Engineering Against Fracture

Spiros Pantelakis · Chris Rodopoulos
Editors

Engineering Against Fracture

Proceedings of the 1st Conference

Springer

Editors

Spiros Pantelakis
Chris Rodopoulos

University of Patras
Dept. Mechanical Engineering
& Aeronautics
265 00 Patras
Greece
pantelak@mech.upatras.gr
rodopoul@mech.upatras.gr

ISBN 978-90-481-8119-3 e-ISBN 978-1-4020-9402-6

Printed on acid-free paper

9 8 7 6 5 4 3 2 1

springer.com

Preface

Within the last thirty years there is a growing acknowledgement that prevention of catastrophic failures necessitates engagement of a large pool of expertise. Herein it is not excessive to seek advice from disciplines like materials science, structural engineering, mathematics, physics, reliability engineering and even economics.

Today's engineering goals, independently of size; do not have the luxury of being outside a global perspective. Survival of the integrated markets and financial systems require a web of safe transportation, energy production and product manufacturing.

It is perhaps the first decade in engineering history that multidisciplinary approaching is not just an idea that needs to materialise but has matured beyond infancy. We can witness such transition by examining engineering job descriptions and postgraduate curricula.

The undertaking of organising a conference to reflect the above was not easy and definitely, not something that was brought to life without a lot of work and commitment. The 1st Conference of Engineering Against Fracture from its conceptual day until completion was designed in a way of underlying the need of bringing all the key players on a common ground that once properly cultivated can flourish. To achieve that the conference themes were numerous and despite their, in principle notional differences, it was apparent that the attendees established such common ground through argumentation. The reader can see this from the variety of research areas reflected by the works and keynote lecturers presented.

A booster to our endeavour was definitely the fact that the conference was also organised to honour the retirement of Prof. Theodore Kermanidis. A figure, which can easily receive the title of the forefather of our laboratory and department.

The editors of these proceedings are in debt to several people including the local and international organising committee and the postgraduate students of the Laboratory of Technology and Strength of Materials, University of Patras. Special thanks should be given to two very special ladies, E. Sotiropoulou and A. Koutsouliakou who without them the conference would have never been a success.

We would also like to acknowledge the help provided by the European Aeronautics Science Network in supporting the organisation of the conference and our sponsors the Technical Chamber of Greece, The University of Patras Research

Committee, the Office for Naval Research Global, the Ministry of Education and Religious Affairs of Greece and Set Point Technologies for financial support.

Special thanks should be given to Ms. Anneke Pot and Ms. Nathalie Jacobs from Springer for their patience and advice in publishing these proceedings.

Patras, Greece
May, 2008

Spiros Pantelakis
Chris Rodopoulos

Contents

Ball-Burnishing and Roller-Burnishing to Improve Fatigue Performance of Structural Alloys

Lothar Wagner, Tomasz Ludian, and Manfred Wollmann

Abstract The HCF response to burnishing of a number of structural materials is compared and contrasted. It is shown that alloys which exhibit marked work-hardening during burnishing respond very beneficially with regard to HCF performance while others which show little work-hardening may even react with losses in HCF strength. Possible explanations for such behavior are outlined in terms of mean stress and environmental sensitivities of the fatigue strengths of the various materials and microstructures.

Keywords Ball-burnishing · Roller-burnishing · Work-hardening · Residual stresses · Fatigue crack nucleation · Fatigue crack propagation

1 Introduction

Mechanical surface treatments such as shot peening and burnishing are often applied to structural metallic materials to improve their HCF performance [1–4]. In general, this improvement can be derived from two contributing factors, namely surface strengthening by the induced high dislocation densities and residual compressive stresses. As opposed to shot peening, burnishing typically results in very smooth surfaces [5, 6]. The influence of burnishing on fatigue life is schematically described in Table 1.

While surface strengthening is able to enhance the resistance to fatigue crack nucleation, micro-crack propagation resistances are detrimentally affected owing to low residual ductility in the cold worked and strengthened surface layer. On the other hand, there is experimental evidence that residual compressive stresses can

L. Wagner (✉), T. Ludian, and M. Wollmann
Institute of Materials Science and Engineering, Clausthal University of Technology, Agricolastraße 6, 38678 Clausthal-Zellerfeld, Germany
e-mail: lothar.wagner@tu-clausthal.de

S. Pantelakis, C. Rodopoulos (eds.), *Engineering Against Fracture: Proceedings of the 1st Conference*,

Table 1 Effects of surface layer properties on the various stages on fatigue life (schematic)

Surface layer properties	Crack nucleation	Micro-crack propagation
Cold work	Retards	Accelerates
Residual compressive stresses	Minor effect	Retards

Table 2 Tensile properties of the various light alloys

	$\sigma_{0.2}$ (MPa)	UTS (MPa)	UTS/$\sigma_{0.2}$	EL (%)
CuZn30	125	310	2,48	67
AISI 304 (1.4301)	270	660	2,44	82
AZ80	235	340	1,45	12
Ti-8.6Al, 10 h 500°C	830	900	1,08	5
Ti-6Al-4 V, EQ, 24 h 500°C	1,060	1,120	1.06	15
Ti-6Al-7Nb, D20/WQ, 24 h 500°C	1,030	1,120	1.09	15
Ti-6Al-7Nb, D20/AC, 24 h 500°C	920	995	1.08	14
Beta C, SHT	840	850	1,01	25
LCB, PS +0.5 h 500°C	1,665	1,730	1,04	3

drastically reduce the growth rate of surface microcracks while crack nucleation resistances are less affected [7–13]. Recently, it was shown that residual tensile stresses which must necessarily be present underneath the near-surface residual compressive stress field, can lead to subsurface fatigue crack nucleation and to losses in HCF strength relative to an electropolished reference [14, 15]. The present investigation is intended to highlight differences in the fatigue response of the various alloys to burnishing.

2 Experimental

The investigation was performed on a number of alloys which tensile properties are listed in Table 2.

Hour glass-shaped fatigue specimens were prepared with a minimum gage diameter of either Ø 4 or Ø 6 mm. In addition, specimens with a circumferential V60° notch and a notch root radius of 0.43 mm ($k_t = 2.5$) were machined (Fig. 1).

Ball-burnishing of the smooth specimens was performed by means of a conventional lathe using a device by which hard metal balls of Ø 3 or Ø 6 mm are hydrostatically pressed onto the rotating specimen surface. Roller burnishing of the notched specimens was performed by a roller element having V55° geometry and 0.35 mm tip radius. Due to roller-burnishing, the notch factor k_t increased from 2.3 to 2.7. Electrolytically polished smooth and notched specimens were prepared to serve as reference. Electropolishing of the notched specimens decreased the

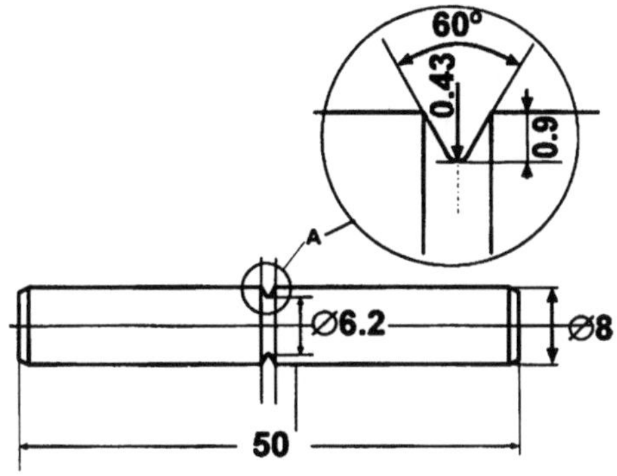

Fig. 1 Geometry of the notched specimens

notch factor from 2.5 to 2.3. The exact k_t value was calculated for each specimen individually. Microhardness-depth profiles and residual stress-depth profiles as determined by the incremental hole drilling method were measured to characterize the process-induced changes in surface layer properties [16]. Fatigue tests were performed in rotating beam loading ($R = -1$) at frequencies of about $50\,s^{-1}$. In addition, axial fatigue tests were done at stress ratios of $R = -1$ and $R = 0.1$ at frequencies of about $90\,s^{-1}$. Some axial fatigue tests at $R = -1$ were performed also in vacuum.

3 Results and Discussion

Examples of the HCF performance of smooth and notched ($k_t = 2.3$) electropolished references are illustrated in Figs. 2 and 3. In addition to nominal stress amplitude σ_a, the maximum stress amplitude at the notch root $\sigma_a \bullet k_t$ of the notched specimens is plotted. On both AISI 304 (Fig. 2a) and CuZn30 (Fig. 2b), the notch factor $k_t = 2.3$ only slightly reduces HCF performance of the smooth ($k_t = 1.0$) specimens. Thus, the HCF strengths of the notched specimens in terms of $\sigma_a \bullet k_t$ are much higher than those of the smooth specimens. On the contrary, results on notched specimens of various magnesium and titanium alloys have demonstrated marked losses in notched HCF performance relative to smooth specimens [17, 18]. Examples on the magnesium alloy AZ80 and the titanium alloy Ti-6Al-4V are illustrated in Fig. 3. Interestingly, the notched HCF strengths of both alloys AZ80 (Fig. 3a) and Ti-6Al-4V (Fig. 3b) in terms of $\sigma_a \bullet k_t$ are the same as those of the smooth conditions.

The effect of notches on the performance of the various alloys is summarized in Table 3 where the fatigue notch sensitivity is expressed as:

$$q = \frac{k_f - 1}{k_t - 1},\ 0 \leq q \leq 1 \text{ and } k_f = \frac{\sigma_{a10^7 smooth}}{\sigma_{a10^7 notched}}.$$

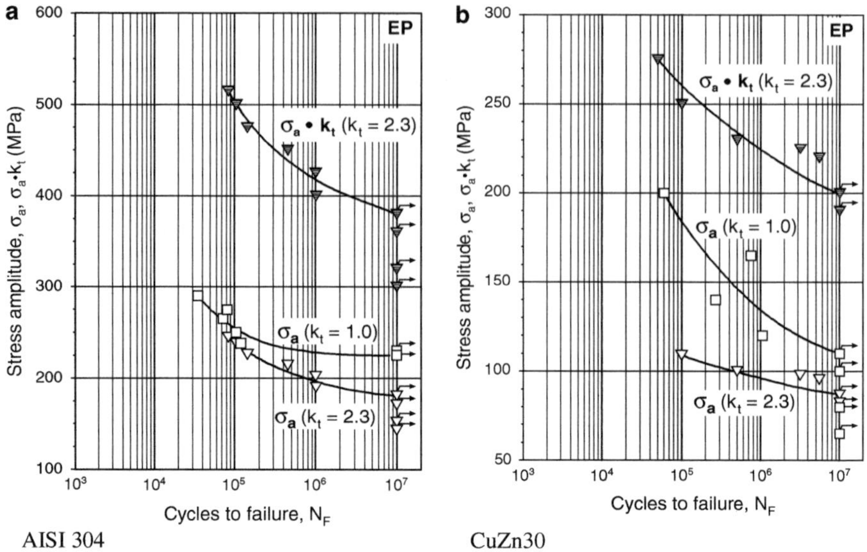

Fig. 2 S-N curves of the conditions EP in rotating beam loading ($R = -1$) in air

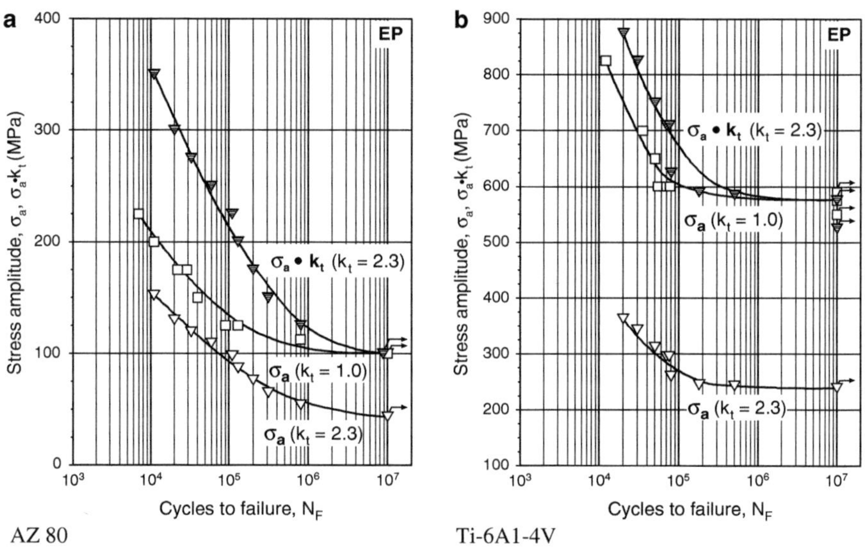

Fig. 3 S-N curves of the conditions EP in rotating beam loading ($R = -1$) in air

Obviously, with the exception of CuZn30 and AISI 304, all other tested alloys are fully (100%) notch sensitive with regard to HCF performance. As seen from Table 1, both CuZn30 and AISI 304 exhibit marked work-hardening capabilities ($UTS/\sigma_{0.2} \geq 2$) which may explain their very low fatigue notch sensitivities.

Table 3 Notch factor k_t, fatigue notch factor k_f and fatigue notch sensitivity q of the various alloys (condition EP)

	k_t	k_f	Q
CuZn30	2.3	1.28	0.25
AISI 304	2.3	1.26	0.20
AZ80	2.3	2.3	1
Ti-6Al-4 V EQ, 24 h 500°C	2.3	2.3	1
Beta C, SHT	2.3	2.3	1
LCB, PS +0.5 h 500°C	2.3	2.3	1

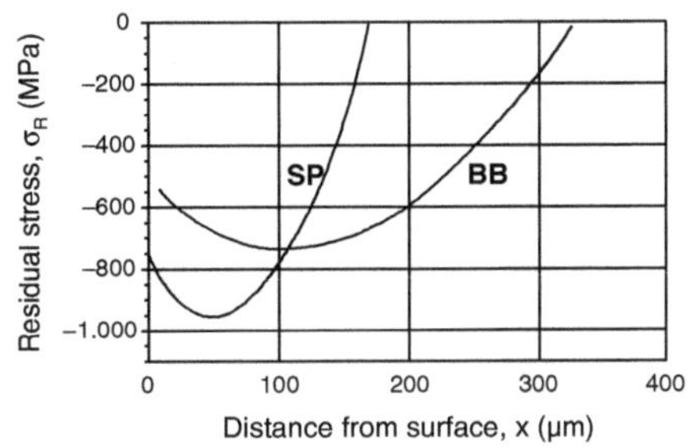

Fig. 4 Compressive residual-stress depth profile after ball-burnishing (BB) and shot peening (SP) in Beta C, SHT

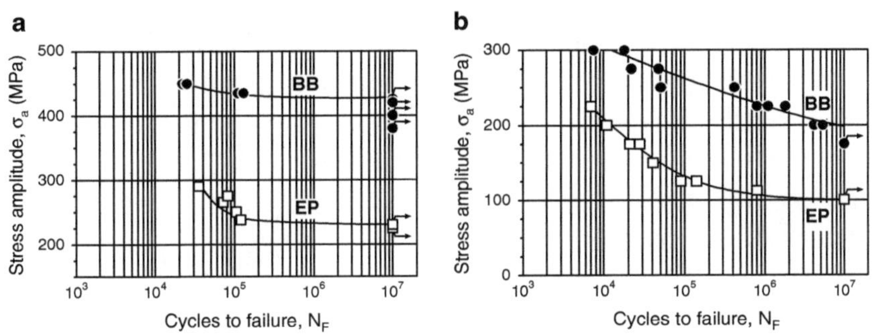

AISI 304 AZ80

Fig. 5 S-N curves in rotating beam loading (R = −1)

An example of the compressive residual-stress depth profile after ball-burnishing of Beta C is illustrated in Fig. 4 where in addition, results after shot peening are also shown.

Ball-burnishing as opposed to shot peening led to deeper penetration depths of residual stresses, however, maximum residual stresses after shot peening were higher than after ball-burnishing.

Both alloys AISI 304 and AZ80 respond very beneficially to ball burnishing (Fig. 5). The enhancements in HCF performance are even superior to those typically observed after optimum shot peening. Similar to results on shot peened HCF specimens, subsurface fatigue crack nucleation was also observed in ball-burnished

HCF specimens while all electropolished specimens failed by surface crack nucleation. Examples of fatigue crack nucleation sites are illustrated in Fig. 6.

From axial HCF tests on ball-burnished specimens it is known that the depths of crack nucleation sites are about the same for a given burnishing pressure. Similar results were reported on shot peened conditions where the depths of subsurface crack nucleation were directly related to the Almen intensity applied. These results clearly indicate that the residual tensile stresses balancing the near-surface compressive stress field (Fig. 4) are not homogeneously distributed over the remaining cross-section [19, 20]. Instead, residual tensile peak stresses must exist at some distances below the compressive stress field. This situation is schematically demonstrated in Fig. 7.

Due to the presence of process-induced residual tensile stresses (Fig. 7), materials with high tensile mean stress sensitivities may respond quite critically to ball-burnishing or shot peening. This can be derived from results on shot peened conditions of Ti-6Al-7Nb [21]. Materials with normal mean stress sensitivity in fatigue as observed on water-quenched duplex microstructures (Fig. 8) also exhibit significant improvements in the HCF performance due to shot peening (Fig. 9). The

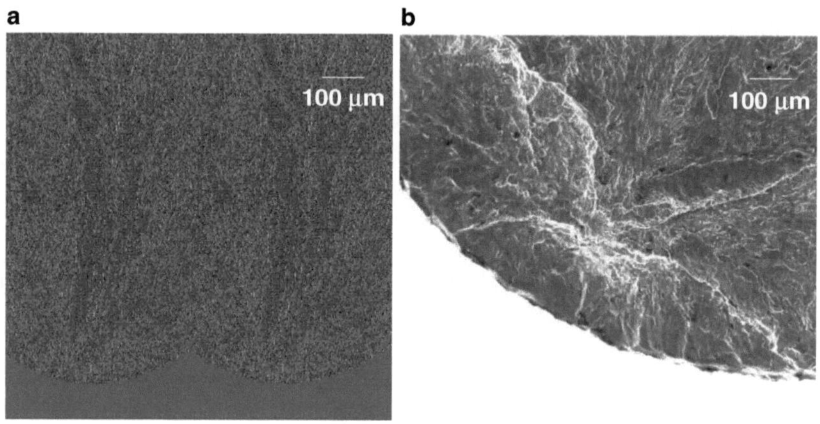

at surface (EP) at subsurface (BB)

Fig. 6 HCF crack nucleation sites in Ti-6Al-4 V

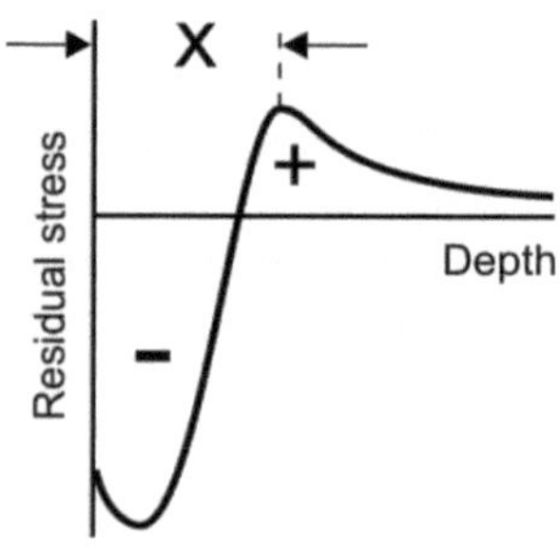

Fig. 7 Residual stress-depth profile after ball-burnishing (schematic)

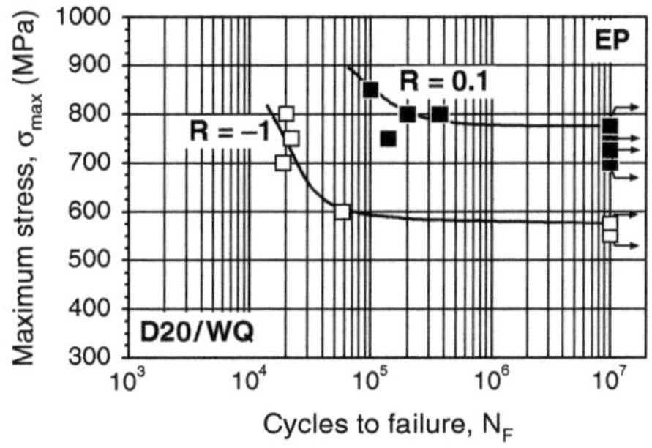

Fig. 8 S-N curves of Ti-6Al-7Nb (EP, D20/WQ)

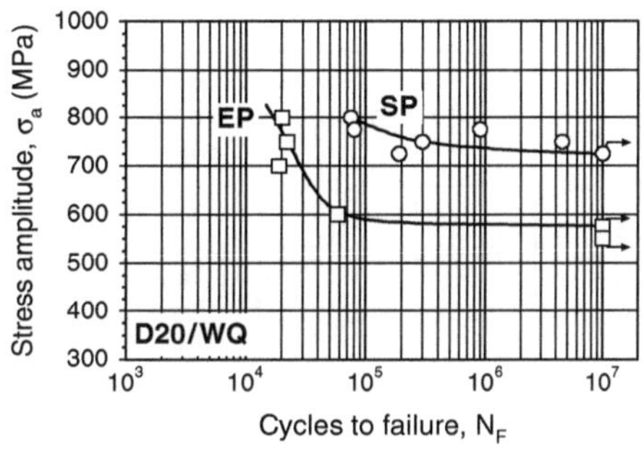

Fig. 9 S-N curves ($R = -1$) of Ti-6Al-7Nb (D20/WQ) after shot peening

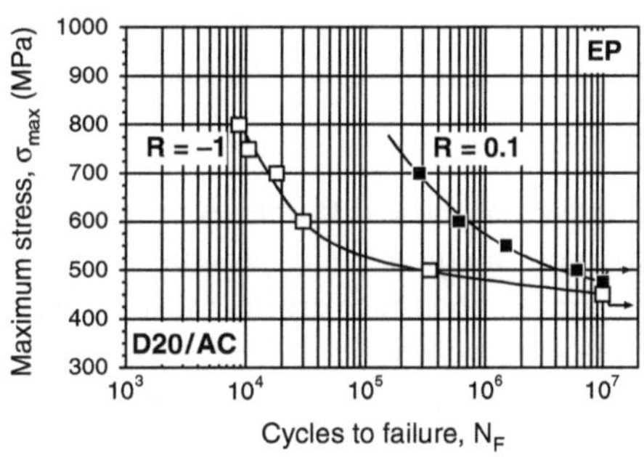

Fig. 10 S-N curves of Ti-6Al-7Nb (EP, D20/AC) in axial loading

latter result is explained by the fact that in these microstructures, the stress amplitude at the endurance limit is hardly reduced by tensile mean stresses.

On the contrary, materials with anomalous mean stress sensitivity in fatigue as found on air cooled duplex microstructures of the same alloy (Fig. 10) exhibit hardly any improvement in HCF performance after shot peening (Fig. 11). The latter is explained by the fact that in these microstructures, an increase in tensile mean stress does not increase the maximum stress at endurance (Fig. 10), i.e., the corresponding stress amplitude at low tensile mean stresses is markedly reduced. Similar results are reported on duplex microstructures in Ti-6Al-4 V [22]. From optical microscopy under polarized light [23] and more recently from EBSD measurements [24] it can be

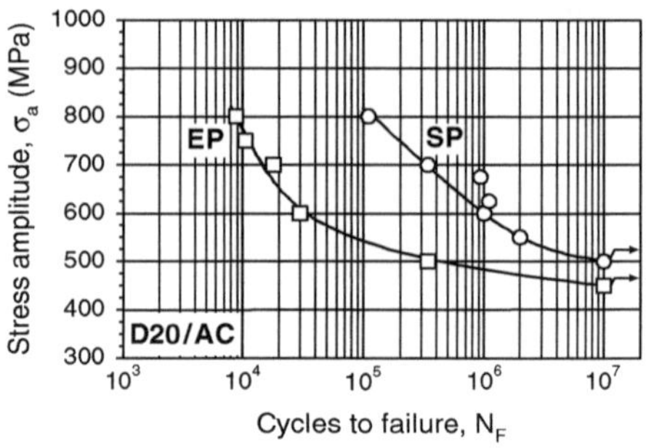

Fig. 11 S-N curves (R = −1) of Ti-6Al-7Nb (D20/AC) after shot peening

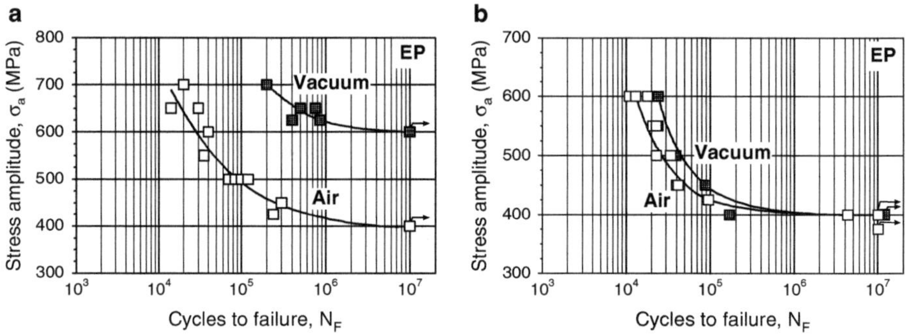

high environmental sensitivity (Ti-8.6Al) low environmental sensitivity (Beta C)

Fig. 12 HCF performance of titanium alloys in axial loading (R = −1)

argued that these differences in mean stress sensitivity of $(\alpha+\beta)$ alloys are related to micro-textures being present in air-cooled conditions.

Since subsurface cracks must nucleate under quasi-vacuum conditions, the environmental sensitivity must be taken into account in order to understand the air HCF performance after ball-burnishing or shot peening. Examples of fatigue performance in air and vacuum of two contrasting alloys are shown in Fig. 12. As seen in Fig.12a, the HCF strength of Ti-8.6Al in vacuum is much higher than the corresponding value in air this indicating high environmental sensitivity of this age-hardened α-alloy. Similar high environmental sensitivities were observed in $(\alpha+\beta)$-titanium alloys such as Ti-6Al-4 V and Ti-6Al-7Nb.

As opposed to α- and $(\alpha+\beta)$-titanium alloys, the HCF strengths of metastable β-titanium alloys such as Beta C or TIMETAL LCB are hardly affected by the air environment as shown in Fig. 12b. This result may be related to the absence of hydrogen embrittlement of the bcc β-phase.

Because of the absence of environmental effects, subsurface quasi-vacuum fatigue crack nucleation in titanium alloys belonging to the metastable β-alloy class is rather easy. Thus, both ball-burnishing and shot peening can lead to marked losses in HCF strength relative to the electropolished reference as demonstrated on TIMETAL LCB and Beta C in Fig. 13.

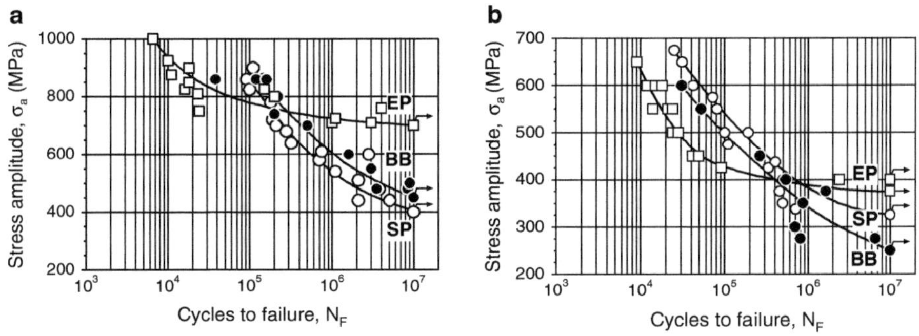

TIMETAL LCB (PS, 0.5 500°C) Beta C (SHT)

Fig. 13 S-N curves of metastable β-titanium alloys in rotating beam loading

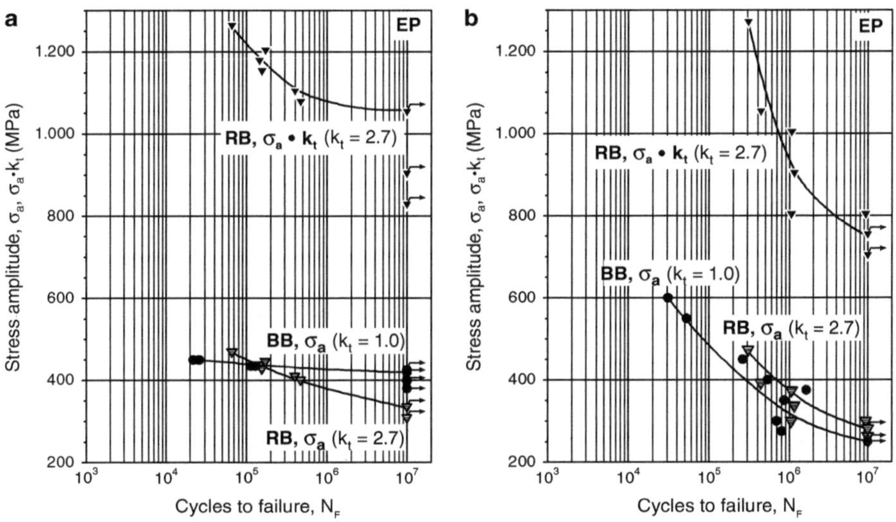

AISI 304 Beta C (SHT)

Fig. 14 S-N curves of the conditions BB and RB in rotating beam loading ($R = -1$) in air

How much the HCF strength in these alloys decreases was shown to depend on the degree of aging as well as on the intensities used in shot peening and burnishing.

The influence of roller-burnishing on notched fatigue performance is illustrated in Fig. 14 where results on AISI 304 (Fig. 14a) are compared with those on Beta C (Fig. 14b). Unlike the effect of notches in the electropolished references (Figs. 2 and 3), roller-burnishing of notched HCF specimens [25] led to enhancements which indicate a full compensation of the notch factor k_t (compare Fig. 14 with Figs. 2 and 3).

Acknowledgements The authors would like to thank the Deutsche Forschungsgemeinschaft (DFG) for financial support through contract WA 692/32-1. They would also like to thank Ms. Johanna Gunkel for fatigue testing and Dr. Sangbong Yi for residual stress measurements.

References

1. Schulze V. (2006), Modern Mechanical Surface Treatment, Wiley-VCH.
2. Wohlfahrt H. (1984), The Influence of Peening Conditions on The Resulting Distributions of Residual Stress, Shot Peening (H. O. Fuchs, ed.) American Shot Peening Society, 316.
3. Wohlfahrt H. (1987), Practical Aspects of the Application of Shot Peening to Improve the Fatigue Behaviour of Metals and Structural Components, Shot Peening (H. Wohlfahrt, R. Kopp and O. Vöhringer, eds.), DGM, Oberursel, Germany, 563.
4. Altenberger I. (2002), Alternative Mechanical Surface Treatments: Microstructures, Residual Stresses & Fatigue Behaviour, Shot Peening (L. Wagner, ed.) Wiley-VCH, ICSP8, Garmisch-Partenkirchen, 421.
5. Mhaede M. H., Wollmann M. and Wagner L. (2008), Influence of Ball-Burnishing on Stress Corrosion Cracking and Corrosion, Fatigue and Corrosion Fatigue of Al 2024 and Al 6082, ICSP10, Tokyo, 505.
6. Wagner L. and Wollmann M. (2008), Shot Peening of Non-Ferrous Alloys to Enhance Fatigue Performance, ICSP10, Tokyo, 355.
7. Burck C. H., Sullivan C. P. and Wells C. H. (1970), Fatigue of a Glass Bead Blasted Nickel-Base Superalloy, Met. Trans. 1A, 1595.
8. Leverant G. R., Langer S., Yuen A. and Hopkins, S. W. (1979), Surface residual stresses, surface topography and the fatigue behavior of Ti-6AI-4V, Met. Trans. 10A, 251.
9. Wagner L. and Lütjering G. (1982), Influence of Shot Peening on the Fatigue Behavior of Ti-Alloys, Shot Peening, Pergamon Press, 453.
10. Hack J. E. and Leverant G. R. (1982), Influence of Compressive Residual Stress on the Crack-Opening Behavior of Part-Through Fatigue Cracks, Residual Stress Effects in Fatigue, ASTM STP 776, 204.
11. Muto Y., Fair G. H., Noble B. and Waterhouse R. B. (1987), The Effect of Residual Stresses Induced by Shot Peening on Fatigue Crack Propagation in Two High Strength Aluminum Alloys, Fat. and Fract. of Eng. Mat. and Struct. 10, 261.
12. Nisitani H. and Fujimura K. (1997), Initiation and Growth Behaviour of a Fatigue Crack in Shot-Peened Steel, Computer Methods and Experimental Methods for Surface Treatment Effects 3, Oxford, Computational Mechanics Publications, (ROYAUME-UNI ed.), 13.
13. Wagner L. and Müller C. (1992), Effect of Shot Peening on Fatigue Behavior in Al-Alloys, J. Materials Manufacturing & Processing, 423.
14. Dörr T. and Wagner L. (1996), Fatigue Behavior of Shot Peened TIMETAL 1100: Effects of Microstructure and Stress Gradient, Surface Performance of Titanium Alloys (J. K. Gregory, H. J. Rack and D. Eylon, eds.) TMS AIME, 231.
15. Kocan M., Rack H. J. and Wagner L. (2005), Fatigue Performance of Metastable Beta Titanium Alloys: Effects of Microstructure and Surface Finish, Beta Titanium Alloys of the 00's (R. R. Boyer, R. F. Denkenberger, J. Fanning and H. J. Rack, eds.), JMEPEG 14, 765.
16. Schwarz T. and Kockelmann H. (1992), VDI Report 940, 99.
17. Gerdes C. and Lütjering G. (1984), Influence of Shot Peening on Notched Fatigue Strength of Ti-6Al-4V, Shot Peening (H. O. Fuchs, ed.) American Shot Peening Society, 175.
18. Gregory J. K., Müller C. and Wagner L. (1993), Bevorzugte Randschichtaushärtung: Neue Verfahren zur Verbesserung des Dauerschwingverhaltens mechanisch belasteter Bauteile, Metall 47, 915.
19. Rodopoulos C. A., Edwards R. E., Curtis S., Romero J. S., Choi J.-H., de los Rios E. and Levers A. (2002), Theoretical Analysis of Beneficial and Detrimental Effects of Controlled

Shot Peening in High Strength Aluminium Alloys, Shot Peening (L. Wagner, ed.) Wiley-VCH, ICSP8, Garmisch-Partenkirchen, Germany, 547.
20. Zheng Y. (2002), Finite Element Simulation of Shot Peen Forming, Shot Peening (L. Wagner, ed.) Wiley-VCH, ICSP8, Garmisch-Partenkirchen, Germany, 554.
21. Holzwarth U., Kiese J. and Wagner L. (1999), Effects of Surface Finishing on Fatigue Performance of the Surgical Implant Alloy Ti-6Al-7Nb, Fatigue Behavior of Titanium Alloys, (R. R. Boyer, D. Eylon and G. Lütjering, eds.), TMS, 323.
22. Lindemann J. and Wagner L. (1997), Mean Stress Sensitivity in Fatigue in Alpha, (Alpha + Beta) and Beta Titanium Alloys, Materials Science and Engineering A 234-239, 1118.
23. Zhang J., Lindemann J. and Wagner L. (2004), Influence of Prior Deformation on the Development of Duplex Microstructures and Fatigue Strengths in Ti-6242, *Ti-2003* Science and Technology (G. Lütjering and J. Albrecht, eds.) Wiley-VCH, Weinheim, 1823.
24. Woodfield A. P., Gorman M. D., Sutliff J. A. and Corderman R. R. (1998), Effect of Microstructure on Dwell Fatigue Behavior of Ti-6242, Fatigue Behavior of Titanium Alloys (R. R. Boyer, D. Eylon, G. Lütjering, eds.) TMS, 111.
25. Gunkel J. (2008), Dauerschwingverhalten des metastabilen austenitischen Stahls X5CrNi18-10 nach dem Festwalzen im glatten und gekerbten Zustand, Diploma thesis, TU Clausthal, IWW, Germany.

Dual Scale Fatigue Crack Monitoring Scheme Considering Random Material, Geometric and Load Characteristics

G.C. Sih

Abstract Effectiveness of the existing fracture control methodology depends on the non-destructive evaluation of defects arising from repeated service of structures. Damage detection is by far being automatic since not all defects are dangerous. The correct diagnostics require not only a knowledge of the size of the prevailing defects but also their likely locations of occurrence from the initial design and/or past experience. What is not known is the time when the defects will become critical since the service conditions change as a rule and the material age in time. These inherent variables can affect the accuracy of non-destructive evaluation which in its strict sense implies certain cut-off scale of small defects that are assumed to be not harmful. Material microstructural effects have been known to affect the structural integrity. The obvious implication is multiscaling, the least consideration of which involves the dual scale of micro/macro. Hence, defect monitoring presents over whelming difficulties because of the entanglement of multi-variables.

Starting with the basic quantities of the local force and displacement that can be detected from the commercially available transducers, continuous records can be made available for the local compliance C and its time rate history, say dC/dt or dC/dN in fatigue with N being the number of cycles. These data can be stored in microprocessors and made available for analyses. The conversion of the force and displacement records to damage by fatigue crack growth is the challenge of this work. Appeal is made to a dual scale micro/macro fatigue crack growth model that has the capability to delineate micro- and macro-cracking. The model makes use of three parameters μ^*, d^* and σ^*. They account for the interaction between the micro/macro effects in terms of the, respective, relative shear modulus μ_{micro}/μ_{macro}, the micro-tip characteristic length d/d_o and the crack surface tightness ratio σ_o/σ_∞ that controls the opening of the fatigue crack. These parameters are assumed to vary in a random fashion as the crack size increase with the repetition of loading.

G.C. Sih
School of Mechanical and Power Civil Engineering, East China University of Science and Technology, Shanghai 200237, China; Department of Mechanical Engineering and Mechanics, Lehigh University, Bethlehem, PA 18015, USA
e-mail: gcs@ecust.edu.cn

S. Pantelakis, C. Rodopoulos (eds.), *Engineering Against Fracture: Proceedings of the 1st Conference*,

Random normal distribution of statistical values of μ^*, d^* and σ^* is investigated for different mean $<\mu>$ and standard deviation $<\sigma>$. The amount of dispersion around the mean is examined for the crack length a and crack growth rate da/dN in addition to the compliance C and its rate dC/dN. It is demonstrated that randomness of the material microstructure has an effect on the local compliance which will increase with crack growth in a stable fashion and then rises more quickly with the number of cycles. The characteristic is reminiscent of the compliance to crack length relation used for determining the energy release rate in fracture mechanics.

The ability to account for micro/macro damage transition is considered to be fundamental for assessing the meaning of the compliance data. Randomness of the material microstructure parameters provides the means for studying the dispersion of the results in terms of fatigue crack growth. The preliminary findings are encouraging and suggest several new directions of research for health monitoring.

Keywords Micro/macro · Fatigue · Cracking · Random properties · Microstructure · Scale transition · Compliance · Health monitoring · Crack growth

1 Introduction

All structures are vulnerable to damage by aging or natural disasters such as earthquakes or tornadoes. Effective methods for inspection and detection of damage are needed to determine whether the structures are worthy of repair and restoration. When the decision entails the loss of life and property such as the collapse of tall buildings or the fracture of bridges, there is the relentless pressure to develop failure control methodologies that can minimize premature failure. It is now a recognized fact that the existing hardware of sensors and transducers must be implemented with the appropriate software. Threshold criterion is needed for developing damage detection models that can account for cumulative effects in terms of measurable parameters with physical meaning. Monitored signal must be supplemented with the correct diagnostic decision.

The work to follow will center on structural damage by fatigue. Accumulation of energy in a unit volume of material that is represented by the sum of the areas enclosed by the hystereses loops [1, 2]. These loops are formed by reversing the cyclic load on a uniaxial test specimen. The size and shape of each loop depend on the previous history. This is in fact the volume energy density function dW/dV, the sum of which when expressed in terms of their mean, can be related directly to the energy density range ΔS at an incremental distance Δr from the location of failure or crack initiation. That is $\Delta r/\Delta N = (1/A)\Delta S$ [3, 4] where A stands for the mean area of the hystereses loops. The more familiar form used in fatigue studies is $da/dN = (1/A)\Delta S$, where the crack length has replaced r while N is the number of cycles. A variety of related forms of da/dN can be found since ΔS can be expressed in terms of parameters that depend on the material microstructure, crack geometry, loading and other factors. Arbitrariness of the form of da/dN will be limited by starting from $da/dN = (1/A)\Delta S$ or the equivalent. While the use of the crack length parameter appears to be a restriction, it can adopt a much broader physical meaning such as the mean characteristic length of a cluster of microcracks. What should be kept in mind is that the final structure instability is dominated by one or very few large cracks. Not all of the cracks are important. Many of them are in fact beneficial as they are self-arresting once the local stress concentrations are relieved. The spirit of fracture mechanics is to trace back from these few damaging cracks to the conditions at a lower scale level so as to determine the probabilities that would favor certain inhomogeneous microscopic conditions to trigger the formation of macrocracks. Not all cracks have the same probability to cause global instability. A sub-set of ΔS would be ΔK which has been used conventionally for treating cracks that propagate nearly straight.

Random normal distribution of the micro/macro material, geometric and loading parameters are invoked for the appropriate mean $<\mu>$ and the standard deviation $<\sigma>$. The former pertains to the average values of the variables $\mu^*(=\mu_{micro}/\mu_{macro})$, $d^*(=d/d_o)$ and $\sigma^*(=\sigma_o/\sigma_\infty)$ while the latter specifies the amount of dispersion around the mean. Simulated are the randomness of the internal material microstructure that in term would cause the microcrack tip to wonder irrespective of the macroscopic load symmetry or skew–symmetry effects. Such a behavior would show up as dispersion of the crack length versus number of cycle relation and transfer to the variation of the compliance as a function of the fatigue cycle. As expected, the dispersion is scale sensitive. It is also indicative of the distance from the crack tip. This is precisely the type of information required in monitoring physical damage.

Detection of crack-like defects should be distinguished from others since crack growth as a rule is accompanied by the release of other forms of energy at the different scale. Acoustic emission when tuned for detecting the nucleation of microcracks is not suitable for finding macrocracks or defect size smaller than the acoustic wave length. Electromagnetic waves are used to detect even smaller inhomogeneities. An a priori knowledge of the scale size with reference to the prevailing non-uniformity is therefore necessary. For large structures, the displacement gage is the basic measurement device for recording the change of distance between two points that may not indicate the presence of surface separation such as that of a crack unless the gages are placed very close to the crack mouth. In practice, the major effort is to find the location of the crack. Conversion of displacement and force history data to local compliance or stiffness can provide surveillance to a larger area but also indicate the tendency of cracks to reach instability. In essence, it is the change of displacement that is calibrated to measure force. This work carries one step further to the compliance, the change of which with the crack length has been recognized as the fundamental parameter for characterizing fracture [5].

2 Micro/Macro Fatigue Crack Growth Model

A modified version of the conventional macrocrack growth model for fatigue has been extended [6, 7] to account for the transition of cracking damage from the micro to macro or vice versa by replacing the ΔK for a macrocrack by ΔK_{micro}^{macro} for micro-/macro-cracking. Unless the cracks take sharp turns, it is adequate to use ΔK^{macro}_{micro} instead of the energy density range ΔS^{macro}_{micro}. The dual scale fatigue model for cracks growing nearly straight can be written as

$$\frac{da}{dN} = M_o(\Delta K^{macro}_{micro})^m \tag{1}$$

where M_o and m are assumed to be constants. Although Eq. (1) was not arrived by fundamental means, the procedure for deriving a differential equation to solve the crack length will be the same. For this purpose, it suffices to start with [6, 7]

$$\Delta K^{macro}_{micro} = \frac{6\sqrt{\pi}(1-\nu_{macro})\mu_{micro}\sqrt{c^2-a^2}}{5d^{0.25}\mu_{macro}}(1-\frac{2\sigma_o}{\pi\sigma_\infty}\sin^{-1}\frac{a}{c})(\sigma_{\max}-\sigma_{\min}) \tag{2}$$

The notations for $d = c - a$ can be found in Fig. 1 where symmetry is assumed to prevail across the x_2-axis. The cyclic restraining stress induced by the oscillating fatigue load $\sigma_\infty(N)$ away from the crack is denoted by $\sigma_o(N)$ as shown in Fig. 1. Anti-symmetry of the condition across the microcrack tip is guaranteed by the mixed boundary condition invoked as an expediency. The motivation for such a representation can be found partly in [8] and is beyond the scope of this discussion. In view of the fact that $a \approx d$, an equivalent but approximate form of Eq. (2) can be used:

$$\Delta K^{macro}_{micro} = \frac{6\sqrt{\pi}(1-\nu_{macro})\mu_{micro}d^{0.25}\sqrt{c+a}}{5\mu_{macro}}(1-\frac{2\sigma_o}{\pi\sigma_\infty}\sin^{-1}\frac{a}{c})(\sigma_{\max}-\sigma_{\min}) \tag{3}$$

The relation $\sigma_{\max} - \sigma_{\min} = 2\sigma_a$ can be further applied to yield

$$\Delta K^{macro}_{micro} = \frac{12\sqrt{2\pi}(1-\nu_{macro})}{5}\sigma_a\sqrt{a}\frac{\mu_{micro}}{\mu_{macro}}d^{0.25}(1-\frac{\sigma_o}{\sigma_\infty}) \tag{4}$$

Take $\nu_{macro} = 0.3$. Equation (4) becomes

$$\Delta K^{macro}_{micro} \approx 4\sigma_a\sqrt{a}\frac{\mu_{micro}}{\mu_{macro}}d^{0.25}(1-\frac{\sigma_o}{\sigma_\infty}) \tag{5}$$

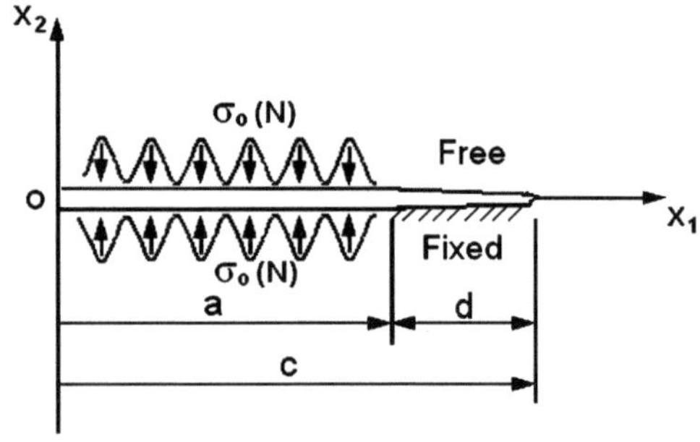

Fig. 1 Micro/macro crack with one-half symmetry

From the definition of the relations

$$d^* = \frac{d}{d_o}, \mu^* = \frac{\mu_{micro}}{\mu_{macro}}, \sigma^* = \frac{\sigma_o}{\sigma_\infty} \tag{6}$$

there results

$$\Delta K_{micro}^{macro} \approx 4d_o^{0.25}\sigma_a\sqrt{a(N)}\ \mu^*(N)\ [d^*(N)]^{0.25}\ [1-\sigma^*(N)] \tag{7}$$

Note that Eq. (7) has the $\sqrt{a}$ dependency. It can be substituted into Eq. (1) to obtain the governing differential equation solving for the half crack length a. Since M_o and m are assumed to be constants which vary with the material, they will not affect the integration for finding a = a(N).

3 Determination of Crack Growth History

Equation (7) is unique; it reveals the interaction of the material microstructure, the microcrack tip reaction and tightness of the crack mouth due to the relative restrain of the material with reference to the applied load. The relation a = a(N) can be obtained by putting Eq. (7) into Eq. (1) to yield the governing differential equation:

$$\frac{da}{dN} = M_o\{4d_o^{0.25}\sigma_a\sqrt{a(N)}\ \mu^*(N)\ [d^*(N)]^{0.25}\ [1-\sigma^*(N)]\}^m \tag{8}$$

Define a new parameter

$$M = M_o[4d_o^{0.25}\sigma_a]^m \tag{9}$$

such that Eq. (8) becomes

$$\frac{da}{dN} = M\{\ \mu^*(N)\ [d^*(N)]^{0.25}\ [1-\sigma^*(N)]\sqrt{a(N)}\}^m \tag{10}$$

The solution of Eq. (10) for a = a(N) depends on the value of m. Numerical integration of Eq. (10) can always be made.

As an illustration, let m = 2 and σ^*, d^* and μ^* to be fixed for each N so that Eq. (10) can be integrated for arbitrarily small segments and written as

$$\frac{da}{dN} = M(\ k_o)^2 a(N) = k\ a(N),\ \text{for m} = 2 \tag{11}$$

such that

$$M = 16M_o(\sigma_a)^2\sqrt{d_o} \tag{12}$$

The contractions

$$k = Mk_o^2\ \text{ and }\ k_o = \mu^*(\ d^*)^{0.25}\ (1-\sigma^*) \tag{13}$$

have been made in Eq. (11) which can be integrated to render

$$a(N) = a_o e^{k(N-N_o)}, \text{ for } m = 2 \tag{14}$$

where a_o and N_o are the initial crack half length and corresponding number of cycles elapsed initially. The range of k_o or k in Eq. (13) can be established by assigning a material with known value of M_o since M can be obtained from the y-intercept extending from the straight line portion of the sigmoidal curve of da/dN. Numerical values of Eq. (11) can thus be found.

In general, m may differ from 2. Equation (10) takes the form

$$\frac{da}{dN} = M(k_o\sqrt{a})^m \tag{15}$$

Integration gives

$$N - N_o = \frac{2a_o}{(m-2)M(k_o\sqrt{a_o})^m}[1 - (\frac{a_o}{a})^{\frac{m}{2}-1}], \ m \neq 2 \tag{16}$$

Solving for the half crack length, the results is

$$a = \left[a_o^{1-\frac{m}{2}} + (1-\frac{m}{2})Mk_o^m(N-N_o)\right]^{\frac{1}{1-\frac{m}{2}}}, \ m \neq 2 \tag{17}$$

Keep in mind that M depends also on the stress amplitude as shown by Eq. (9). The parameter k_o has been defined by the second of Eqs. (13). The oscillatory character of k_o depending on the randomness of the micro/macro parameters governing the stiffness, microcrack tip, restraining character of the material will be discussed subsequently.

4 Physical Model of Random Material Properties

For small cracks that are comparable with the characteristic length of the material microstructure, it is necessary to consider the inhomogeneity of the material since it will influence the crack growth behavior.

4.1 Generation of Random Numbers

The parameters k_o in the second of Eqs. (13) will be assumed to vary randomly with the number of fatigue cycles such that $\sigma^* = \sigma^*(N)$, $\mu^* = \mu^*(N)$ and $d^* = d^*(N)$ take discrete values for different N. Figure 2a–c show, respectively, the randomly generated values of σ^*, μ^* and d^* with the identical probability in their ranges [9] which are known from previous work [6, 7]. For example, consider $s^* = 0$, $d^* = 1$ and $\mu^* = 1 \sim 5$. Hence, the second of Eqs. (13) gives $1 \leq k_o \leq 5$. Moreover, $\sigma^* = 0.9$, $d^* = 10$ and $\mu^* = 1 \sim 5$ yield $0.178 \leq k_o \leq 0.889$ which establishes the range for generating the random values of σ^*, μ^* and d^* in Fig. 2a–c.

Since the material will degrade with time or number of cycle N, it is anticipated that the range for σ^*, μ^* and d^* will not be constant with N but it is expected to decrease with N. The second of Eqs. (13) shows that k_o will also decrease with N. Over a range of about two million cycles, six segments of k_o can be considered such that each successive segment has decreasing amplitude as given in Table 1. Displayed in Fig. 3 are the oscillatory character of k_o with N. Randomness variations of σ^*, μ^* and d^* or k_o may not have same probabilities. This requires the consideration of random normal distribution with different mean $<\mu>$ and standard deviation $<\sigma>$ in each of the six segments.

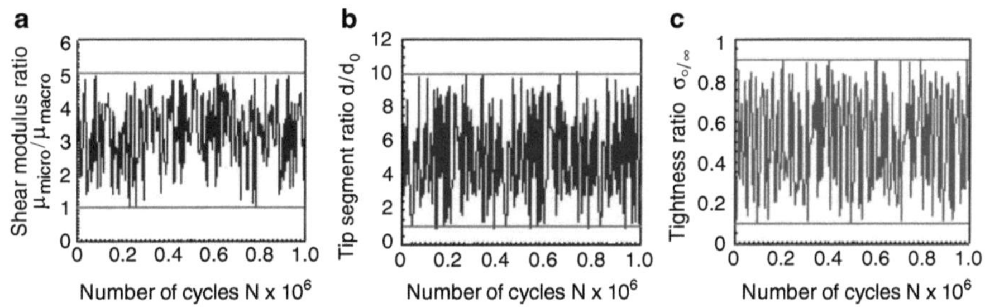

Shear modulus ratio vs cycle number

Tip segment ratio vs cycle number

Tightness ratio vs cycle cycle number

Fig. 2 Randomly generated values [9] for micro/macro stiffness, geometric and material restraining stress

Table 1 Segmented random variations of k_o with cycle number N

Segment no.	Range of $N \times 10^6$	Range of k_o
1	$1.4 \sim 1.45$	$2.1 \sim 4.1$
2	$1.45 \sim 1.5$	$2.3 \sim 3.8$
3	$1.5 \sim 1.6$	$2.5 \sim 3.2$
4	$1.6 \sim 1.7$	$2.8 \sim 3.1$
5	$1.7 \sim 1.9$	$2.9 \sim 3.1$
6	$1.9 \sim 2.0$	$2.95 \sim 3.05$

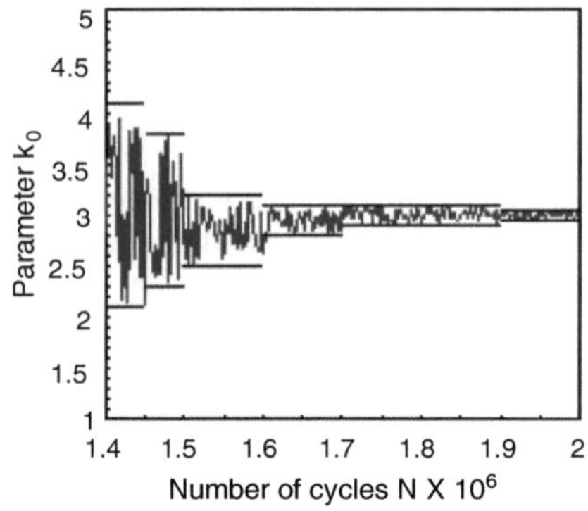

Fig. 3 Random variations of parameter k_o with cycle number N (Table 1)

Table 2 Segmented range of N for normal distribution with different $<\mu>$ and $<\sigma>$

Segment no.	Range of $N \times 10^6$	Mean $<\mu>$	Standard deviation $<\sigma>$
1	$1.4 \sim 1.45$	3.0	0.8
2	$1.45 \sim 1.5$	3.0	0.5
3	$1.5 \sim 1.6$	3.0	0.2
4	$1.6 \sim 1.7$	3.0	0.09
5	$1.7 \sim 1.9$	3.0	0.05
6	$1.9 \sim 2.0$	3.0	0.03

4.2 *Normal Distribution with Mean $<\mu>$ and Standard Deviation $<\sigma>$*

In order for the randomness of σ^*, μ^* and d^* to have different probabilities, different mean $<\mu>$ and standard deviation $<\sigma>$ will be selected for the six segments as N is increased while the range of the amplitude of k_o will be kept as those in Table 1. To be specific, a fatigue life of two million cycles or $N = 2 \times 10^6$ will be considered. The crack initiation period is taken as 80% of the total fatigue life with the crack growth being the remaining 20% of the total life. For The purpose of discussion, numerical results will be found for $a_o = 5\,\text{mm}$, $N_o = 1.4 \times 10^6$ while different m and M values may be considered.

Making use of the $<\mu>$ and $<\sigma>$ in Table 2, another set of σ^*, μ^* and d^* will be found [9] with the same limits as those in Fig. 2a–c but their oscillations will differ. From the second of Eqs. (13), k_o is also found to fluctuate with the mean and standard deviation in Table 2. The results are displayed in Fig. 4. The difference of the oscillations in Figs. 3 and 4 is obvious. The assumed trend of the material degradation combined with the geometric and crack tightness properties represented by k_o is such that the amplitude of variations become decreasingly small in a random fashion as the fatigue life is approached. The specifics can be adjusted by altering the

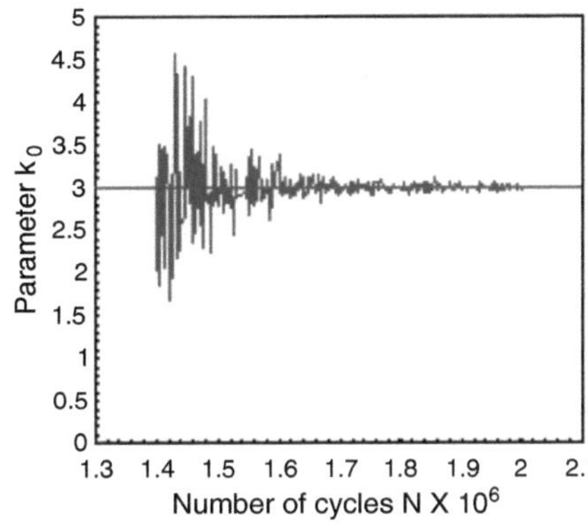

Fig. 4 Oscillation of segmented k_o versus N for normal distribution (Table 2)

limits of σ^*, μ^* and d^* and the values of $<\mu>$ and $<\sigma>$. Much of this knowledge will depend on the interaction of the fatigue crack with the governing parameters that will change during the life of the structure.

4.3 Change of Local Compliance with Fatigue Crack Growth

Compliance or its reciprocal the stiffness reflects the integrity of a structure. The change of compliance is indicative of the presence of damage and/or deterioration of the material properties. The change of the local compliance C in a specimen can be identified with the growth of a crack-like defect, the behavior of which has been used to develop the fracture control methodology for the inspection of aircrafts. The monolithic nature of the aircraft structure raises the sensitivity of the changes in the relation between dC and da where a denotes the crack length. In fact, the rate dC/da is known to increase progressively until it reaches a critical value when the onset of rapid fracture takes place under stress controlled situations. Strain or displacement controlled conditions are usually invoked in design to avoid catastrophic failure. The addition of redundant members in civil engineering structures is made for this purpose.

Consider a representative block of material that contains damage in the form of a line-crack of length 2a. The particular shape of the defect or configuration of damage can be reflected by the dC/da versus time or load cycle N relation. Moreover, there is no loss in generality by assuming a line crack 2a for the sake of illustration. The object is therefore to seek the relation between $C = C(N)$ and $a = a(N)$ or dC/da as a function of N to da/dN. The local displacement $u = u(N)$ and force $P = P(N)$ are presumably monitored to reflect the effect of the crack. For a block of width b, height h and thickness B, a relation between the equivalent strain u/h and equivalent stress P/bB can be written [10] as

$$\frac{u}{h}=\frac{1-\nu_{macro}^2}{E_{macro}}\left(1+\frac{2\pi a^2}{A_o}\right)\frac{P}{bB} \tag{18}$$

where $A_o = bh$ with E_{macro} and ν_{macro} being the macroscopic elastic modulus and Poisson's ratio. The compliance C of the block is defined as

$$C = \frac{u}{P} \tag{19}$$

It follows that

$$C = \frac{h}{bBE_{macro}}(1-\nu^2_{macro})(1+\frac{2\pi a^2}{A_o}) \tag{20}$$

yields a parabolic relation for $C = C(a)$. It will be shown subsequently that the compliance C will not depend on the shape of the representative block but only on the size of the area A_o. Differentiating both sides of Eq. (17) with N, there results

$$\frac{dC}{dN} = \Lambda_o a \frac{da}{dN} \text{ or } \frac{dC/dN}{\Lambda_o} = a\frac{da}{dN} \tag{21}$$

where

$$\Lambda_o = \frac{4\pi(1-\nu^2_{macro})}{b^2 BE_{macro}} = \varphi\frac{1-\nu^2_{macro}}{E_{macro}} \tag{22}$$

For the rectangular block in Fig. 5, $\varphi = 4\pi/b2B$. In general, φ may be regarded as the shape factor that can be applied to any shape of the representative block. Now, integration of Eq. (21) give

$$\int_{C_o}^{C} dC = \Lambda_o \int_{a_o}^{a} a da \tag{23}$$

and hence

$$\frac{C - C_o}{\Lambda_o} = \frac{1}{2}(a^2 - a_o^2) \tag{24}$$

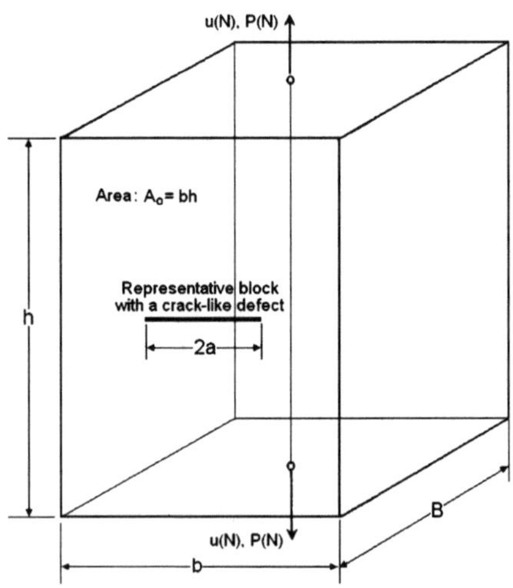

Fig. 5 Representative block containing damage by cracking

where C_o and a_o are the initial compliance and crack half length, respectively. While a_o can be arbitrary, C_o can be determined from Eq. (20) as

$$\frac{C_o}{\Lambda_o} = \frac{a_o^2}{2} + \frac{A_o}{4\pi} \tag{25}$$

Note that C_o depends on the area A_o of the representative block or the size of the region under investigation. Combining Eqs. (24) and (25), there results

$$\frac{C}{\Lambda_o} = \frac{a^2}{2} + \frac{A_o}{4\pi} \tag{26}$$

The area A_o gives a translation to the $C = C(a^2)$ relation. It will not affect $(dC/dN)/\Lambda_o$. Once $a = a(N)$ and da/dN are known, $(dC/dN)/\Lambda_o$ can be plotted as a function of N. The shape of the block in Fig. 5 is contained in Λ_o which does not affect the character of the curve for dC/dN as a function of N. Hence, $(dC/dN)/\Lambda_o$ applies to any representative block while a can be regarded as an equivalent damage length.

4.4 Randomness of Micro/Macro Modulus, Geometric and Restraining Parameters for Different Materials

Recall from Eq. (15) that the two parameters M and m need to be specified such that Eq. (17) can be solved for the crack length a with the same initial values of $a_o = 5\,\text{mm}$ and $N_o = 1.4 \times 10^6$. Equation (9) shows that M also depends on the microscopic length d_o and the stress amplitude σ_a which will be fixed and hence M can be regarded to depend only on M_o for the form of da/dN in Eq. (1). For the sigmoidal fatigue crack growth curve that possesses the three regions I, II and III, both M_o and m can be determined from the test data from the straight line extending from region II. Summarized in Table 3 are four sets of M and m values referred to, respectively, as I, II, III and IV. Material type I with $m = 2$ corresponds to the da/dN relation in [11] such that crack growth satisfies self-similarity while material type III pertains to the frequently referred to fourth power law in [12] when the dual

Table 3 Four sets of material related parameters for Eq. (15)

Material type	Y-intercept M	Slope m
I	4.6×10^{-7}	2
II	3.9×10^{-8}	3
III	3.7×10^{-9}	4
IV	3.91×10^{-10}	5

scale ΔK^{macro}_{micro} is replaced by the single scale ΔK for a macrocrack. It is worthwhile pointing out that the sigmoidal curve will still be retained using ΔK^{macro}_{micro} if the crack surface tightness adjustment is not assumed to depend on crack growth.

4.5 Crack Growth Relations

Using Eq. (14) for $m = 2$ and Eq. (17) for $m = 3$, 4 and 5, the curves in Fig. 6a–d are obtained. The range of crack growth has been set at approximately from 5 to 70 mm. Except for the case of self-similar crack growth with $m = 2$, the other three curves have a similar trend that is the random oscillation increases rapidly when the rate of crack growth towards the end of the fatigue life of two million cycles. Of particular interest is the different behavior of the curves in Fig. 6a for self-similar and Fig. 6c non self-similar crack growth, respectively.

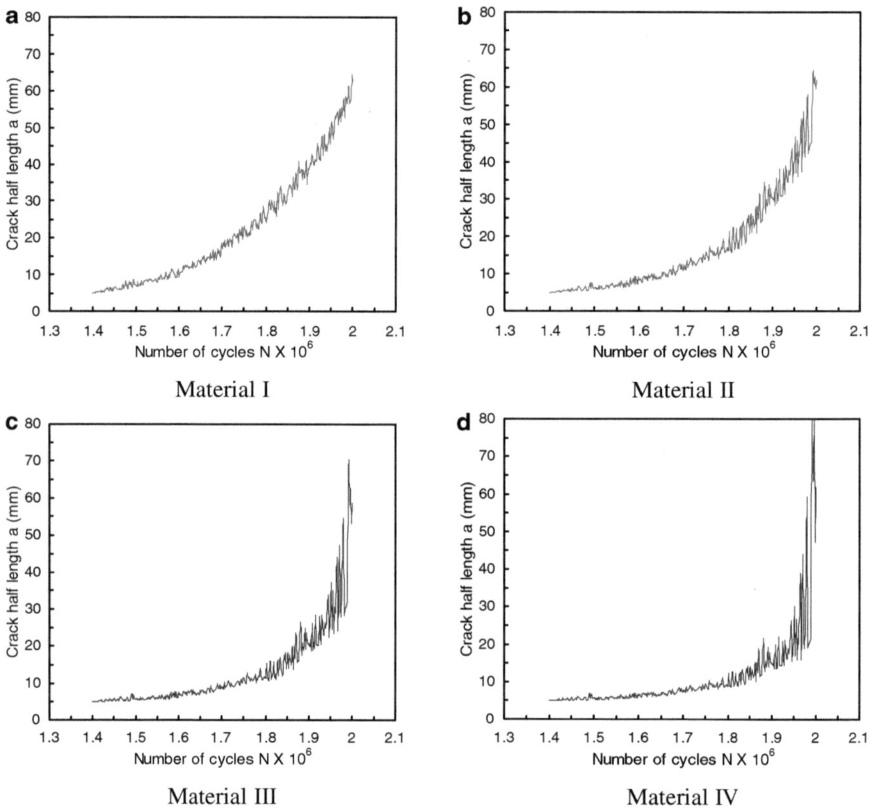

Fig. 6 Randomness of crack growth for materials I, II, III and IV

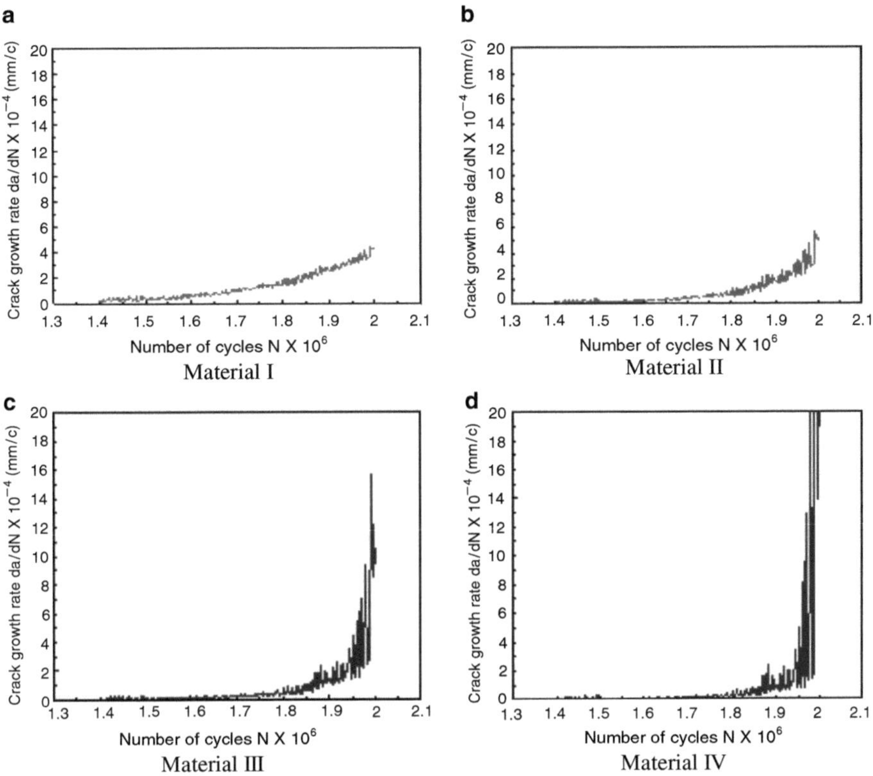

Fig. 7 Randomness of crack growth rate for materials I, II, III and IV

4.6 *Crack Growth Rate Relations*

Displayed in Fig. 7a–d are the crack growth rate da/dN of the curves in Fig. 6a–d. They are of the order of 10^{-4} mm/c for metals. The rise of da/dN for self-similar crack growth is relatively slow reaching a maximum of 4×10^{-4} mm/c. An increase of more than four time are shown for the results of da/dN in Fig. 7c, d when N reached 2×10^6 cycles.

4.7 *Compliance Versus Cycle*

The rise of the compliance C with N from Eq. (24) is different from that of a with N. Figure 8a, b can be compared to those for a = a(N) in Fig. 6a, b.

These results are not affected by the shape of the representative block chosen in Fig. 5 because the factor Λ_o in Eq. (22) that governs this behavior has been normalized with referenced to C.

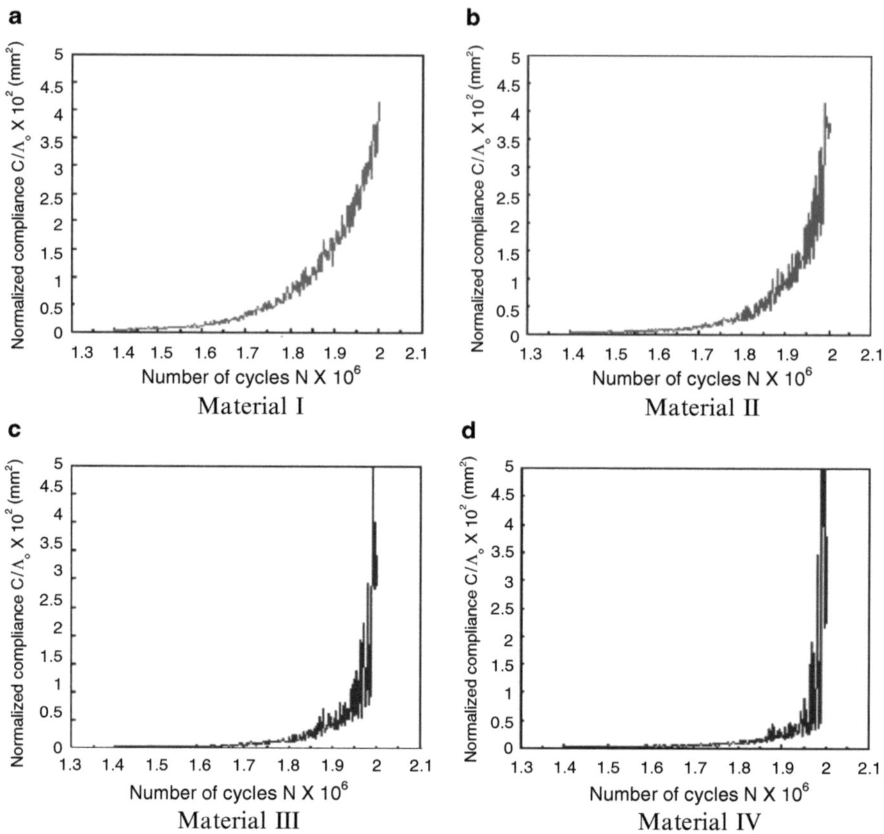

Fig. 8 Randomness of compliance for materials I, II, III and IV

4.8 Rate Change of Compliance Versus Cycle

Plotted in Fig. 9a–d are the rate change of the compliance with N. The violent oscillation of dC/dN for large N of material types III and IV in Fig. 9c, d may be exaggerated. Such behavior can be checked from specimen tests.

4.9 Rate Change of Compliance with Crack Length

The curves in Fig. 10a–d show the plots of $(dC/da)/\Lambda_o$ versus the number of cycles N. They can be obtained from Eqs. (21). The implication from fracture mechanics is that the rise of the slope of $(dC/da)/\Lambda_o$ gives an indication that the crack growth is tending towards instability. Since Λ_o has been factored out, the curves apply to any shape of the representative block in Fig. 5. The foregoing presumption is based on

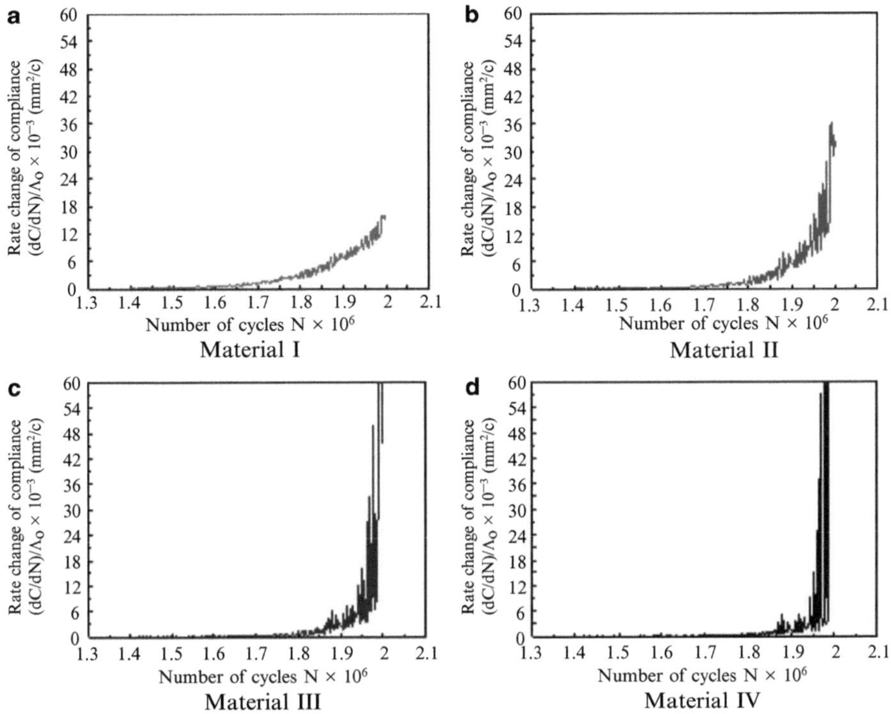

Fig. 9 Randomness of compliance rate for materials I, II, III and IV

idealized laboratory conditions that may not coincide with the field conditions. Adjustments are always needed when connecting field and laboratory results. One of the conditions needs to be studied is the critical value of dC/da corresponding to the onset of crack growth. Figure 10 shows that dC/da would depend on the mean value $<\mu>$ and standard deviation $<\sigma>$ depicted for the micro/macro parameters. Nevertheless the trend of the data should be relevant to the tendency of the crack-like defect to become unstable. Time and experience will provide more precise interpretation of the $(dC/da)/\Lambda_o$ versus time data.

5 Concluding Remarks and Future Work

As a first attempt to simulate microscopic effects with random oscillations, the corresponding fatigue crack growth characteristics are derived. They also exhibit oscillations although the transition from micro to macro behavior that the model is capable of analyzing was not studied because of limited time and space of this writing. Much of the results presented are manifestation of scaling adjustment for measuring laboratory data, say for stress and strain. Smooth curves are usually

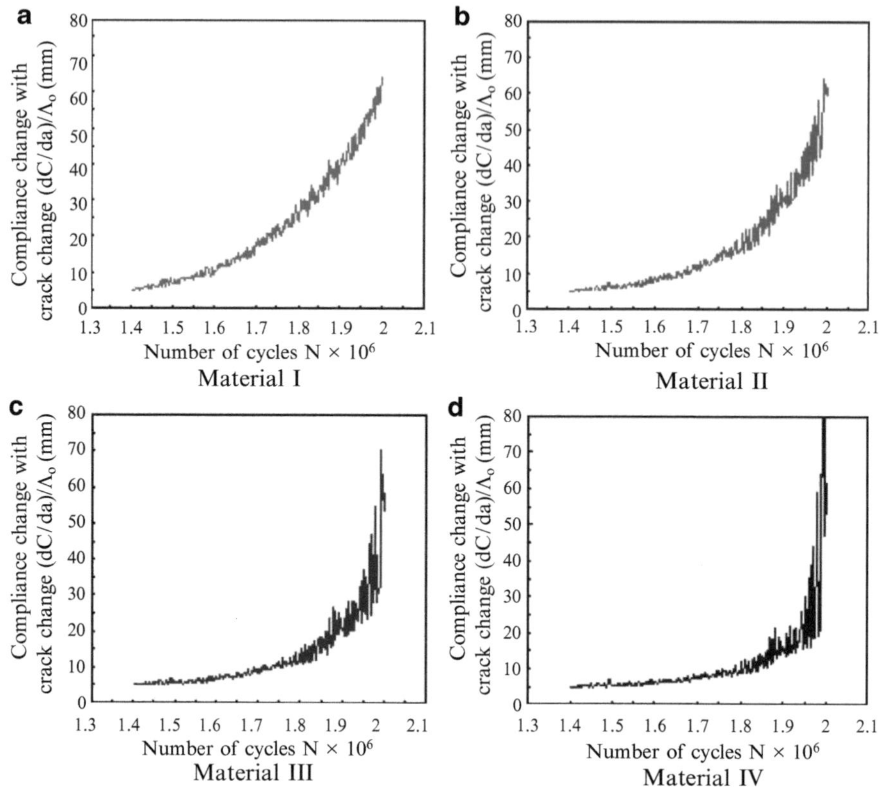

Fig. 10 Randomness of compliance change versus crack length change for materials I, II, III and IV

adjusted for monitoring macroscopic phenomena while microscopic effects would be oscillatory in character. Normal distribution is used for generating random values of the micro/macro stiffness, micro-length and material restraining parameters. Required are the appropriate selection of the mean $<\mu>$ and standard deviation $<\sigma>$ for a given range of the crack growth behavior and the local compliance that has been chosen as the quantity to be monitored for a structure. From the view point of fracture mechanics damage by micro/macro cracking can be best reflected by the change of the compliance with the crack size. Although the line crack model was used in the analysis, the formulation shows that the shape of the representation block can be normalized such that the results are valid regardless of the actual shape of the damage. It suffices to consider the resulting change in the compliance with time or the number of cycles in fatigue. Morphology of defects and/or inhomogeneities are reflected the compliance and its rate change with N. Structural integrity becomes threatened only when the change of local compliance begins to affect the global instability. Such an assessment can be made by finding the local and global stationary values [13] of the energy density function dW/dV that is expressible in terms of

$\Delta S/\Delta r$. The increment change ΔS at an increment distance Δr away from the damage in question is a more general quantity to use for estimating fatigue crack growth da/dN.

The main object of this work was to explore the application of fracture mechanics. This was accomplished by relating the change of compliance with the line crack as an indicator for the damage of the structure. Randomness of the material microstructure and micro/macro geometric parameters was shown to play a role in the interpretation of the data. Consistency in multiscaling of size and time is found to be critical. The preliminary findings provide a way to filter out the microscopic effects that can interrupt the macroscopic signals that are not always obvious. The exclusion of results at a pre-determined scale can greatly simplify model development.

References

1. Sih G. C. and Jeong D. Y., Hysteresis loops predicted by isoenergy density theory for polycrystals. Part I: fundamentals of non-equilibrium thermal–mechanical coupling effects, Theoretical and Applied Fracture Mechanics, 2004, 41: 233–266.
2. Sih G. C. and Jeong D. Y., Hysteresis loops predicted by isoenergy density theory for polycrystals. Part II: cyclic heating and cooling effects predicted from non-equilibrium theory for 6061-T6 aluminum, SAE 4340 steel and Ti–8Al–1Mo–1V titanium cylindrical bars, Theoretical and Applied Fracture Mechanics, 2004, 41: 267–289.
3. Sih G. C., Fatigue crack growth and macroscopic damage accumulation, in: Sih G. C. and Zorski H. (Eds.), Defects and Fracture, Martinus Nijhoff, Boston, MA, 1980, 53–62.
4. Sih G. C., Mechanics of Fracture Initiation and Propagation, Kluwer, Boston, MA, 1991.
5. Standard test method for plane-strain fracture toughness of metallic materials, Annual Book of ASTM Standards, Part 10, American Society for Testing and Materials, Philadelphia, PA, 1981.
6. Sih G. C. and Tang X. S., Dual scaling damage model associated with weak singularity for macroscopic crack possessing a micro/mesoscopic notch tip, Journal of Theoretical and Applied Fracture Mechanics, 2004, 42(1): 1–24.
7. Sih G. C. and Tang X. S., Simultaneous occurrence of double micro/macro stress singularities for multiscale crack model, Journal of Theoretical and Applied Fracture Mechanics, 2006, 46(2): 87–104.
8. Molent L. and White P., Private communication, on: Microcrack tip opening related to fatigue striations in aluminum alloy that the opposing microcrack surfaces are untouched: Defence Science and Technology Organisation, Australia, October 2006.
9. XU S. L., FORTRAN Program Collections of Common Computational Methods, Tsinghua University Press, Beijing, China, 2003.
10. Sih G. C. and Baisch E., Angled crack growth estimate with overshoot and/or overload, Journal of Theoretical and Applied Fracture Mechanics, 1992, 18: 31–45.
11. Liu H. W., Fatigue crack propagation and applied stress range, ASME Transaction, Journal of Basic Engineering, 1963, 85D(1): 115–122.
12. Paris P. C., The growth of cracks due to variations in load, Ph.D. dissertation, Department of Mechanics, Lehigh University, 1962.
13. Chue C. H. and Wei Y. D., Failure initiation sites and instability of structural components with localized energy density, Journal of Theoretical and Applied Fracture Mechanics, 1991, 15(1): 163–177.

Exploitation of the TRIP Effect for the Development of Formable, Fracture and Fatigue Resistant Steels for Automotive Applications

Gregory N. Haidemenopoulos

Abstract The present paper summarizes recent and on-going work on the exploitation of TRansformation-Induced Plasticity (TRIP) in an effort to develop automotive steels which would possess high-strength combined with high formability while these steels could also exhibit high fracture and fatigue resistance. Especially for the automotive industry, the driving force for these developments is the vehicle weight reduction, which could eventually lead to lower fuel consumption combined with reduced greenhouse gas emissions.

The discussion starts with the modelling and characterization of the retained austenite stability (the transforming phase) as well as the modelling of the transformation kinetics, i.e. evolution of transformation with plastic strain.

Enhancement of formability is discussed next. Constitutive micromechanical modelling has been employed for the calculation of forming limit diagrams (FLD) for these complex steels, an issue of great practical importance for the optimisation of stretch-forming and deep-drawing operations.

Fracture resistance can be considerably increased by the TRIP effect. A review is made of the "transformation toughening" arising from metastable austenitic dispersions in ultrahigh strength steels and the high fracture toughness achieved in this class of materials. The potential of the TRIP effect in increasing fatigue strength has not yet received considerable attention. However once we understand the fatigue behaviour of these materials, new applications, e.g. long products made of TRIP steel, might emerge for automotive applications.

Keywords TRIP steels · Formability · Austenite stabilization · Transformation toughening

---

G.N. Haidemenopoulos
Department of Mechanical Engineering, University of Thessaly, Volos, Greece
e-mail: hgreg@mie.uth.gr

S. Pantelakis, C. Rodopoulos (eds.), *Engineering Against Fracture: Proceedings of the 1st Conference*,

1 Introduction

Low-alloy TRIP steels constitute a relatively new class of steels that exhibit excellent combinations of strength and cold-formability, making them particularly suitable for sheet-forming applications in the automotive industry. These steels possess a multiphase microstructure containing ferrite, bainite and retained austenite. During cold-forming operations, such as stretch-forming and deep-drawing, the retained austenite transforms to martensite under the action of the applied stresses and strains. This mechanically-induced martensitic transformation of the retained austenite is responsible for the TRansformation-Induced Plasticity (TRIP) effects found in these materials. The major effect of TRIP is plastic flow stabilization which leads to significant improvements in ductility, formability and fracture toughness. The objective of this paper is to review current work in TRIP steels and discuss the issues of formability, fracture and fatigue resistance in these steels.

2 Austenite Stabilization

TRIP effects in the steels under consideration are influenced significantly by *austenite stability*. If austenite stability is low, then the austenite will transform very early during the forming operation, without beneficial effects in formability. On the other hand, if austenite stability is too high, then the austenite might not transform at all during the forming operation. Thus, an optimum level of stability is required in order to maximize formability. Austenite stability depends on the following factors: (a) *chemical composition*, (b) *austenite particle size*, (c) *matrix strength* and (d) *stress state*. These factors correspond to different scale levels, as depicted in Fig. 1. Chemical composition refers to the atomic scale. In order to increase stability, the austenite should be enriched with carbon or other stabilizing elements by atomic diffusion. Austenite particle size refers to the micro-scale or the so-called microstructure scale. In order to increase stability, the austenite particle should be made finer, in order to reduce the probability of finding heterogeneous nucleation sites in the particles which would trigger the martensitic transformation.

The strength of the matrix also refers to the micro-scale. High strength destabilizes the austenite due to the higher mechanical driving force contribution to the total driving force for the martensitic transformation. In addition, it is well established that strength is controlled by microstructure via the various strengthening mechanisms. Finally, the stress state refers to the macro-scale. Its effect originates from the interaction of the transformational volume change with stress triaxiality. High stress triaxiality destabilizes the austenite. For example, consider an austenite particle, with a certain size and chemical composition, which could resist transformation in uniaxial tension. The same particle would transform readily in the triaxial stress field ahead of a crack tip. This is of particular importance when designing TRIP steels for forming applications, due to the complex stress states found in these processes. The austenite stability should be tuned for maximum TRIP interactions

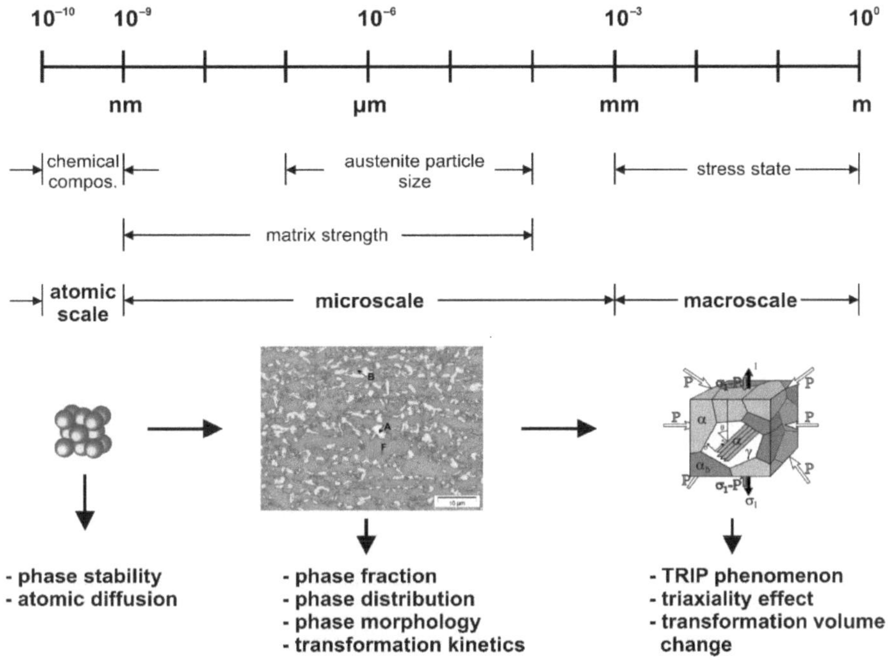

Fig. 1 The parameters and scale-levels affecting austenite stability

for the particular stress state. It is apparent that a model for the stability of austenite, accounting for all these factors, is needed. Since the parameters of the model come from different size scales, the model itself should possess a multiscale character, in the sense that it will incorporate these parameters in the same equation. The first step towards developing such a model is to identify a single parameter to characterize stability. In this work, the M_s^σ temperature has been chosen for this task, in the same way that the M_s temperature is used to characterize the stability of austenite against transformation on cooling. The M_s^σ temperature can be defined by considering the mechanically-induced martensitic transformation of austenite [1] in Fig. 2. Spontaneous transformation occurs on cooling the austenite to the M_s temperature. This transformation is triggered at pre-existing nucleation sites in the austenite. The same sites can operate also above the M_s temperature under the action of an externally applied stress.

The higher the temperature, the higher the required stress. This *stress-assisted* transformation is denoted by line AC. At point C and at the M_s^σ temperature, the applied stress reaches the yield stress of the austenite. Above M_s^σ new potent nucleation sites, which are produced by the plastic deformation of austenite, trigger the *strain-induced* transformation. Thus the M_s^σ temperature defines a boundary between the temperature regimes where separate modes of transformation dominate: below M_s^σ the transformation is stress-assisted and above M_s^σ the transformation is strain-induced. Near the M_s^σ temperature both modes operate. Due to transformation

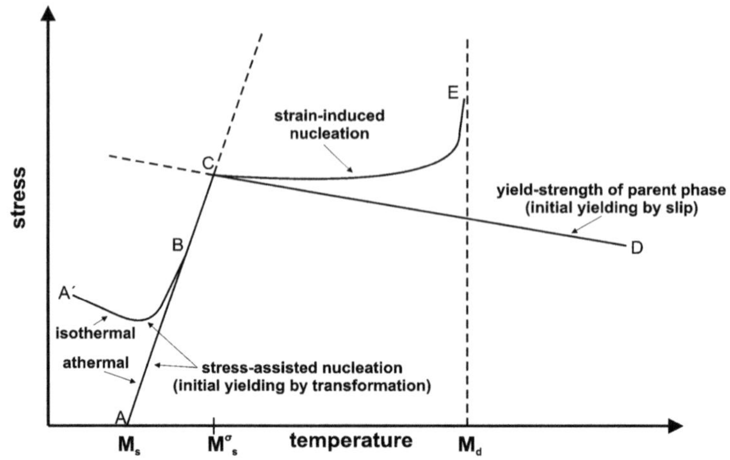

Fig. 2 Stress-temperature interactions in TRIP steels [1]

plasticity the observed yield stress follows the stress for stress-assisted transformation below the M_s^σ. A reversal in the temperature dependence of the flow stress provides a convenient determination of the M_s^σ temperature. Actually this technique has been used successfully for the determination of the M_s^σ temperature in steels containing austenitic dispersions, either as retained austenite in martensitic steels [2], or more recently for retained austenite in low-alloy TRIP steels [3]. Having defined the M_s^σ temperature as the single parameter characterizing stability, the model objective is to develop an expression for the M_s^σ as a function of the four stability parameters defined previously. The model is based on the fact that, for the case of stress-assisted transformation, the applied elastic stress aids the transformation kinetics by modifying the effective potency distribution of pre-existing nucleation sites. The model details are given elsewhere [4]. The final result is a general expression for the M_s^σ temperature as:

$$M_s^\sigma = (a_1 + a_2 X_C)^{-1} \left\{ \begin{array}{l} b_1 + b_2 X_C + b_3 X_{Mn} + b_4 X_C X_{Mn} + \dfrac{c_1}{\ln(c_2/V_p)} + \\ \sigma_y [d_1 + d_2(\sigma_h/\bar{\sigma})] \end{array} \right\} \tag{1}$$

In Eq. (1), X_i denotes the mole fraction of alloying elements (i.e. the chemical composition) in austenite, V_p is the mean austenite particle volume, σ_y is the matrix yield strength and $\sigma_h/\bar{\sigma}$ is the ratio of the hydrostatic to the von Mises equivalent stress, which characterizes the triaxiality of the stress state. The constants (a_i, b_i, c_i, d_i) in Eq. (1) depend on the specific steel composition and, as stated earlier, are given in [4]. The key point is, however, that Eq. (1) provides an analytical expression for the M_s^σ temperature as a function of the chemical composition of austenite, the austenite particle size, the yield strength and the stress state. It is interesting that parameters from three size scales (atomic, micro and macro) enter this equation.

As stated in the introduction, TRIP effects and the associated formability enhancement depend directly on the amount and stability of retained austenite.

A large research effort over the last years has focused on establishing the suitable heat-treatments for obtaining stable retained austenite in the microstructure. A typical heat-treatment for the production of a TRIP steel involves intercritical annealing to produce a ferrite-austenite mixture, followed by cooling to an intermediate temperature and holding for the isothermal transformation of austenite to bainite. During the bainitic transformation, carbon is rejected from the forming bainite to the austenite. This carbon stabilizes the remaining austenite against martensitic transformation on cooling to room temperature. Optimisation of the design of new alloy compositions and/or processing routes, necessitate the clarification of the effect of chemical composition and heat-treatment conditions on the evolution of the bainitic transformation. A very important modelling effort, based on the principles of the "nucleation-control" approach, was recently presented by Azuma et al. [5]. This model takes simultaneously into account all the aforementioned metallurgical processes that may occur during the bainitic transformation. The model can predict the overall transformation kinetics, with and without the precipitation of cementite. In order to apply the model, driving forces for ferritic bainite and cementite have to be calculated for the upper and lower bainite regimes. The model has been applied [6] for the calculation of the volume fraction as well as the carbon content of retained austenite as a function of chemical composition of the steel and heat treatment conditions (bainite isothermal transformation time and temperature). Results are shown in Fig. 3 for a Fe–0.6C–1.5Mn–1.5Si (in mass%) steel treated at 400°C. The results indicate a significant carbon enrichment of retained austenite.

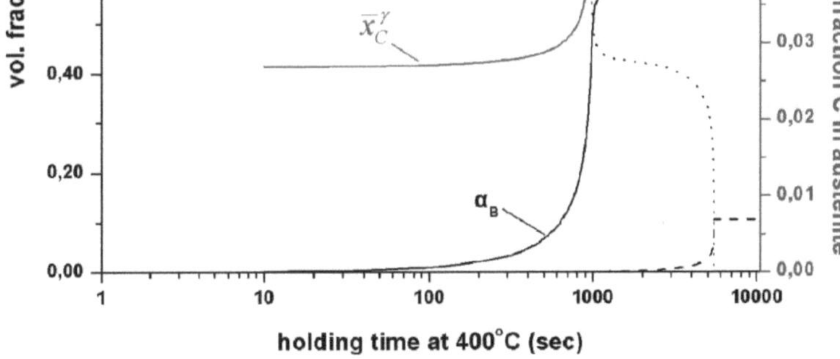

Fig. 3 Variation of volume fraction and carbon content (X_C) of retained austenite (γ) as a function of holding time during the bainitic transformation [6]

3 Kinetics of the Mechanically-Induced Transformation

As already explained with the aid of Fig. 2, transformation-induced plasticity can occur by two distinct mechanisms, i.e. stress-assisted and strain-induced transformation of retained austenite to martensite. In the stress-assisted regime martensite forms on pre-existing nucleation sites, whereas in the strain-induced regime new and more potent nucleation sites are created by plastic deformation of austenite. As the steel is stressed and deformed, retained austenite will transform to martensite by the simultaneous operation of both mechanisms. The stress-assisted mechanism prevails at stresses lower than the yield-strength of austenite, whereas the strain-induced mechanism prevails after the yield-strength has been surpassed. The total volume fraction of retained austenite transforming to martensite (f) can be expressed in the following form:

$$f = f(\sigma)_{stress} + f(\sigma, \varepsilon)_{strain} \tag{3}$$

In Eq. (3), $f(\sigma)_{stress}$ and $f(\sigma, \varepsilon)_{strain}$ denote the contributions of the stress-assisted and strain-induced mechanisms, respectively.

A model described in [7] gives the expression of Eq. (3) as

$$f = 1 - \exp\left\{-V_p\left[N_v^o \exp\left(\frac{2\gamma_s \alpha_{stress}}{A\rho}\right) + N\left[1-\exp\left(-k\varepsilon^n\right)\right]\exp\left(\frac{2\gamma_s \alpha_{strain}}{B\rho}\right)\right]\right\} \tag{4}$$

Equation (4) establishes a relation between the vol. fraction of retained austenite transforming to martensite (f) and true plastic strain (ε). It is, therefore, possible to predict the kinetics of the mechanically-induced transformation as a function of plastic strain. Figure 4 depicts calculated (lines) and experimentally measured (symbols) "$f - \varepsilon$" kinetics for two typical low-alloy TRIP steels. The nominal chemical composition of steel "TRIP 1" was 0.20C–1.40Mn–0.50Si–0.70Al (in mass%), whereas steel "TRIP 2" had the same composition with a small addition of 0.03 mass% Nb. As shown, model results display good agreement with experimental measurements. It should also be noted that a significant amount of retained austenite is transforming in the elastic region, due to the stress-assisted mechanism.

4 Enhancement of Formability

Based on micromechanical considerations, a constitutive model was developed [8] for the mechanical behavior of TRIP steels. In particular, based on previous work for dual-phase steels [9], constitutive equations were developed for four-phase TRIP steels, which consist of a ferritic matrix with dispersed bainite and austenite, the later transforming gradually to martensite as the material deforms plastically. Details of the model can be found in [8]. This constitutive model was used to predict the form of forming limit diagrams for TRIP steels. A sheet made of TRIP steel is considered, which is deformed uniformly on its plane in such a way that the in-plane

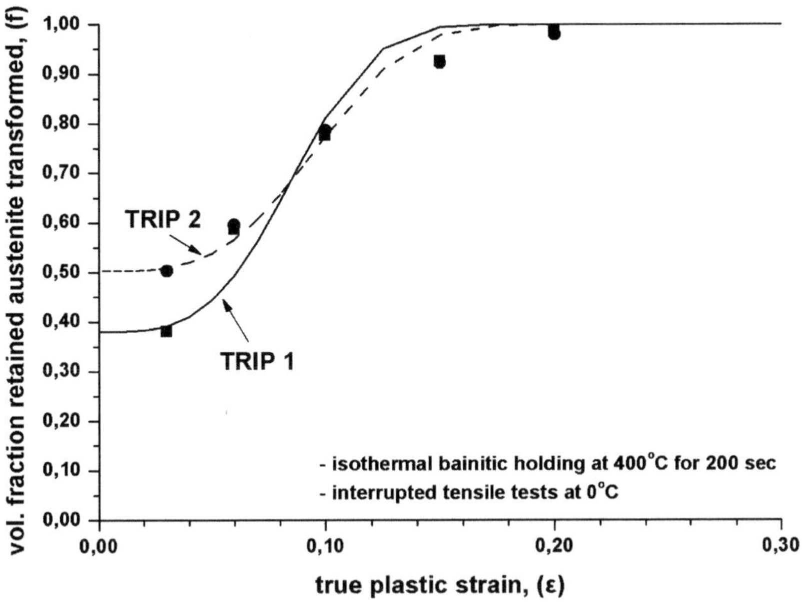

Fig. 4 Volume fraction retained austenite transformed to martensite as a function of plastic strain [7]

principal strain increments increase in proportion. The possibility of the formation of a neck in the form of a narrow straight band is studied, and the corresponding forming limit diagram is constructed. The sheet is assumed to contain a small initial inhomogeneity (imperfect sheet) and necking results from a gradual localization of the strains. The inhomogeneity is in the form of straight narrow band (neck) of reduced thickness $H^b < H$. Figure 5 shows forming limit curves obtained for imposed proportional straining ρ for two different values of the initial thickness imperfection, namely $H^b/H = 0.999$ and $H^b/H = 0.99$. The two solid curves correspond to the TRIP steel, whereas the dashed curves are for the non-transforming steel. The TRIP phenomenon increases the necking localization strains. A comparison of the model predictions with available experimental data is also presented in Fig. 5. An "Erichsen" universal sheet metal testing machine was employed for the experiments. A hemispherical punch with a diameter of 50 mm was used and the punch velocity was set to 1 mm/s. The agreement between the model prediction and the experimental data is reasonable.

5 Enhancement of Fracture Toughness

Transformation toughening is one of the major consequences of the TRIP effect in steels. Transformation toughening has been studied extensively in the past for Ultrahigh-strength secondary-hardening martensitic steels based on the Fe–Ni–Co

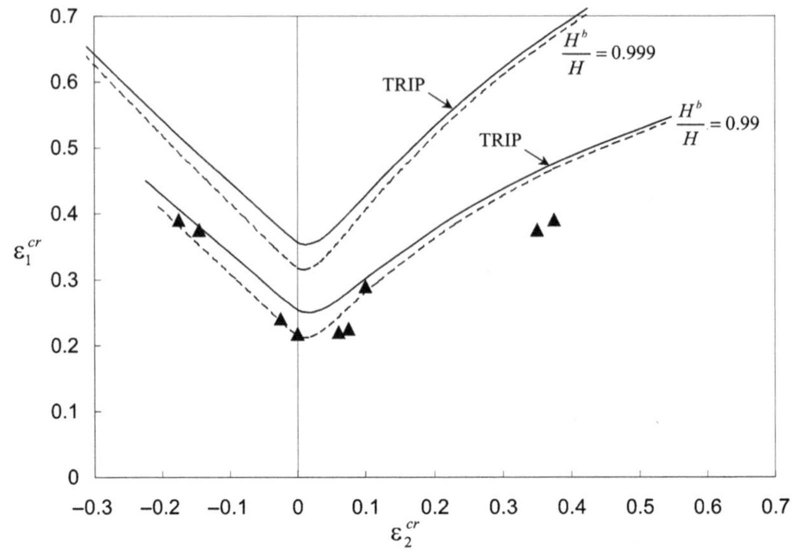

Fig. 5 Forming limit curves for two different values of initial thickness inhomogeneities H^b/H. Solid lines correspond to the TRIP steel, whereas the dashed lines are for a non-transforming steel. Dark triangles are experimental data [8]

system. Details of this research can be found in [10, 11]. Research was focused on the AF1410 steel the composition of which is Fe–0.16C–14Co–10Ni–1Mo–2Cr (in mass%). The standard tempering treatment of this steel is performed at 510°C for 8 h. In order to get TRIP effects, an austenite dispersion should be formed. For this purpose, a two-step heat treatment was applied consisting of heating at 600°C for 15 min followed by the normal tempering of 510°C for 8 h. The purpose of this heat treatment was to increase the driving force for austenite precipitation and enhance the diffusivity of Ni. The combination of these effects produced an ultrastable intralath austenitic dispersion with a high degree of Ni-enrichment, the nickel content reaching almost 30 mass%. This compositional enrichment combined with the very small size of the austenite particles (10–20 nm) made this dispersion extremely stable. Tensile testing and J_{IC} fracture toughness testing was performed for the two-step treatment as well as for the standard 510°C/8 h treatment. The yield strength, UTS, uniform strain, fracture strain, J_{IC} fracture toughness and the volume fraction of precipitated austenite (determined by X-ray diffraction) are listed in Table 1. The two-step treatment possessed higher uniform ductility and fracture strain relative to the standard treatment. The J_{IC} fracture toughness increased about 25%. This increase is higher than the percentage increase in the tensile properties indicating that the TRIP effects associated with intralath austenite interact more with the crack tip stress state rather than the stress state corresponding to uniaxial tension.

These types of ultra-high strength steels (e.g. steel AerMet 100), possessing transformation toughening effects have been used with success by the automotive

Table 1 Tensile properties and JIC fracture toughness of standard and two-step tempering of AF1410 steel [10, 11]

Tempering treatment	510°C/8h	650°C/15min ± 510°C/8h
Yield strength (MPa)	1,362	1,370
UTS (MPa)	1,576	1,573
Uniform strain	0.096	0.111
Fracture strain	1.21	1.35
$J_{IC}(\text{kJ/m}^2)$	201	250
Volume fraction austenite	0.06	0.15

industry in the drive shafts of racing cars. However due to cost reasons these steels are not yet used in mass production.

Transformation toughening effects in low-alloy multiphase TRIP steels have not been studied extensively. In a recent investigation [12] the failure mechanisms of low alloy TRIP steels were studied in unnotched and notched tensile tests to identify the effect of microstructure on damage initiation and development. It was found that the major parameters affecting fracture mode in automotive TRIP steels are (a) the initial TRIP microstructure (especially if martensite is present where it accelerates crack initiation), (b) dislocation density (by affecting stability of retained austenite), (c) neighbourship conditions (e.g. austenite within bainite colonies leads to easy crack path through bainite sheaves), (d) stress state (triaxiality destabilises austenite and accelerates damage development). The interactions are highly complex and therefore more work is needed to gain insight in the design of damage tolerant automotive TRIP steels.

6 Implications on Fatigue Resistance

The fatigue resistance of low-alloy TRIP steels has not been an issue up to now, probably because the TRIP steels are used under static loads or at the extreme case of high-strain rate loading for crash resistance. TRIP–fatigue interactions are not well understood and contradicting results appear in the literature. Cyclic hardening and retardation of fatigue crack growth has been associated with compressive residual stresses arising from the transformation of retained austenite [13, 14]. In low-cycle fatigue, at low strains TRIP effects seem to delay crack initiation while at high strains the martensite formed during fatigue accelerates crack growth [15]. Questions arise as to the role of austenite stability and amount on fatigue behaviour and the interaction of strain-induced transformation on cyclic plasticity and fatigue crack initiation. Work is currently underway to shed light on the effect of austenite stability on the cyclic plasticity mechanisms associated with fatigue crack initiation [16]. A thorough understanding of these issues is a prerequisite for a reliable design of TRIP steels for fatigue resistance.

Acknowledgments The author is grateful to Dr. A.Vasilakos, Dr. A. Katsamas and Professor N. Aravas for their contribution in various aspects of research in TRIP steels. Financial support for this research has been granted from EU through the ECSC/RFCS programs and from the Greek Ministry of Education through the IKYDA program.

References

1. Olson, G.B., Mechanically-Induced Phase Transformations in Alloys, Encyclopedia of Materials Science and Engineering (ed.) Bever, M.B., Pergamon Press, Cambridge, MA, 2929–2932 (1986).
2. Haidemenopoulos G.N., Grujicic M., Olson G.B., and Cohen M., "Transformation Microyielding of Retained Austenite", Acta Metall., Vol. 37, No. 6, pp. 1677–1682 (1989).
3. Vasilakos A.N., Papamantellos K., Haidemenopoulos G.N., and Bleck W., "Experimental Determination of the Stability of Retained Austenite in Low-Alloy TRIP Steels", Steel Res., Vol. 70, No. 11, pp. 466–471 (1999).
4. Haidemenopoulos G.N and Vasilakos A.N., "Modelling of Stability in Low Alloy Triple-Phase Steels", Steel Res., Vol. 67, No. 11, pp. 513–519 (1996).
5. Azuma, M., Fujita, N., Takahashi, M., Senuma, T., Quidort, D., and Lung, T., "Modelling Upper and Lower Bainite Transformation in Steels", ISIJ Int., Vol. 45, No. 2, pp. 221–228 (2005).
6. Katsamas, A. and Haidemenopoulos, G.N., Research Work in Progress, University of Thessaly, Volos, Greece (2008).
7. Haidemenopoulos, G.N., Katsamas, A., and Aravas, N., "Stability and Constitutive Modelling in Multiphase TRIP Steels", Steel Res. Int., Vol. 77, No. 9–10, pp. 720–726 (2006).
8. Papatriantafillou, I., Agoras, M., Aravas, N., and Haidemenopoulos G., "Constitutive Modeling and Finite Element Methods for TRIP Steels", Comput. Meth. Appl. Mech. Eng., Vol. 195, pp. 5094–5114 (2006).
9. Stringfellow R.G., Parks D.M., and Olson G.B., "A Constitutive Model for Transformation Plasticity Accompanying Strain-Induced Martensitic Transformation in Metastable Austenitic Steels", Acta Metall. Mater., Vol. 40, pp. 1703–1716 (1992).
10. Haidemenopoulos, G.N., Dispersed-Phase Transformation Toughening in Ultrahigh-Strength Steels, Doctoral thesis, Department of Materials Science and Engineering, Massachusetts Institute of Technology, Cambridge, MA (1988).
11. Haidemenopoulos, G.N., Olson, G.B., and Cohen, M., Dispersed-Phase Transformation Toughening in Ultrahigh-Strength Steels, Proceedings of 34th Sagamore Army Materials Conference on Innovations in Ultrahigh-Strength Steel Technology (eds.) Olson, G.B., Azrin, M., and Wright, E.S., pp. 549–592 (1990).
12. Papaefthimiou, S., Failure Mechanisms of Multiphase Steels, Doctoral thesis, IEHK, RWTH Aachen, Germany (2005).
13. Sugimoto, S.I., Kobayashi, K.I., and Hashimoto, S.I., "Low-Cycle Fatigue Hardening of TRIP-Aided Dual Phase Steels", J. Jpn. Inst. Met., Vol. 54, No. 12, pp. 1350–1357 (1990).
14. Yokoi, T., Kawasaki, K., Takahashi, M., and Mitzui, M., "Fatigue Properties of High Strength Steels Containing Retained Austenite", JSAE Rev., Vol. 17, No. 1, pp. 210–212 (1996).
15. Hu, Z.Z., Ma, M.L., Liu, Y.Q., and Liu, J.H., "The Effect of Austenite on Low Cycle Fatigue in Three-Phase Steel", Int. J. Fatigue, Vol. 19, No. 8–9, pp. 641–646 (1997).
16. Dimakarakou, E., Austenite Stabilization, Cyclic Deformation and Fatigue of Low-Alloy TRIP Steels, Doctoral thesis Research in Progress, Department of Mechanical Engineering, University of Thessaly, Volos, Greece (2008).

Hybrid Metal Laminates for Low Weight Fuselage Structures

Marco Pacchione and Elke Hombergsmeier

Abstract Metal Laminates (MLs), combined with selective reinforcements and bonded design principles, is a hybrid structural concept suitable for use in primary fuselage structures. Following a screening of the static, fatigue and damage tolerance performances of thin MLs coupons an investigation, focused on identification of fuselage areas best suited for ML application has been performed. After a weight optimization analysis large panel tests have been defined and tested to investigate the two main structural design requirements: the post-buckling shear-compression response, and the F&DT behavior including large damage capability, of thin ML shells.

Keywords Metal laminates · Adhesive bonding · Selective reinforcements · Damage tolerance

1 Introduction

Metal laminates (MLs) are produced by adhesive bonding of a number of thin sheets to obtain a larger thickness. The ML concept is not new, Schijve, in [1], based on investigations performed at Delft University in the 1970s, suggested that MLs could significantly improve the damage tolerance performances of aircraft structures due to four main advantages in comparison to monolithic materials.

M. Pacchione (✉)
Generic Design and Research, Airbus Deutschland GmbH, Kreetslag 10, 21129 Hamburg, Germany
e-mail: marco.pacchione@airbus.com

E. Hombergsmeier
EADS Innovation Works Germany – Metallic Structures, P.O. Box 80 04 65, 81663 Munich, Germany
e-mail: elke.hombergsmeier@eads.net

S. Pantelakis, C. Rodopoulos (eds.), *Engineering Against Fracture: Proceedings of the 1st Conference*,

1. Thickness effect: fatigue crack growth in thin sheets is somewhat slower than in thick sheets and plates (plane stress/plane strain).
2. Peak load delay: under variable-amplitude loading high peak loads produce larger plastic zones in thin sheets and longer crack growth delays.
3. Crack arrest: many cracks start as part through cracks and the initial growth also occurs in the thickness direction. Adhesive layers will retard crack growth in the thickness and as a result crack growth in the width direction will also be reduced as a consequence of restraint on crack opening by adjacent non-cracked laminas.
4. Fracture toughness: K, for a thin sheet (plane stress) is higher than a valid K_{Ic} for a thick plate (plane strain).

Metal laminates concepts for fuselage structure applications have been investigated in frame of the EU Project (6th Framework) DIALFAST – Development of Innovative and Advanced Laminates for Future Aircraft STructures. A description of the complete project is given in [2].

2 Current Challenges for a Metallic Fuselage

General Airbus strategy is to have a strong and clear competition between metallic and composite design approaches in order to push both technologies to their frontier, with the final goal of ensuring delivery of the most efficient and advanced structure to the customers.

New aluminium alloys have potential for weight savings although are not expected to directly contribute to cost savings. Advanced bonded design, has potential to deliver weight savings and extension of the inspection intervals compared to existing metallic aircraft structures. MLs in a bonded design shall be considered as an additional possibility to improve material properties and design to achieve additional weight savings. In MLs, dissimilar materials for skin laminas, selective reinforcements and doublers can be mixed to tailor the mechanical performances; furthermore a layered configuration decreases material costs (buy to fly ratio) of expensive alloys and minimize chemical/mechanical milling operations.

Cost reduction is not the main attribute for bonded structures; to make them affordable it is essential to implement a cost aware design. MLs to be used as skin material for fuselage panels may result in an affordable manufacturing process following recommendations like:

- Reduction of lay-up efforts keeping the number of laminas to the minimum
- Use of single waffle plate instead of separate doublers
- No forming for low thickness sheets (usually less than 2 mm), curvature to be induced by the bonding process
- Simultaneous bonding of several parts (stringers, window frames, doublers, etc.)
- No joggling for stringers

2.1 Design Drivers for a Bonded ML Fuselage

Fuselage adhesive bonded panels have the highest Fatigue and Damage Tolerance (F&DT) allowable stresses among the different metallic manufacturing technologies available today, see [3]. The bonded design is the most suited for the upper part of an aircraft fuselage, which is typically dominated by F&DT requirements.

Metal bonding as production process for panels in lower fuselage areas has never been used in Airbus fuselage to avoid risk of bondline corrosion in the presence of bilge fluid. The bondline corrosion is today scientifically understood, and with use of unclad aluminium sheets, the adhesive bonding process is technically possible even in the bilge area. Bonding can deliver a significant advantage in structural areas, like the lower shells or areas with large opening for doors and windows, where a combination of static stability and F&DT is required.

Other important criteria for fuselage panels are a long durability and resistance to small impacts (hail, runway debris, etc.) to minimize maintenance burden to the customers.

The ML structural concept is expected to contribute to all the aspects mentioned above improving the design in areas dominated by F&DT as well as static stability, increasing the fatigue life and the impact resistance.

3 Experimental Investigations

A range of coupon tests has been carried out in order to screen the basic mechanical performance as well as the durability of the proposed MLs. The materials tested were produced by adhesive bonding of two, and in some cases three, thin aluminium sheets, as shown in Fig. 1. The alloys that were investigated are:

- 2024 T3 clad, 1.6 mm and 0.8 mm, (reference material)
- 2524 T3 clad 0.8 mm and bare 0.7 mm
- 7475 T7 clad 0.8 mm and 0.4 mm
- Ti64 0.8 mm (only for local reinforcements)

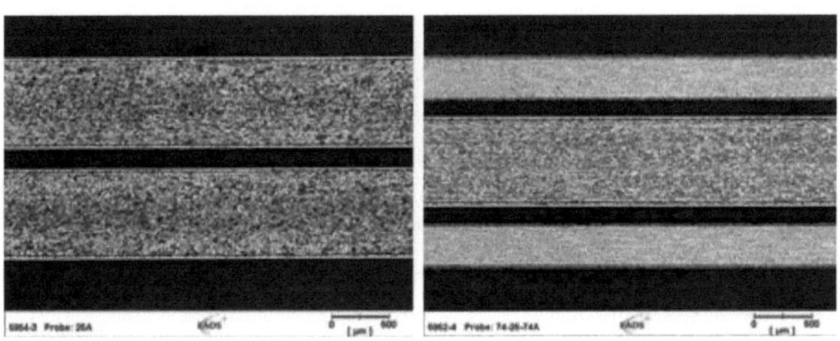

Fig. 1 Example of MLs: two laminas of 2524 alloy, 0.8 + 0.8 mm thick (*left*) and three laminas of 7475-2524-7475 alloys 0.4 + 0.8 + 0.4 mm (*right*)

All aluminium alloys were treated with CAA and primer BR127.

The reference adhesive used for the investigation was the FM 73M.03 film, 0.1 mm thick. Three alternatives were investigated:

- EA 9696NW.03 (can be cured at vacuum without autoclave)
- EA 9380 (paste adhesive to be cured at a "low" temperature of 70 °C)
- EA 9686NW.03 ("high" maximum applicable temperature of 148 °C)

All the tested products are not experimental adhesives and have the potential to be applied within a few years.

The tensile tests results have shown that regarding pure static performances no significant advantages shall be expected from MLs. The scatter of the static tensile test results was rather low and no influence of the different adhesives on the static strength values was found. The static strength is according to the metal volume fraction and the bare condition has a positive effect on the static tensile strength.

The blunt notch strength (open hole specimens under static tensile loading) was 5–10% lower than monolithic alloys. This result, attributed to the layered composition, was confirmed in all MLs and a clear understanding of the reason was not found. In a similar way the bearing strength was between 1% and 7% lower than the comparable monolithic alloy.

Impact tests were performed at TU-Delft with the objective to determine the maximum allowable impact energy. All test results, plotted in Fig. 2, are more or less on one line. The relationship between the impact energy and the dent depth times thickness can be explained by the fact that for a given dent depth the increase in impact energy is by first approximation proportional to the thickness of the impacted sheet. In conclusion:

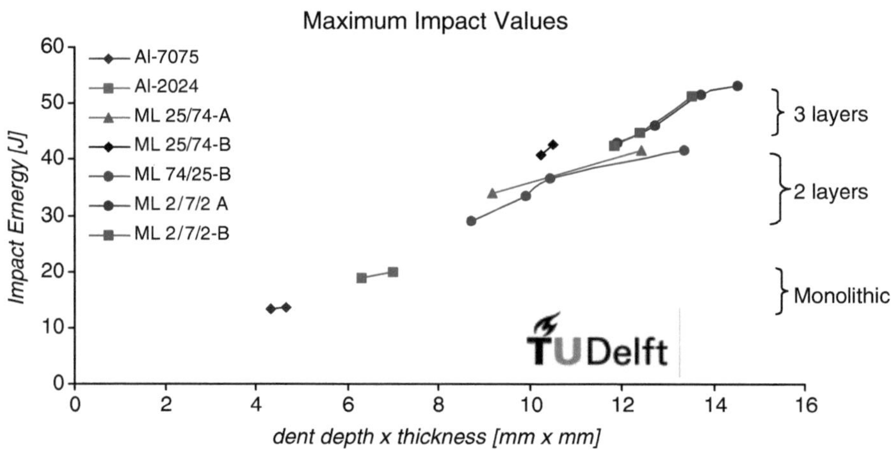

Fig. 2 Test results showing better impact properties for MLs in comparison to monolithic alloys (specimens size 125 × 125 mm)

- MLs had far better impact properties than comparable monolithic aluminium alloys.
- MLs consisting of three layers had slightly better impact properties than the ones with two layers.
- There was no significant difference between different adhesives (FM-73 or EA 9696).

Open hole fatigue test coupons, with dimensions of 330 × 40 mm, were tested at Linköping University and compared to two references:

- Monolithic 2.54 mm thick 2024T3, clad with CAA (Airbus standard)
- MLs 3 × 0.7 mm 2524T3 bare with PAA and primer BR127 (Saab standard)

All the fatigue test data fell within the limits defined by the SAAB and Airbus references:

- The monolithic 2024T3 reference had the lowest fatigue life;
- ML with two layers of 2524 increased the maximum stress at 1E5 cycles about 15% in comparison to monolithic 2524T3 (Fig. 3);
- Fatigue stress increase up to 25% were measured for a ML with three laminas.

Important drivers for the fatigue life of MLs were the presence or not of the clad layer, and the surface anodisation process (CAA is more detrimental than PAA). In Fig. 3, for clarity, only a limited set of the SN data have been plotted.

Crack propagation tests were carried out on M(T) specimens loaded at constant amplitude according to ASTM E 647 at EADS-IW Ottobrunn. There was no reduction of the propagation rate of through cracks in laminates compared to the monolithic sheets of the same metal and thickness. Part through cracks (surface

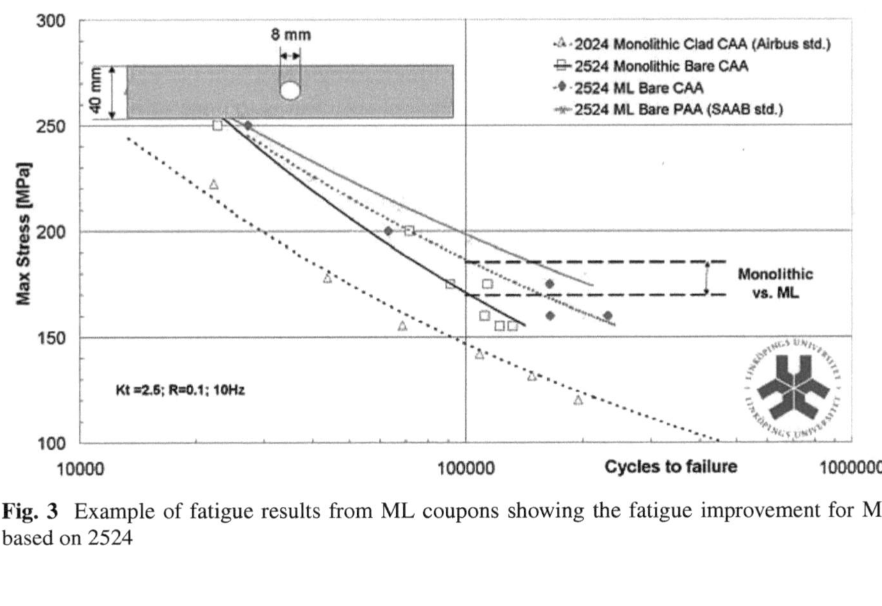

Fig. 3 Example of fatigue results from ML coupons showing the fatigue improvement for ML based on 2524

cracks) and spectrum (overloads) were not tested. A slight influence of different adhesives and aluminum doubler material on the crack propagation behavior was found.

Similar conclusions were obtained from crack propagation and residual strength tests of stiffened flat panels with seven stringers. ML sheets didn't provide significant improvements on the crack propagation behavior whereas the residual strength was slightly increased compared to the monolithic panels. The test results revealed that the use of local reinforcement is crucial for the panel performance, on the other hand, changing the aluminum alloy (2024, 2524) and the thickness of the doubler (0.6, 0.7 and 1.2 mm) didn't affect significantly the results. A large improvement was obtained with Titanium reinforcements. Ti–6Al–4V doublers had the best crack arresting capabilities for crack propagation and residual strength tests. Bonded Ti doublers between the stringers extended the crack growth period across the first bay of about 2.7 times compared to the case of aluminum doublers. The residual strength increased 10% using Ti doublers under the stringers instead of aluminum ones.

4 Fuselage Structural Design with MLs

The coupon tests have shown that a ML skin, in comparison to a monolithic design, offers better fatigue properties (mainly related to the better material properties of the thin foils in comparison to the monolithic sheets) and the impact resistance is significantly higher. On the other hand the coupon tests have clearly shown that MLs bring no improvement of the static properties, slight or negligible improvement in terms of reduction of through crack growth rate (under constant amplitude) and minor improvement in residual strength. In the ML structural concept have been introduced selective reinforcements to achieve a significant fuselage skin thickness reduction and maintain sufficient damage tolerant characteristics. To fulfil requirements like the large damage capability in longitudinal direction (so called two bay crack over broken frame) an extremely efficient solution is the use of high strength glass fibres embedded in the skin as selective reinforcement.

For areas dominated by static strength requirements the possibility to reduce the weight (e.g. skin thickness) is linked to the exploitation of higher efficiency under compression shear loading. For these areas the goal is to trade off the increased load carrying capabilities from bonded stringers with reduced skin and stringer sections, thus achieving a weight saving.

Following an analytical, as well as a numerical investigation, large panel tests were designed to investigate the two principal structural requirements: post-buckling stability under shear-compression and the F&DT behavior, including large damage capability. The large panels have been manufactured and produced using only qualified materials and processes. This means that all the results obtained could be transferred to real manufacturing without needs of new qualifications and a reduced development effort.

5 Static Stability of Metal Bonded Structures

It is know that bonded stiffened panels are very effective in terms of static stability under compressive loads. In [4] are compared test results from large panels having the geometry of Airbus Long Range aircraft: in one case the stringers are riveted, in the other case the stringers are adhesively bonded; the performances increase range from 20% in pure compression to 8% in pure shear. In DIALFAST project, instead of assessing the failure load increase, it was decided to maximize and validate the weight saving achievable from a bonded design in comparison to a riveted one. Looking at the possible weight saving, in general it can be stated that a thin bonded structure will show a larger benefit than a thick bonded structure. For this reason a large diameter fuselage structure was investigated given that the results could subsequently be extended to smaller fuselages having lower skin thickness.

The original design of a side panel shell from A330-300 aircraft was optimized minimizing the weight and keeping the same load carrying capability. The steps taken to arrive from the benchmark to the final optimized configuration were:

1. Switch of stringer-skin connection to bonding
2. Reduction of skin (2024T3) thickness from 1.6 to 1.2 mm
3. Introduction of a doubler below stringer (0.6 mm thickness)
4. Reduction of stringer thickness from 2.0 to 1.6 mm

In Fig. 4 are given the interaction curves of the original shell design and of the optimized design. The comparison shows that both panels achieve similar failure loads; nevertheless the weight of the optimized bonded design is about 20% lower taking into account the weight of stringers, skin and doublers.

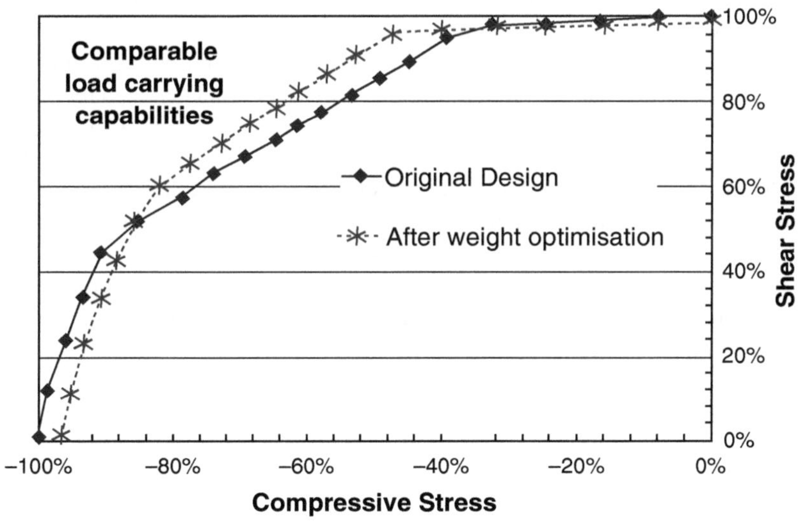

Fig. 4 Analytical interaction curve of the original design and optimized panel

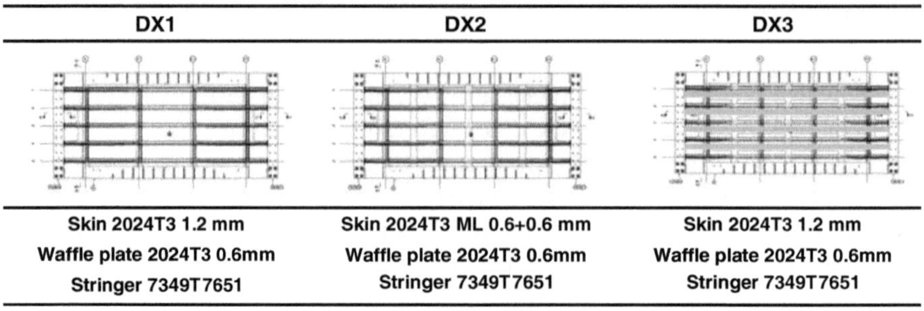

Fig. 5 Sketches of the adhesive bonded panels tested under shear compression

The main driver for the improvement in performance is the bonded connection of stringers and skin. The doubler bonded below the stringers (waffle plate) yields only slight improvements. Stringer reduction in thickness also generates substantial weight savings, while compression/shear allowable loads remain similar: only slightly lower than reference for pure compression.

Three curved panels were tested to validate the predicted shear-compression interaction curve. The panels, with five stringers and four frames, had a 2024T3 1.2 mm thick skin, a 0.6 mm thick waffle plate of the same material and bonded stringer 1.6 mm thick from 7349T7651 material. Sketches are given in Fig. 5. The first panel, DX1, was designed reproducing the configuration of minimum weight; the waffle plate increases the skin thickness only below the stringers. The second and the third panels were reinforced increasing the extension of the waffle plate. The main drivers for the additional material were to retard the initial buckling of the very thin skin (with circumferential reinforcement) and to quantify the potential benefit under shear compression of design features mainly needed to improve crack propagation properties (e.g. longitudinal reinforcements in the waffle plate). The DX2 waffle plate included a circumferential strip of material between the frames, the DX3 panel, in addition, had a longitudinal strip of material between the stringers. In all the three cases the waffle plate consisted of 0.6 mm thick 2024T3.

The DX2 panels had a Metal Laminates skin made from two foils of 2024T3 each 0.6 mm; the DX3 panels included glass fibers strips, in longitudinal direction, located between the stringers and below the waffle plate. The glass prepreg was intended only as a crack retardation feature and was included in the static test only to prove that the prepreg would not debond during postbuckling. The prepreg has lower peeling strength compared to standard bonding film and in postbuckling regime very high peeling force are induced from the panel deformation.

The panel DX1 and DX2 have been tested under a moderate shear load; the DX3 panel with the additional doubler between the stringers was tested at high shear load. Test results vs. predicted failure loads are given in Fig. 6. The graphs show the strain gauge (rosette back to back) readings on the skin at the center of each panel. All tests have validated the analysis method. The analysis, performed using nominal material properties, shows a difference of about 6–7% on the safe side.

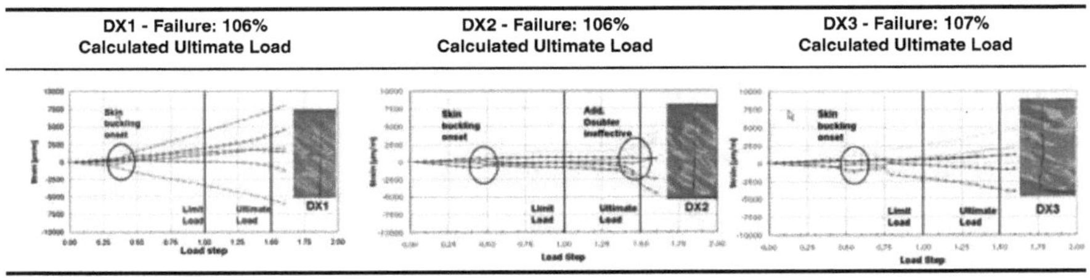

Fig. 6 Test results of the three panels. The strain gauge readings show the onset of buckling

The DX1 panel, as expected, had a very early skin-buckling onset at 0.35LL without any detrimental effect on ultimate load carrying capability. The stringer deformation was very low, which means that the waffle plate was effective. Despite the very low initial buckling load the panel exceed the predicted failure load by about 6%.

In the second test, the panel DX2 had a skin buckling 30% higher than DX1 but the failure load did not change (as expected). The ML skin did not suffer any debonding during postbuckling. Test validates that doubler located between stringers helps for retardation of onset of buckling but not for post-buckling failure. The additional doubler between the stringers can be used in case of fatigue-buckling issue to reduce the occurrences of buckling events.

The third panel, DX3, had a similar skin buckling as the panel DX2, the final failure load was not higher than predicted. The additional doubler in longitudinal direction was not effective for post-buckling failure load increase. The glass prepreg used below the waffle plate did not debond during postbuckling confirming the possibility to introduce selective reinforcements, if required from damage tolerance considerations.

In conclusion, the DX1 configuration was the most effective in term of low weight and reduced complexity, and was confirmed that with small weight penalty it is possible to retard the occurrence of initial buckling or to introduce local selective reinforcements.

6 Curved Shell Under Pressurization and Biaxial Loads

To validate the ML structural concept for areas dominated by F&DT requirements a large curved panel with seven frames and ten stringers has been tested under biaxial load and internal pressurization. The design, sketched in Fig. 7, was representative of an Airbus Single Aisle upper front fuselage. The main features in the panel design were the extreme low skin thickness (0.8 mm in the pockets) and the use of selective reinforcements embedded in the skin to act as crack arrester. The low skin thickness causes a relative fast crack propagation rate and the structural concept developed with ML rely on strong crack arrester at the stiffeners to stop, or turn, any propagating crack. The crack turning of long damages would prevent an explosive decompression of the pressurized fuselage since the damage would results in

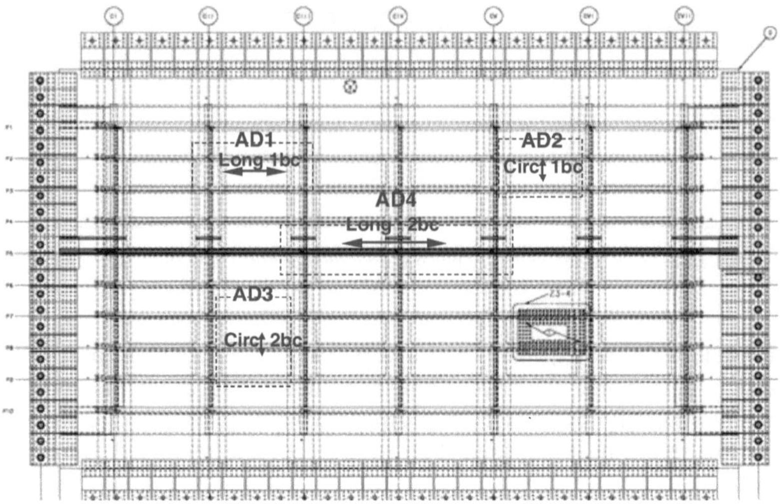

Fig. 7 Large shell tested to validate the ML concept under F&DT requirements. Sketched the location of the artificial damages (ADs) introduced during the test

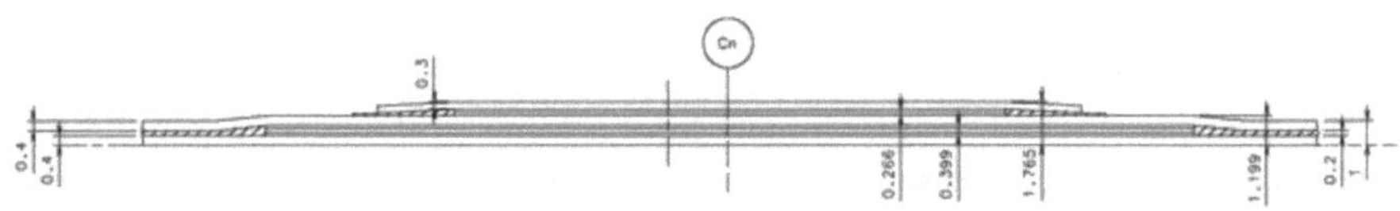

Fig. 8 Longitudinal section of the skin below the circumferential frames showing the selective reinforcements embedded in the skin

flapping of large opening with subsequent loss of internal pressure i.e. crack driving force. The crack turning phenomena is an important fail-safe measure for a pressurized fuselage, although it is quite difficult to achieve with a classic design. The use of selective reinforcements embedded in the skin was motivated by the requirement of fulfilling the large damage capability criteria with a very limited weight increase. Previous experience with curved panel test having monolithic skin and FML crack arresters, see for example [5], was used to size the selective reinforcements.

The panel skin was a metal laminate constituted of two layers of 2024T3 unclad, each one 0.4 mm thick. A waffle plate of 2024T3 unclad 0.3 mm thick was used to increase the skin thickness below stringers and frames. The metal bonding was based on the AF163 system with curing temperature of 120°C; all metal foils were anodized with CAA. The aluminum thickness was 0.8 mm in the skin pockets and 1.1 mm below stringers and frames.

The reinforcements were considered necessary only below the frames in circumferential direction in order to be able to fulfill the longitudinal large damage capability criteria: "two bay crack over broken frame". The reinforcement consisted of five layers (each one 0.133 mm thick) of S2 glass prepreg FM94, as sketched in Fig. 8. Three layers were embedded in the skin and two layers were between the skin and the waffle plate. The prepreg was embedded in the metal laminate during the lay-up process; it was a manual procedure, quite labor intensive.

With a skin thickness of 0.8 mm the nominal pressurization stress in the pockets is about 140 MPa (calculated from ΔP R/t thus neglecting stiffeners contribution, using the ΔP = 564 mbar, R = 1,975 mm). This high stress value poses significant challenges for reparability and, in addition, a thickness of 0.8 mm makes impossible to install countersunk fasteners due to knife edge effects. For these reasons, to validate the reparability was introduced a cutout with dimension of 200 × 90 mm and an external riveted patch. The external repair, visible in Fig. 7, was constituted of an outer doubler made of 2024-T3 clad with a thickness of 1.2 mm; surface treatment CAA. The installation was made with four fasteners rows in both circumferential and longitudinal direction. The critical first rivets row was located in the skin pocket where the thickness is only 0.8 mm.

The frames were metal sheet 2024-T42 with thickness of 1.2 mm, and were connected to the skin by means of riveted shear clips. The frames used had the minimum dimensions among all the frames currently in use in the single aisle aircrafts. The selective reinforcement embedded in the skin was the key concept that allowed the use of light frames. In a standard design very strong frames are essential to keep stable the skin when a longitudinal skin crack reach dimensions up to two frame bays (one bay is 533 mm in a Single Aisle aircraft). A skin with embedded reinforcements made of glass prepreg can arrest such long damages without the need of strong circumferential frames therefore attaining a weight saving.

The extruded stringers were made of high strength 7349T76511 alloy with thickness 1.2 mm. The stringers were adhesively bonded to the skin in a second bonding cycle. Due to the smooth thickness transition in the area of selective reinforcements no joggling of the stringer was necessary.

A limit of the ML concept based on foils with thickness as low as 0.4 and 0.3 mm is the narrow width of the aluminum coils, which have a maximum width of 1.5 m. To create large fuselage shells it is necessary to splice, in longitudinal direction, several aluminum foils. Two different concepts of longitudinal splices have been investigated in the curved panel: a bonded splice between stringer P6 and P7 and a Friction Stir Welded (FSW) splice at stringer P5. The bonded splice design, given in Fig. 9, was based on the design principles developed for FMLs, and can be considered a state of the art solution. The alternative concept for the longitudinal splicing consists of the butt-welding of each thin foil. The laminate was built using welded foils as sketched in Fig. 10; the welds were located below the stringer foot, where the stresses are lower. In comparison to a standard bonded splice the main advantages from the welded splice are the simplified design and, during manufacturing,

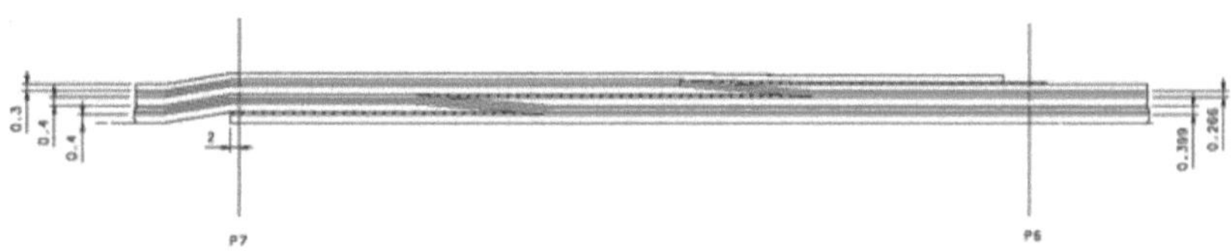

Fig. 9 Drawing of the bonded metal laminate splice

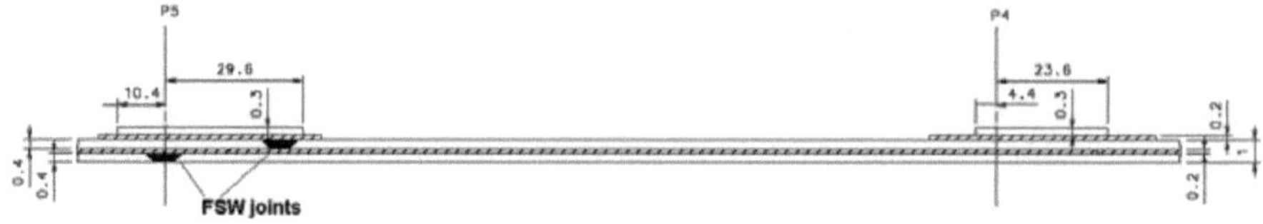

Fig. 10 Drawing of the FSW metal laminate splice

Fig. 11 Microsection of the FSW butt weld of 0.4 mm 2024T3

a reduced layup time. A minor weight saving is an additional benefit. Figure 11 shows a microsection of the high quality welded joint obtained for 2024T3 foils 0.4 mm thick.

The skin was loaded in both longitudinal and circumferential directions by independent actuators; the frames were loaded separately with a separate actuators. The constant amplitude fatigue testing was performed with a stress ratio for longitudinal and circumferential direction of about 1:1 and due to the already high design stresses, no load increase factor was used. At 1 ΔP load the hoop membrane stress in the skin pocket, measured by strain gauges, was about 115 MPa; the bending stress (30 MPa difference in the stress for inner and outer skin surface) was quite low.

After the fatigue testing (60,000 cycles at constant amplitude loading) were introduced four artificial damages: a sketch is given in Fig. 7.

The longitudinal one bay crack AD1 was propagated under spectrum loading from a starting size of 30 mm and reached the reinforcements below frames in about 6,000 flights. Once the crack reached the selective reinforcements below the frames the propagation speed reduced significantly and both crack tips started to turn in circumferential direction, see Fig. 12. After complete tuning of both crack tips the damage was repaired.

The circumferential one bay crack AD2 with an initial dimension of 30 mm in the center of the bay was not propagating. After 26,000 load cycles the damage was artificially extended to 100 mm. The crack started propagating but the speed slow down reaching the stringer. After additional 9,000 load cycles, with the crack tips reaching the stringers, the crack was stop drilled.

The circumferential two bays crack over broken stringer AD3 was propagated under spectrum loading. The crack has shown a not expected tendency to turn to a 45° direction in the early stage of the propagation as shown in Fig. 13. The longitudinal load was increased in order to retard the crack turning without success. The crack reached the stringer in about 25,000 cycles; crack tip 2 went parallel to stringer and was stop drilled; when crack tip 1 reached the speed of 1 mm/cycles it was stop drilled as well and the damage repaired.

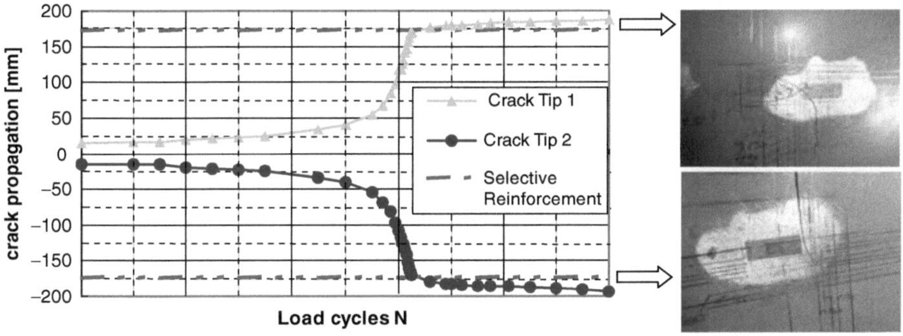

Fig. 12 Artificial damage AD1 – longitudinal one bay crack: the selective reinforcement proved very effective in arresting the crack

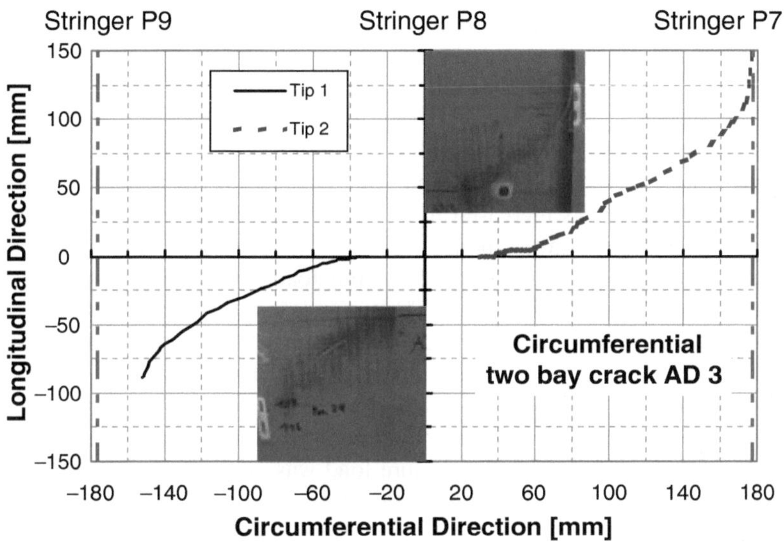

Fig. 13 Artificial damage AD3 – circumferential two bays crack showing crack deviation at the next stringer

The longitudinal two bay crack AD4 was introduced with a skin through cut of 60 mm; the frame was left intact. The crack growth rate was very slow as shown in Fig. 14. After 37,000 cycles, and a natural crack extension of only few millimeters, the frame was saw cut: no significant increase in crack speed of the skin was measured. The selective reinforcement below the frame was able to completely retard the skin crack propagation even with a broken frame. To have the crack propagating the damage was artificially extended twice, the second time cutting completely all the fibers below the frame. With the selective reinforcement completely cut the crack propagation rate was very fast although the crack never got instable. Reaching the selective reinforcement of the next frame both crack tips slow down and began to deviate in circumferential direction.

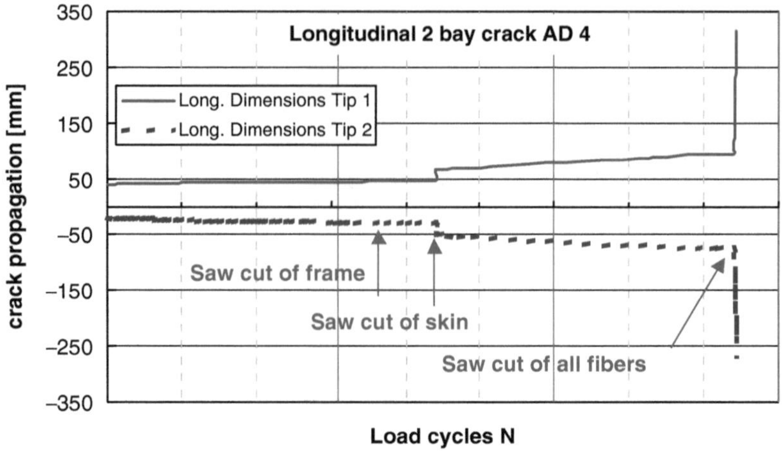

Fig. 14 Longitudinal two bays crack. Crack propagation is very slow when the crack is in the selective reinforcement area, even with broken frame

To conclude the investigation a residual strength test was performed with the two bay longitudinal crack (and broken frame) of AD4. During the residual strength test the longitudinal load was increased in phase with pressure up to ΔP load and the skin stress ratio long/circ was about 1/1; for higher pressure the longitudinal load was kept constant to the 1 ΔP level. The limit load for a residual strength test is 1.15 ΔP to be considered as ultimate. The test was stopped for 30 s at 1.15 ΔP and then continued increasing pressure and circumferential loads up to the final failure. During the test the crack extended in stable way in a predominantly circumferential direction and slow down approaching the foot of the bonded stringers. The panel failure occurred at 1.97 ΔP when the skin opened cutting the three stringers above the crack, as shown in Fig. 15. The failure load was almost twice the fatigue pressurization load and 70% higher than the target load. Such high pressurization load, applied in combination with a large damage, cannot be reached during real aircraft service and was tested only to assess the limit of the structure.

At the end of the test the panel had cumulated 144,000 cycles and no natural damages where detected in the external riveted repair, in the FSW splice and in the bonded splice.

The weight of the metal laminates curved panels has been compared to the weight of an equivalent monolithic fuselage panel able to fulfill similar F&DT structural requirements (inclusive of the large damage capability). Figure 16 shows the weight split for skin, frames and stringers, as well as the total weight in comparison to monolithic references based on 2524T3 and Al-Li 2198T8 skins.

The weight saving potential from the ML concept is in the range of 10–13% in comparison to a metallic monolithic design.

Fig. 15 The curved panel at the end of the F&DT test. It is visible the large opening caused by the crack flapping

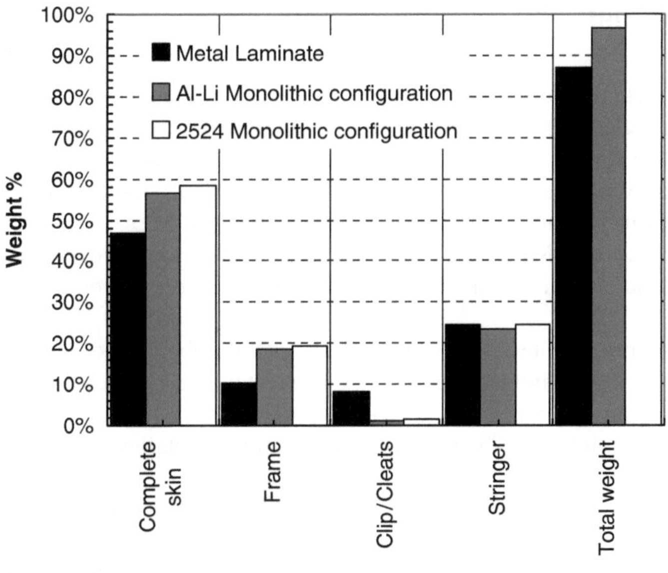

Fig. 16 Weight comparison for the curved ML shell and monolithic references

6.1 Manufacturing Aspects of ML

The study was completed including an assessment of the main manufacturing costs for ML skins. The cost was only estimated for the skin, based on figures from the Airbus internal accounting system, assuming a cylindrical shell with the lay-up and reinforcements configuration of the test panel. The process sequence assumed for ML skin was:

- Decoiling, cutting and contour milling of the aluminum foils
- FSW of the foils
- Chemical treatment and application of bond primer
- Panel lay-up in the mould and curing in autoclave
- Lay-up of stringers and second curing cycle
- C-Scan inspection
- Panel milling
- Final assembly

The design to cost investigation has show that a ML skin can be up to four times more expensive than a monolithic skin; for complex shells the cost difference is expected to be lower.

The comparison of the two splicing concepts investigated, bonded and friction stir welded, has shown a slightly higher cost for the welded splice.

7 Conclusions

A fuselage structural concept consisting of adhesively bonded laminated skin with waffle plate, selective reinforcements and bonded stringers has been developed and validated with several coupons as well as large stiffened panel tests.

Coupons tests have shown that regarding static properties (tensile, blunt notch, bearing strength) no advantages shall be expected from MLs. Conversely, the tests results revealed that MLs deliver higher fatigue life, improved impact resistance properties, and are suited for the introduction of local reinforcements in a bonded design.

Shear-compression test results have confirmed that bonded structures offer high potential for weight savings in comparison to a standard riveted design, in particular weight savings opportunities up to 20% exists for fuselage areas dominated by shear-compression loads. The main improvement was obtained from the bonding of the stringers to the skin.

The F&DT curved panel test has confirmed that a long fatigue life and high damage tolerance can be achieved with a hybrid metal laminate concept. The low thickness skin concept is based on the superior fatigue life from MLs and the high damage tolerance characteristics provided by integrated selective reinforcements. Crack turning has been experienced for both one bay crack and two bays longitudinal crack scenarios. The reparability of a very thin ML skin (0.8 mm) has been proven by means of an external riveted repair with four rows of fasteners. The weight comparison with metallic monolithic riveted design shows a potential weight saving in area dominated by F&DT in the range of 10–13%.

For simple cylindrical shells the cost of a ML skin is definitely higher than the cost of a monolithic skin. The weight advantages shall be traded off with the additional costs. The ML structures investigated are based on standard qualified materials and can be manufactured with today's technologies without requiring much further validation.

Acknowledgments The metal laminate investigation has been performed in frame of WP2 of the DIALFAST project (European Specific Targeted Research Project – 6th Framework). Partners involved were Airbus-D, EADS-IW Ottobrunn, TU-Delft, ISTRAM and Linköping University. The financial support of the EU 6th Framework is acknowledged.

References

1. J. Schijve, H.T.M. van Lipzig, G.F.J.A. van Gestel, A.H.W. Hoeymakers (1979) Fatigue properties of adhesive-bonded laminated sheet material of aluminum alloys, Engineering Fracture Mechanics Vol. 12. pp. 561–579
2. E. Hombergsmeier (2006) Development of advanced laminates for aircraft structures. ICAS, Hamburg
3. H.-J. Schmidt (2005) Damage tolerance technology for current and future aircraft structure. Plantema Lecture, ICAF, Hamburg
4. M. Pacchione, J. Telgkamp (2006) Challenges of the metallic fuselage. ICAS, Hamburg
5. M. Pacchione, J. Telgkamp, N. Ohrloff (2008) Design of pressurized fuselage structures under consideration of damage tolerance requirements, 40. Tagung des DVM - BR, Stuttgart

Multifunctional Materials Used in Automotive Industry: A Critical Review

Konstantinos Salonitis, John Pandremenos, John Paralikas, and George Chryssolouris

Abstract Nowadays, advanced materials and related processes in the automotive industry, are more widely used, leading to an effort towards reducing weight and fuel consumption. The use of such advanced materials and technologies tends to increase the cost. Multifunctional materials (MFMs) and related processing technologies aim at overcoming this increase of cost by exploiting the high level of functional integration. MFMs are designed so as to meet specific requirements through tailored properties. The use of such materials, in the automotive body construction, can help reduce produced parts, lightweight design, high level of integration of functionalities, advancements in mechanical properties of structures etc. In the current study, there is a clear definition of MFMs, and a critical view of such materials used both in the automotive body construction and in other industrial applications.

Keywords Multifunctional · Materials · Functions · Automotive body

1 Introduction

The increasing competition in global market has pushed manufacturers to develop new concepts and forms of their products. The development of the materials used, plays a major role in the formation of new concepts and manufacturing technologies. The Materials created by intimately integrating different types of materials as well as functions for implementing new applications that can be proved both technically challenging and economically favorable [14, 24], are called multifunctional materials. This kind of materials promises to provide a new level of functionality, adaptability, and tailorability for future engineered systems [42, 50]. The number of

K. Salonitis, J. Pandremenos, J. Paralikas, and G. Chryssolouris (✉)
Laboratory for Manufacturing Systems and Automation,
Department of Mechanical Engineering and Aeronautics, University of Patras, Patras, Greece
e-mail: {kosal, jpandrem, jparali, xrisol}@lms.mech.upatras.gr

S. Pantelakis, C. Rodopoulos (eds.), *Engineering Against Fracture: Proceedings of the 1st Conference*,

parts can be reduced and a lightweight design can be easily achieved. In addition to this, a higher level of functional integration can be achieved and finally, the overall cost can be reduced. Recent research studies focus on structural materials incorporating different functions, into their processing techniques, as well as into coatings of metallic surfaces to improve their properties and enhance them with functionalities [17, 47, 59, 66]. Materials that are truly integrated this is the true promise of multifunctional materials for performance tailored structures [45].

The entire automotive industry needs to reduce fuel consumption and therefore, the emission of their products by the methods of weight (without losses in performance) savings and technological improvements [13, 38]. Great potential weight savings can be obtained in the vehicle structure, with a body-in-white (BIW) providing the largest contribution. In order to reap the benefits of weight-savings in cost and fuel economy, the functional properties of a vehicle's body-in-white (BIW), including its requirements for passive safety, service life, dynamic stiffness, acoustics and comfort must be improved [6]. MFMs could be distinguished in materials that combine properties and functions unaffected from external stimuli and the time domain, in materials that sense the process information and respond (smart materials) [18], as well as in affordable composite materials [24, 52].

The use of advanced materials can also affect design methods and concepts in the automotive industry. Modularity is a major trend, which is used for simplifying the management of complex systems [54]. Modularity in design (MID), in use (MIU) and in production (MIP) are the three main categories, which provide flexibility, speed and expanded design capability and reduced cost (labor, engineering efficiency, scale economics and lower). These design techniques allow the use of advanced materials, such as MFMs, and related processing technologies without any significant cost increase.

2 Materials Used in Automotive Industry

The functionality requirements of vehicles have increased [62], and the materials help these requirements to be met. In order for advanced materials to be used in mass-produced consumer products such as automobiles, they must exhibit a number of general characteristics [38, 62], such as affordable cost and tooling, the supply of raw materials to be plentiful and readily available, and the highly reliable parts and assemblies to be recyclable.

2.1 Advanced Steels and Pre-coated Steels

Different types of steels still remain the most commonly used materials for vehicles. They offer a wide variety of material characteristics such as thermal, chemical or mechanical resistance, ease of manufacture and durability [76]. High strength

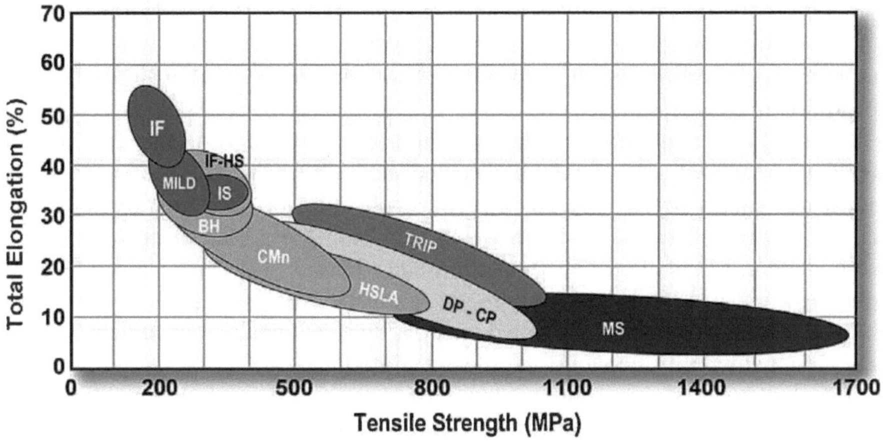

Fig. 1 Schematic of AHSS steels (shown in color) compared to low strength steels (dark grey) and traditional HSS (light grey) [2]

(HSS) and advanced high strength (AHSS) steel grades are increasingly used in high-volume production for parts such as sheets or profiles, which are assembled by advanced manufacturing techniques (Fig. 1). Steels are initially classified as hot and cold rolled [2], which are then categorized according to their production method and quality (commercial CQ, drawing DQ, deep drawing DDQ, etc.) for formability. In a deeper view, automotive steels can also be defined in a number of different ways [38,77]. The most common steels used by automotive industries are (i) by metallurgical designation, (ii) based on mechanical properties and (iii) based on total elongation. The most common and evolving steel types are: mild, bake-hardenable (BH), isotropic, carbon-manganese (CM), high strength low alloy (HSLA), dual phase (DP), transformation induced plasticity (TRIP), complex phase (CP), martensitic (Mart), ferritic-bainitic (FB), twinning induced plasticity (TWIP), hot formed (HF), post forming heat treatable (PFHT) and evolving AHSS steels [37].

In response to the automotive industry's demands for better functional steels, the steel industry develops new types. TRIP and TWIP steels are both categorized as AHSS. The Evolving AHSS steels are designed to reduce density, improve strength and/or increase elongation. TWIP steels have become very important for automotive applications. They have a great ability to absorb the crass energy extremely quickly [60]. TRIP steel can resist high stresses without deforming. TWIP steel deforms with low stresses, but does not break until the strain reaches around 90%. Salzgitter AG [60] launches on the market the evolution of TRIP steel as HSD (high strength and ductility) steel. This type of steel is extremely light due to the low density of its alloy elements. The TWIP/TRIP steels can be used in thinner sheets that result in 10–30% weight savings in the automotive structures. Arcelor and ThyssenKrupp have developed a new super-high strength steel with TWIP for weight reduction and improved crash resistance named X-IP1000 [15, 51]. This austenitic product, which is based on high manganese alloy metallurgy, has a tensile strength greater than 1,000 MPa for a total elongation superior to 50%. X-IP steels exhibit the best combination of strength and ductility over the whole range of ductile steel grades for automotive applications, such as forming of complex shapes and exploiting energy absorption in passive safety.

The need of the automotive industry to improve corrosion resistance has brought about an increased use of metallic coated steel sheets in place of cold rolled (uncoated) sheets formerly used [2]. The application of multifunctional coating can be classified into the traditional four-step model used for electroplating: (1) pre-treatment, (2) plating, (3) passivation and (4) seal [47]. Major advantages of applying precoated steels in the automotive industry, are the enhanced corrosion resistance, decorative finish, controlled friction, thermal shock management, ease of operation, paintability, economic, and weldability [75].

The challenges of managing increased loads and additional functionalities while keeping weight down has resulted in a significant transformation in the use of steel and the blend of several steel grades. The designs of modern cars can be divided into four main groups: monocoque, unibody, spaceframe, mix of unibody/spaceframe [33]. Steel unibody has for years been the dominant method of producing vehicle bodies in the automotive industry. BMW (3-Series) managed to improve the body's functionality and to increase the average minimum yield strength and significant weight saving. Blanding of steel grades has also been used by Mercedes-Benz (M-Class) [30, 32].

2.2 *Aluminum Alloys*

The European automotive industry has more than doubled the average amount of aluminum used for the construction of passenger cars. Determining the right alloy for the body structure and hang-on panels has been the subject of considerable development effort and most of the activity is concentrated on a relatively small number of alloys [49]. One of the main advances of aluminum is its availability in a large variety of semi-finished forms. Such semis are very suitable for mass production and innovative solutions in the form of compact and high integrated parts that meet the high demands of performance, quality and cost efficient manufacturability [34]. The main aluminum alloy classes for automotive sheet application are the non heat treatable Al-Mg (EN 5000 series) and the heat treatable Al-Mg-Si (EN 6000 series) alloy system. Undoubtedly, aluminum is more expensive to be manufactured than steel is [12]. Aluminum alloys offer a wide range of properties that can be engineered precisely to the demands of specific automotive applications, through the choice of alloy, temper and fabrication process. The key characteristics of aluminum in the automotive use are lightweight, high strength to weight ratio, resilience, corrosion resistance, forming and fabricating, joining, crashworthiness, cold resistance, recyclability, thermal conductivity and reflectivity [69].

Audi with the use of ASF (Audi Space Frame [7, 40, 56]), has improved A8 handling and safety performance and achieved a high-strength aluminum framework, which the larger sheet aluminum elements were integrated into and performed multiple load-bearing functions [6]. Audi has also simplified the production line by placing body production and final assembly in one building and improved body stiffness (improved passenger comforts [7, 8], provided less vibrations, as well as maximum stability with minimal mass) [7, 40]. Novelis Fusion^{TM} technology [53] is a breakthrough process that simultaneously casts multiple alloy layers into a single aluminum rolling ingot. The whole process can be used as an automotive sheet with both strength and improved formability to enable new design options.

2.3 Aluminum Foams

The aluminum foam is a new class of materials with low densities and novel physical, mechanical, thermal, electrical and acoustic properties. It offers potential for lightweight structures, for energy absorption, and for thermal management, as it is recyclable and non toxic [5]. Foamed aluminum can be produced by two methods, powder metallurgy and melt foaming. The properties (energy absorption, thermal, sound insulation and mechanical strength) depend upon the history of the foaming process [64]. Aluminum foams can be classified according to their structure, into *open-cell* (can form an interconnected network) and *closed-cell* (The pores can be sealed) foams. The applications of the aluminum foams related to the automotive industry, and their allocation to the automotive body depends on their structure. Major multifunctionalities offered by aluminum foams are strain isolation, mechanical damping, vibration control, NVH improvements, energy absorption, energy and thermal management, excellent strength and stiffness to weight ratios and filtering [5].

Aluminum foams with open-celled structures have a wide range of applications (filtration, separation, heat or mass exchange, sound and energy absorption [39, 70, 71]) in functional structures. Closed cell aluminum foam offers a unique combination of properties such as low density, high stiffness and energy absorption [19]. Cymat [16] has introduced and commercialized the Stabilized Aluminum Foam (SAF), a revolutionary material with numerous automotive applications [16].

Another type of aluminum foam is that of the Advanced Pore Morphology (APM) [25]. This type consists of small volume metallic foam elements (shaped spherical) which are expanded in bulk/mass production. Joined to each other in a separate process step the foam elements form the APM foam part. This kind of aluminum foam has pore morphology, and mechanical properties adjustable independent of the part's geometry and volume. In addition, it has maximum volume of the largest pores for complete part volume guaranteed, homogeneous, and multi-phase and/or gradient pore structure. Mono-, bi- or multi-modal combinations/mixtures of APM foam elements with different properties (density, geometry, matrix material) are possible [67].

2.4 Magnesium Alloys

Magnesium is a quite attractive material for automotive use, primary because of its light weight, 36% lighter than aluminium and 78% lighter than iron. When alloyed, Mg has the highest strength-to-weight ratio of all the structural metals. Magnesium alloys are currently used relatively in small quantities for auto parts, generally limited to die castings (e.g. housings). Mg sheet can be used in body non-structural and semi-structural applications, while extrusions could be used in such structural applications as spaceframes [27]. The key properties and functions of magnesium, in automotive applications, are lightweight, reduction of number of parts, recyclability, excellent mechanical properties at high temperatures [20].

The development of new alloys has helped acceptable properties to be achieved [27]. Mg may have lower fabrication and joining costs. Its substitution by lightweight materials may enable secondary weight savings, and lifetime fuel costs to be reduced, so the total life-cycle cost of a Mg part may actually be lower than that made from another material. Magnesium is quite easy to form and join. Several drawbacks (most amenable to technical or institutional fixes) with the most important one the material's physical properties and unstable price [72] have limited the growth of Mg usage in automobiles [26, 48]. A considerable application is the Magnesium space-frame (VW 1-Liter Car) [33, 65] consumed fuel just 1 l per 100 km by using magnesium that is lightweight, has high strength to weight ratio and excellent die-casting properties. The space frame is constructed with composite welding and adhesive bonding of different magnesium alloys.

2.5 Sandwich Materials

Stressed skin sandwich panels are among the most efficient constructions, based on stiffness per unit weight basis. Combined with natural sound dampening and thermal attenuation properties, sandwich panels can offer significant improvements in reducing vehicle weight, reducing parts and streamlining the body assembly process. The greater the thickness is in the core material, the greater the increases are in the bending/flexural strength and especially in stiffness [29]. Functions achieved using combinations of different materials and core materials in sandwich structure are: weight saving, high strength to weight ratios, customization, noise damping, resistant to harmful chemicals and absorption of shock and impact loads [58].

Tribond®, patented from ThyssenKrupp, is a three layer hot strip produced by firmly bonded layers made from different steel grades with correspondingly different properties [67]. In addition to the "wear-resistant/ductile/wear-resistant" combination, other combinations of properties are also conceivable. Aluminum foams could be used as cores into multifunctional sandwiches. In particular, structural efficiency, cost requirements, and improvements in manufacturing processes have created a growing interest in aluminum foams and in layered composite materials such as aluminum foam sandwiches (AFS). This foam, produced with aluminium powder and a foaming agent, can be molded into complex 3D geometry [48]. Arcelor has developed a three layer composite Usilight steel/polymer sandwich [31]. With the use of this composite, preliminary studies suggest a typical total system weight and cost reduction up to 20% – while achieving noise reduction levels up to 50% [4]. Metal-plastic-metal laminates, the so called MPM-sheets, have been in the market and possessed by pressing for many years. BONDAL® in Mercedes-Benz A-Class [35, 74] managed to balance the vehicles' weight, to reduce manufacturing costs and insulate airborne, structural born noise and vibration.

2.6 Plastics, Composites and Nano-composites

Plastics have low density, are mostly economic to be produced and their characteristics can be fitted to specific demands. Their wide variety of properties can be created through chemical or physical material combinations [76]. Polymeric materials are characterized by long chains of repeated molecule units known as "Polymers" [22]. These long chains intertwine to form the bulk of the plastic. Most plastics can be classified as either thermoplastic (the polymer chains are only weakly bonded) or thermoset (adjacent polymer chains form strong cross links).

SMC is a type of fibre reinforced plastic, which primarily consists of a thermosetting resin, glass fibre reinforcement, and filler. Additional ingredients, such as low-profile additives cure initiations, thickeners, and mould release agents are used for enhancing the performance or processing of the material. Advantages of SMC versus steel automotive constructions are considerable weight savings, corrosion resistance, reduced tooling cost, resistance to minor impact, surface finish, as well as compatibility with steel body assembly [63].

Fiber reinforced composites are made of fibers and a matrix. Fibers are the reinforcement and the main source of strength, while the matrix 'glues' in shape all the fibers together and transfers the stresses between the reinforcing fibers. The primary function of the fibers is to carry the loads along their longitudinal directions, while those of the matrix are to transfer stresses between the reinforcing fibers (hold fibers together) and protect the fibers from mechanical and/or environmental damages. Common fiber reinforced agents include Aluminum, Aluminum oxide, Aluminum silica, Asbestos, Beryllium, Beryllium carbide, Beryllium oxide, Carbon (Graphite), Glass (E-glass, S-glass, D-glass), Molybdenum, Polyamide (Aromatic polyamide, Aramid), e.g. Kevlar 29 and Kevlar 49, Quartz (Fused silica), Natural fibers, etc. [21, 43, 57]. Common resin materials include Resin Matrix, epoxy, phenolic, polyester, polyurethane and Vinyl Ester [21].

Long-Fiber reinforced Thermoplastics (LFT) have gained an increasing market share in the automotive industry in the last decade. The LFT-D process differs fundamentally from the GMT and LFT-G processes in that it avoids the need for semi-finished products manufactured by the material supplier [41]. It is therefore a cost effective measure. Further advantages of this technology are high output-capacity and flexibility in fiber and polymer-modification in accordance with the customers' needs.

People are turning to nanomaterials as a new tool to improve properties and gain multifunctionality [55]. Multifunctional composites with application in the automotive sector can be classified into three main categories: (i) Nanocomposites (carbon nanotubes, nanoclays), (ii) Multi-functional composites (SMA and Piezo-materials) and (iii) Magnetic nanocomposites (metal nanoparticles). Nanocomposites are a novel class of polymeric materials exhibiting superior mechanical, thermal, and processing properties suitable for replacing metals in automotive [28]. Nanocomposite plastic parts offer great advantages (weight and energy savings). Nanocomposite-based parts provide stiffness, strength, and reliability comparable with better metals. They offer corrosion resistance, noise dampening and enhanced modulus, thermal and dimensional stability [28, 55]. With the use of Fillers or nanofillers, the heat-distortion temperature of materials increased and the material could withstand high or elevated temperatures. However, nanofillers are not expected to impart many of the property losses of conventional fillers. Nanoclays compounds, in automotive applications, have advantages including reduced weight, a wider processing window, improved colourability, improved toughness and rheological properties and improved scratch and mar resistance [28]. Mercedes Benz nano-based clearcoats [11] are resistant to scratch and mar, provide less stress concentration and reduce damage potential. In addition to this, Hummer H2 cargo bed nanocomposites [10,61,68], reduce cargo's bed weight, improve scratch protection, enhance the vehicles' overall efficiency and improve rheological properties. Menzolit's Advanced SMC for 2004 SLR McLaren Mercedes [23, 46] can be molded similar to standard SMC and exceeds by far the properties of standard. Great weight reduction (even lighter than the same panels made of aluminum), excellent paint surfaces, improved energy absorption during crash and mechanical properties just a little below high performance composites are some additional advantages. BMW 1 series multifunction tray from D-LFT [9] has improved the part performance and reduced the complexity in moulder's operations offering a unique product in the market. Finally, Maserati Quattroporte uses nanotechnology in the engine bay [44] and achieves reduced weight, increased mechanical performance (especially at high temperatures) and improved surface appearance.

3 Multi-material Approach in Automotive Industry

Several attempts have been made in the automotive industry for e exploiting the advantages of each material category in the automotive body. The application of the multi-material approach offers opportunities for combination of functions, properties and processing technologies into a single body structure, with significant advantages.

Lamborghini uses a hybrid solution with combination of stampings and space-frame elements in Gallardo's body. Cast, extruded and sheet components are integrated into the body as modules or individual components [36]. New Steel Body (NSB®) Thyssen Krupp [1,73] offers enormous weight savings, better utilization of

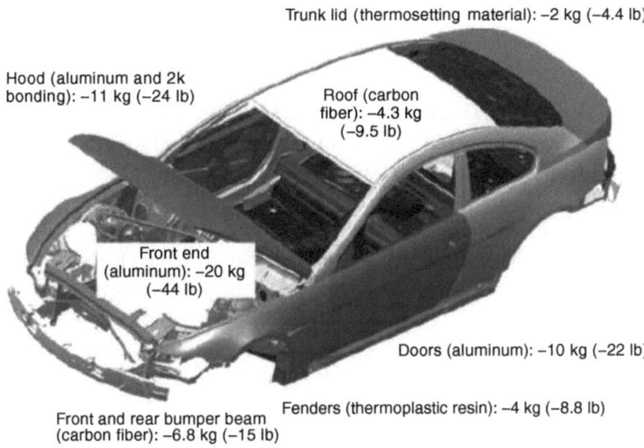

Fig. 2 Mixed-material approach used in the BMW M6 sports car [32]

packaging space, thanks to the absence of flanges, enhanced torsional stiffness and crash performance, uniform stretching of the material, low tooling costs and very high part accuracy. Additionally, Arcelor's Body Concept (ABC) [3] achieves cost reduction, improved pedestrian and passenger safety and it finally demonstrates the steel solution advantages (mass savings, improved vehicle strength and reduction in fuel consumption). BMW M6 mixed-material approach [12, 30, 32] has reduced the vehicles' weight, offered advanced car crashworthiness and achieved optimum vehicle balance. BMW researchers have made use of advanced materials (steel grades, aluminium, and plastic and carbon fibre) in this project (Fig. 2).

4 Conclusions

The major multifunctional materials used and the trends for evolving materials in the automotive industry have been presented in the current study. The term "multifunctional" has been given from the industrial point of view, and major materials used in the automotive body constructions have been identified. Industrial applications, and multifunctionality achieved have been described, in order for an emphasis to be given to the applicability of advanced materials in the automotive industry. The higher cost of advanced material could be overcome using advanced processing, "intelligent" choice and application of such materials, as well as a high level of integration of functions into the automotive structure. Several applications of the mix-material approach have been described, as this trend has become a more common way of thinking for automotive manufacturers and suppliers.

Acknowledgement The work reported in this paper was partially supported by CEC/FP6 NMP Programme, "Integration Multi-functional materials and related production technologies integrated into the Automotive industry of the future" (FP6–2004-NMP-NI-4–026621).

References

1. Adam, H., Osburg, B., Patberg, L., Gruneklee, A., Floth, T., Hinz, M., 2004, NSB® NewSteel-Body – Auto body weight reduction with steel, ThyssenKrupp techforum, July
2. American Iron and Steel Institute & Auto/Steel Partnership, 2002, Automotive steel design manual, Revision 6.1
3. Arcelor Auto Newsletter, 2005, The ABC project enters the limelight, No. 9, January
4. Arcelor – Steel Solutions for a Better World, Quiet Steel, URL: http://www.arcelor.com/fcse/prg/selfware.pl?id_sitemap = 37&language = EN, http://www.quietsteel-europe.com/body/en/material.php
5. Ashby, M. F., Evans, A. G., Fleck, N. A., Gibson, L. J., Hutchinson, J. W., Wadley, H. N., 2000, Metal Foams: A Design guide, Butterworth Heinemann, Oxford, UK
6. Audi of America Press Site, 2003, AUDI and Weight-Saving Construction – From Pioneer in Aluminum To High-Volume Production, URL: http://media.audiusa.com/article_display.cfm?article_id = 9101
7. Audi of America Press Site, 2003, Latest-Generation AUDI Space Frame Advances Design, Safety and Performance Standards, URL: http://media.audiusa.com/article_display.cfm?article_id = 8994
8. Automotive Engineering International Online: Global Viewpoints, 2006, Audi A2 production, URL: http://www.sae.org/automag/globalview_02–00/01.htm
9. Borealis, Shaping the Future with Plastics, PP for d-LFT, URL: http://www.borealisgroup.com/public/customer/automotive/dlft/MainPage.jsp#
10. Birch, S., 2004, "Dwarf" in Greek, the Word Nano has Big Implications, Material Innovations, AEI Magazine, August 2004, URL: http://www.aei-online.org
11. Birch, S., 2004, Mercedes Makes a Scratch Hit, Material Innovations, AEI Magazine, February 2004, URL: http://www.aei-online.org
12. Carle, D., Blount, G., 1999, The suitability of aluminium as an alternative material for car bodies, Materials and Design, 20:267–272
13. Cole, G., Glove, A., Jeryan, R., Davies, G., 1997, Steel World 21:75–83
14. Committee on Materials Research for Defense After Next, 2003, Materials Research to Meet 21st Century Defense Needs, National Research Council
15. Cugy, P., Hildenbrand, A., Bouzekri, M., Cornette, D., Goglu, S., Hofmann, H., 2006, A Super-High Strength FE-MN-C Austenitic Steel with Excellent Formability for Automotive Applications, URL: www.msm.cam.ac.uk/phase-trans/2005/LINK/92.pdf
16. Cymat (CYM) foam products, Aluminium Foam Technology Applied to Automotive Design, URL: http://www.cymat.com/Cymat_Foam_products0.htm
17. Department of Trade and Industry, 2006, Report on Key Technology Area: Advanced Materials, URL: http://www.dti.gov.uk/files/file27985.pdf
18. Department of Trade and Industry, 2006, The Technology Programme: Smart Materials and Related Structures, URL: http://www.dti.gov.uk/files/file18875.pdf
19. Deqing, W., Xiangjun, M., Weiwei, X., Ziyuan, S., 2006, Effect of processing parameters on cell structure of an aluminum foam, Materials Science and Engineering, A 420:235–239
20. Duffy, L., 1996, Magnesium alloys – An introduction, Materials World, 4(3):127–130
21. EFUNDA, Engineering Fundamentals, Fiber Reinforced Composites, URL: http://www.efunda.com/formulae/solid_mechanics/composites/comp_FRC_intro.cfm
22. EFUNDA, Engineering Fundamentals, Polymer Materials - Overview, URL: http://www.efunda.com/materials/polymers/science/polymer_science_intro.cfm
23. European Alliance for SMC/BMC, News from the Member Companies, Carbon Fibre (CF) Reinforced SMC Used in Series Production, URL: http://www.smc-alliance.com/members/members_news/members_news.html
24. Flatau, A. B., Chong, K. P., 2002, Dynamic smart and structural systems, Engineering Structures, 24:261–270
25. Fraunhofer IFAM, Advanced Pore Morphology (APM) Foams, URL: http://www.ifam.fhg.de/index.php?seite=/2801/leichtbauwerkstoffe/metallschaeume/apmschaum/&lang=en

26. Friedrich, H., Schumann, S., 2001, Research for a “new age of magnesium” in the automotive industry, Journal of Materials Processing Technology, 117, 276–281
27. Gaines, L., Cuenca, R., Stodolsky, F., Wu, S., 1996, Potential Automotive Uses of Wrought Magnesium Alloys, Automotive Technology Development Conference, Detroit, Michigan
28. Garces, J. M., Moll, D. J., Bicerano, J., Fibiger, R., McLeod, D. G., 2000, Polymeric Nanocomposites for Automotive applications, Advanced Materials, 12(23):1835–1839
29. Greve, B., 2004, Sandwich panel construction for lightweight vehicle designs, ThyssenKrupp Budd Technology and Innovation Centre, Auburn Hills, MI
30. Gehm, R., 2005, M-Class Shows Its Metals, Material Innovations, AEI Magazine, September 2005, URL: http://www.aei-online.org
31. Gehm, R., 2006, Arcelor Serves Up Steel Sandwich, Material Innovations, AEI Magazine, April 2006, URL: http://www.aei-online.org
32. Gehm, R., 2006, BMW Mixed It Up in BIW, Material Innovations, AEI Magazine, April 2006, URL: http://www.aei-online.org
33. Herold, H., Pieschel, J., Juttner, S., 2005, The Application of High Alloyed Steel in Vehicle Structures Welding Solutions, URL: http://www.msm.cam.ac.uk/phase-trans/2005/LINK/115.pdf
34. Hirsch, J., 2006, Automotive trends in aluminum, the European perspective: Part One, Key to Metals, URL: http://www.key-to-metals.com/Article135.htm
35. Hoch-Baumann, C., 2005, The A-Class features a Bondal® dash panel, ThyssenKrupp information, Compact, Issue 1
36. Hocher, U., Hage, M., Keller, T., 2007, Product and process engineering as a key to lightweight construction, ThyssenKrupp techforum, December
37. International iron & steel institute Committee on Automotive Applications, 2006, Advanced high strength steel (AHSS) application guidelines, Version 3, URL: http://www.worldautosteel.org
38. Jambor, A., Beyer, M., 1997, Technical report: New cars – new materials, Materials and Design, 18(4 & 6):203–209
39. Jiang, B., Zhao, N. Q., Shi, C. S., Li, J., 2005, Processing of open cell aluminum foams with tailored porous morphology, Scripta Materialia 53:781–785
40. Kochan, A., 2000, Audi moves forward with all-aluminium cars, Assembly Automation, 20(2):132–135
41. Krause, W., Henning, F., Tröster, S., Geiger, O., Eyerer, P., 2003, Reinforced thermoplastic components LFT-D — A process technology for large scale production of fiber, Journal of Thermoplastic Composite Materials 16:289–302
42. Liu, G. R., Han, X., Lam, K. Y., 2000, Material characterization of FGM plates using elastic waves and an inverse procedure, Journal of Composite Materials 35(11)11:954–971
43. Marsh, G., 2003, Next step for automotive materials, Materials Today 6(4):36–43
44. Maserati of Minneapolis Blog, Cutting Edge Nanotechnology – Hidden in Plain View, URL: http://maseratiofminneapolis.blogspot.com/2006/07/cutting-edge-nanotechnology-hidden-in.html
45. Matic, P., 2003, Overview of multifunctional materials, Smart Structures and Materials, Proceedings of SPIE 5053:61–69
46. Menzolit, Member of the Pastal Group, Carbon Fibre Reinforced SMC: Menzolit Advanced SMC, URL: http://www.menzolit.de/products/smc/carbon_smc/carbon_smc.html
47. Mertens, M., 1998, Multifunctional coatings in the automotive finishing industry, Metal Finishing 96(5):10–14
48. Metal Foams, FOAMINAL®, Fraunhofer-Institute for Manufacturing and Advanced Materials IFAM, URL: www.ifam.fraunhofer.de
49. Miller, W. S., Zhuang, L., Bottema, J., Wittebrood, A. J., Smet, P. D., Haszler, A., Vieregge, A., 2000, Recent development in aluminium alloys for the automotive industry, Materials Science and Engineering A280:37–49
50. Momoda, L. A., 2004, The future of engineering materials: Multifunction for performance-tailored structures, The Bridge, National Academy Engineering 34(4):18–21

51. Newsletter Arcelor Auto, 2005, News of steel in the automotive industry, No. 10, June
52. Noor, A. K., Venneri, S. L., Paul, D. B., Hopkins, M. A., 2000, Structures technology for future aerospace systems, Computers and Structures 74:507–519
53. Novelis Fusion Technology, 2007, URL: http://www.novelis.com/Internet/en-US/AboutUs/ResearchTech/fusion/AdditionalInformation/fusionbackground.htm
54. Pandremenos, J., Paralikas. J, Salonitis, K., Chryssolouris, G., 2008, Modularity Concepts for the Automotive Industry: A Critical Review, CIRP Design Conference
55. Markarian, J., 2005, Automotive and packaging offer growth opportunities for nanocomposites, Plastics Additives & Compounding 7(6):18–21
56. Ponticel, P., 2004, Alcoa Recast for AUDI A8 Role, Material Innovations, AEI Magazine, January 2004, URL: http://www.aei-online.org
57. R&G, 2006, Composites Materials Handbook, Edition B, URL: http://www.r-g.de
58. Reis, E., 2006, Lightweight Composite Panel, Ezine Articles, URL: http://ezinearticles.com/?Lightweight-Composite-Panel&id=313020
59. Roebben, G., Watjen, U., Lamberty, A., 2005, Certified reference materials for the characterization of multifunctional materials, Materials Science Forum 492–493:635–640
60. Schroder, T., 2004, Cooking steel for the car of tomorrow, Iron research, Max Planck Research
61. Science Technology & Society, the University of Texas @ Austin, Societal Implications of Nanotechnology 2005, "What About There? Nano Products", URL: http://sts.utexas.edu/projects/nanomodules/2NanoProducts.pdf
62. Sherman, A. M., Krause, A. R., Friedman, P. A., Steenkamer, A. A., Houston, D. Q., 2006, Automotive body materials, Encyclopedia of Materials: Science and Technology, 415–421
63. SMC Automotive Alliance, 1991, SMC design manual - Exterior body panels, The Composites Institute of the Society of the Plastics Industry, Washington, DC
64. Song, Z., Zhu, J., Ma, L., He, D., 2001, Evolution of foamed aluminum structure in foaming process, Materials Science and Engineering A298:137–143
65. Successful Drive with 1-Litre Car of Volkswagen, With Surface-Treated Magnesium Space Frame, URL: http://www.ahc-surface.com/ahce/news/one-litre-car.htm
66. Taylor, S. R., Sieradzki, K., 2003, The development of a multifunctional aerospace coating: Considerations in the use of the nano-dimensional materials, Progress in Organic Coatings 47:169–173
67. Tamler, H., Becker, J., Wunderlich, R., Friedrich, K., Rademacher, P., 2006, Tribond – hot rolled clad strip, customized steel composite material from coil, ThyssenKrupp Techforum, Issue 1
68. The A to Z of materials, 2004, Hummer H2 SUT Benefits from Lightweight High Performance Nanocomposite Materials, URL: http://www.azom.com/details.asp? ArticleID=2563
69. The Aluminium Association, 2001, Aluminium: The Corrosion Resistant Automotive Material, Publication AT7
70. The UK's National Measurement Laboratory NPL, Applications of Metal Foams, URL: http://www.npl.co.uk/materials/metal_foams/applications.html
71. The UK's National Measurement Laboratory NPL, General Information on Metal Foams, URL: http://www.npl.co.uk/materials/metal_foams/general_info.html
72. ThyssenKrupp Stall AG, 2004, ZE-Mg surfaces. The new generation of zinc alloy coatings: greater protection against corrosion; enhanced technological properties, Division Auto
73. ThyssenKrupp Steel, Publications, NSB NewSteelBody Brochure, URL: http://www.thyssenkrupp-steel.com/auto/en/publikationen/index.jsp?catcode=tkscsauto.pub.auto_informationen
74. ThyssenKrupp Steel, Highlights, Quiet Please! With BONDAL®, URL: http://www.thyssenkrupp-steel.com/auto/en/highlights/highlights_detail.jsp?cid=1001781
75. Vernor, J., Challinor, C., 1997, Precoated steels for automotive industry, Materials World 5(3):136–137
76. Wallentowitz, H., Leyers, J., Parr, T., 2003, Materials for Future Automotive Body Structures, Business Briefing: Global Automotive Manufacturing & Technology
77. Wick, C., Benedict, J., Veilleux, R., 1984, Tool and manufacturing engineers handbook - Volume 2 Forming, 4th Edition, SME, Dearborn, MI

Session I
Engineering Properties

Effect of Talc Filler Content on Poly(Propylene) Composite Mechanical Properties

Lubomir Lapcik Jr., Pavlina Jindrova, and Barbora Lapcikova

Abstract This presentation examines the effect of the micro/nano-sized talc filler on the physico-chemical and mechanical properties of the filled poly(propylene) (SABIC PP 108MF10, SABIC PP 33 MBTU [Saudi Basic Industries Corporation]) composite matrix. A range of mechanical properties were measured (tensile test, bending test, fracture toughness, notched impact strength (at the ambient temperature and −20°C), strain at break, impact strength), as well as micro-hardness testing and thermal stability from 40°C to 600°C measured by thermal analysis DTA and TGA. It was found that increasing filler content lead to a concomitant increase mechanical strength and toughness. The observed increase in tensile strength ranged from 15% to 25% (maximum tensile strength at break was found to be 22 MPa). The increase in strength and toughness simultaneously lead to higher brittleness reflected in the decrease of mean impact strength from the initial $18\,kJ/m^2$ (for the virgin PP sample) to $14\,kJ/m^2$, i.e. a 23% decrease. Similar dependency was also obtained for the samples conditioned at −20°C (decrease of 12.5%). It was found, that with increasing degree of filling of the talc-PP composite matrix the thermooxidative stability was increased, the highest magnitude was obtained for the sample 20 wt% (482°C decomposition temperature cf. 392°C for virgin polymer).

Keywords Composites · Poly(propylene) · Talc · Mechanical properties

1 Introduction

In the beginning of the third millennium polymeric materials represent the most important segment of all materials, based on production and usage volumes. In comparison with metals, polymers can offer better processing, lower density,

---

L. Lapcik Jr. (✉), P. Jindrova, and B. Lapcikova
Institute of Physics and Materials Engineering, Tomas Bata University in Zlín, Nad Stráněmi 4511, CZ-760 05 Zlín, Czech Republic
e-mail: lapcik@ft.utb.cz, pavlina.jindrova@seznam.cz, lapcikova@ft.utb.cz

S. Pantelakis, C. Rodopoulos (eds.), *Engineering Against Fracture: Proceedings of the 1st Conference*,

higher strength-to-weight ratio, greater resistance to corrosion and often a better price/performance ratio [1, 2]. Current increased demand for application of synthetic polymers in the automotive and aeronautic industries is evident, mainly in utilization of poly(ethylene), poly(propylene) (PP), polycarbonate or polyamide components of the interior, exterior or other functional parts of the vehicle [3].

Minerals first served as additives to polymer systems as a cost reducing technology – increased mineral addition lead to a decreased of polymer requirement [4]. Exploitation of mineral and polymer processing and chemistry has lead to minerals used as a functional additive, bringing specific properties to the final composite [4]. Often these mechanical and physicochemical properties are optimized with the use of nanostructured or nanocomposite materials. Richard P. Feynman proposed in the late 1950s that materials whose structure will be possible to control on extremely small dimensions will exhibit new and surprising properties [5]. His forecast is now being realized.

This work looks at the reinforcement of polypropylene with the addition of the mineral talc. Talc is a phyllosilicate mineral with a tri-octahedral layered structure, $Mg_3Si_4O_{10}(OH)_2$. The idealized crystal structure contains a layered structure of a 'sandwich' of magnesium oxide (brucite-like) octahedra between tetrahedra of silica [6]. This leads to a neutrally charged system, with all vacancies satisfied with no net surface charge. As a result, the lamellar platelets are only held together by van der Waals forces, which lead to talc being the softest mineral, defined as 1 on Mohs scale. Furthermore, the main sites for chemical attack or amphiphilic reaction on the talc surface are mineralogical defects and platelet edges [7–9].

The aim of this paper was to better understand and characterize basic physicochemical and material data of highly filled mico/nano composite matrix of PP-talc bases.

2 Experimental

2.1 Materials

Poly(propylene) SABIC PP 108MF10 natur. copolymer (lot no. 6091218L23) (sample A) and talc filled poly(propylene) copolymer with ethylene propylene diene elastomer EPDM SABIC PP 108MF10 and SABIC PP compound 33 MBTU (lot no. 0060271436) (sample B) (Saudi Basic Industries Corporation) were used in this study. Filled samples of different talc content (5, 10, 15, 20, 25 and 30 wt%) were prepared by homogenizing pure and filled PP in microblender Brabender at 180°C, 185°C and 195°C. Samples were then injection moulded on a Demag NC4 Ergotech 50-200 system. The processing parameters for samples A and B are summarized in the Table 1.

Table 1 Processing parameters used for the samples A and B preparation by injection moulding of the pure and filled poly(propylene) (SABIC). Parameters kept constant for all samples

Sample	Injection pressure (bar)	Injection rate (mm/s)	*Dye* temp. (°C)	Screw resistance (bar)	RPM (1/min)	Melt temp. (°C)	jet temp. (°C)
A, B, all talc comp.	845	50	60	150	150	238	230

2.2 *Methods*

2.2.1 Tensile Strength Testing

Tensile strength (σ) tests were performed on universal tensile test machine Tiratest 2160 according to CSN EN ISO 527-1 and 527-2 standards. Applied deformation rate was 50, 200 and 500 mm/min. Measurements were performed at ambient laboratory conditions (25°C and 50% relative humidity). Five separate readings of results for each sample gave an average representative maximum load value and elongation at break (ε).

2.2.2 Bending Strengths Testing

Bending strengths tests were performed on Tiratest 2160 universal test machine according to CSN EN ISO 178-64 0607 standard (a three-point flexural test). Applied deformation rate was 5 mm/min. Support span length was adjusted to 60 mm. The determined maximum deformation was set to be 6 mm (i.e. 1.5 sample thicknesses), as the sample were not broken during testing. Five separate readings of results for each sample gave an average representative value.

2.2.3 Microhardness Measurements

Surface micro-hardness was measured according to the Czech standard CSN EN ISO 2039-1 (Plastics – Determination of the hardness – Part 1: Sphere impressing method) by use of the Rockwell-Rapid instrument. Measurements were performed at 25°C and 42% of relative humidity. The sphere was impressed into six different positions of the test surface at defined preload (132.4 N) applied for 10 s. Results are shown in Fig. 2.

2.2.4 Fracture Toughness

Fracture toughness was determined on Charpy's hammer (0.4 kPm hammer) at 25°C and at −20°C temperature for each material under study. Charpy fracture toughness is defined as the kinetic energy of the hammer used for the fracture of the tested specimen normalised to the specimen original cross-section. Each measurement was repeated five times.

2.2.5 Notched Impact Toughness

Notched impact toughness was determined according to TL 526 31 standard on Charpy's hammer (0.4 kPm hammer) at 25°C and at −20°C temperature for each material under study. Charpy notched impact toughness is defined as the kinetic energy of the hammer used for the fracture of the notched tested specimen recalculated to the specimen original cross-section. Each measurement was repeated five times.

2.2.6 Thermal Stability

Thermal stability of the samples was characterized by means of thermogravimetry (TG) and differential thermal analysis (DTA) methods. Measurements were performed on DTG 60/60H Differential Thermogravimetric Analysator (Shimadzu) in the air atmosphere. Measurement parameters: heating rate 10°C/min, starting temperature 40°C/min, ending temperature 600°C/min.

3 Results and Discussion

Results of the mechanical tensile tests of samples A and B (for varying filler concentrations) are shown in Fig. 1. Obtained results confirm expected behaviour, i.e. the gradual increase of tensile strength with increasing filler concentration accompanied with respective decrease of the elongation at break. Similar trends were obtained for all measured deformation rates. Observed behaviour suggests that with increasing filler content, samples do not preserve their original polymer chain rearrangement mobility, most probably due to the creation of specific inter particle–polymer interactions. Talc is a strong reinforcing filler due to its thin platy or flaky nature and possesses the capability to orientate to the polymer flow during processing [10]. The increase in tensile strength for the 50 mm/min deformation rate was ranging from 16.68 to 21.93 MPa, a 31.5% increase. Corresponding to the latter increase of the tensile strength the 98% decrease of the elongation at break was detected (decrease from the original 265% down to 5%). With increasing deformation rate the loss of elongation was not so great – 94% for the 200 mm/min deformation rate this was 94% and 86% for 500 mm/min. The obtained increase in tensile strength for virgin sample A for increasing deformation rates suggests successful pseudo-plastic deformation of the amorphous polymer chains part.

With increasing filler content the higher crystal like morphologies were created, reflected in the latter mention increase of the tensile strength, suggesting

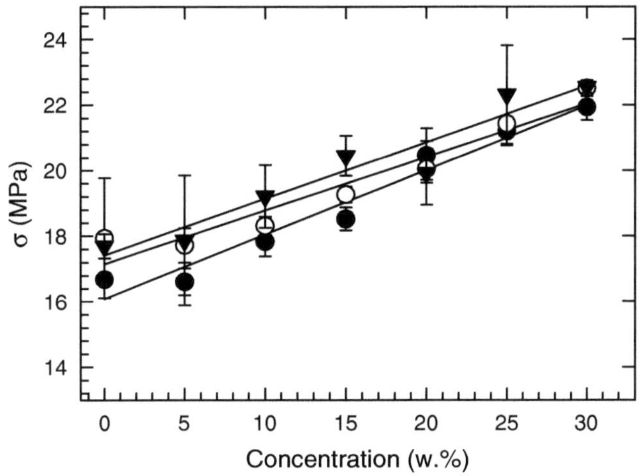

Fig. 1 Talc filler concentration dependence of the tensile strength for samples A, B and C. Applied deformation rates: samples A and B (SABIC): full circle – 50 mm/min, empty circle – 200 mm/min

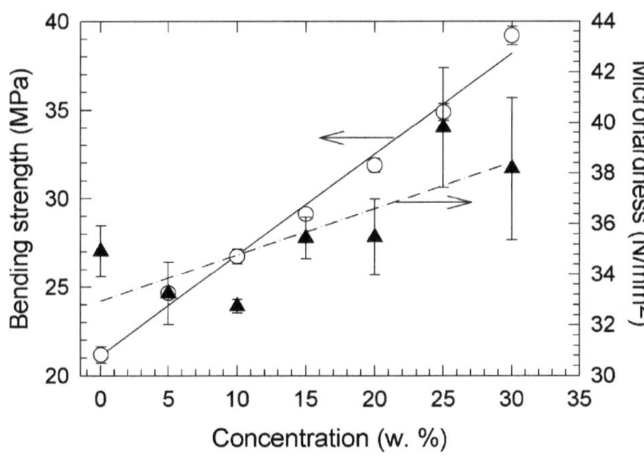

Fig. 2 Talc filler concentration dependence of bending strength (empty circles) and of Microhardness (full triangles) of samples A and B (SABIC)

substitution of PP by the largely more rigid filler. Filler thus can restricting the mobility and deformability of the matrix by introduction of a mechanical restraint [10]. This effects can be attributed also to the fact, that talc is a strong crystallization agent which is affecting nucleation processes in PP matrix, which might change the crystalline structure from weaker β spherulites into the stronger α spherulites. As confirmed by thermal analysis data (see Table 2), our samples T_m were in the range of 170°C, which is typical for α modification (T_m for β modification is at approximately 152°C) [11–13]. Similar trend was confirmed also in the corresponding increase of the bending strength from original 21.18–39.20 MPa for 30 wt% talc concentration, i.e. increase of 85% (see Fig. 2). As shown in Fig. 2 with increasing filler concentration microhardness gradually increased by approximately 9% (from original 34.94–38.18 N/mm^2). With increasing hardness a corresponding increase in brittleness may be expected, thus measurements of both fracture toughness and notched impact toughness were performed at ambient temperature and at −20°C.

In the case of fracture toughness both dependencies at the ambient temperature as well as at −20°C follow the same gradual linear decrease pattern having a gradient of $-0.1\,kJ/m^2$ and the abscissa at $16.7\,kJ/m^2$ (for 25°C) and $23.7\,kJ/m^2$ (for −20°C) respectively (Fig. 3). Different dependencies were obtained for notched impact toughness measurements for SABIC PP, where at 15 wt% (for 25°C) and at 5 wt% talc concentration (for −20°C) vigorous decrease of impact toughness was found suggesting a step change in samples brittleness. These processes might be accompanied by filler active suppressing of the polymer matrix ability to undergo a plastic deformation process. This triggers change of the failure mode of the PP matrix from ductile to brittle. Such a strong increase of brittleness corresponds to the change of the spatial arrangement of the filler particles in the composite matrix, reflected in thermal gravimetric data (Table 2). Here the strong increase

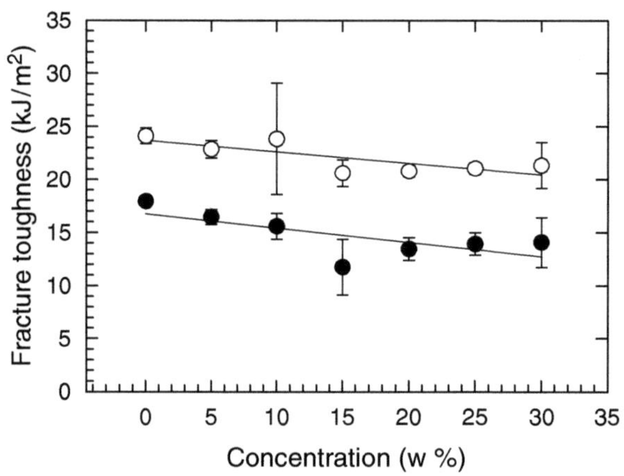

Fig. 3 Talc filler concentration dependence of fracture toughness of unnotched (sample A, B (SABIC) full circle: at +25°C, empty circle: at −20°C)

Table 2 Results of the thermal analysis of sample A and B (SABIC)

Sample/talc conc. (wt%)	T_m(°C) DTA	TG	
		T_{start}(°C)	Weight loss (%)
A/0	172.6	260.5	100
B/5	171.4	298.5	93.5
B/15	170.4	288.7	84.2
B/30	171.5	275.4	67.6

T_m – Melting point temperature, T_{start} – Starting temperature of decomposition in decomposition starting temperature point was found in the case of 20 wt% talc concentration, from the original 260.5–303.6°C suggesting presence of the strong intermolecular or inter particle bonds.

4 Conclusions

It was found in this study, that with increasing degree of filler content increased mechanical strength and toughness of the poly(propylene)/talc composite matrix was found. This effect was ascribed to the influence of the excellent dispersion and high order (crystal like lattice) of microparticles (their spatial configuration) in the 3D PP matrix and to their mutual synergistic effect resulting in the creation of the specific bonds between individual particles and the polymer matrix. Observed increase in the tensile strength was ranging between 15% to 25% (maximum tensile strength at break was found 22 MPa, SABIC). Simultaneously with this the increase brittleness was obtained, what was reflected by the decrease of the mean impact toughness from the initial $18\,\text{kJ/m}^2$ (for the virgin PP sample) to $14\,\text{kJ/m}^2$, i.e. 23% decrease. Similar dependency was obtained also for the samples conditioned at −20°C (decrease of 12.5%). It was found, that with increasing degree of filling of the talc-PP composite matrix the thermooxidative stability was increased, the highest magnitude was obtained for the sample 20 wt% (virgin unfilled sample showed decomposition temperature at 392°C, sample 20 wt% at 482°C).

Acknowledgements Authors would like to express their gratitude for partial financing of this research by Cadence Innovation, Liberec company (member of Ventura, USA) (Project No.: HS260006020) and to Ministry of Education, Youth and Physical Training of the Czech Republic (Grant VZ MSM7088352101).

References

1. Mazumdar SK. Composites Manufacturing. Materials, Product, and Process Engineering. Boca Raton, FL/London/New York/Washington, DC: CRC Press, 2002.
2. Lapcik L, Jr, Raab M. Materials Engineering II. Text Book, 2nd Edition. Zlin: Tomas Bata University, 2004.
3. Lapcikova B, Lapcik L, Jr, Smolka P, Dlabaja R, Hui D. Application of Radio Frequency Glow Discharge Plasma for Enhancing Adhesion Bonds in Polymer/Polymer Joints. J Appl Polym Sci 2006; 102: 1827–1833.
4. Rothon RN, Hancock M. Particulate Filled Polymer Composites, 2nd Edition, Ed., R.N. Rothon. Shrewsbury, UK: Rapra Technology, 2003.
5. Feynman R. There is Plenty of Room at the Bottom. Talk at the California Institute of Technology, 29 December 1959.
6. Rayner JH, Brown G. Crystal-Structure of Talc. Clay Clay Miner 1973; 21: 103–114.
7. Douillard JM, Salles F, Henry M, Malandrini H, Clauss F. Surface Energy of Talc and Chlorite: Comparison Between Electronegativity Calculation and Immersion Results. J Colloid Polym Sci 2007; 305: 352–360.

8. Mitsuda T, Taguchi H. Formation of Magnesium-Silicate Hydrate and Its Crystallization to Talc. Cement Concrete Res 1977; 7: 223–230.
9. Alonso M, Gonyales A, Desaja JA, Escalona AM. Quality-Control of Mineral Impurities in Industrial Talcs by Thermogravimetric Analysis. Thermochim Acta 1991; 184: 125–130.
10. Abu Bakar MB, Leong YW, Ariffin A, Ishak ZAM. Mechanical, Flow and Morphological Properties of Talc- and Kaolin-Filled Polypropylene Hybrid Composites. J Appl Polym Sci 2007; 104: 434–441.
11. Wang T, Liu D, Keddie JL. An Alternative Approach to the Modification of Talc for the Fabrication of Polypropylene/Talc Composites. J Appl Polym Sci 2007; 106: 386–393.
12. Varga J. Supermolecular Structure of Isotactic polypropylene. J Mater Sci 1992; 27: 2557–2579.
13. Rybnikar F. Transition of Beta to Alpha Phase in Isotactic Polypropylene. J Macromol Sci Phys 1991; B30: 201–223.

Fracture Properties of Polypropylene Reinforced with Short Glass Fibres: The Influence of Temperature, Notch Length and Geometry

Petr Elisek and Miroslav Raab

Abstract Commercial grades of isotactic polypropylene TATREN TI 924 and SLOVALEN PH 61 GF 30 were used for preparation of specimens with 0%, 10%, 20% and 30% of short glass fibres reinforcement ratio.

Injection-moulded specimens were notched and exposed to impact bending tests with instrumented Charpy pendulum MFL Systeme. For each type of specimen the measurement was repeated ten times for the assessment of the dynamic modulus of elasticity (res. dynamic yield strength) and five times for the fracture toughness evaluation.

Fracture toughness characteristics were evaluated in temperatures range from −40°C to 50°C with 10°C increment. The influence of notch length and geometry on fracture toughness characteristics were evaluated at the temperatures of (20 ± 1)°C.

Keywords Isotactic polypropylene · Short glass fibre · Charpy impact testing · Dynamic modulus of elasticity · Dynamic yield strength · Notch geometry

1 Introduction

For thermoplastic polymers, both ductile and brittle modes are possible, and many of these materials show a ductile-to-brittle transition. Factor that favour brittle fracture are a reduction in temperature, an increase in strain rate, the presence of a sharp notch, increased specimen thickness, and, in addition, a modification of the polymer

P. Elisek
M.Sc., Ph.D. - Tomas Bata University, Faculty of Technology, TGM 275, 762 72 Zlín, Czech Republic

M. Raab
Professor RNDr. CSc. – Institute of Macromolecular Chemistry, Academy of Sciences of the Czech Republic, Heyrovského nám. 2, 162 06 Prague, Czech Republic
e-mail: kavcr@kav.cas.cz

S. Pantelakis, C. Rodopoulos (eds.), *Engineering Against Fracture: Proceedings of the 1st Conference*,

structure (chemical, molecular, and/or super molecular). Glass reinforced polymers are brittle at relatively low temperatures; as the temperatures is raised, they become ductile in the vicinity of their glass transition temperatures and experience plastic yielding prior to fracture.

Prior to the advent of fracture mechanics as scientific discipline, impact testing techniques were established so as to ascertain the fracture characteristics of materials. It was realized that the results of laboratory tensile tests could not be extrapolated to predict fracture behaviour [1].

The Charpy impact testing is the most widespread method for determination of predisposition for the brittle behaviours of material which was designed and still used to measure the impact energy, sometimes also referred to as notched impact strength [1].

2 Materials and Experimental Technique

2.1 Materials

Two basic materials were used for the present study: neat polypropylene TATREN TI 924 (Slovnaft, Inc, Bratislava, Slovakia) and the same basic polymer containing 30% glass fibers SLOVALEN PH 61 GF 30 (Amida, Ltd, Czech Republic). The properties of the materials, as given by the producers, are summarised in Table 1.

Other tested samples with 10 and 20 wt% short glass fibres, were prepared by blending of the basic materials in the rates of 2:1 and 1:2 (TATREN TI 924:SLOVALEN PH 61 GF).

2.2 Specimen Preparation

All specimens were prepared by injection moulding using DEMAG ERGOtech 50-200 system machine (Fig. 1). According to .SN EN ISO 179 two types of test

Table 1 Properties of materials [2, 3]

Property	Unit	Test method	Materials	
			SLOVALEN PH 61 GF 30	TATREN TI 924
Glass fibres content	%	ISO 3451	30 ± 2	0
Density	g/cm^3	ISO 1183	1.10	0.90
Tensile strength	MPa	ISO 527	80	28
Elongation	%	ISO 527	4	11
Tensile modulus	MPa	ISO 527	4,900	1,200
Flexural modulus	MPa	ISO 178	5,300	1,100
Charpy notched impact strength 23°C	kJ/m^2	ISO 179	4	2

Fig. 1 DEMAG ERGtech 50-200 [4]

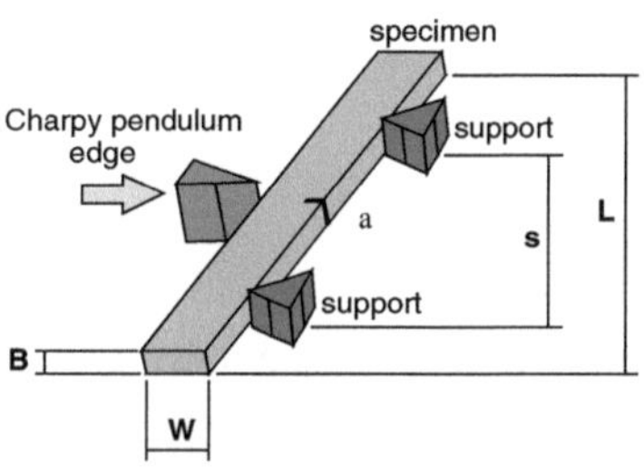

Dimension	Description	Size (mm)
B	Thickness of the specimen	4
W	Width of the specimen	10
L	Length	60
a	Notch length	4
s	Span of supports	40

Fig. 2 Test specimen

specimens were prepared – unnotched (for the assessment of dynamic yield strength and dynamic modulus of elasticity) and notched for the determination of K_{Jd} or J_{Id}.

All specimens were conditioned to achieve stress relaxation and to obtain the desired temperature for a minimum duration of 24 h. The layout of the test as well as the specimen dimensions are shown in Figs. 2 and 3.

Dimensions, radius of notch bases, geometry of notches and their preparation were proposed for the determination of the influence of notch depth, geometry and preparation on fracture toughness of specimens. Specimens were prepared according to ISO 179 and ISO 179-2 [5, 6]. The size and shapes of specimens are shown in Fig. 3. The first type of notch was made by slow pushing of a razor blade into the specimen with using special machine with micrometric shift. The depths of notches were used 1, 2, 3 and 4 mm.

The second type of notch was prepared by Power-Driven Notchvis-code 6951.000 apparatus using exchangeable knife with one side blade. The speed

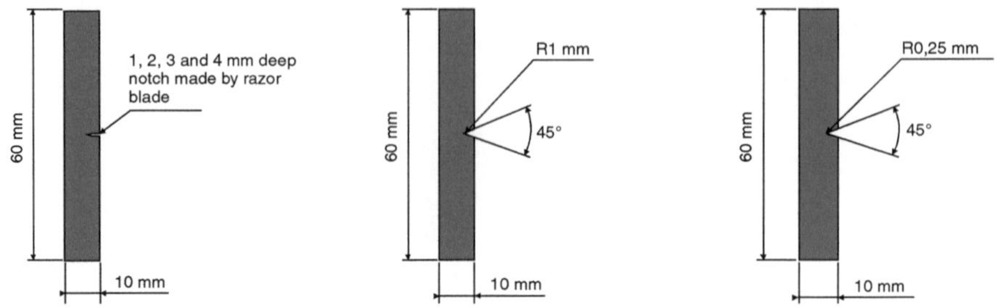

Fig. 3 Dimensions and geometry of notches

of knife is variable and rate is between 12–42 m/min. We used cutting speed rate 12 m/min. Notchvis-code 6951.000 apparatus is designed and built to meet the international standards [7]. Depths of notches were prepared from 1 to 4 mm, step 1 mm, radius 0.25 mm.

The last type of notch with base radius 0.25 and 1 mm was prepared by milling machine PC16CNC. Shift of milling was set-up on 500 mm/min and knife rotation speed on 1,200 rpm. Under these conditions specimens with V-type notches and deeps 2 and 4 mm were prepared.

2.3 *Impact Testing*

At impact testing the specimen with prismatic profile placed horizontally on two supports is exposed to an impact in perpendicular direction at a high nominally constant rate. The acquired stress-deflection curve represents the behaviour of the tested specimen during the test.

Charpy impact tests were carried out on an instrumented Charpy pendulum MFL Systeme (Fig. 4) used for the impact testing of dynamic fracture toughness. The hammer deflection of 20° was used giving a rate of $0.95\,m\,s^{-1}$ and energy $E = 4.7\,J$. The specimens were measured according to standard specification ISO 179-2 [6].

A minimum of ten specimens were broken for the assessment of the dynamic modulus of elasticity (i.e. dynamic yield strength) and another five for the fracture toughness evaluation. The specimens were impact tested at the temperatures of from −40°C to 50°C, the step being 10°C. Thus, the dependence of waveform of the force-deflection curves on temperature was obtained.

Linear elastic fracture mechanics can determine fracture characteristics of brittle plastic materials, but this method is not applicable for tough materials, where the concept of J-integral can be applied. Initiation of the unstable crack propagation is determined by the critical value of the energy. The characteristic measured quantity is FGY [N] detected from the intersection of 5% secant with test trace.

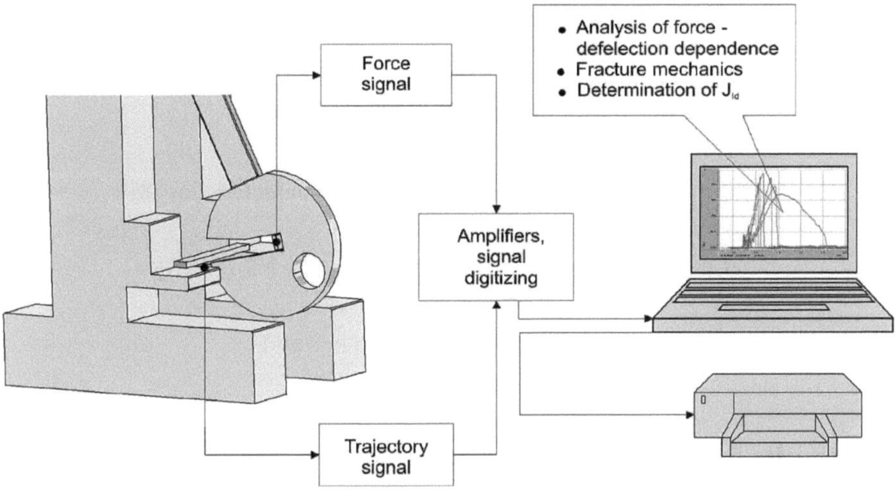

Fig. 4 Scheme of Charpy instrumented pendulum

The "provisional" value of the J-integral is given as [8]:

$$J_{Qd} = \eta_{el} \cdot \frac{A_{el}}{B \cdot (W - a_0)} + \eta_{pl} \cdot \frac{A_{pl}}{B \cdot (W - a_0)} \cdot \frac{W - a_{eff}}{W - a_0} \quad (1)$$

$$\eta_{el} = \frac{2 \cdot F_{GY} \cdot s^2 \cdot (W - a_0)}{f_{GY} \cdot E_d \cdot B \cdot W^3} \cdot f^2_{(a/W)} \cdot (1 - \nu)^2 \quad (2)$$

$$\eta_{pl} = 2 - \frac{(1 - a_0/W) \cdot (0.892 - 4.476 \cdot a_0/W)}{1.125 + 0.892 \cdot a_0/W - 2.238 \cdot (a_0/W)^2} \quad (3)$$

a/W ratio (notch length/specimens width) is calculated as [8]:

$$f_{(a/W)} = \frac{3}{2} \cdot (a/W)^{0.5} \cdot \left(\frac{1.99 - a/W \cdot [1 - a/W] \cdot \left[2.15 \cdot 3.93a/W + 2.7\,(a/W)^2\right]}{(1 + 2a/W) \cdot (1 - a/W)^{3/2}} \right) \quad (4)$$

Dynamic modulus of elasticity (i.e. dynamic yield strength) is obtained as [8]:

$$E_d = \frac{F_{GY} \cdot s^3}{4 \cdot B \cdot W^3 \cdot f_{GY}} \qquad \sigma_d = \frac{3 \cdot F_{GY} \cdot s}{2 \cdot B \cdot W^2} \quad (5,6)$$

Where:

F_{GY} [N] … the force value of the point of intersection of 5% secant with test trace

f_{GY} [mm] … deflection corresponding to F_{GY}

E_d [MPa] … dynamic modulus of elasticity

σ_d [MPa] ... dynamic yield stress
a_{eff} [mm] ... effective length of crack
ν [-] ... Poisson's constant
$f_{(a/W)}$ [-] ... function of test piece geometry and (a/W)
s, B, W, a ... explanation see above in Fig. 1

The rules for the validity of the calculated toughness characteristics [8]:

$$B, a_0, (W - a_0)\rangle \varepsilon \cdot \frac{J_{Qd}}{\sigma_d}, \varepsilon = 256.0 \cdot J_{Qd}^{-0.96} \Rightarrow J_{Qd} = J_{Id} \tag{7}$$

The previous condition must be fulfilled for J_{Qd} to be valid.

3 Results and Discussion

The philosophy of this work was verified effect of reinforcement ratio of glass fibres, temperature, deep, geometry and preparation of notch on fracture behaviour of PP composite.

Figures 5 and 6 show dependence of waveform of force-deflection curves on temperature for isotactic polypropylene TATREN TI 924 and SLOVALEN PH 61 GF 30. Characteristic curves of these two commercial grades there are shown for illustration at temperature interval from −30°C to 50°C, the step being 20°C.

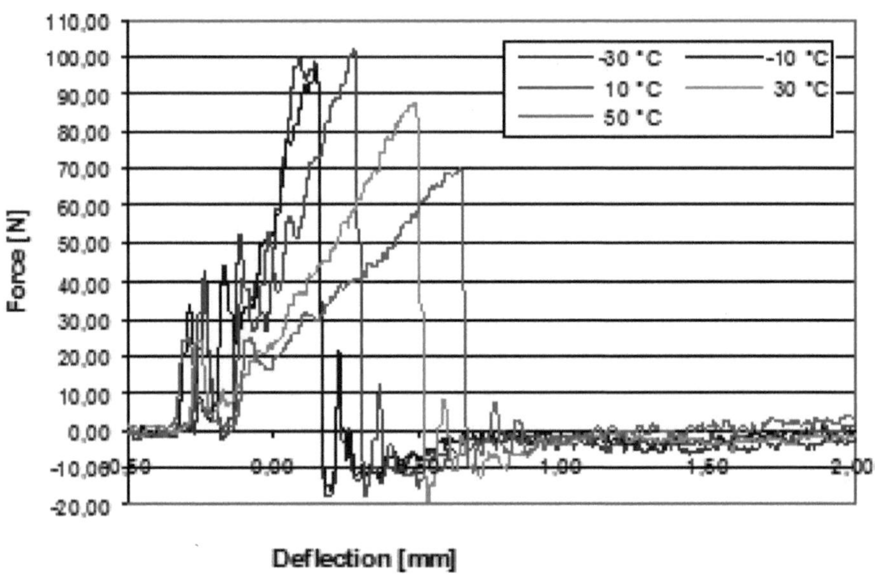

Fig. 5 Dependence of waveform of force-deflection curves on temperature (PP homopolymer)

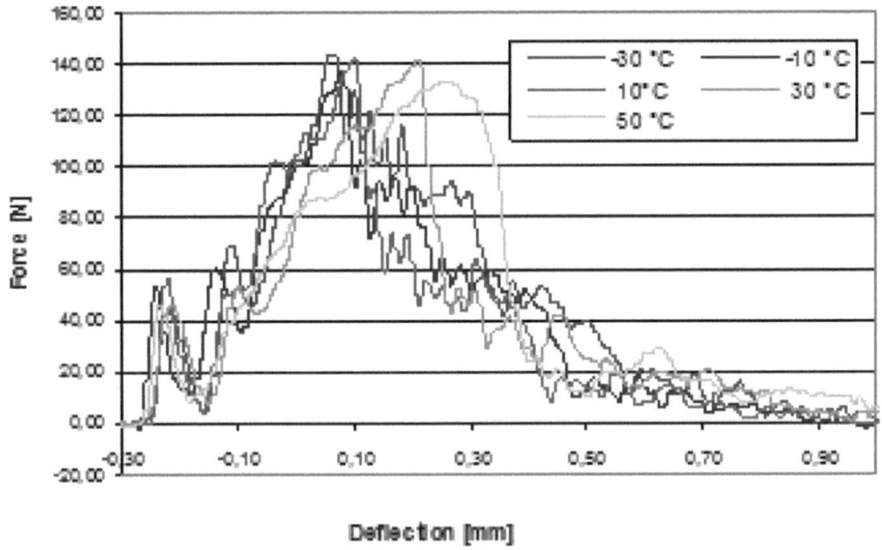

Fig. 6 Dependence of waveform of force-deflection curves on temperature (PP with 30% GF)

As can be seen, polypropylene homopolymer and glass filled composite are brittle in at relatively low temperatures; as the temperatures is raised, the materials become ductile as can be determined from the waveform of curves. For the 30% GF composite achieved as high as 140% of value of force necessary to initiate failure in pure polypropylene used in test.

Firstly, diametrically different increasing of stress in both basic types of tested materials – pure and reinforced PP and consequential process of their failure is evident. Another important piece of knowledge is that with growing up ratio of filling (10%, 20% and 30% of GF in polypropylene matrix) increase value of force necessary to initiate of crack. It follows that short glass composites used in test are preferable for force stressed application than pure matrix.

Moreover, dynamic modulus of elasticity and dynamic yield strength indicate high dependence on temperature and reinforcement ratio (notch depth 4 mm). Both quantity decreases with increasing temperature and increase with increasing reinforcement ratios as can be seen in Figs. 7 and 8.

Figures 9 and 10 show, that dynamic modulus of elasticity and dynamic yield strength significantly depend on reinforcement ratio of glass fibres in isotactic polypropylene. Value of Ed grows up with increasing of short glass fibres volume. Volume increment 10 wt% of GF (on 10, 20 and 30 wt% of GF in polypropylene) means that Ed grows in average about $(65 \pm 9)\%$. This means that addition of 30 wt% GF in neat isotactic polypropylene TATREN TI 924 cause growth of dynamic modulus of elasticity about $(195 \pm 9)\%$, i.e. $(2{,}200 \pm 100)$ MPa. Dynamic yield strength indicates identical dependence – increasing of δd with increasing reinforced ratio of short GF, in average about $(22 \pm 8)\%$.

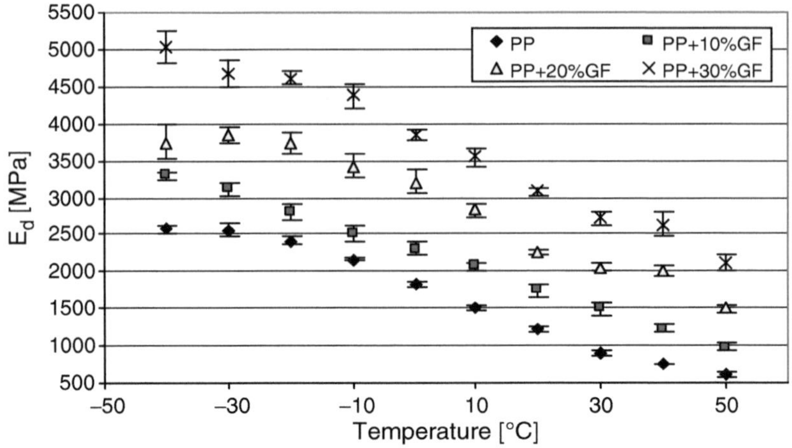

Fig. 7 Dependence of Ed on temperature and short glass reinforcement ratio of polypropylene composite

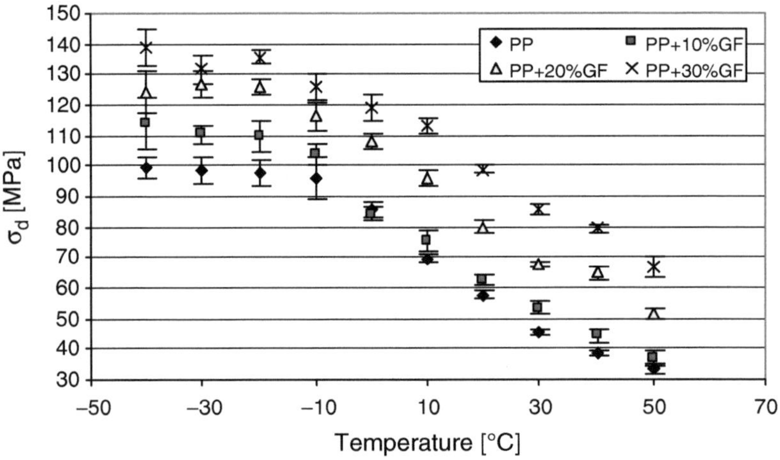

Fig. 8 Dependence of σ_d on temperature and short glass reinforcement ratio of polypropylene composite

Figure 11 shows dependence of KJd on geometry, depth, preparation of notches and reinforcement ratio of short GF.

The first important piece of knowledge is fact that fracture toughness KJd, as can we see in Fig. 11, significantly depends on reinforcement ratio of short GF in isotactic polypropylene. The quantity of fracture toughness increases with growing up ratio of filling (10, 20 and 30 wt% of GF in polypropylene matrix). At the same time fracture toughness KJd decreases with growing up length of notch.

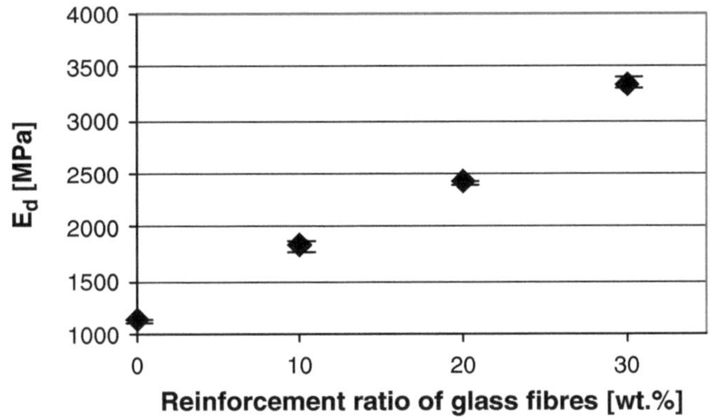

Fig. 9 Dependence of E_d and σ_d on reinforcement ratio of short GF

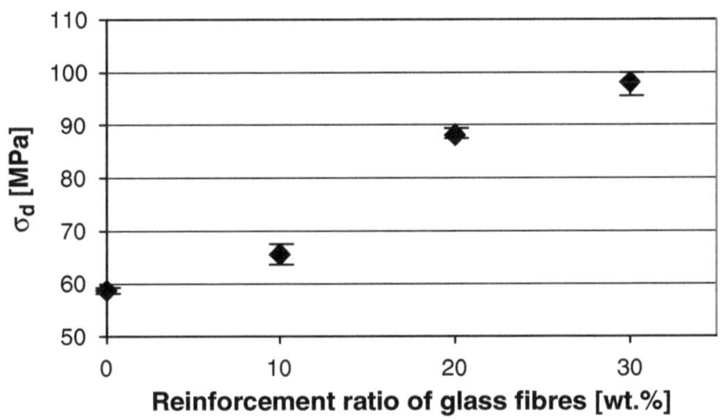

Fig. 10 Dependence of σ_d on reinforcement ratio of short GF

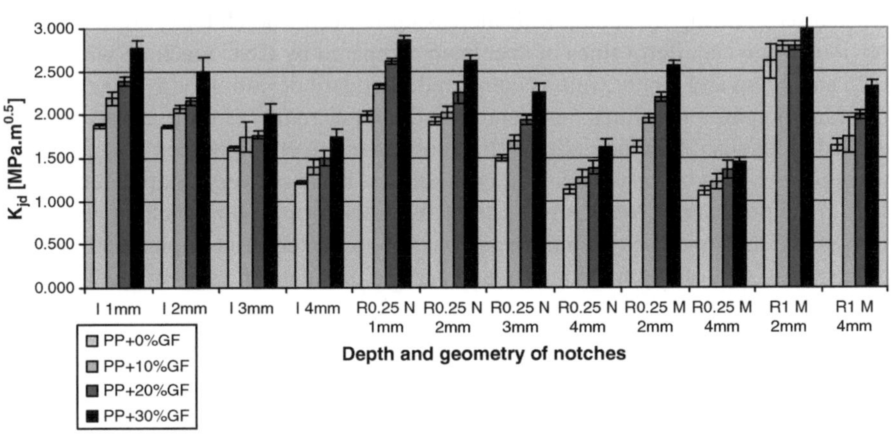

Fig. 11 Dependence of KJd on geometry, depth, preparation of notches and reinforcement ratio of short GF (I – razor blade, N – Notchvis, M – milling machine, R – radius of notch base [mm])

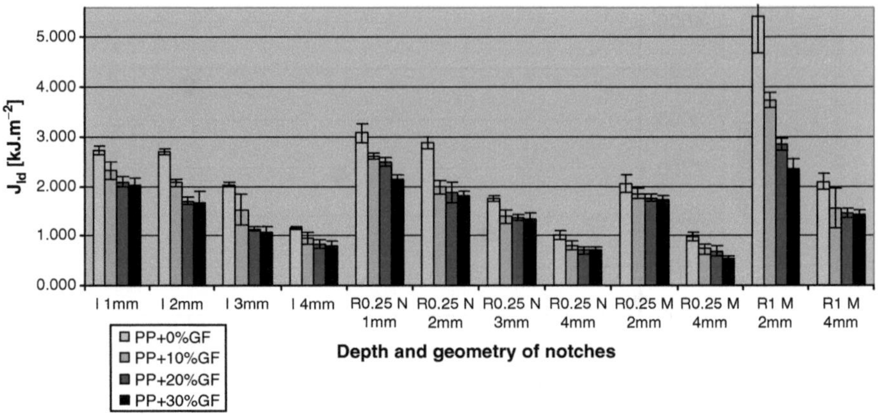

Fig. 12 Dependence of J_{Id} on geometry, depth, preparation of notches and reinforcement ratio of short GF (I – razor blade, N – Notchvis, M – milling machine, R – radius of notch base [mm])

Moreover, measurement shows that preparation of notch and appropriate select of apparatus is important too. As an example we can show specimens with 2 and 4 mm depth of notch, prepared by Power-Driven Notchvis [9] (R0.25 N 2 and 4 mm) and milling machine PC16CNC [4] (R0.25 M 2 and 4 mm). Both prepared type of specimens have the same radius 0.25 mm. The higher values of fracture toughness are given by Notchvis prepared specimens. It caused by strict implementation and precise hold size of depth and geometry of the notch. Another important argument is better smooth surface of the notch made by Notchvis and heat modified surface made by rotating knife on CNC milling machine.

Dependence of J_{Id} on geometry, depth, preparation of notches and reinforcement ratio of short GF in isotactic polypropylene matrix can be seen in Fig. 12. Values of J-integral decrease with increasing deep of notch again. In contrast to previous figure, value of this property is decreasing with reinforcement ratio too. Once again we can see growing up values with increasing radius of notch base. Above all it is obvious when acquired values of specimens prepared by CNC machine with radius 0.25 and 1 mm and depth 2 mm is compared. Standard deviation of average of CNC machine prepared specimens especially with higher radius (1 mm) is very high. It caused by the fact, that the radius of the crack is relatively high, then it is necessary to prepare crack with very precise smooth surface, because every small defect of surface in the presence of lower stress can evoke crack propagation. Notch preparation using CNC machine is quick and cheap, but not guarantee strict implementation and precise hold size of depth, geometry and surface of the notch as mentioned above.

4 Conclusion

In this paper the fracture properties of neat isotactic polypropylene and the same isotactic polypropylene reinforced by 10, 20 and 30 wt% of short glass fibres has been described. Especially influence of reinforcement ratio of GF, temperature, deep, geometry and preparation of notch was determined. Instrumented Charpy pendulum, data evaluation software and methods of fracture mechanics were used for characterization and interpretation of measured data.

Firstly high influence of dynamic modulus of elasticity and dynamic yield strength on reinforcement ratio was affirmed. Every increment 10 wt% of GF caused that material's properties Ed and σd were increased in average about $(65 \pm 9)\%$ and $(22 \pm 8)\%$. High dependence of Ed and σd on temperature and reinforced ratio was found. Both quantities decreased with temperature and increased with increasing reinforcement ratio.

Moreover, importance of precise production and appropriate manner of production of notch was given by the measurement of fracture toughness and comparing of values for specimens with different machine method of notch preparation. The higher values of fracture toughness were found for notch prepared by special apparatus, by reason of better accuracy dimensions and better notch surfaces.

Fracture mechanics properties – fracture toughness K_{jd} and J-integral J_{Id} were measured and determined using fracture mechanics methods and Instrumented Charpy pendulum. Both quantities were decreased with increasing of notch's deep. Further the fracture toughness K_{jd} was grown up with increasing ratio of filling and notch radius. On the other side J_{Id} integral was decreasing with reinforcement ratio but it grown up with increasing notch base radius as fracture toughness.

The beneficial effect of the reinforcement on the behaviour of the material under service conditions is evident. The established complex relation between reinforcement ratio, notch length, notch radius and resulting fracture toughness could be used for material selection for particular technical applications.

Acknowledgements Partial support of the Academy of Sciences of the Czech Republic (project AVOZ 40500505) and of the Ministry of Education, Youth, and Sports of the Czech Republic (research grant MSM 265200015) is gratefully acknowledged.

References

1. Callister W.D.: Materials science and engineering: an introduction, 4th edition, 1997 U.S.A. ISBN 0-471-13459-7.
2. www.amida.cz.
3. www.slovnaft.pl.
4. http://www.ft.utb.cz/czech/upmt/.
5. ČSN EN ISO 179: Plasty – Stanovení rázové houževnatosti metodou Charpy, Český normalizační institut, 1998.
6. ČSN EN ISO 179-2: Plasty – Stanovení rázové houževnatosti metodou Charpy, Část 2: Instrumentovaná rázová zkouška, Český normalizační institut, 2000.
7. ESIS Technical Committee: A linear elastic fracture mechanics (LEFM) standard for determining K_c and G_c, Testing protocol, 1990.
8. Grellmann W., Seidler S., Hesse W.: Prozedur zur Ermittlung des Risswiderstandsverhaltens aus dem Instrumentierten Kerbschlagbiegeversuch Institut für Polymerwerkstoffe e.V., Merseburg, Deutschland, 1993.
9. www.ceast.com.

Influence of Width of Specimen on Tensile Properties of NiCo Thin Film

Jun-Hyub Park, Joong-Hyok An, Yun-Bae Jeon, Yun-Jae Kim, and Yong-Hak Huh

Abstract This paper represents the results of tensile test for NiCo thin films, to investigate the size effect on tensile properties. An axial loading tensile tester developed by authors was used to measure the mechanical characteristics of thin film materials. The tester has a load cell with maximum capacity of 0.5 N and a non-contact position measuring system based on the principle of capacitance micrometry. The "dog-bone" type specimen was designed and fabricated by electroplating process. The 50 and 150 μm width of reduced section respectively were designed on 6 in. wafer. The length of reduced section is 1,000 μm and the radius of the blending fillet is 1,000 μm to minimize a stress concentration of the specimen. The thickness is 10 μm. The elastic moduli of NiCo thin film were about 18 ± 4.1 GPa for 50 μm width and about 148 ± 8.4 GPa for 150 μm width. The ultimate tensile strength for the width of 50 and 150 μm were $2{,}431 \pm 87.8$ and $2{,}348 \pm 93.3$ MPa, respectively. However, the difference between tensile strengths is a little, as 3.3%.

Keywords NiCo thin films · *Micro-tensile Stress–Strain* Curves · MEMS applications

1 Introduction

Thin films are widely employed in microelectromechanical system (MEMS) and integrated microelectronic components. One of the most important items in reliability of MEMS is to determine relevant mechanical properties of the MEMS component,

J.-H Park
Department of Mechatronics Engineering, Tongmyong University, Busan, Korea

J.-H. An, Y.-B. Jeon, and Y.-J. Kim
Department of Mechanical Engineering, Korea University, Seoul, Korea

Y.-H. Huh (✉)
Department of Metrology for Quality Life, Korea Research Institute of Science and Standards, Science Town, Taejon, 305-600, Korea
e-mail: yhhuh@kriss.re.kr

S. Pantelakis, C. Rodopoulos (eds.), *Engineering Against Fracture: Proceedings of the 1st Conference*,

such as elastic and plastic tensile properties, fatigue properties and so on. One notable point is that the size of components in MEMS devices is in the order of microns, and that their mechanical properties are not same as those for bulk materials, as mechanical properties in micro-scale depend on the size and fabrication process [1]. Thus, it is essential to perform mechanical (tensile and fatigue) tests in order to obtain relevant mechanical properties.

Several mechanical testing methods for micron-sized specimens have been proposed [2]. Some notable techniques include the microscale tension test, axi-symmetric plate bend test, micro-beam bend test, M-test, wafer curvature measurements, dynamic (resonant) tests, fabrication of passive strain sensors, and Raman spectroscopy. Srikar et al. [3] critically reviewed these methods and suggested a rational approach for the selection of test techniques for the design of micro-systems. The round-robin test at five different Japanese institutions, for single crystal silicon, polysilicon, thin film nickel and thin film titanium, suggested that, in tensile testing of micron-sized specimens, the method of gripping samples in fact is very important [4]. In this respect, the authors proposed a new specimen, test procedure, and test machine for tensile testing of micron-sized specimens [5] and reported test results [6].

The NiCo thin film has recently become a widely used material for MEMS devices, such as a probe tip. Especially, when the film is used as probe tip in wafer probe inspection system, the film is repeatedly applied by mechanical and thermal stresses and then is often failed. Therefore, the mechanical properties of the film are necessary to design a reliable structure of probe tip. In this paper, tensile tests on NiCo thin film were performed to obtain the tensile properties. To investigate the width effects and wafer effects, specimens with two different widths (50 and 150 μm) were tested.

2 Experimental Procedures

2.1 Design of Specimens

All specimen dimensions could not be in compliance with a standard such as ASTM because of some difficulties in fabrication. However, the shape of the conventional "dog-bone" tensile test specimen was designed based on ASTM E466-96. The 29 and 22 specimens with 50 and 150 μm wide reduced sections, respectively, were designed on a 6 in. wafer. The length of a reduced section was 1,000 μm, and the radius of the blending fillet was 1,000 μm to minimize stress concentration of the specimen. The width and length of the grip section were 4,000 and 3,500 μm, respectively.

As noted, one important issue to be resolved in tensile and fatigue testing of micron-sized specimens is the gripping method. In the present work, the specimen suggested by the authors [5] is used and is shown in Fig. 1. The suggested specimen has two holes at the grip ends with an 850 μm diameter for pins to secure the specimen to the tester so that the UV bond is no more than necessary when mounting

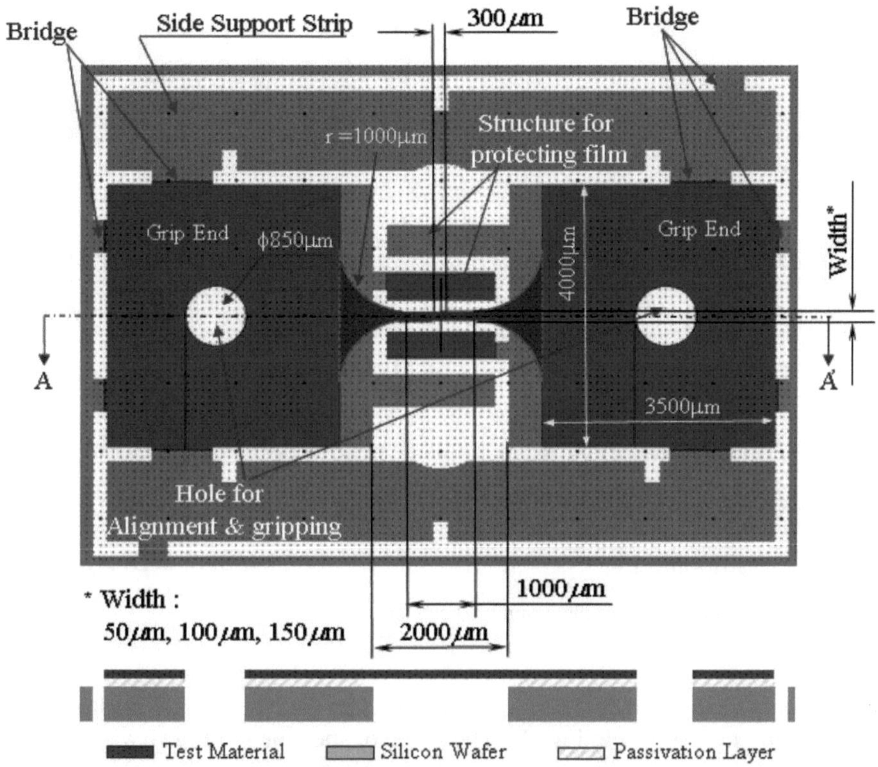

Fig. 1 Schematic diagram of specimen

the specimen. The distance from the center of the hole to the edge is 1,700 μm. It should also be noted that, in fabricating the specimen, the surroundings of the specimen, including the side support strips, were etched except for six bridges. This makes it possible to extract the specimen easily from the wafer without any damage. Moreover, by etching between the side support strip and test film in advance to remove the side support strip after mounting the specimen, except for the four bridge, it is possible to avoid sawing that could damage the test film. As shown in Fig. 1, structures are connected with the grip ends of the specimen so as to protect the test film from fracture due to residual stress of very brittle passivation layers, such as SiO_2 or Si_3N_4.

2.2 Fabrication of Specimens

The fabrication processes of a specimen are schematically shown in Fig. 2. A single-side polished 6 in. silicon wafer with (100) surface was used as a substrate. The 100 nm thick pure aluminum was deposited by sputtering process and the Al film

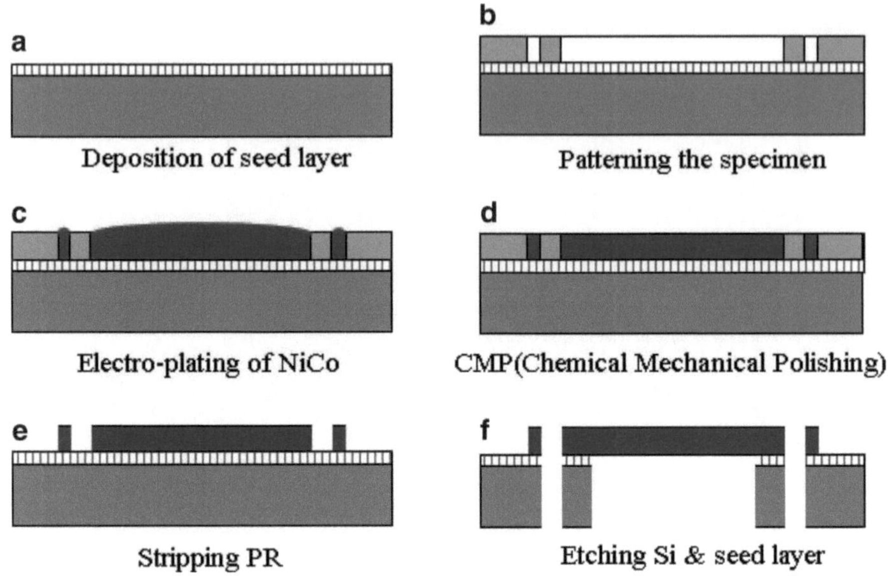

Fig. 2 Fabrication processes of the designed specimen

on the front-side plays the role of a seed layer when the NiCo was deposited by the electro-plating, as shown in Fig. 2a. The shape of specimen of Fig. 1 was patterned using mask and developed to make the mold, after PR(Photo Resist) was coated in the target thickness of specimen (Fig. 2b), and the NiCo film to be tested was deposited by electro-plating process (Fig. 2c). The deposited layer was polished out by CMP(chemical mechanical polishing) until the target thickness was reached, as shown in Fig. 2d. The PR layer was stripped, as shown in Fig. 2e. The open area below the gage length in the test film was patterned on the backside, and the silicon substrate was dry-etched by the deep RIE (reactive ion etching) process until reaching the Al layer, as shown in Fig. 2f. The Al layer was etched by a wet etching process to make a free standing specimen without any damage on the NiCo film, as shown in Fig. 2f.

By arranging various sizes of specimen on one wafer, it is possible to investigate the effect of specimen width on mechanical properties of thin films without fabrication effects.

2.3 *Experimental Procedure*

Tensile properties were determined by using the microtensile-testing machine developed by the authors, equipped with a load cell for measuring load, a capacitance sensor for measuring displacement between grips and CCD camera for measuring extension of gage length. A detailed description about the microtensile-testing machine has been described previously [5]. The range and rated output of the loadcell are 500 mN and 0.1 mV/V/g, respectively. The range and nominal scale factor of capacitance sensor are 250 μm and 25 μm/V, respectively. The specimen (Fig. 1) was mounted to the specially designed grips and was aligned using pins and bolts.

We developed a program with a closed-loop control function that is able to control a constant load or displacement rate during the tensile test. The extension between grips was measured using a capacitance displacement sensor and load was measured using loadcell. And then the signals were recorded into a computer using an A/D converter.

3 Results and Discussion

3.1 Micro-tensile Stress–Strain Curves

For the axial tensile tests, the 50 μm width specimens were chosen from two different wafers which were fabricated in the same plating bath at the same time. The average thickness of NiCo films of wafer No. 2 was 7.23 μm and the average thickness of NiCo films of wafer No. 3 was 8.66 μm. The 150 μm width specimens were chosen from wafer No. 2. The length of the reduced section is 1,000 μm. The three samples for each case were tested in atmospheric air at room temperature. The rate of load application was constant, 11.0 MPa/s. Load and extension between grips were measured and saved to the PC simultaneously using AD converter with a 200 Hz sampling rate. And also the extension of gage length was directly measured using CCD camera with a 1 Hz sampling rate during tensile test.

Figure 3 shows the measured load–displacement curves. Tensile properties, such as elastic modulus, yield strength and ultimate tensile strength, should be

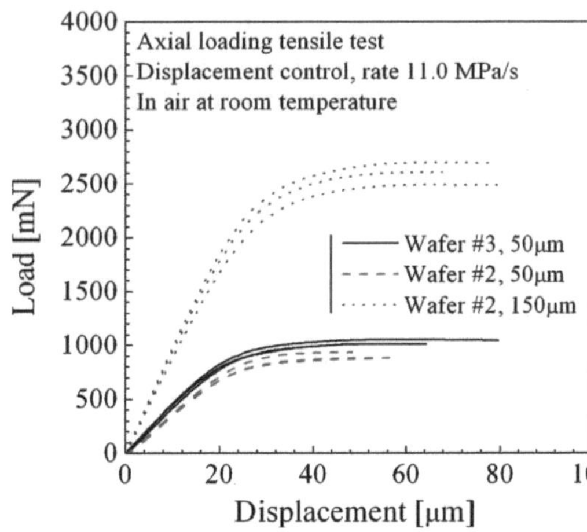

Fig. 3 Typical curves of the measured force–displacement with various samples for NiCo thin films

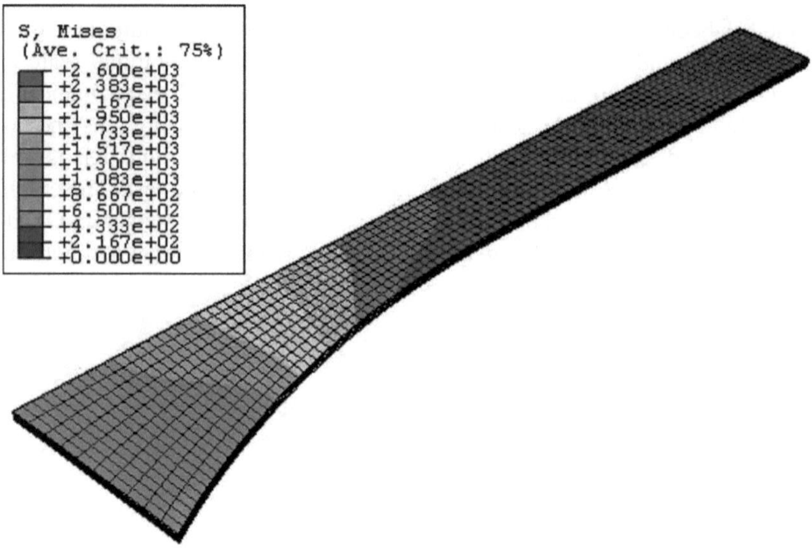

Fig. 4 Model and stress contour of FE analysis

obtained from the curves. However, for obtaining the properties, the measured load–displacement curve should be transformed into an engineering stress–strain curve. The engineering stress is calculated using the measured force and the initial area of gage section of specimen and the engineering strain is calculated using the extension of gage length and the initial length of gage length.

In this study, the extension between grips and the extension of gage length was measured, simultaneously. To obtain the mechanical properties, the extension between grips was primarily used because the capacitance sensor has higher resolution and sampling rate than the method using CCD camera. Therefore, the extension of gage length was calculated by multiplying the measured the extension between grips by the ratio of the extension of gage length (300 μm) to the extension of distance between grips (2,000 μm). The ratio was calculated using FE analysis. Figure 4 shows that the specimen was modeled and analyzed for a linear elastic FE method in order to calculate the ratio. The ratio was previously calculated for each width, 0.185 for 50 μm and 0.166 for 150 μm width. It was easily found that the thickness of specimen and elastic modulus of material do not influence the ratio. The method is applicable only to specimens without stress concentration before plastic deformation occurs.

Figure 5 shows the engineering stress–strain curves converted using the ratio. From the engineering stress–strain curves, tensile properties were obtained, as shown in Table 1. To verify that the main tensile properties of a material, such as elastic modulus, yield strength and tensile strength, could be extracted from the converted engineering stress–strain curve practically without question, the two extension measurement methods were compared, as shown in Fig. 6.

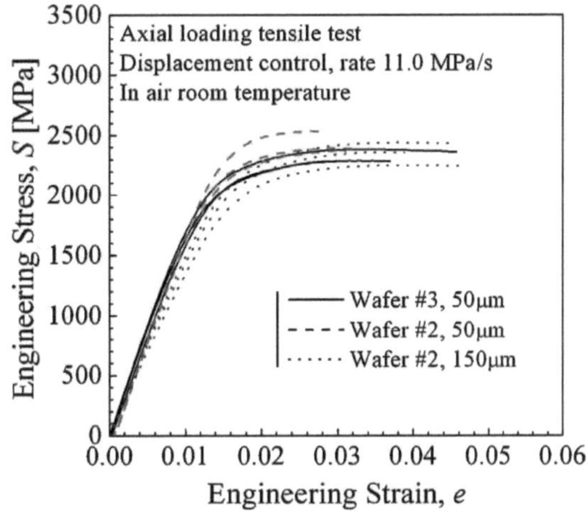

Fig. 5 Typical stress–strain curves with various samples for NiCo thin films

Table 1 Results of tensile test of NiCo thin films

Wafer no.	Width (μm)	Elastic modulus (GPa)	0.2% offset yield strength (MPa)	Ultimate tensile strength (MPa)
3	50	177 ± 5.5	1,930 ± 66.7	2,318 ± 54.6
2	50	183 ± 4.1	2,013 ± 152.0	2,431 ± 87.8
2	150	148 ± 8.4	1,943 ± 73.0	2,348 ± 93.3

3.2 Tensile Behavior

The stress–strain curves indicated the difference in both the slope of the curves and the fracture strains between the widths of specimen at the same wafer No. 2, as shown in Fig. 5. The difference in the slopes reflects the difference in elastic modulus, as can be seen in Table 1. Table 1 and Fig. 5 show that there is difference in the elastic modulus of 50 μm width samples between wafers. The elastic moduli of NiCo samples are 177 GPa for 50 μm width of wafer No. 3,183 GPa for 50 μm width of wafer No. 2 and 148 GPa for 150 μm width of wafer No. 2, which is less than the bulk value of 207 GPa of the pure Ni [7] but which is between the elastic moduli of 113 ∼ 202 GPa of the LIGA Ni MEMS structure [4,7]. The 150 μm width samples at the wafer No. 2 exhibits more ductile fracture than the 50 μm width samples at the wafer No. 2, as shown in Fig. 5. The higher elastic modulus of the narrower specimen is attributed to the differences in the width of the two kinds of samples. The location of sample in 6 in. wafer was different along with width. The 50 μm width samples were located in the top of wafer and the 150 m width samples were located in the center of wafer. There is a little difference in the elastic modulus along with wafers, as shown in Table 1 and Fig. 5.

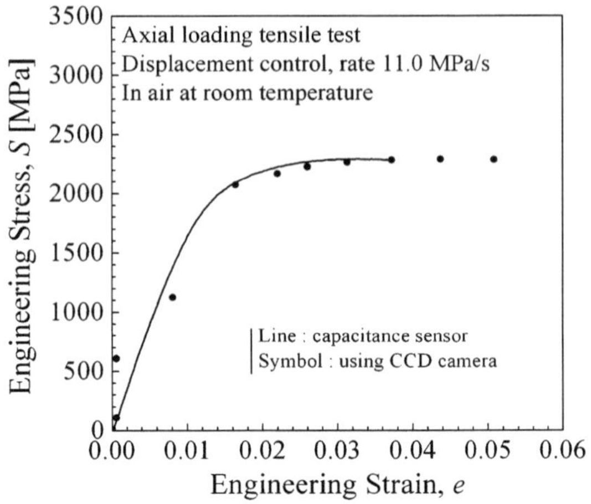

Fig. 6 Comparison of the two extension measurement methods

There are a little difference in the yield strength and ultimate tensile strength along with wafer and width, as shown in Table 1 and Fig. 5. The samples have the yield strength of 1,930 $\sim$ 2,013 MPa, which is much greater than the 59 MPa of Ni bulk and the strength of 323 $\sim$ 620 MPa for the LIGA Ni MEMS structure [7]. Furthermore, the samples' ultimate tensile strength of 2,318 $\sim$ 2,431 MPa is much greater than the 317 MPa of Ni bulk [7] and the strength of 540 $\sim$ 2,180 MPa for the LIGA Ni MEMS structure [7]. It has been reported that the yield strength and ultimate tensile strength depend on the fabrication process and the film thickness and that they are increased by the precipitates such as NiCo. It is well known that stiff precipitates, NiCo in the present case, block dislocation movement through a soft matrix, thereby increasing the yield strength [8]. This alloy effect is potentially of practical importance for MEMS applications because one can improve the strength of a Ni membrane while only slightly diminishing its conductivity [9].

Typical SEM photomicrographs of the deformed gauge sections, and fracture surfaces of the 50 μm tensile specimen samples, are presented in Fig. 7a, b. The general appearance of the necking area is apparent in part (a). Ductile dimpled fracture was observed in the necking area (Fig. 7b). Fracture surfaces were parallel to the maximum shear stress directions. The tensile fracture surfaces were generally associated with inclined surfaces, as shown in Fig. 7a, b. In the case of the 150 μm width samples, a similar range of deformation and fracture modes were observed in Fig. 8a, b. Failure occurred by ductile dimpled fracture but the size of dimple is smaller than the 50 μm width samples.

4 Summary and Conclusions

Micro-mechanical properties of NiCo thin films with various widths and wafers were evaluated by axial loading tensile tests. The tensile stress–strain curves for

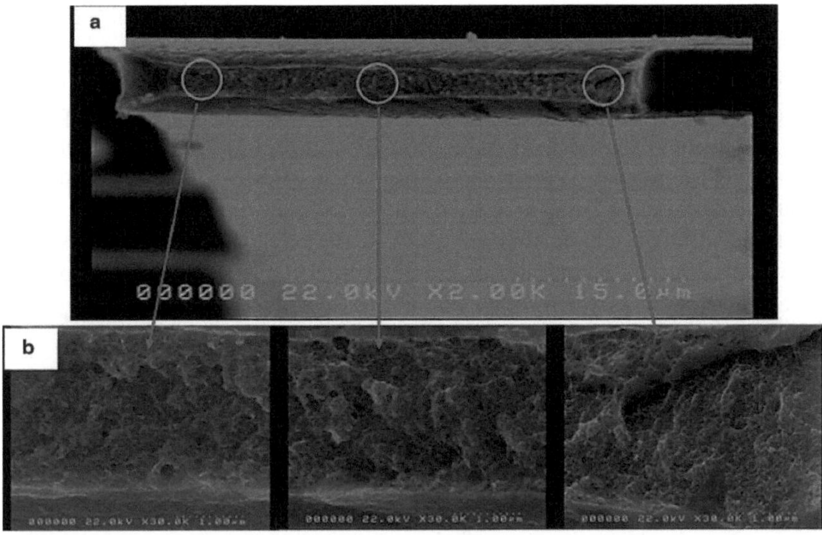

Fig. 7 Fracture surface of 50 µm width NiCo sample tested under tensile loading. (**a**) Overall view, (**b**) high magnification of the neck area

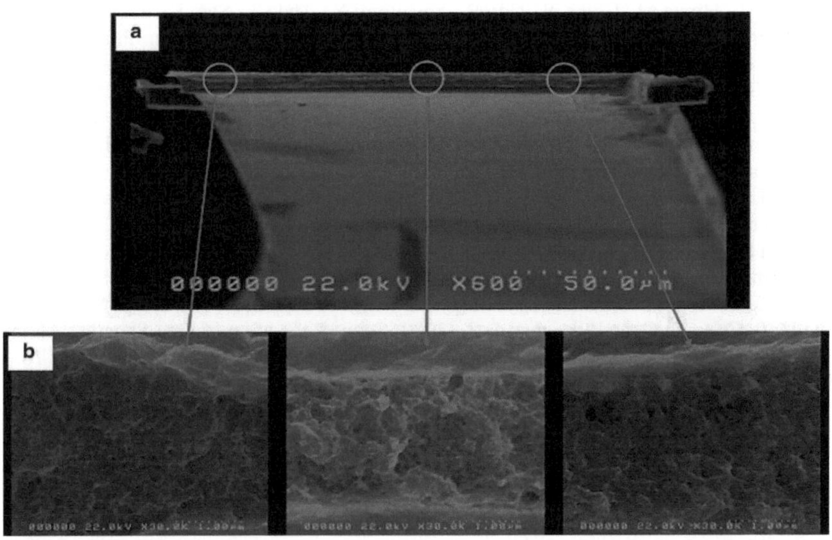

Fig. 8 Fracture surface of 150 µm width NiCo sample tested under tensile loading. (**a**) Overall view, (**b**) high magnification of the neck area

the films were determined measuring tensile strains during tensile loading with two methods, capacitance sensor and CCD camera. The elastic moduli of NiCo thin films are about 180 GPa for 50 μm width and 148 GPa for 150 μm width. The 50 μm width samples were located in the top of wafer and the 150 μm width samples were located in the center of wafer. There is little difference in the elastic modulus along with wafers. The samples have the yield strength of 1,930 ∼ 2,013 MPa and the ultimate tensile strength of 2,318 ∼ 2,431 MPa. There are little differences in the yield strength and ultimate tensile strength along with wafer and width. It was verified that the alloy effect is potentially of practical importance for MEMS applications because the strength of a LIGA Ni membrane was improved by the precipitates, NiCo.

References

1. Hardwick D. A. (1987) The mechanical properties of thin films: a review, Thin Solid Film, 154: 109–124.
2. Larsen K. P., Rasmussen A. A., et al. (2003) MEMS device for bending test: measurement of fatigue and creep of electroplated nickel, Sensors and Actuators A, 103: 156–164.
3. Srikar V. T. and Spearing S. M. (2003) A critical review of microscale mechanical Testing Methods used in the design of micromechanical systems, Experimental Mechanics, 43(3): 238–247.
4. Tsuchiya T., Hirata M., Chiba N., Udo R., Yoshitomi Y., Ando T., Sato K., Takashima K., Higo Y., Satome Y., Ogawa H. and Ozaki K. (2003) Cross comparison of thin film tensile-testing methods examined with single crystal silicon, polysilicon, nickel and titanium films, Proceedings of IEEE 16th Annual Conference on Microelectromechanical Systems (MEMS), 666–669.
5. Park J-H., Kim C. Y., Choa S. H., Lee C. S., Che W. S. and Song J. H. (2005) New structures and techniques for easy axial loading test of static and fatigue properties of MEMS materials, Key Engineering Materials, 297–300: 545–550.
6. Park J-H., Myung M. S. and Kim Y-J. (2008) Tensile and high cycle fatigue test of Al-3%Ti thin films, Sensors and Actuators A: Physical (submitted).
7. Allameh S. M., Loua J., Kavishe F., Buchheit T. and Soboyejo W. O. (2004) An investigation of fatigue in LIGA Ni MEMS thin films, Materials Science and Engineering, A 371: 256–266.
8. Dieter, G. E. (1988) Mechanical Metallurgy, SI metric ed., McGraw-Hill Book, New York, 20.
9. Lee H. J., Cronella G. and Bravman J. C. (2000) Stress Relaxation of Free-Standing Aluminim Beams for Micorelectromechanical Systems Applications, Applied Physics Letters, 76 (23): 3415–3417.

Session II
Fatigue Damage Physics and Modeling

Dynamic Properties and Fatigue Failure of Aircraft Component

Sergei Kuznetsov, Eriks Ozolinsh, Ilmars Ozolinsh, Igors Pavelko, and Vitalijs Pavelko

Abstract The basic purpose of given article is consideration of the problems connected with application of a method of concentrated weights in the tasks of mechanical systems dynamics at non-classic internal and external constraints. The method of the concentrated weights is convenient means of the analysis of dynamic properties of elastic mechanical systems. It has relative simplicity of definition of the parameters of equivalent discrete system, the clearness of computing algorithms and provides comprehensible accuracy of definition of the lowest natural frequencies. Doubtless advantage of a method is convenience of modeling of non-classic constraints of fastening and internal constraints between elements of complex systems. Such problems arise at the decision of practice tasks of the analysis of dynamics of real systems. The method is used for the analysis of vibrations of a beam with variable parameters at presence of elastic supporting of the beam and the attached additional concentrated weight. A result of dynamic analysis has a good correlation with the fatigue damages observed at fatigue test of a beam.

Keywords Aircraft · Dynamics analysis · Fatigue damage · Fracture mechanics

1 Introduction

This article is connected with 6FP Euro project AISHA. The basic purpose of the project-development of the continuous monitoring system of a technical condition integrated into a structure. Progressive methods and means of the control over use of ultrasonic technology are developed. In thin-walled structures it uses properties of elastic Lamb waves. The final stage provides carrying out of full-scale fatigue tests on components of real aviation structures for demonstration of working capacity and

S. Kuznetsov, E. Ozolinsh, I. Ozolinsh, I. Pavelko, and V. Pavelko (✉)
Aviation Institute of Riga Technical University, Latvia
e-mail: s.v.kuznetsov@btv.lv, ilmars_o@inbox.lv, eriks_ozolins@inbox.lv, Igors.Pavelko@rtu.lv, Vitalijs.Pavelko@rtu.lv

S. Pantelakis, C. Rodopoulos (eds.), *Engineering Against Fracture: Proceedings of the 1st Conference*,

Fig. 1 Stand of dynamic testing of the helicopter MI-8 tail beam (1 – the tail beam, 2 – support, 3 – mechanical vibrator, 4 – motor, 5 – V-belt gear)

efficiency of methods and means of the non-destructive testing. The object of testing is the helicopter MI-8 tail beam (Fig. 1). It is Al alloy typical aircraft component.

At planning experiment there are the problems with a choice of an optimum mode of the tests combining realization of demanded distribution of stresses, duration of tests and relative simplicity of excitation of mechanical load. In this connection there was a necessity of the analysis of dynamic characteristics of the testing object.

There is a plenty of methods of the dynamic analysis of mechanical system [1–4]. Exact analytical methods have the limited application and, mainly, their role is the reference decisions for an estimation of accuracy of decisions by the approximated methods. Among the last the visible place is occupied by Ritz's methods [5], various versions of energetic methods [6], and also methods of replacement of system with continuously parameters by system with the limited number of the concentrated weights [6–8]. Universal numerical method of finite elements (FEM) is also some one of this group method [9].

There are many methods of the dynamic behavior analysis of complex elastic structure and usually special estimation of accuracy by comparison with precise analytical solution shows the rational selection of model parameters allows obtaining of acceptable result of simulation. But modeling of boundary conditions is often no adequate. Therefore real experimental data is unacceptable difference from results of simulation in similar cases [10, 11].

In presented research there are two aims. First of them is the method of the concentrated weights application for definition of dynamic characteristics of real aircraft component with continuously distributed weight. Second (and it is main aim) is correct simulation of non-classic internal and external constraints at application of a method of concentrated weights.

2 The Method of Concentrated Weights

In connection with the relatively big length of a beam in comparison with diameter of its cross-section the beam schematization of such structure is admissible at the analysis of its dynamic characteristics. The equation of movement in this case looks like

$$m(z)\frac{\partial^2 v}{\partial t^2}+\frac{\partial^2}{\partial z^2}\left(EJ_x\frac{\partial^2 v}{\partial z^2}\right)=q(z,t) \tag{1}$$

where $v(z,t)$ is a deflection of a beam, $m(z)$ is the intensity of the weight distribution, EJ_x is bending stiffness of the cross-sections of a beam, and $q(z,t)$ is intensity of the distributed external load. If $q(z,t)\equiv 0$, the free movement without participation of external forces is realized.

Generally, if the mass and stiffness of the cross-sections is distributed non-uniformly, the analytical decision cannot be received. Therefore the approximated method of the concentrated weights is used for the purpose of this research. For definition of natural frequencies and forms the actual beam with continuously distributed weight must be replaced by a weightless beam with the same bending stiffness, but with finite number of the concentrated weights in finite number of nods $(k+1)$. For this purpose the beam has been sheared into finite number k parts. For each of them weight M_i and its center coordinate z_{ic} was defined and then it was distributed between nods $(i-1)$ and i those are finite nods of the part. From a condition of static equivalence there are the next expressions

$$M_i=\Delta M_{i-1}+\Delta M_i$$

$$M_i\cdot z_{ic}=\Delta M_{i-1}\cdot z_{i-1}+\Delta M_i\cdot z_i=\int_{z_{i-1}}^{z_i} z_i\cdot m(z)dz$$

where ΔM_{i-1} and ΔM_i are the lots of weight of the part, attached to units $(i-1)$ and i accordingly.

If the beam is sheared into enough plenty of parts within the limits of one part distribution of weight can be accepted linear and then last two equations can be written down so

$$M_i=\Delta M_i+\Delta M_{i-1}\cong\frac{m_i+m_{i-1}}{2}\cdot\Delta z_i$$

$$M_i z_{ic}=\Delta M_{i-1}z_{i-1}+\Delta M_i z_i\cong\frac{m_i z_i+m_{i-1}z_{i-1}}{2}\cdot\Delta z_i$$

or in the dimensionless form

$$\overline{M}_i=\Delta\overline{M}_i+\Delta\overline{M}_{i-1}\cong\frac{\overline{m}_i+\overline{m}_{i-1}}{2}\cdot\Delta\bar{z}_i,$$

$$\overline{M}_i\cdot\bar{z}_{ic}=\Delta\overline{M}_{i-1}\cdot\bar{z}_{i-1}+\Delta\overline{M}_i\cdot\bar{z}_i\cong\frac{\overline{m}_i\cdot\bar{z}_i+\overline{m}_{i-1}\cdot\bar{z}_{i-1}}{2}\cdot\Delta\bar{z}_i$$

where $\Delta\overline{M}_i=\Delta M_i/(m_0 l)$, $\Delta\bar{z}_i=\Delta z_i/l$

As a result relative weights can be expressed in the dimensionless form so

$$\Delta\overline{M}_i = \overline{M}_i \cdot \frac{\overline{z}_i - \overline{z}_{ic}}{\Delta\overline{z}_i}, \quad \Delta\overline{M}_{i-1} = \overline{M}_i \cdot \frac{\overline{z}_{ic} - \overline{z}_{i-1}}{\Delta\overline{z}_i} \tag{2}$$

Where

$$\overline{z}_{ic} = \frac{\overline{m}_i \cdot \overline{z}_i + \overline{m}_{i-1} \cdot \overline{z}_{i-1}}{2M_i} \cdot \Delta\overline{z}_i$$

It is obvious, that the total relative weight in the nod i is $\overline{\boldsymbol{m}}_i = \overline{\boldsymbol{M}}_i + \overline{\boldsymbol{M}}_{i+1}$.

On the basis of a principle of superposition the displacement Δi in the nod i $(i = 0, 1, \ldots k)$ as result of the action of system of the vertical forces P_j concentrated in all nods, is defined by the following sum

$$\Delta_i = \sum_{j=1}^{n} \delta_{ij} \cdot P_j = -\sum_{j=1}^{n} \delta_{ij} \cdot m_j \cdot \Delta_j'' \tag{3}$$

where δ_{ij} is the factor of elastic compliance.

In the second part of Eq. (3) the force P_j is replaced by the inertial force that appears in case of natural vibrations. These vibrations are harmonious with frequency ω. Therefore

$$\sum_{j=1}^{n} \delta_{ij} \cdot m_j \Delta_j'' + \Delta_i = 0 \qquad \Delta_i = A_i \cdot Sin\omega t$$

where A_i is amplitude of vibration of the nod i.

As a result for definition of amplitudes of free vibrations the system of k the linear homogeneous algebraic equations turns out

$$\delta_{i1} m_1 A_1 + \ldots + \left(-\frac{1}{\omega^2} + \delta_{ii} m_i\right) A_i + \ldots + \delta_{in} m_n A_n = 0 \ldots (i = 1, \ldots, k) \ldots \tag{4}$$

For a greater generality this system of the equations is convenient for writing down in a dimensionless view

$$\overline{\delta}_{i1} \overline{m}_1 A_1 + \ldots + (-x + \overline{\delta}_{ii} \overline{m}_i) A_i + \ldots + \overline{\delta}_{in} \overline{m}_n A_n = 0 \tag{5}$$

where

$$x = \frac{EJ_{x0}}{m_0 l^2} \cdot \frac{1}{\omega^2}, \qquad \overline{\delta}_{ij} = \frac{\delta_{ij}}{l^3/(EJ_{x0})}, \qquad m_j = \frac{m_j}{m_0 l}$$

It is known that non-trivial decision of system (5) exists, if the determinant of coefficients of the equations is equal to zero.

$$Det\{[\overline{\delta}_{ij} \overline{m}_j] - x[I]\} = 0 \tag{6}$$

Roots x_k of the Eq. (6) define natural frequencies of vibrations.

$$\omega_k = \frac{1}{l^2} \cdot \sqrt{\frac{EJ_{x0}}{x_k m_0}} \tag{7}$$

After their definition the natural forms can be received too.

Verification of definition of dynamic characteristics by a method of the concentrated weights was carried out calculation for a console beam of constant cross-section and uniformly distributed weight. It is established, if there is $k = 8 \ldots 10$, then at least first two natural frequencies practically coincide with their exact values.

3 The Analysis of Dynamic Characteristics of a Beam Under Ideal Boundary Conditions

The method described above leads the numerical analysis of natural frequencies and forms of a beam of the helicopter at perfect conditions of supporting: a jamming at one tip. In Fig. 2 the scheme of the test system is presented. The thin-walled tail beam of the helicopter has an elastic attachment 1 to the motionless tip, in the middle part of a beam the additional weight 2 (about 94 kg) is attached.

On the free tip of a beam the mechanical vibrator 3 in weight about 50 kg is installed. If to accept, that connections of a beam with the basis and additional weight with a beam are absolutely rigid, then the relative distribution of weight among nods is in Table 1. Own weight of the beam, the attached weight in an middle zone of a beam and weight of the vibrator consist common weight.

In Fig. 3 results of calculation of first two natural frequencies and forms are shown. In connection with large bending stiffness of a beam and small weight it has high frequencies. The second form of vibrations is obeyed by the requirements of planned tests: the maximal curvature bended axes of a beam so, and the maximal stress take place in the middle part of a beam.

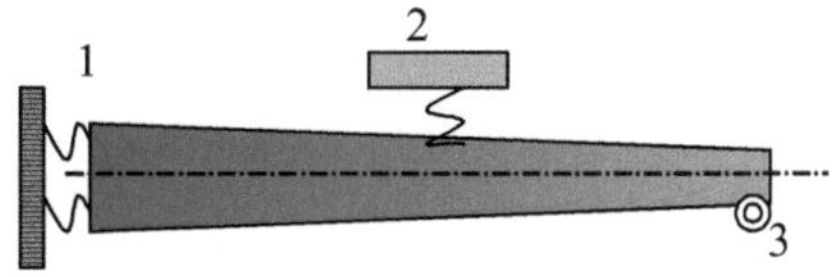

Fig. 2 The beam has elastic support 1 and the mass 2 also has the same kind of connection with beam, 3 is the vibrator

Table 1 Mass distribution among the nods of system

Weight number	0	1	2	3	4	5
m/m_0	1	0.043	0.063	0.061	0.243	0.241
Weight number	6	7	8	9	Total	
m/m_0	0.049	0.041	0.039	0.220	1.00	

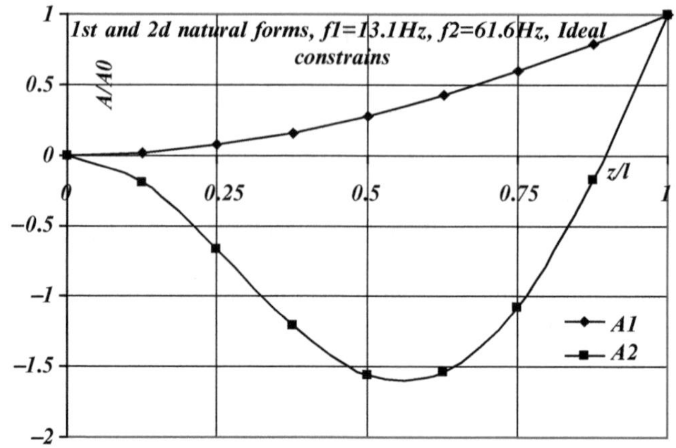

Fig. 3 First and second natural form of vibration

4 The Account of Influence of Non-classic Internal and External Constraints

Two types of deviations from ideal boundary conditions are examined:

- Elastic fastening
- An elastic attachment of the additional concentrated weight

Let the characteristic of the elastic hinge looks like $M_0 = K\theta$, where K is factor of elastic hinge stiffness, θ is the angle of turn in the hinge, M_0 is the moment in the hinge.

At action of single force in nod there is moment $M_0 = 1z_j$ causing rigid turn of a beam about the axis of the hinge $\theta_j = M_0/K = z_j/K$. As a result in section j there will be the additional deflection caused by an elastic compliance of the hinge

$$\delta_{0ij} = \theta_j z_i = \frac{z_i z_j}{K}$$

or in the dimensionless form

$$\overline{\delta}_{0ij} = \frac{\delta_{0ij}}{l^3/(EJ_{x0})} = \frac{\overline{z}_i \overline{z}_j}{Kl/(EJ_{x0})} \tag{8}$$

So that to consider influence of elastic compliance hinge fastenings it is necessary to calculate a total compliance $\overline{\delta}_{\Sigma ij} = \overline{\delta}_{ij} + \overline{\delta}_{0ij}$. The subsequent calculation of frequencies and forms remains the same.

The account of elasticity of an attachment of the additional concentrated weight can be shown also to updating a matrix of an elastic compliance. Let the additional weight m_a is attached in a nod k by means of elastic connection with stiffness C. It is obvious, that connection of additional weight increases number of degrees of

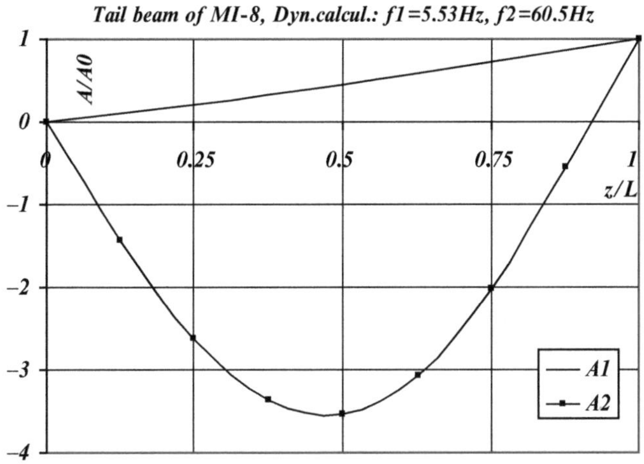

Fig. 4 First and second natural frequencies and forms of the tail beam of helicopter at non-classic connections realized in dynamic experiment

freedom and, means, the number of the equations of movement will be equal $n+1$. Thus

$$\delta_{i,n+1} = \begin{cases} \delta_{ik}, \ldots\ldots\ldots\ldots i \neq k \\ \delta_{kk} + 1/C, \ldots\ldots i = k \end{cases} \tag{9}$$

Both of the kinds of elastic connections are available in the dynamic system examined in p. 3. For the real object prepared for test, the approximated estimation of stiffness of connections has been executed. Their relative values are equal accordingly

$$\overline{K} = \frac{K}{EJ_{x0}/l} = 2.0 \quad \overline{C} = \frac{C}{1/\delta_{kk}} = 1.0$$

Figure 4 presents results of calculation of first two natural frequencies and forms of a beam in structure of this dynamic system. The natural form value for additional concentrated weight is equal – 1.42. It means the displacement of additional concentrated weight in 1.42 times more than the displacement of free tip of the beam and has opposite phase of vibrations. It is visible, that on the first form of moving of a beam are defined mainly by an angular compliance of a trailer support. The axis of a beam is kept practically direct that testifies about small effect of bending. The second natural frequency and the form in this case is more similar to the first frequency and the form of vibrations at a rigid jamming. Nevertheless this form is much more complex.

The lead analysis shows exclusively strong influence of conditions of fastening of a beam and presence of elastic connections with other objects to dynamic characteristics, in particular, to natural forms and frequencies of vibrations.

5 The Estimation of Nominal Stress of a Beam Under its Form Bended Axes

Let $v_i(z)$ is the form i of the bended axes of a beam. Then the bending moment $M(z)$ in this section is defined by curvature of this axis in the given section

$$M(z) = EJ_x \frac{d^2 v_i}{dz^2} \tag{9}$$

and the maximal direct stress are defined under the formula

$$\sigma_{\max} = \frac{M(z)}{W(z)} \tag{10}$$

where $W(z)$ is the moment of resistance of section.

If amplitude of moving of the free end of a beam to designate A_0, then the formula (2) of the maximal direct stress can be wrote in next form

$$\sigma_{\max} = \frac{EJ_x(z)}{W(z)} \frac{d^2 v_i}{dz^2} = \frac{1}{2l^2} EDA_0 \frac{d^2 \overline{v}_i}{d\overline{z}^2} \tag{11}$$

where J_x(z) is the moment of inertia of section, D and l is external diameter of section and length of a beam, E is the module of elasticity of a material of a beam. The form of vibrations is expressed in the relative form $\overline{v}_i = v_i/A_0$ as a function of relative coordinate of section $\overline{z} = z/l$.

It is obvious the constant $\sigma_0 = \frac{1}{2l^2} EDA_0$ has the same dimension as the stress and its value is defined by the amplitude of moving of the free tip of a beam. As a result

$$\overline{\sigma}_{\max} = \frac{\sigma_{\max}}{\sigma_0} = \frac{d^2 \overline{v}_i}{d\overline{z}^2} \tag{12}$$

Thus, the maximal stress distribution along beam longitudinal axis at a bend of a beam under the form i is completely defined by the relative curvature of bended axes. For definition of absolute value of stress it is necessary to set amplitude of moving of the free tip of a beam and to calculate a scale constant σ_0.

In [12] first two natural frequencies and forms of a tail beam of helicopter Mi-8 are obtained at the conditions of supporting those are approximately the same as at dynamic tests. It allows leading an estimation of the maximal direct stress in the sections of a beam. With this purpose the second form presented in the relative form was approximated by a polynomial of the fifth degree. There is enough high accuracy of approximation of the form. However it is known that differentiation of a polynomial can lead to greater mistakes for derivatives. Therefore it was carried out also approximation by a polynomial of higher degree. Comparison of results really confirms this statement, if to consider the form on all length of a beam. However, significant differences are characteristic for zones of small stress, near to the tips of a beam. In the most intense zone the difference of stresses at the fifth and sixth degrees of a polynomial are insignificant.

In Fig. 5 the corrected results of calculation of relative pressure in sections of a beam are presented. It is visible, that the maximum of pressure is approximately on three quarters of length of a beam.

Dynamic tests at frequency of the exciting force equal to the second natural frequency of beam were caused with initiation of fatigue damages to a zone of predicted maximal stresses. The example of a fatigue crack in a frame of a beam is presented on Fig. 6.

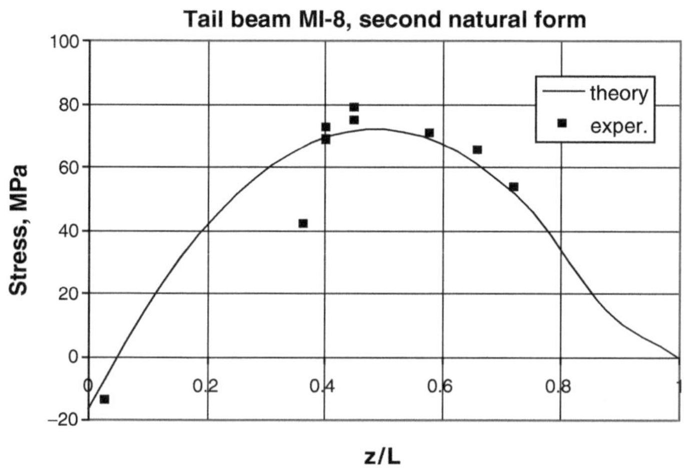

Fig. 5 Comparison theoretical and experimental results for the dynamic stress in a skin of tail beam of helicopter

Fig. 6 The fatigue cracks in a frame as a result of dynamic overload induced by additional concentrated weight in middle part of the tail beam

6 Conclusions

The method of the concentrated weights is convenient means of the analysis of dynamic properties of elastic mechanical systems. It has relative simplicity of definition of the parameters of equivalent discrete system, the clearness of computing algorithms and provides comprehensible accuracy of definition of the lowest natural frequencies. Doubtless advantage of a method is convenience of modeling of non-classic constraints of fastening and internal constraints between elements of complex systems. Such problems arise at the decision of practical problems of the analysis of dynamics of real systems. Presented calculation for the real testing system demonstrates very important effect of real boundary conditions to dynamic properties of elastic systems.

Acknowledgments This research was induced by the authors' participation in 6FP research project AISHA (Aircraft Integrated Structural Health Assessment). AISHA has brought together major European aircraft manufactures, research and academic institutions in order to provide effective integrated system of continues detection of damages in the aircraft structure.

References

1. Timoshenko, S. P., Young, D. H. and Weaver, W., Vibration Problems in Engineering, 4th edn, Wiley, New York, 1974, 444 pp.
2. Tse, E. S., Morse, I. E. and Hinkle, R. T., Mechanical Vibrations, Theory and Applications, 2nd edn, Allyn & Bacon, Boston, 1983.
3. Mei, C. Differential Transformation Approach for Free Vibration Analysis of a Centrifugally Stiffened Timoshenko Beam. Journal of Vibration and Acoustics, April 2006, Volume 128, Issue 2, pp. 170–175.
4. Romeo, F. and Luongo, A. A Transfer-Matrix-Perturbation Approach to the Dynamics of Chains of Nonlinear Sliding Beams. Journal of Vibration and Acoustics, April 2006, Volume 128, Issue 2, pp. 190–196.
5. Zirkelback, N. L. and Ginsberg, J. H. Ritz Series Analysis of Rotating Shaft System: Validation, Convergence, Mode Functions, and Unbalance Response. Journal of Vibration and Acoustics, October 2002, Volume 124, Issue 4, pp. 492–501.
6. James, M. L., Smith, G. M., Wolford, J. C. and Whaley, P. W., Vibration of Mechanical and Structural Systems, Harper & Row, New York, 1989.
7. Newland, D. E., Mechanical Vibration Analysis and Computation, Longman, New York, 1989.
8. Collar, A. R. and Simpson, A., Matrices and Engineering Dynamics, Ellis Horwood, Chichester, England, 1987.
9. Huebner, K. H., The Finite Element Method for Engineers, Wiley, New York, 1975.
10. Close, C. M. and Frederick, D. K., Modeling and Analysis of Dynamic Systems, Houghton Mifflin, Boston, MA, 1978.
11. Smith, J. D., Vibration Measurement and Analysis, Butterworth-Heinemann Ltd, Oxford, Butterworths, 1989.
12. Pavelko V., Some problems of dynamics of mechanical systems at the non-classic internal and external constraints. The article in this issue of RTU Proceedings, 2006, pp.105–112.

Investigation of the Fatigue Behaviour of the Structural Magnesium Alloy AZ31

Spiros G. Pantelakis, Apostolos N. Chamos, and Vasilis Spiliadis

Abstract The fatigue behaviour of wrought magnesium alloy AZ31 has been evaluated experimentally under constant amplitude fatigue tests for both parent and pre-corroded specimens. The S-N curve of the parent material exhibits a very smooth transition from low to high cycle fatigue regime indicating a strong stress sensitivity. Crack initiation occurs already at early stage of the fatigue damage accumulation process. A transgranular initiation fracture is observed followed by an intergranular mode of propagation. The presence of pitting due to corrosion exposure facilitates essentially the onset of fatigue cracks and, hence, reduces the fatigue life of the corroded specimens appreciably. The effect of existing corrosion pits on the fatigue life increases with decreasing fatigue stress amplitude. Fractographic analysis revealed twin marks in the initiation area of the pre-corroded material, while the fast fracture area is characterized as quasi-cleavage.

Keywords Magnesium alloy · Fatigue · Crack initiation · Corrosion

1 Introduction

Permanent objective for the aeronautical industry is the weight reduction of fuselage parts, systems and interior components. Magnesium is the lightest structural engineering metal, and therefore, particularly attractive for applications where weight saving is of major concern. Improvements in mechanical properties, corrosion resistance [1,2] and the development of advanced manufacturing processes have led to increased interest in magnesium alloys and, therefore, several works are in progress to develop wrought magnesium alloys suitable for aerospace applications [3, 4].

S.G. Pantelakis (✉), A.N. Chamos, and V. Spiliadis
Laboratory of Technology & Strength of Materials, Department of Mechanical Engineering and Aeronautics, University of Patras, Rion, Greece
e-mail: pantelak@mech.upatars.gr, hamosa@mech.upatras.gr, spiliadisv@edrasis.gr

S. Pantelakis, C. Rodopoulos (eds.), *Engineering Against Fracture: Proceedings of the 1st Conference*,

Amongst them, AZ31 represents one of the most popular wrought alloys usually supplied in thin sheet form; diverse uses include satellite components, military applications, door inner automotive components, computer cases and cameras [5]. The major problems limiting the use of wrought magnesium alloys in aircraft structure applications are the high corrosion susceptibility and the reduced fatigue life [6, 7] as compared to other competitive aircraft alloys like the 2xxx and 6xxx aluminum alloy series.

It is worth mentioning that although the high corrosion susceptibility of magnesium alloys is well known, the effect of prior corrosion damage on the mechanical performance of magnesium alloys has not been faced in the scientific literature. Up to now, only the mechanical behaviour during corrosion exposure has been investigated, e.g. [8, 9]. Yet, for components operating over long period in corrosive environments, investigations on specimens subjected first to corrosion exposure and then mechanical loading may represent a more relevant case, as corrosion damage accumulates gradually, under or without loading, during the service life of the component under consideration. When the component is subjected to mechanical loading during service, it involves already a certain degree of prior corrosion damage.

The present work aims to contribute to the investigation of the fatigue behaviour of the wrought magnesium alloy AZ31. For the investigation, constant amplitude fatigue tests were performed on both, parent and pre-corroded specimens and the S-N curves of the alloy were derived. Crack initiation and short crack growth have been monitored for the case of the parent material using the replication technique. Corrosion induced degradation of the fatigue resistance is discussed. The derived results were supported by fractographic analysis on selected specimens.

2 Material and Experimental Procedure

A hot rolled magnesium alloy AZ31 in the form of thin plate with nominal thickness of 2.0 mm was used for the present study. The material is characterized as high-purity (hp) as the concentration of the contaminants (Fe, Cu and Ni) was held under certain limits. After the rolling procedure the material was subjected to annealing (O-temper) at 300°C for 30 min.

For all tests performed, specimens and testing procedure were according to the ASTM specifications. Fatigue specimens were cut according to ASTM E466 specification with their long axis parallel to the rolling direction, having a stress concentration factor of 1.0 (Kt = 1.0, un-notched). Displayed in Fig. 1 is the configuration of the fatigue specimen.

For the corrosion exposure of the fatigue specimens, the accelerated salt spray fog environment has been used. The salt spray tests were conducted according to ASTM B117 specification. The corrosion exposure time was selected to 6 h, based on the results of the metallographic examination and tensile properties degradation obtained from a previous work of the authors [10]. It should be underlined that in

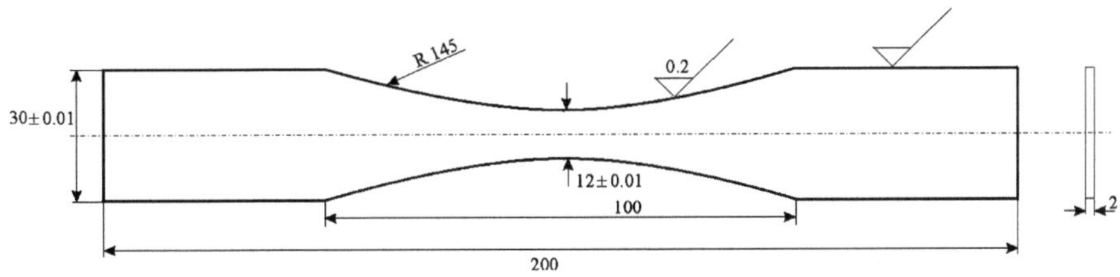

Fig. 1 Fatigue specimen configuration

practical applications magnesium alloys are not used as bare materials but are protected against corrosion by using coatings. However, as the possibility of damage of the coating layers can not be excluded, the corrosion resistance of the bare material should be also investigated. The salt spray test involved in this study is an accelerated laboratory corrosion test and, hence, causes a corrosion attack which is much more severe as compared to the expected corrosion in real service conditions. According to [11], it has been found that 27 years industrial atmospheric exposure is roughly equivalent to 10 days salt fog exposure. Yet, this outcome is based only on measurements of the corrosion rate of some magnesium alloys and is not correlated with the degradation of the mechanical performance due to corrosion exposure.

For the testing procedure a servo-hydraulic MTS machine with a capacity of 100 kN was used. Data were obtained under constant amplitude tests at a stress ratio, R, of 0.1. For the parent material, crack initiation and short crack propagation were monitored with replication technique. The test was periodically interrupted and replicas were taken and then crack length was measured using a microscope.

After the experiments, a basic fracture surface analysis has been performed in selected specimens. Standard metallographic preparation has been employed. The etching solution for the AZ31 alloy acetic picral consisted of 5 ml acetic acid, 6 g picric acid, 10 ml H_2O and 100 ml ethanol.

3 Results and Discussion

3.1 Microstructure and Tensile Properties

A typical image of the microstructure of the alloy under investigation is shown in Fig. 2. The alloy shows equiaxed structure with an average grain size of 14.5 μm. No precipitations corresponding to the β-phase (Mg17Al12) were observed at the grain boundaries. Furthermore, no twins are observed after the annealing treatment.

The average tensile properties of the alloy under investigation are summarized in Table 1. The values fit well to the range of values given in DIN-EN 10002 for the same alloy.

More details about the tensile behaviour of the alloy may be found in [12].

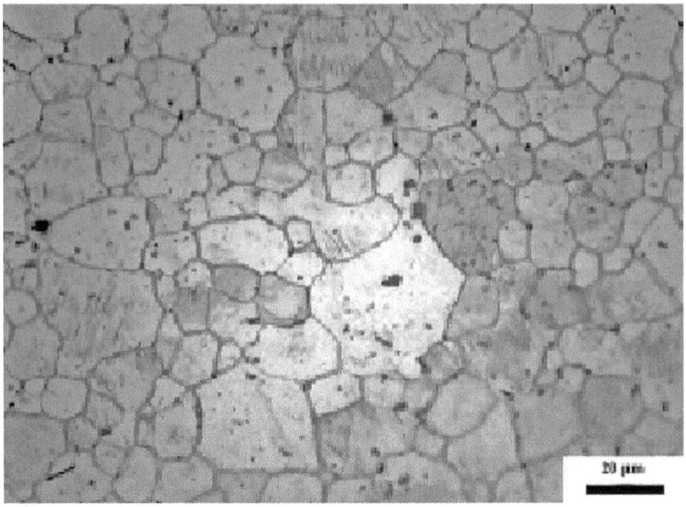

Fig. 2 Microstructure of the AZ31hp-O alloy

Table 1 Tensile properties of the AZ31hp-O alloy

Property	$R_{p0.2}$ (MPa)	R_m (MPa)	A_f (%)	$W(MJ/m^3)$
AZ31hp-O	165	263	19.45	48.00

3.2 *Fatigue Behaviour of the Parent Material*

The fatigue data obtained from the fatigue tests performed were used to derive the S-N curve and also to obtain the fatigue strength data of AZ31 alloy. The experimental data obtained are plotted by means of maximum stress (linear scale) versus cycles to failure (logarithmic scale) in Fig. 3. As it may be seen in the above figure, the results exhibit a very smooth transition from the low to high cycle fatigue regime, which is a typical behaviour for Mg alloys. This very high stress sensitivity is an essential disadvantage for applications where fatigue resistance is of major concern. Assuming the above results may be fitted by a non-linear analysis (Weibull function), the curve drawn in Fig. 3 has been derived. The four point Weibull equation is defined as:

$$\sigma_{max} = P_1 + \frac{P_2 - P_1}{\exp\left(\frac{\log N_f}{P_3}\right)^{P_4}}$$

where P1 and P2 stand for fatigue endurance limit and ultimate tensile strength and P3 and P4 are regression analysis coefficients.

To better understand the fatigue behaviour of the alloy under investigation, crack initiation and short crack growth observations on the specimen tested at 160 MPa

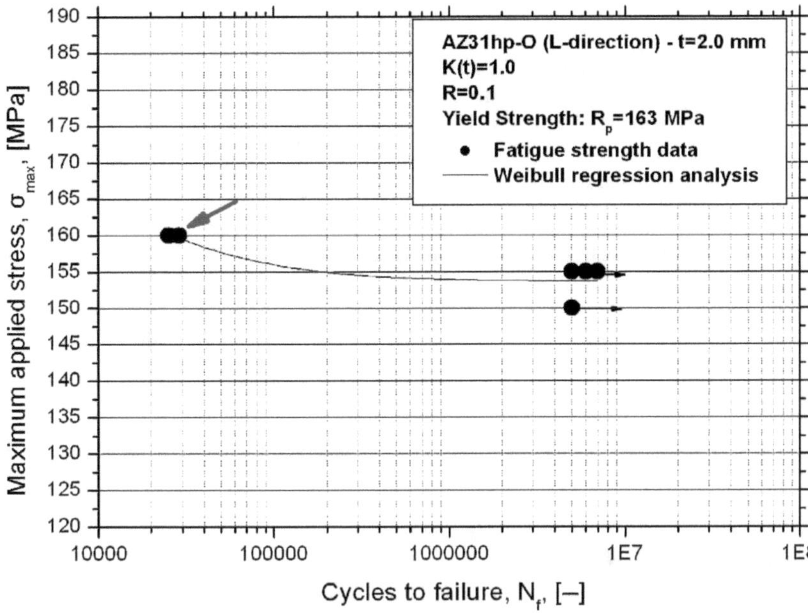

Fig. 3 S-N curve of the AZ31hp-O alloy

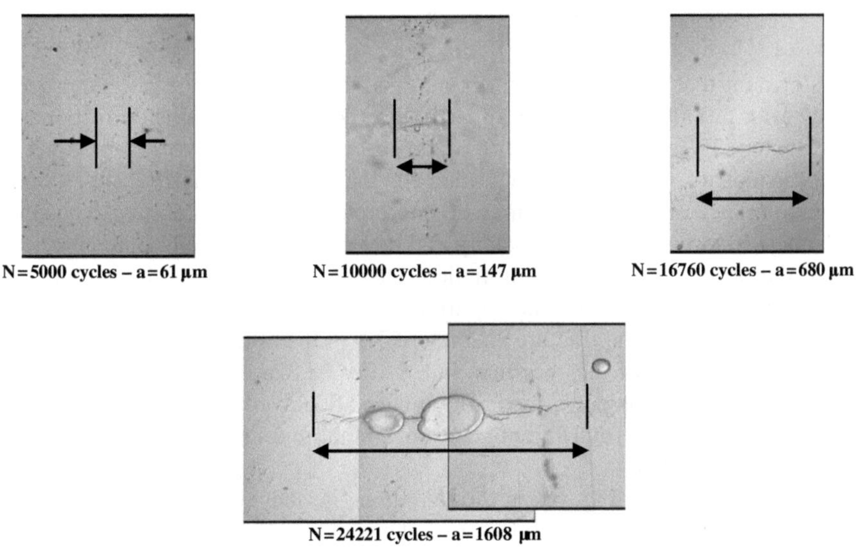

Fig. 4 Successive observation of the evolution of the fatigue crack

maximum stress, indicated with the arrow in Fig. 3, were made using the replication technique. The test was periodically interrupted and replicas were taken from the surface of the material. Selected successive observations may be seen in Fig. 4.

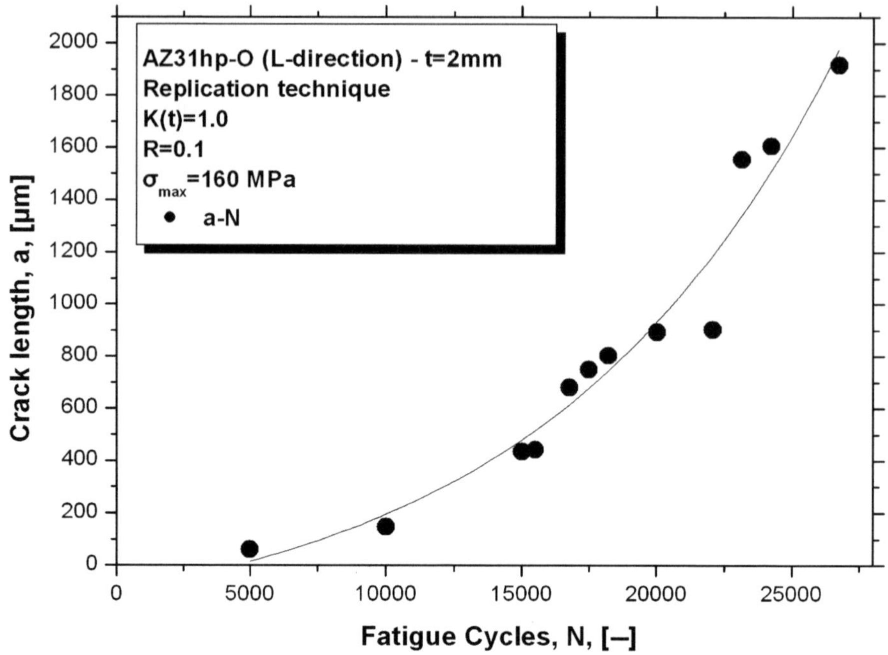

Fig. 5 Short crack growth evolution of the AZ31hp-O alloy

It is evident that crack initiation occurs at an early stage of the fatigue process. Already after 5,000 cycles a crack of 61 μm has been detected; this corresponds to approximately 17% of the fatigue life of the specimen. The above observations are also consistent to other works performed for the AZ31 material [6, 7]. The poor damage tolerance of the material may be interpreted as a lack of ability to accumulate damage in terms of cyclic plasticity at the early stages of the fatigue process; it leads to the observed early crack initiation. Displayed in Fig. 5 is the evolution of the fatigue crack from a short crack up to the fracture of the specimen.

In order to evaluate the fracture characteristics of the alloy under investigation, a fractographic analysis of the specimen failed after 28,828 cycles at 160 MPa maximum applied stress was performed. The images displayed in Fig. 6 show fracture characteristics referring to different stages of the fatigue life of the specimen, as it was pointed out in [12].

As it may be seen in Fig. 6a, in the initiation site there is an arc-shaped region of flat transgranular fracture with linear marks which could be striations. However, some marks appear in pairs and could be traces of twin boundaries (Fig. 6b). The size of the transgranular section at the direction normal to the applied load is of the order of 100 μm. The fatigue crack propagation area (Fig. 6c) is characterized by a fracture surface with flake-like appearance. The flake boundaries are clearly grain boundaries, therefore the fracture surface is intergranular. Crack branching (Fig. 6d) can also be observed in the propagation area.

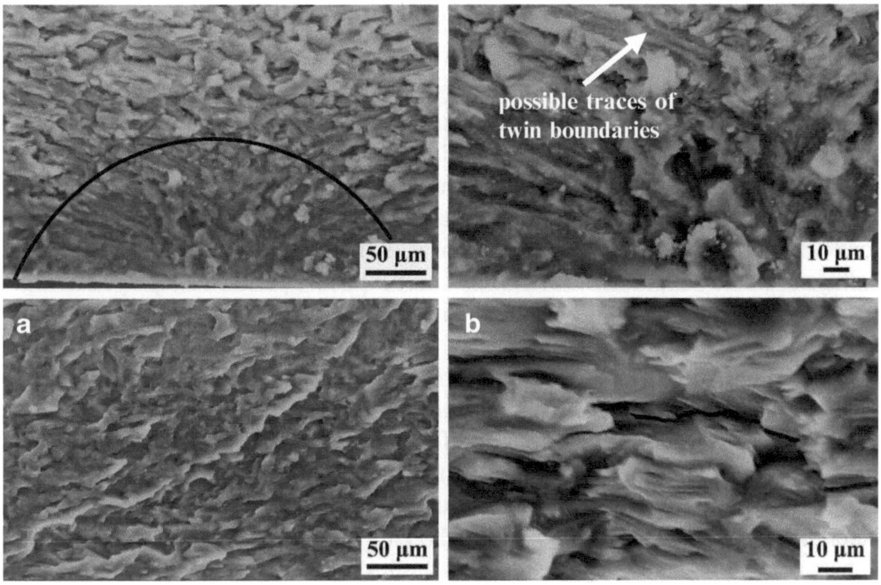

Fig. 6 Fracture characteristics of a fatigue specimen

As it has been discussed in [12], in order to have homogeneous deformation in a stressed material, five independent slip systems should be active. Magnesium alloys, due to their hexagonal structure possess four independent systems in room temperature. The required fifth system is provided by the activation of twinning. Twinning activation in the alloy under investigation occurs at stresses above the yield strength. Therefore, in the investigation of the fatigue behaviour in this work, where the stresses are below the yield strength of the material, twinning deformation is not evident and might occur locally at preferred oriented grains or locations of stress concentrations. This is consistent to the fractographic analysis presented above, where a few traces of possible twin boundaries are observed in the initiation area.

Furthermore, the observations on the collected replicas revealed a crack of 61 μm after 5,000 fatigue cycles. By taking into account the theory of short cracks suggested by Miller [13], crack initiation may be considered when the defect reaches a magnitude between: $d/2 < a < d$, where d is the grain size. Recall that the grain size of the material under investigation is 15 μm. It is reasonable to assume that a significant part of the 5,000 fatigue cycles is consumed for the crack to increase from 15 to 61 μm. Therefore, one could reasonably assert that crack initiation occurs at the very early fatigue stage. The above observations may lead to the conclusion that the material can hardly accumulate damage in terms of plasticity at the first stage of the fatigue process, resulting in a very early crack initiation and consequently poor damage tolerance of the material, compared to other structural alloys. Still, further investigation is needed to understand the fundamentals of fatigue damage accumulation in rolled magnesium alloys.

3.3 Effect of Pre-exposure on the Fatigue Behaviour

The effect of pre-existing corrosion damage on the fatigue behaviour of AZ31 material has been assessed by exposing the respective specimens in a salt spray chamber for 6 h prior to fatigue loading. As it has been observed in [10], the maximum pit depth and pitting density for 6 h exposure were 325 μm and 82 samples/100 mm^2 respectively. Furthermore, the residual ductility of the material in terms of elongation to fracture and strain energy density was of the order of 65%. Hence, corrosion damage is already appreciable after only short exposure periods. Displayed in Fig. 7 is the S-N curve of the pre-corroded material, compared to respective one of the parent material.

As expected, the corrosion attack results in a significant drop of the material's fatigue life. The presence of corrosion pitting facilitates essentially the onset of fatigue cracks and, hence, reduces the fatigue life of the corroded specimens appreciably. The effect of existing corrosion pits on the fatigue life increases with decreasing fatigue stress amplitude. As a result, the shape of the S-N curve of the pre-corroded material is no longer smooth as it was the case for parent material but appears a higher inclination. The fatigue endurance limit drops from 155 MPa for the parent material to 80 MPa for the pre-corroded specimens, which corresponds to an almost 50% reduction. At this point, it should be mentioned that a qualitatively similar behaviour, after corrosion exposure, has also been observed for several structural aluminum alloys of the 2xxx, 6xxx and 7xxx series.

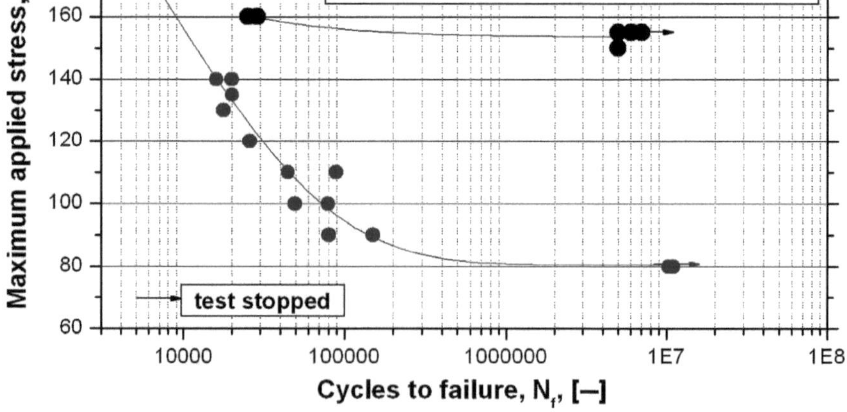

Fig. 7 S-N curves of parent and pre-corroded AZ31hp-O alloy

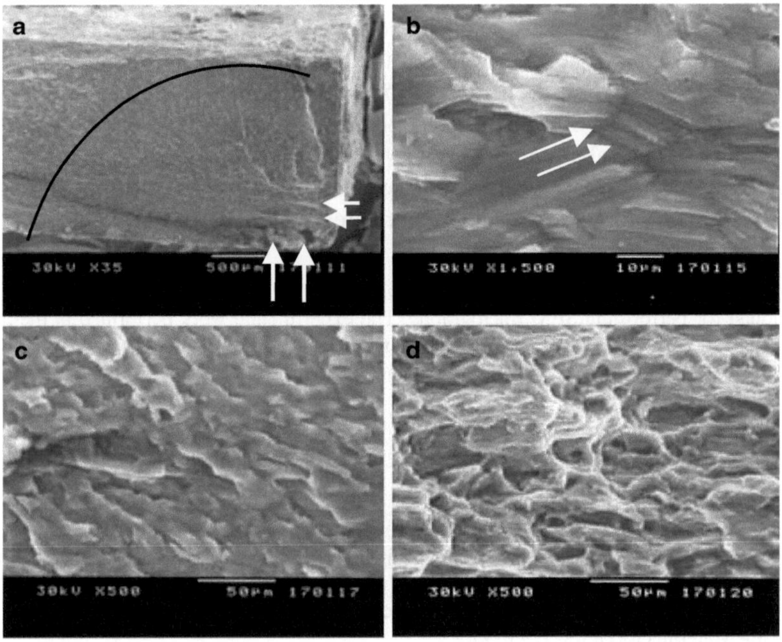

Fig. 8 Fracture characteristics of a pre-corroded fatigue specimen

The fracture characteristics of the pre-corroded specimen failed after 2.6E4 cycles under 120 MPa maximum stress are displayed in Fig. 8. The curved line in Fig. 8a marks the fatigue area of the specimen. The initiation site of the crack is difficult to be defined due to extensive corrosion damage. The arrows mark the corrosion sites and corrosion products. The ridges on the fracture surface (white lines) lead to the initiation site. By observing a magnification of the area between these ridges (Fig. 8b) some twin-like features seems to appear. Finally, at the propagation area (Fig. 8c) flakes are observed as in the case of the parent material, while the fast fracture area is characterized as quasi-cleavage (Fig. 8d).

4 Conclusions

- The S-N curve of the parent material exhibits a smooth transition from the low to high cycle fatigue regime, indicating a very high stress sensitivity.
- Crack initiation occurs at a very early stage of the fatigue process. This could be attributed to the lack of ability of the material to accommodate damage in terms of plasticity.
- A transgranular mode of crack initiation and an intergranular mode of crack propagation have been observed.

- Pre-existing corrosion damage facilitates essentially the onset of fatigue cracks and, hence, reduces the fatigue life of the corroded specimens appreciably. The fatigue endurance limit drops from 155 to 80 MPa, which corresponds to an almost 50% reduction. Similar qualitatively characteristics have also been observed for several structural aluminum alloys.
- Fractographic analysis revealed some twin marks in the initiation area of the pre-corroded material, while the fast fracture area is characterized as quasi-cleavage.

Acknowledgments The authors are grateful to Professor G.N. Haidemenopoulos and Dr. E. Kamoutsi for performing SEM fractography. The experimental work performed in the present work is part of the AEROMAG Project "Aeronautical Application of Wrought Magnesium" (Project no AST4-CT-2005-516152), which was supported by the European Union.

References

1. M. Avedesian and H. Baker (1999). Magnesium and Magnesium Alloys. ASM Specialty Handbook, ASM International, Metals Park, OH.
2. G. Song and A. Atrens (1999). Corrosion mechanisms of magnesium alloys. Advanced Engineering Materials Reviews, 1, 11–33.
3. Anonymous. AEROnautical application of wrought MAGnesium (AEROMAG). EC/FP6 Growth/Aeronautics, 2005–2008.
4. Anonymous. Integrated design and product development for eco-Efficient production of low-weight aeroplane equipment (IDEA). EC/FP6-IP/Aeronautics, 2005–2008.
5. Wrought Magnesium Alloys. www.magnesium-elektron.com
6. K. Tokaji, M. Kamakura, Y. Ishiizumi and N. Hasegawa (2004). Fatigue behaviour and fracture mechanism of a rolled AZ31 magnesium alloy. International Journal of Fatigue, 26, 1217–1224.
7. S. Ishihara, Z.Y. Nan and T. Goshima (2007). Effect of microstructure on fatigue behaviour of AZ31 magnesium alloy. Materials Science and Engineering A, 468–470, 214–22.
8. C. Blawert, W. Dietzel and A. Atrens (2005). A study on stress corrosion cracking and hydrogen embrittlement of AZ31 magnesium alloy. Material Science and Engineering A, 399, 308–317.
9. Z.Y. Nan, S. Ishihara and T. Goshima (2008). Corrosion fatigue behaviour of extruded magnesium alloy AZ31 in sodium chloride solution. International Journal of Fatigue, 30, 1181–1188.
10. Sp.G. Pantelakis, N.D. Alexopoulos and A.N. Chamos (2006). Effect of salt spray corrosion on the tensile behaviour of wrought magnesium alloy AZ31. In Proceedings of the 7th International Conference on Magnesium Alloys and Their Applications, Dresden, Germany, 743–748.
11. D. Tawil (2004). The Principles of Magnesium Corrosion Protection, Magnesium Elektron. In website: www.magnesium-elektron.com/data/downloads/Corrosion%20Protection%20Principles.pdf.
12. A.N. Chamos, Sp.G. Pantelakis, G.N. Haidemenopoulos and E. Kamoutsi (2008). Tensile and fatigue behaviour of wrought magnesium alloys AZ31 and AZ61. Fatigue and Fracture of Engineering Materials and Structures (to be published).
13. K.J. Miller (1987). The behaviour of short fatigue cracks and their initiation Part I – A review of two recent books. Fatigue and Fracture of Engineering Materials and Structures, 10, 75–91.

Session III
Engineering Properties II

Coulomb Failure Surfaces in Ductile Non Linear Elastic Materials

Nikos P. Andrianopoulos, Vassilios M. Manolopoulos, and Ioannis T. Dernikas

Abstract In the present work a theoretical approach to the problem of failure of non-linear elastic materials is described. This approach is based on the Principle of Conservation of Energy – the total elastic strain energy density in this case. This Principle is equivalent to the path-independence of strain energy, the same not being true for its components, i.e. dilatational and distortional densities. This way, an intimate quantitative relationship is established between the secant bulk, K_S, and shear, G_S, elastic moduli of the material. Experimental evidence from the open literature in case of a ductile material supports clearly the present predictions.

Keywords Coulomb criterion · Non linear elasticity · Ductile materials

1 Introduction

The development of failure criteria outside Fracture Mechanics seems, nowadays, out of fashion. Microscopic approaches are considered more fruitful, due to the necessity of developing new materials. But, practical applications require answers in terms of macroscopic level i.e. Continuum Mechanics. The limited interest for macroscopic considerations is, additionally, explained in the view of Coulomb's statement that any material fails when '*a proper combination of shear and normal stresses is realized*'. This statement is logically perfect (what else?) but lacks a quantitative description as far as it is based on experimental data and arbitrary assumptions for the shape of the failure envelope. Coulomb's postulate, simply, reflects the only two geometrically permissible and discrete modes of failure activated during loading, i.e. slip caused by shear and cleavage caused by normal stresses.

N.P. Andrianopoulos (✉), V.M. Manolopoulos, and J.T. Dernikas
Department of Mechanics, Faculty of Applied Sciences, National Technical University of Athens, GR-157 73, Hellas
e-mail: nandrian@central.ntua.gr, manolobill@gmail.com, johnder2004@yahoo.gr

S. Pantelakis, C. Rodopoulos (eds.), *Engineering Against Fracture: Proceedings of the 1st Conference*,

However, pressure dependence of failure surfaces expresses an inherent property of materials. Many attempts have been focused to this problem, due to its high technological interest. Some of them [9, 11–13, 18] comply with theoretical requirements, the remaining being more or less empirical [6, 7, 15–17, 19–21] and based on rich experimental evidence especially for brittle, non linear elastic materials, like stones, marbles etc.

Along the lines of the as above Coulomb's 'doctrine', a rigorous theoretical approach was, recently, introduced [4, 5]. The general case of a non-linear elastic material was considered and the total elastic strain energy density, T, was anticipated as the characteristic quantity for the respective conservative field $T\,[T_V(I_1,I_2),\ T_D(I_1,I_2)]$. Thus, T is a path independent quantity and, along a closed path $\gamma = \gamma(I_1,I_2)$ equals to zero. Here T_V, T_D are the dilatational and distortional parts of T and I_1, I_2 the first strain and second deviatoric strain invariants, respectively.

However, path independence of T does not hold necessarily for its parts T_V, T_D. Then, either T_V or T_D arrives first at a critical value, say $T_{V,0}$ or $T_{D,0}$, depending on the type of loading. Obviously, T_V, T_D are measures of normal and shear stresses and the *proper combination* of Coulomb and the respective failure surfaces are easily obtained through conservation of elastic strain energy density, T.

The so-obtained failure surfaces are closed, a finding not accompanying Coulomb's criterion. It is known that Coulomb's criterion results to open surface for high negative pressures and problematic predictions for positive pressures (tension cut-off and other tailored assumptions). Failure surfaces are unique when elastic properties of the material (secant shear G_S and bulk K_S moduli) depend only on second deviatoric I_2 and first strain I_1 invariants, respectively. In the general case, when elastic moduli depend on both strain invariants, failure surfaces are not unique. Various non linear elastic constitutive equations, especially those of higher order, result to a great variety of failure surfaces, through the simple mechanism described here. This finding may be of technological interest. Here, the study is focused in the behaviour of ductile materials and experimental evidence indicates promising agreement with theoretical predictions.

2 Theoretical Considerations

2.1 Theoretical Basis

The behaviour of an isotropic (linear or not) elastic material, in the general case of loading, can be described in terms of the secant elastic moduli $K_S(I_1,I_2)$ and $G_S(I_1,I_2)$, with I_1 and I_2 the first strain and the second deviatoric strain invariants [8, 10]. Furthermore, the elastic strain energy density, T, stored into the material consists of two parts, T_V representing the cost (storing capacity) for volume changes and T_D for shape changes, respectively, given by:

$$T = T_V + T_D = \int_0^{\varepsilon_{ij}} \sigma_{ij} d\varepsilon_{ij} = \int_0^{I_1} \frac{1}{2} K_S(I_1, I_2) dI_1^2 + \int_0^{I_2} 2G_S(I_1, I_2) dI_2 \tag{1}$$

It can be easily seen that when:

$$K_S = K_S(I_1), \quad G_S = G_S(I_2), \tag{2}$$

i.e. when the integrals at the right hand side of E q. (1) depend only on the current values of I_1 and I_2, respectively, the elastic strain energy densities T_V and T_D are path independent. In the simplest case of linear elastic materials, where $K_S = K_0^* = \text{const}$ and $G_S = G_0^* = \text{const}$, both dilatational and distortional strain energy densities are given by the well-known expressions:

$$T_V = \frac{1}{2} K_0^* I_1^2, \; T_D = 2G_0^* I_2 \tag{3}$$

Equation (3), when either T_V or T_D takes a critical value $T_{V,0}$ or $T_{D,0}$, respectively, indicate failure of the material. So, a failure criterion for the general case of non-linearity can be introduced, having the form:

$$\left.\begin{aligned} T_V &= \int_0^{I_{1,0}} \frac{1}{2} K_S(I_1, I_2) dI_1^2 = T_{V,0} \text{ for failure by cleavage} \\ T_D &= \int_0^{I_{2,0}} 2G_S(I_1, I_2)\, dI_2 = T_{D,0} \text{ for failure by slip} \end{aligned}\right\} \tag{4}$$

The second of Eq. (4) is, exactly, the von Mises criterion in strain energy density terms. The first of them is a simple and obvious addendum to the von Mises criterion in the sense that it provides the missing boundary for hydrostatic or nearly hydrostatic loading paths. In case of linear materials, where Eq. (3) hold, condition $T_D = T_{D,0}$ forms in the stress space the well-known von Mises failure cylinder. The first condition, $T_V = T_{V,0}$, forms two planes normal to the hydrostatic axis at equal distance from the centre of the coordinate system. So, the failure surface is a closed cylinder with its surface corresponding to slip and its two circular bases corresponding to cleavage. Equation (4) form the so-called T-criterion [2, 3, 22], according to which *brittle failure (cleavage) occurs when T_V exceeds a critical value $T_{V,0}$ and shear failure (slip) occurs when $T_D$exceeds a critical value $T_{D,0}$*, respectively.

Things are complicated in the more general case, of a non-linear, elastic, isotropic material with elastic moduli depending on both strain invariants, i.e.:

$$K_S = K_S(I_1,\, I_2),\; G_S = G_S(I_1,\, I_2) \tag{5}$$

Now, path independence of the two integrals at the right hand side of Eq. (1) is not guaranteed. In fact, volume changes do not depend only on hydrostatic stresses/strains and distortions do not depend only on stresses/strains deviations. There is an interaction between volumetric and deviatoric response through K_S and G_S. The mechanical properties of the material, expressed by K_S and G_S, depend on the loading path. Exaggeratedly speaking, we deal with a different linear elastic material at each loading step. Under these conditions, the only available fundamental assumption for the problem is that the total elastic strain energy field is conservative and it is derived by a potential function V. It implies that T in Eq. (1) keeps the same difference in its value between two points, independently of the path connecting these points. However, this is not true for either one of the two integrals in the right hand side of Eq. (1).

To quantify the failure behaviour of a material described by Eq. (5), the evaluation of upper integration limits, $I_{1,0}$ and $I_{2,0}$, for the quantities I_1 and I_2 appearing in Eq. (4) is required. If so, one of the two integrals takes the respective critical value $T_{V,0}$ or $T_{D,0}$, before its pair reaches its critical value, indicating that the material failed by either cleavage or slip, respectively. The study is based on the principle of conservation of the total strain energy density T or, equivalently, on the path independence of this quantity. In other words, it is asked for a vector field $\mathbf{F}\left[P(x,y),Q(x,y)\right]$ to satisfy the condition:

$$\oint_{\gamma} Pdx + Qdy = 0 \tag{6}$$

along a closed path $\gamma = (I_1, I_2)$. In the present case, the conservative vector field is:

$$\mathbf{F} = [P(I_1,I_2), Q(I_1,I_2)] = \mathbf{F}\left[I_1 K_s\left(I_1,I_2\right), 2G_S(I_1,I_2)\right] \tag{7}$$

as it can be obtained from Eq. (4). Path independence of T is satisfied when:

$$\frac{\partial P}{\partial I_2} = \frac{\partial Q}{\partial I_1} \text{ or } \frac{\partial\left(I_1 K_S\right)}{\partial I_2} = 2\frac{\partial G_S}{\partial I_1} \tag{8}$$

Here, Eq. (8) constitutes a relationship between secant elastic moduli and strain invariants, imposed by the principle of conservation of total elastic strain energy density. In fact, Eq. (8) establishes an interaction between secant elastic moduli, in a manner depending on loading path. Finally, one of the functions $K_S\left(I_1,I_2\right)$ or $G_S(I_1,I_2)$ being experimentally given, Eq. (4) can be evaluated for a given loading path.

2.2 Theoretical Application

It is assumed that the bulk secant modulus is related to strain invariants I_1 and I_2 with:

$$K_S(I_1,I_2) = K_0^* e^{-\left(a_1 I_1 + \alpha_2 \sqrt{I_2}\right)} \tag{9}$$

The exponential type of this function is arbitrarily chosen, in the light of common experimental evidence for many non-linear elastic materials. However, it could have any other form, depending on specific materials [8, 10, 14]. In this equation, α_1, α_2 are two suitable real parameters depending on the material. The presence of the term $\sqrt{I_2}$ instead of I_2 in Eq. (9) is justified by the natural necessity to represent the same physical entity as I_1 (strain). This way, the 'sensitivity' of the material to normal and shear strains is balanced. In addition, the second deviatoric strain invariant, I_2, is a non-negative quantity. Here, it must be positive. For that, it is assumed that at the initial steps of loading the material is linear elastic and the lowest allowable values for I_1, I_2, in case of non-linearity, are, say, λ_1, λ_2. A proper selection of these constants allows for the easy description of the material from nearly full linear to nearly full non-linear behaviour (high or low values of λ_1, λ_2, respectively). Then, Eqs. (8 and 9) give: $\frac{\partial G_S}{\partial I_1} = \frac{\partial (I_1 K_S)}{2\partial I_2} = -\frac{K_0^* \alpha_2}{4} \frac{I_1}{\sqrt{I_2}} e^{-(\alpha_1 I_1 + \alpha_2 \sqrt{I_2})}$, which, after integration, gives:

$$G_S(I_1, I_2) = \frac{K_0^* \alpha_2}{4\alpha_1 \sqrt{I_2}} \left[\left(I_1 + \frac{1}{\alpha_1} \right) e^{-(\alpha_1 I_1 + \alpha_2 \sqrt{I_2})} + f(I_2) \right] \equiv G_0^* Z(I_1, I_2) \tag{10}$$

where $f(I_2)$ is an integration function. Function $Z(I_1, I_2)$ is introduced in order for G_S to take the same form as K_S in Eq. (9). It is, then, allowed by inspection of Eq. (10) to put:

$$G_0^* = \frac{K_0^* \alpha_2}{4\alpha_1} \tag{11}$$

$$Z(I_1, I_2) = \frac{1}{\sqrt{I_2}} \left[\left(I_1 + \frac{1}{\alpha_1} \right) e^{-(\alpha_1 I_1 + \alpha_2 \sqrt{I_2})} + f(I_2) \right] \tag{12}$$

The real constants K_0^*, α_1 and α_2 and the integration function $f(I_2)$ can be evaluated, without loss of generality, by considering the cases of simple tension, where:

$$\sqrt{I_2} = \frac{1}{\sqrt{3}} \frac{1+\nu}{1-2\nu} I_1 \tag{13}$$

(ν being the Poisson's ratio of the material) and torsion where: $I_1 = 0$.

Equation (13) was obtained by combining the relations that give the definition of the first strain invariant $I_1 = \varepsilon_1 + \varepsilon_2 + \varepsilon_3$ and the second deviatoric strain invariant $I_2 = \frac{1}{\sqrt{6}} \left\{ (\varepsilon_1 - \varepsilon_2)^2 + (\varepsilon_2 - \varepsilon_3)^2 + (\varepsilon_3 - \varepsilon_1)^2 \right\}$ for the case of simple tension.

Consequently, plotting the experimental points of a simple tension and torsion test and using non linear regression theory, the real constants K_0^*, G_0^* and α_2 can be calculated. We also find that:

$$f(I_2) = e^{-a_2 \sqrt{I_2}} \left(\sqrt{I_2} - \frac{1}{a_1} \right) \tag{14}$$

Therefore:

$$\left.\begin{array}{l} P = K_S I_1 = K_0^* I_1 e^{-(\alpha_1 I_1 + \alpha_2 \sqrt{I_2})} \\ Q = 2G_S = 2G_0^* \left(1 + \frac{1}{\sqrt{I_2}}\left(I_1 + \frac{1}{a_1}\right) e^{-(\alpha_1 I_1 + \alpha_2 \sqrt{I_2})} + e^{-a_2\sqrt{I_2}}\left(\sqrt{I_2} - \frac{1}{a_1}\right)\right) \end{array}\right\} \quad (15)$$

Then, the required for the description of the non-linear elastic material parameters are three, namely K_0^*, G_0^* and α_1, since α_2 is related to K_0^*, G_0^* and α_1 through Eq. (11). Finally, through Eqs. (4 and 8), it is obtained that:

$$T_V = \int_{\lambda_1}^{I_{1,0}} K_s I_1 dI_1 = \frac{K_0^*}{a_1^2}\left[(1 + a_1\lambda_1) e^{-(a_1\lambda_1 + a_2\sqrt{I_2})} - (1 + a_1 I_{1,0}) e^{-(a_1 I_{1,0} + a_2\sqrt{I_2})}\right] = T_{V,0} \quad (16)$$

$$T_D = \int_{\lambda_2}^{I_{2,0}} 2G_s(I_1, I_2) dI_2 = \int_{\lambda_1}^{\bar{I}_1} 2G_s(I_1, I_2) F'(I_1) dI_1 = T_{D,0} \quad (17)$$

where $F'(I_1) = \frac{dF(I_1)}{dI_1}$ is the derivative of the stretching path. Thus, for a given stretching path $F(I_1)$ and material (K_0^*, G_0^*, α_1) with known failure properties ($T_{V,0}$, $T_{D,0}$) experimentally determined, the complete failure surface can be derived.

3 Application in Case of a Mild Steel

A series of experiments [1] in mild steel under various loadings was used. Experimental constitutive equations $\sigma_{eq} = f(\varepsilon_{eq})$ and $p = f(\theta)$ are plotted in Fig. 1, the

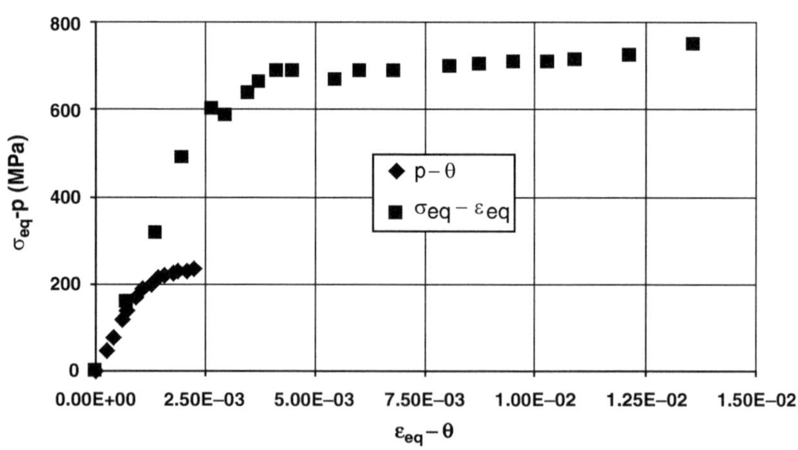

Fig. 1 Constitutive equations $(\sigma_{eq}, \varepsilon_{eq})$ and (p, θ) for a mild steel

points being merged from all the experiments. There, $(\sigma_{eq}, \varepsilon_{eq})$ are the equivalent von Mises stress/strain and (p, θ) the hydrostatic pressure/strain, respectively.

It is clear that a unique curve for each pair of quantities is obtained, allowing the evaluation of unique, as well, values of the mechanical properties of the material. These properties are gathered in Table 1.

It is evident from this figure (Fig. 2) that tangent bulk modulus K_S is quite sensitive to the first strain invariant I_1, showing slight variation with the second deviatoric strain invariant I_2. The opposite is true for secant shear modulus G_S, a reasonably expected result.

In Table 2 eight combinations of simple tension and torsion for the mild steel, under study, are presented [1]. These loading paths range from simple tension to simple torsion, passing through intermediate mixed tension–torsion or torsion–tension paths.

Table 1 Mechanical properties of a mild steel

Tensile yield strength (MPa)	K_0^* (GPa)	G_0^* (GPa)	a_1	a_2	$T_{V,0}$ (MPa)	$T_{D,0}$ (MPa)
604.8	194.6	77.8	399.6	197.9	0.20	1.17

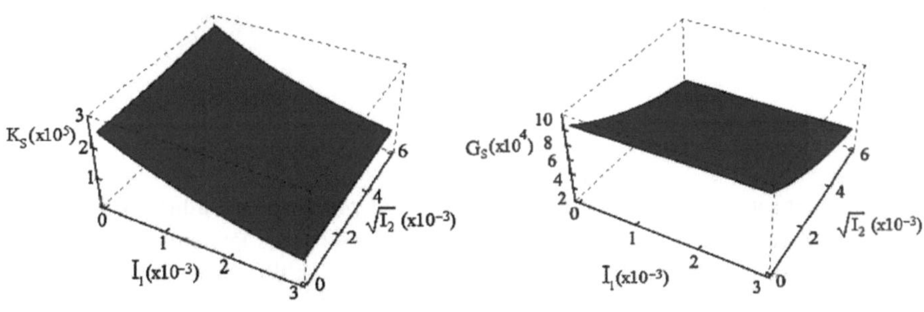

Fig. 2 Three-dimensional plots of the constitutive equations $K_S(I_1, I_2)$ and $G_S(I_1, I_2)$

Table 2 Experimental loading paths

Line	Experiment
OD	Tension – virgin specimen
OAH	25% of yield stress → torsion
OBF	50% of yield stress → torsion
OCE	75% of yield stress → torsion
OL	Torsion – virgin specimen
OPG	25% of yield torque → tension
ONJ	50% of yield torque → tension
OMK	75% of yield torque → tension

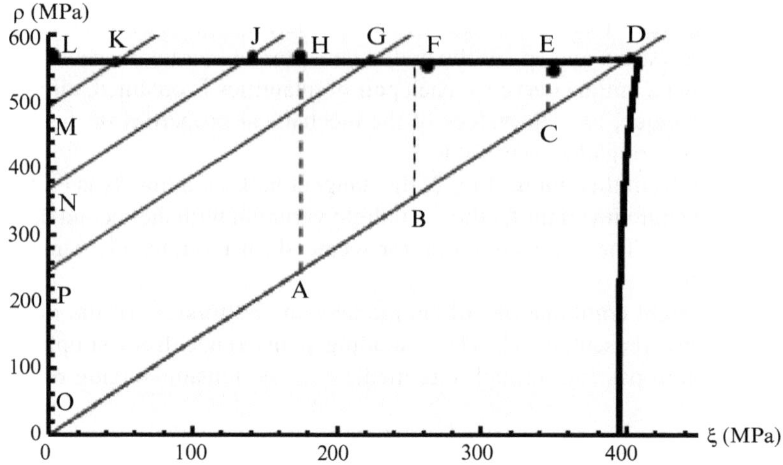

Fig. 3 Failure surface in the Haig–Westergaard space (ξ, ρ)

The respective eight experimental failure points D, E, F, G, H, J, K, and L belong to the curve $T_D = T_{D,0}$, as it is evident in Fig. 3. In other words, the specific material failed by slip for all the paths. It is interesting to observe that triaxiality ratio p/σ_{eq} increases from point L to point D and a little higher value of this ratio would drive to failure by cleavage i.e. to transition of D to the curve $T_V = T_{V,0}$.

4 Conclusions

In the present work a theoretical approach to the problem of failure of non-linear elastic materials is described. This approach is based on the Principle of Conservation of Energy – the total elastic strain energy density in this case. This way, an intimate quantitative relationship (Eq. (8)) is established between the secant bulk, K_S, and shear, G_S, elastic moduli of the material. This relationship is qualitatively equivalent and explains the Coulomb's postulate of "*a proper combination of normal and shear stresses*" for a material to fail. In fact, this desired *combination* is a matter of energy conservation. As dilatational and distortional strain energy densities must have an upper limit, defining the maximum capacity of the material to afford volume or shape changes before failure, limits can be drawn to obtain the respective closed failure surface. Experimental evidence in case of a typical ductile material is in favor of the present predictions. In addition, some interesting properties of the present approach are worth to mention. Namely, the present approach quantifies and rationalizes Coulomb's criterion by a strong and simple theoretical basis and generalizes it for any type of material, brittle [5] or ductile. Contrarily to Coulomb's criterion, failure surfaces are closed and, especially in case of brittle materials, they are not unique, reflecting a common experimental finding (pressure dependence).

References

1. Ali ARM, Hashmi MSJ (1999) Theoretical and experimental results of the elastic–plastic response of a circular rod subjected to non-proportional combined torque and tension loadings. Proc. Inst. Mech. Eng., 213, Part C
2. Andrianopoulos NP (1993) Metalforming limit diagrams according to the T-criterion. J. Mat. Proc. Tech., 39: 213–226
3. Andrianopoulos NP, Boulougouris VC (1994) Failure by fracture or yielding in strain hardening materials according to the T-criterion. Eng. Fract. Mech., 47: 639–651
4. Andrianopoulos NP, Boulougouris VC (2004) Loading path dependence of failure surfaces. Seventh International Congress on Mechanics, HSTAM 2004, Chania
5. Andrianopoulos NP, Manolopoulos VM, Boulougouris VC (2007) Considerations on failure surfaces in brittle, non linear elastic materials. Eighth International Congress on Mechanics, HSTAM 2007, Patras
6. Balmer GG (1949) Shearing Strength of Concrete Under High Triaxial Stress. Computation of Mohr's Envelope as a Curve Structural Research Laboratory Report SP-23, Denver, CO
7. Bresler B, Pister KS (1958) Strength of concrete under combined stresses. ACI J., 55: 321–345
8. Chen WF, Han DJ (1988) Plasticity for Structural Engineers. Springer, New York
9. Desai CS (1971) Nonlinear analysis using spline functions. J. Soil Mech. Found. Div., ASCE, 97: 1461–1480
10. Desai CS, Siriwardane HJ (1984) Constitutive Laws for Engineering Materials. Prentice-Hall, Englewood Cliffs, NJ
11. Hansen JB (1963) Discussion of "Hyperbolic Stress-Strain Response: Cohesive Soils", by Kondner, R.L. J. Soil Mech. Found. Div., ASCE, 89: 241–242
12. Hsieh SS, Ting EC, Chen WF (1982) A plasticity fracture-model for concrete. Int. J. Solids Struct., 18: 181–197
13. Kondner RL (1963) Hyperboloid stress-strain response: cohesive soils. J. Soil Mech. Found. Div., ASCE, 89: 15–143
14. Kotsovos MD, Pavlovic MN (1999) Ultimate Limit-State Design of Concrete Structures: A New Approach. Thomas Telford, London
15. Kupfer H, Hilsdorf HK, Rusch H (1969) Behavior of concrete under biaxial stresses. ACI J., 66: 656–666
16. Launay P, Gachon H (1972) Strain and ultimate strength of concrete under triaxial stresses special publication SP-34. ACI J., 1: 269–282
17. Mills LL, Zimmerman RM (1970) Compressive strength of plain concrete under multiaxial loading conditions. ACI J., 67: 802–807
18. Nelson I, Baron ML (1971) Application of variable MODULI models to soil behaviour. Int. J. Solids Struct., 7: 399–417
19. Ottosen NS (1977) A failure criterion for concrete. J. Eng. Mech. Div., ASCE 103(EM4): 527–535
20. Richart FE, Brandtzaeg A, Brown RL (1928) A Study of the Failure of Concrete under Combined Compressive Stresses. University of Illinois, Engineering Experimental Station Bulletin, 185
21. Tasuji ME, Slate FO, Nilson AH (1978) Stress-strain response and fracture of concrete in biaxial loading. ACI J., 75: 306–312
22. Theocaris PS, Andrianopoulos NP (1982) The T-criterion applied to ductile fracture. Int. J. Fracture, 20: R125–R130

Effect of Stress on the Fire Reaction Properties of Polymer Composite Laminates

A.E. Elmughrabi, A.M. Robinson, and A.G. Gibson

Abstract A small-scale loading frame was used to apply tensile and compressive stresses to glass vinyl ester and glass polyester laminates in a cone calorimeter under a heat flux of 75 kW m^{-2}. It was found, for the first time, that stress has a small but significant effect on the fire reaction properties. Increasing tensile stress increased heat release rate and smoke production, while shortening the time-to-ignition. Compressive stress had the reverse effect. This was attributed to the fact that tensile stress promotes the formation of matrix microcracks, facilitating the evolution of flammable volatiles. This hypothesis is further supported by the observation that stress has the greatest effect on the early heat and smoke release peaks, with a lower effect on the final 'run-out' values.

Stress rupture (time-to-failure) curves were produced for tension and compression. In tension, the behaviour was fibre dominated, with times-to-failure being roughly ten times those in compression. Compressive failure involved resin dominated local fibre kinking, initiated near to the rear face of the specimen. The failure time was determined by a significant proportion of the specimen reaching its glass transition temperature.

Keywords Composite laminates · Fire reaction · Cone calorimeter · Stress

1 Introduction

Fibre reinforced polymers are used because of a range of useful properties, including high specific strength and resistance to corrosion. Applications include transport and infrastructure as well as the energy and chemical industries. A significant problem, however, is their poor fire performance, which is related to the combustibility of the organic matrix [1–7]. Thermosetting polymers loose much of their stiffness and

A.E. Elmughrabi, A.M. Robinson, and A.G. Gibson (✉)
Centre for Composite Materials Engineering, University of Newcastle upon Tyne
e-mail: {awad.elmughrabi, newrail, a.g.gibson}@ncl.ac.uk

S. Pantelakis, C. Rodopoulos (eds.), *Engineering Against Fracture: Proceedings of the 1st Conference*,

strength when they reach the glass transition temperature, which often lies in the range from 80°C to 150°C. Higher temperatures, typically in excess of 400 °C result in decomposition, producing products that may burn. These products release heat that can contribute to fire growth [1–20]. In addition, smoke and toxic fumes may be released, limiting visibility, hindering escape and thus constituting a further hazard.

Fire behaviour of materials and components is categorized in terms of *fire reaction* and *fire resistance* [1]. *Fire reaction* properties relate to the response of the material, often in the early stages to fire. Important measured quantities are time-to-ignition, heat release and smoke evolution. Of these, heat release is regarded as the most significant in determining the hazard posed by a particular material. *Fire resistance* properties, on the other hand, relate to the integrity of structures in fire and to the ability of barrier walls and partitions to prevent the flow of heat from one compartment of a structure to another. Fire resistance tests are generally larger in scale than fire reaction tests, and they are often required to characterise performance over periods of an hour or more, sometimes under load. Often, however, useful information can be gained from the results of testing composites after fire exposure [8–11]. Recently it has been suggested, by a number of workers [1, 11–15] that fundamental information about fire resistance can be gained from relatively low cost, small-scale tests on samples under constant load, subject to constant one-sided heat flux.

Returning to heat release, the principal means of characterising this, and several other fire reaction properties, is the now cone calorimeter [19, 20], shown in Fig. 1.

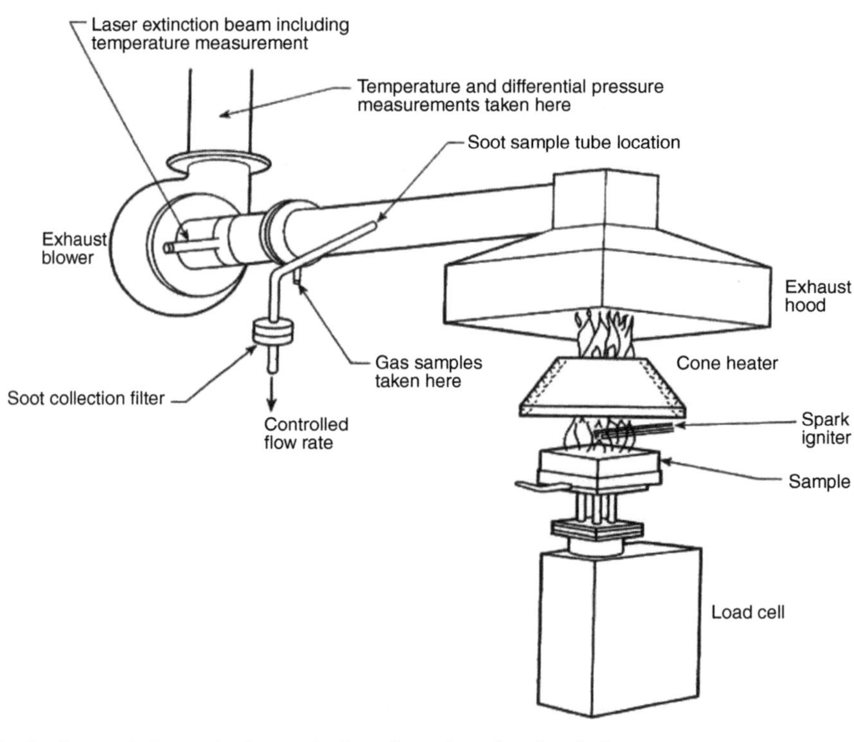

Fig. 1 Cone calorimeter in the standard configuration, showing the key components

The key advance embodied in this instrument, developed in the 1980s, is the use of the 'oxygen consumption principle', an empirical law which observes that the heat released on combustion is proportional to the oxygen consumed. This enables heat release to be inferred from oxygen concentration measurements, more accurately and conveniently than by direct calorimetric methods. The oxygen consumption principle holds for a broad range of materials, including polymer-based ones: its validity and limitations have been discussed elsewhere [19–21].

The cone calorimeter, Fig. 1, now underlies many standards for fire characterisation and qualification of materials, relating to a wide range of applications. An electrically heated cone element produces a near-constant heat flux over the $100\,mm^2$ surface of the test specimen. All the combustion products are drawn, with an excess of air, through the hood extraction system. The air flow rate, temperature and oxygen concentration are measured, enabling a continuous record of heat release to be calculated. Several fire reaction parameters can be measured simultaneously in the cone calorimeter. For instance, a laser is used to determine the obscuration level in the exhaust gas and hence the smoke generated and, in the standard form of the instrument, shown in Fig. 1, the sample is mounted on a load cell, enabling mass loss and 'effective' heat of combustion to be measured. It is also possible, with the addition of appropriate sensors, to measure the evolution of key gases, including CO_2 and CO.

There have been many recent studies of the heat conduction and ablative behaviour of composite structures in fire, including [1, 10–18, 22–28]. Many real composite structures are thermally 'thick', which means that, when they are heated or exposed to fire one side, there can be a significant temperature difference through the thickness of the material. One important factor that limits and determines the transmission of heat is the decomposition of the polymeric matrix, into a mixture of solid char, along with liquid and solid components. This reaction is strongly endothermic, and can significantly delay transmission of heat through the structure, at least up the point where the decomposition of the resin is complete. It is this effect that is employed in composite and elastomeric passive fire protection for steel structures. The key requirement to describe this type of behaviour quantitatively is a thermal model that takes into account the thermal decomposition behaviour of the material, as well as the transport law (i.e. Fourier's Law) determining the transmission of heat through the structure. The most effective model of this type was developed by Henderson and co-workers [22, 23] and the use of several variants of this have been subsequently reported [24–28].

A number of studies have discussed the behaviour of composite structures under load in fire, including the post-fire residual properties [1,8–11], as well as the failure of loaded structures during fire [1, 12–18]. It has been found possible to generate 'stress rupture' curves (i.e. curves relating the applied stress to the time-to-failure) for structures in fire – similar to the relationships more commonly employed for other types of environmental exposure, albeit with a shorter timescale.

Hitherto it has been assumed that thermal and mechanical effects on composites are separable. In other words, if the temperature and decomposition history of the material can be modelled using, say the Henderson Equation [22, 23] approach, and if properties can be modelled as a function of temperature and decomposition, then it should be possible to model the behaviour of the structure by summing, in an appropriate fashion, the contributions of its parts. The assumption is equivalent to regarding the *fire reaction* behaviour of a structure as independent of the presence of stress. To date, however, there have been no reports on this effect, so the purpose of the present investigation was to investigate the validity of this assumption.

We will describe the characterisation of composite laminate samples in a modified cone calorimeter under tensile and compressive loading, with the aim of measuring the effect of stress on fire reaction properties.

2 Materials and Experimental

2.1 Composite Materials

The composite samples used in this investigation were vinyl ester and polyester-based laminates of the type commonly employed in marine structures. The materials were 11.3 mm thick vinyl ester and 11.6 mm thick polyester glass plates, prepared by hand laminating onto a flat mould. The laminates were supplied by the Composite Technology Centre (CTC) of VT Halmatic, Portchester, a manufacturer of large marine craft. The reinforcement was non-crimp stitched E-Glass fabric, with equal numbers of layers in the 0° and 90° directions. 24 plies in total of the glass fabric were used in each laminate. The resins were Dow 411-45 vinyl ester and Scott Bader Crystic 489 isophthalic polyester, cured with organic peroxides according to the manufacturers' specifications. The fibre weight fractions were 0.64 and 0.62, respectively for the vinyl ester and polyester laminates, corresponding to volume fractions of 0.46 and 0.44. The respective void contents were 4.6% and 2.6%. The laminates were not post-cured, which is quite common for composite parts used in large marine structures. They were, however, stored at ambient temperature for approximately 12 months prior to testing.

2.2 Tensile and Compression Measurements under Constant Heat Flux

The loading rig shown in Fig. 2, was designed to apply stress to composite samples, while exposing one side to a constant heat flux in a cone calorimeter (as supplied by FTT Ltd, East Grinstead, West Sussex, UK). Some compromises were required to enable the rig to fit in the restricted space under the cone element in the standard type of instrument. It was necessary, for instance, to remove the mass loss measuring load cell, seen in Fig. 1, so no mass loss measurements were possible in the tests reported here. The 25 mm gap between the cone element and the sample was kept the same as in a conventional test.

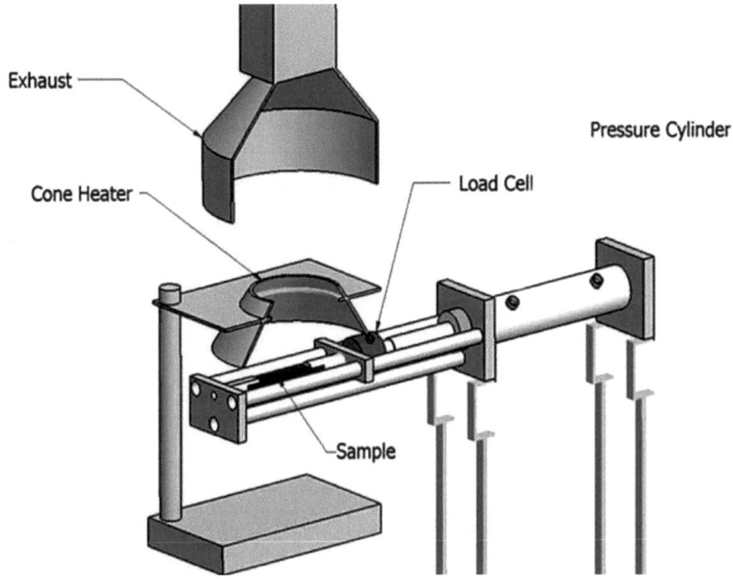

Fig. 2 Loading frame for applying tensile or compressive stress to samples under constant heat flux in the cone calorimeter

As can be seen in Fig. 2, the rig consisted of a pressure cylinder and loading frame comprising three 25 mm diameter steel bars. An electronic load cell was fitted on the pressure cylinder plunger to measure the load applied to the samples. To minimise thermal effects this was situated just outside the area affected by the radiant heat flux from the cone element. The limited working area under the cone restricted the space available for sample grips, so it was necessary to reduce the exposed area of the sample from the standard 100 mm square, down to 100 × 30mm. The grips are shown in Fig. 3. The tensile specimens were given a 15 mm corner radius at either end of the gauge length. The small difference in area resulting from this was allowed for when computing the heat release per unit area. The sample edges were insulated with Kaowool, to ensure as far as possible that heat was applied to only one side, giving near one-dimensional heat flow in the sample.

The change in sample dimensions will have had the effect of decreasing the accuracy of the fire reaction measurements and there may also have been some effects due to the increased area of the sample edges, compared to those in a conventional test. Nevertheless, tests carried out with zero load on the sample produced very similar heat release profiles to those measured with conventional sample dimensions. It would, nevertheless, be desirable, for future instruments of this type, to modify the design a little, in order to permit full-size specimens to be placed under the cone heater.

Figure 4 shows the loading frame in place under the cone calorimeter heating element. A heat flux of 75 kW m^{-2} was employed in this study, corresponding to a fairly severe fire, equivalent to a minimum source temperature of about

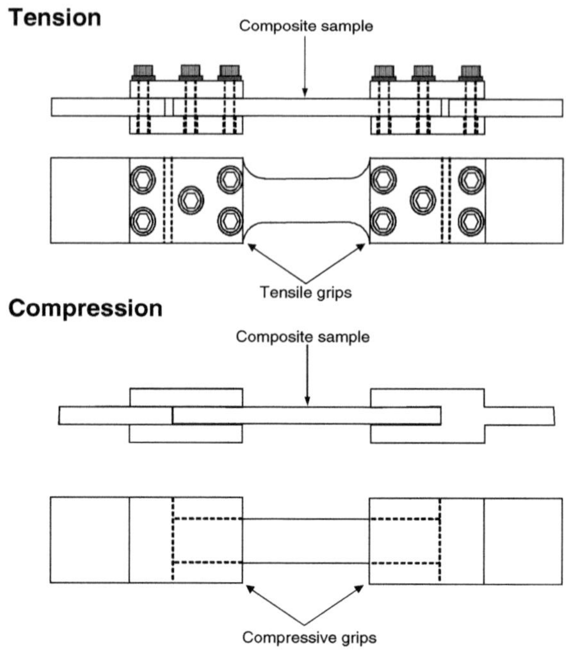

Fig. 3 Specimen grips for tension and compression. In each case the heat-exposed sample dimensions were 100 mm long by 30 mm wide

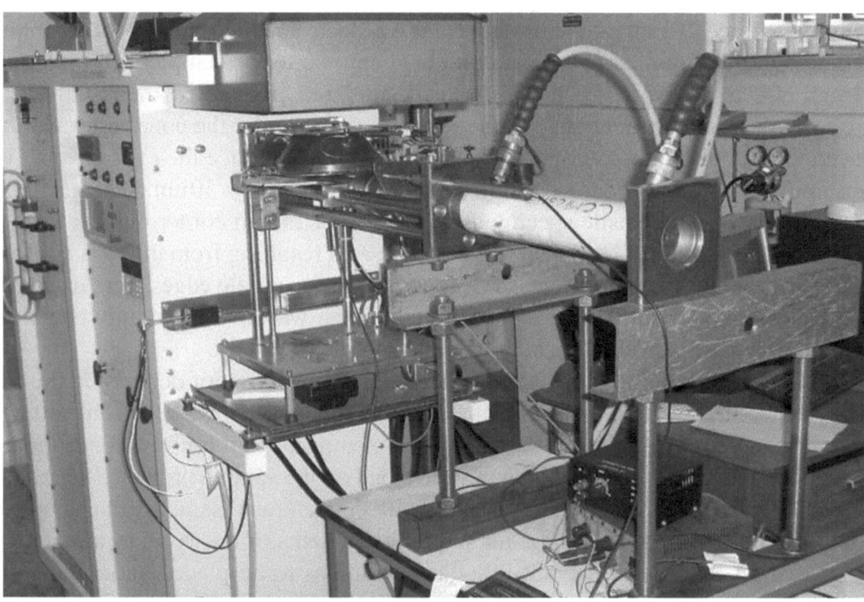

Fig. 4 Loading frame in place under the cone calorimeter heating element

800 °C, according to the Stefan-Boltzmann relationship. As with conventional cone calorimeter measurements a spark igniter was employed, to ensure the ignition of evolved decomposition products. The time elapsed to visible ignition of these products was recorded.

The samples were held under load until mechanical failure occurred, the time-to-failure being defined as the time at which the laminate could no longer support the externally applied load. Heat release rate and smoke production rate were measured simultaneously by the cone calorimeter. The peak values of heat release and smoke generation rate were reported. It is also common to report average values of these parameters, computed over a period, often of 300 s. In the present study, however, this was not possible, as the samples failed after different time periods depending on the stress level. Instead, values of the heat and smoke release rate immediately prior to failure were recorded. These will be referred to as the minimum heat release rate and the minimum smoke production rate.

3 Results and Discussion

3.1 Mechanical-Thermal Effect on Heat Release Rate and Time-to-Ignition

The full heat release rate profiles, up to sample failure, for glass vinyl ester and polyester are shown in Figs. 5 and 6, respectively, as functions of the applied tensile stress at constant heat flux. The main effect of increasing tensile stress was to shorten the time taken to failure, an effect that will be discussed later. It can also be seen that increasing the tensile stress has a small, but significant effect on the heat release rate: increasing it. In the case of compressive stress, shown in Figs. 7 and 8, the failure times were shorter and the heat release decreased slightly with increasing stress.

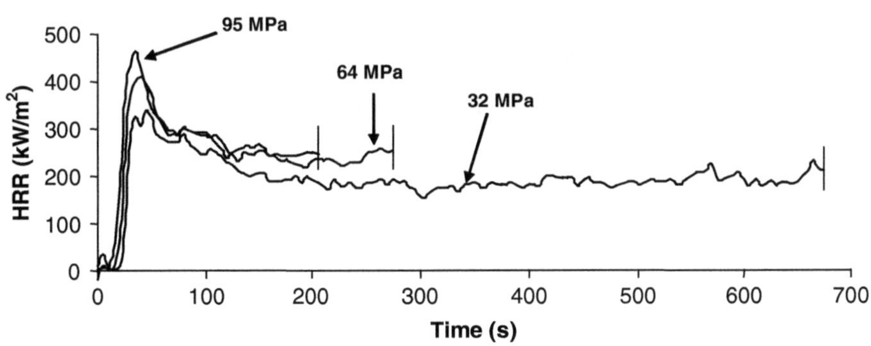

Fig. 5 Full heat release rate profiles for glass vinyl ester laminates under tensile stress at a constant heat flux of 75 kW m^{-2}

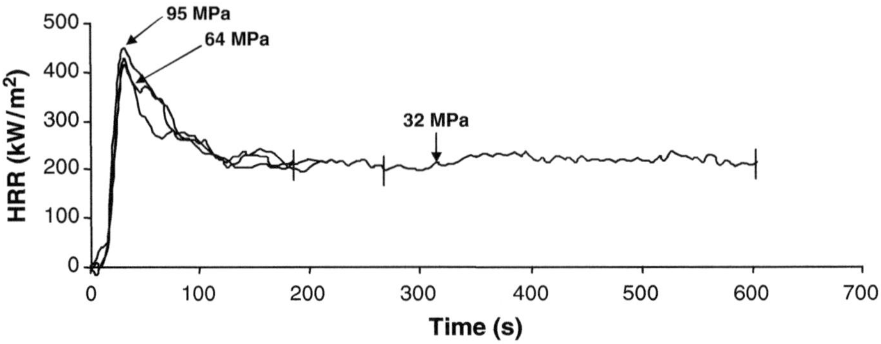

Fig. 6 Full heat release rate profiles for polyester laminates under tensile stress at a constant heat flux of 75 kW m^{-2}

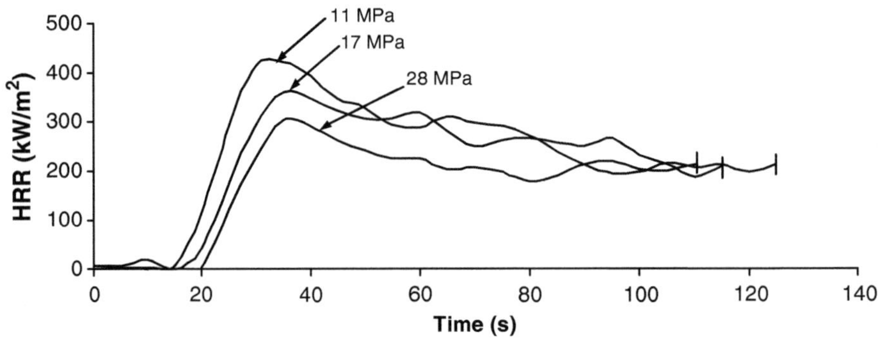

Fig. 7 Full heat release rate profiles for glass vinyl ester laminates under compressive stress at a constant heat flux of 75 kW m^{-2}

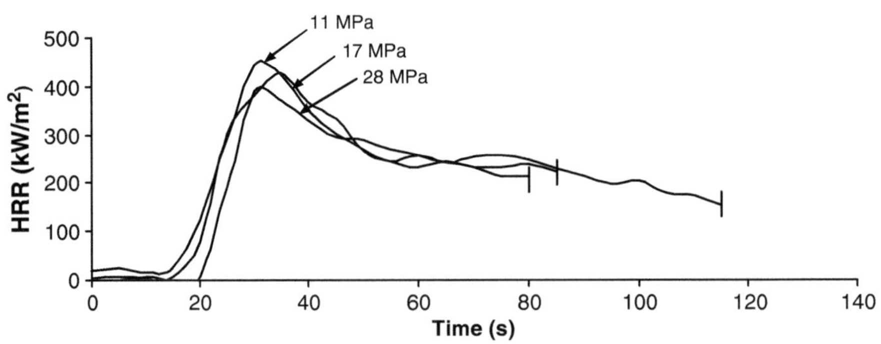

Fig. 8 Full heat release rate profiles for glass polyester laminates under compressive stress at a constant heat flux of 75 kW m^{-2}

All the heat release rate profiles showed similar characteristics: a short delay, corresponding to the observed time-to-ignition, a rapid increase up to the peak heat release rate, then a slow decline to a steady, almost constant 'minimum' release rate that lasted up to the point of sample failure.

It is interesting to note that there was little change in the heat release profiles at the point of sample failure, either in tension or compression. This may be because, when failure occurred, the peak heat release rate had already passed, with the heat release rate having falling to the lower almost steady value mentioned above. At this stage a significant proportion of the resin will have already been burnt out from the sample surface layers.

The heat release profiles and the size of the stress effect can be seen to be fairly similar for the vinyl ester and the polyester-based laminates. This is probably to be expected, as both resins are based on 'solvent monomer' systems, in which styrene is an important component. In the present case the heat release rate of the vinyl ester samples was slightly lower than that of the polyester, which is in contrast to previous work [1], where the reverse was reported. It is probable that the heat release of these resins depends to some extent on the particular grade of vinyl ester or polyester and, most especially, on the actual styrene content and chemical differences between the species present in the main chain backbone. As reported previously [1], the thermal decomposition characteristics of vinyl ester and polyester resins are similar, in that they both generally decompose almost completely into volatile products, leaving only a small percentage of char, often of the order of 4% of the original resin mass. This almost complete decomposition is again likely to be due to the resin styrene content. Such behaviour is in marked contrast to resins such as phenolic, which contain complex highly crosslinked aromatic species, that can produce a substantial yield of carbonaceous char on decomposition [1, 3, 8, 9, 13, 14].

The vinyl ester is a little more thermally stable than the polyester [1]. For instance, thermogravimetric analysis (TGA) at a heating rate of 25 °C min^{-1} in N_2 shows a maximum decomposition rate at 450 °C for vinyl ester, compared to 425 °C for polyester [1].

The details of the important initial heat release peaks are difficult to distinguish in Figs. 5–8, so the information was re-plotted over a shorter timescale of 120 s, in Figs. 9 and 10, for the two resin types, with the data for tensile and compressive stress combined. There were no tensile failures within 120 s, but some compressive samples did fail within this period. This has been indicated in Figs. 9 and 10 by placing a cross on the curves in question at the point of failure. Despite a certain level of scatter, which is fairly typical for this type of measurement, a clear pattern can be discerned, in which the release profiles decrease fairly progressively in magnitude going from the highest tensile stress, down to the lowest compressive one. This decrease is accompanied by an increase in the induction time for the onset of heat release, which is roughly the same as the observed time to ignition.

The physical processes involved when composites are exposed to heat have been discussed by Mouritz and co-workers [1, 16–18]. First the laminate surface is heated by the incoming heat flux. The surface temperature increases, resulting in decomposition of the resin, producing volatile, often flammable, by-products, which generate

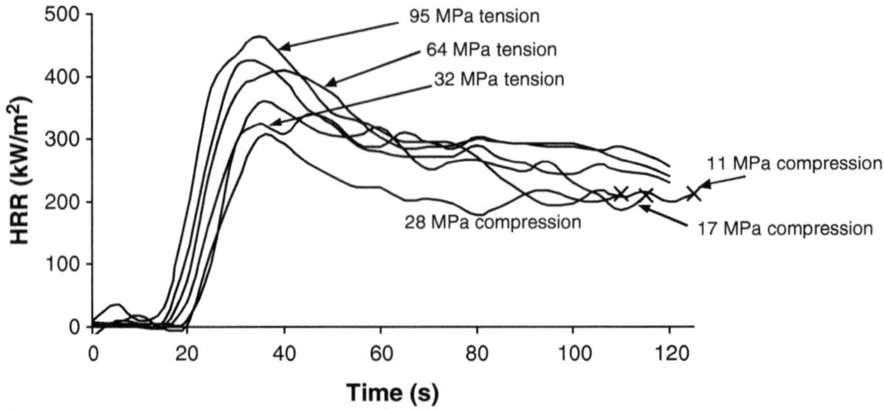

Fig. 9 Heat release rate profiles over 120 s for glass vinyl ester laminates under tensile and compressive stress, at a constant heat flux of $75\,kW\ m^{-2}$

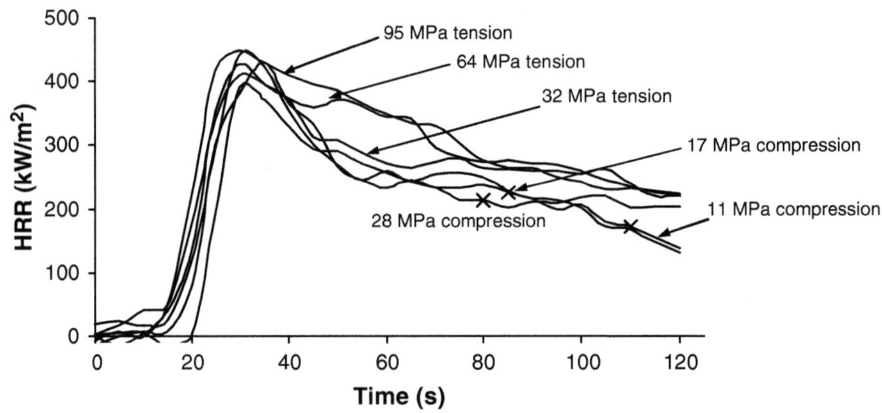

Fig. 10 Heat release rate profiles for glass polyester laminates under tensile and compressive stress, at a constant heat flux of $75\,kW\ m^{-2}$

a pressure within the resin phase [22,23]. This leads eventually to the formation of a network of cracks in the resin phase, permitting the easier escape of volatiles. When the laminate surface temperature reaches a particular value, depending on the heat flux [1,16–18], the volatile evolution rate is sufficient to allow ignition and sustained combustion to occur. This critical surface temperature for the onset of ignition has been estimated for vinyl ester resin to be about 600 °C for the present heat flux of $75\,kW\ m^{-2}$ [1,3,16–18].

The present results show the onset of ignition to be significantly affected by stress. The most probable mechanism is that tensile stress promotes the formation of resin cracks, shortening the time-to-ignition. Conversely, compressive stress can be expected to inhibit cracking, giving the opposite effect.

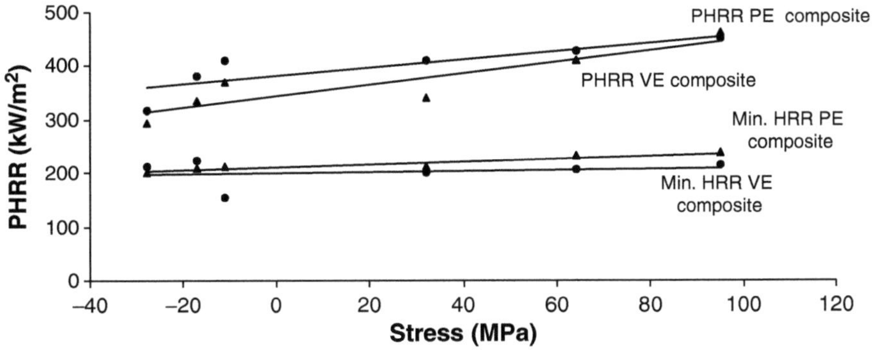

Fig. 11 Peak and minimum heat release rates versus stress for glass vinyl ester and glass polyester laminates at a heat flux of 75 kW m^{-2}

As mentioned, the laminates in this study are thermally 'thick': which means that there can be a significant temperature difference between the heat-exposed and rear surfaces. A 'front' of decomposing resin passes through the thickness of the laminate, generating flammable volatile products that migrate to the laminate surface, where they burn. The speed at which this decomposition front passes through the laminate is controlled to a certain extent by heat transfer, and the endothermic nature of the resin decomposition process. The heat release profiles demonstrate that an almost steady-state condition is achieved, with near constant release rate, which persists through to the point of mechanical failure.

Figure 11 shows the peak heat release rates and minimum release rates at failure as a function of stress. The stress effect is clear on the peak heat release rate curves, again emphasising the increased release rate for tensile stress and *vice versa* for compressive. The effect can be seen to be rather less for the minimum heat release at failure. This is in keeping with the hypothesis that stress predominantly affects matrix cracking and the initial heat release response. It is probably to be expected that the heat release rate during the later stages of exposure would be less significantly affected.

Figure 12 shows the variation of the time-to-ignition also with stress, which is progressive, decreasing with increasing tensile stress and increasing with compressive stress. Both resin types behave similarly, with the vinyl ester showing slightly longer times-to-ignition.

3.2 Mechanical-Thermal Effect on Smoke Evolution

Figures 13 and 14 show the effect of stress on the smoke generation profiles over a 120 s period. These are rather similar in shape to the heat release profiles already discussed, albeit with greater superimposed fluctuations. The vinyl ester smoke generation is somewhat greater than that for the polyester laminates. As with heat

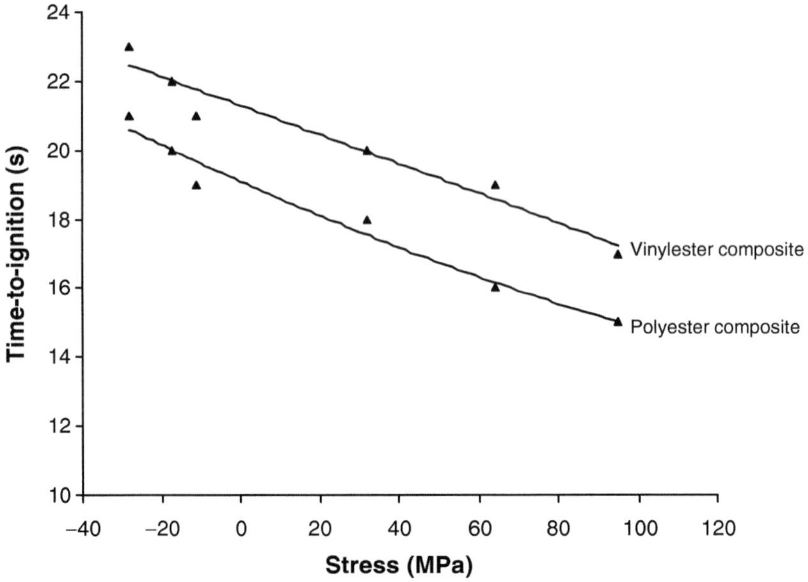

Fig. 12 Time-to-ignition versus stress for glass vinyl ester and glass polyester laminates, at 75 kW m^{-2}

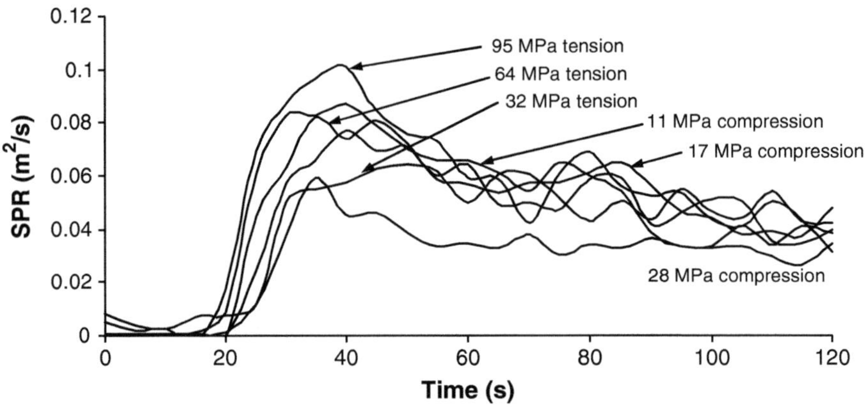

Fig. 13 Smoke production rate profiles for glass vinyl ester laminates under tensile and compression stress, at 75 kW m^{-2}

release, a significant effect of stress can be observed. Figure 15 shows the peak smoke generation rate and the minimum smoke generation rate at sample failure. As with heat release the stress effect can be seen, the curves being rather similar to those in Fig. 11 for peak heat release rate. The close relationship between heat release and smoke generation for composite systems of this type has been discussed previously [3]. However, the final value of smoke production rate at failure,

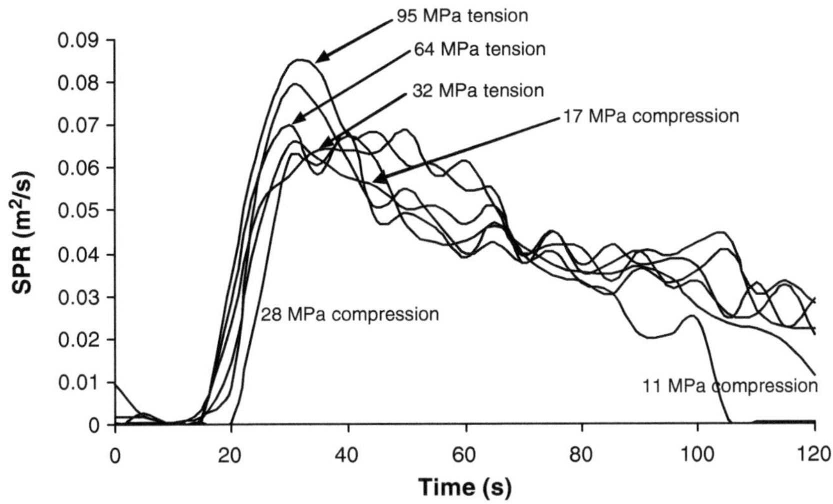

Fig. 14 Smoke production rate profiles for glass polyester laminates under tensile and compression stress, at 75 kW m^{-2}

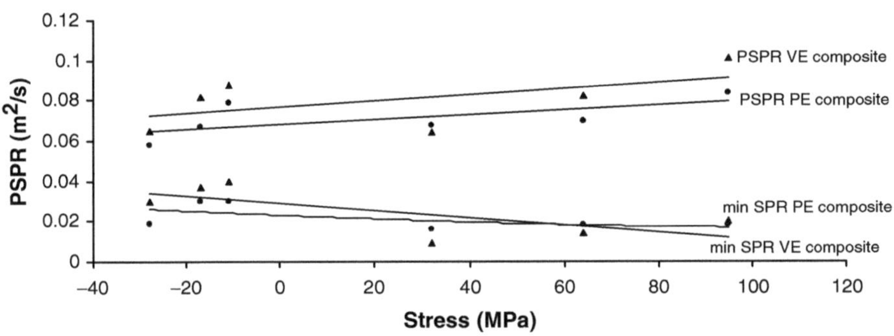

Fig. 15 Peak and minimum smoke production rate versus stress, for glass vinyl ester and glass polyester laminates at 75 kW m^{-2}

which shows quite a lot of scatter, actually declines a little with increasing stress. The probable reason may be that much of the smoke is evolved during the initial peak, so increasing this peak will result in a decline smoke during the later stages of exposure.

3.3 Time-to-Failure

Figures 16 and 17 show 'stress rupture' plots of under tensile and compressive stresses, respectively. These relate the time-to-failure, shown on the horizontal axis, to the applied stress. It can be seen that in both tension and compression the failure

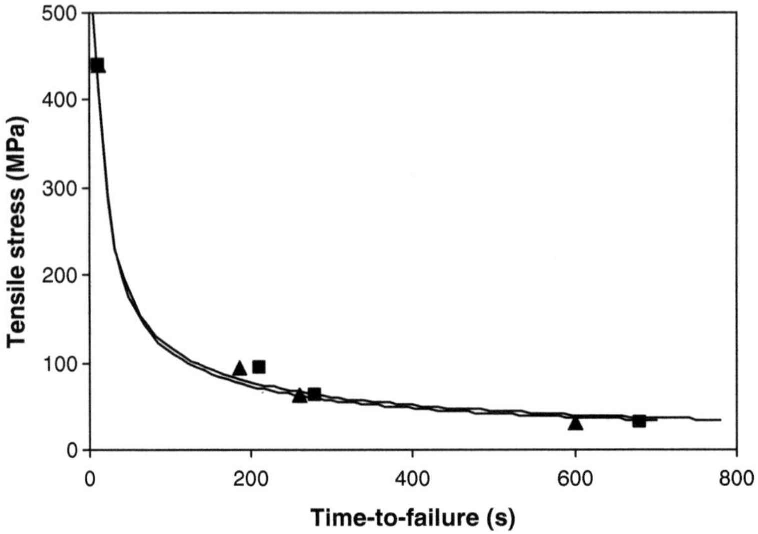

Fig. 16 Tensile stress rupture curves (stress vs time-to-failure) for glass vinyl ester laminates (squares) and glass polyester laminates (triangles) at a heat flux of $75\,kW\,m^{-2}$

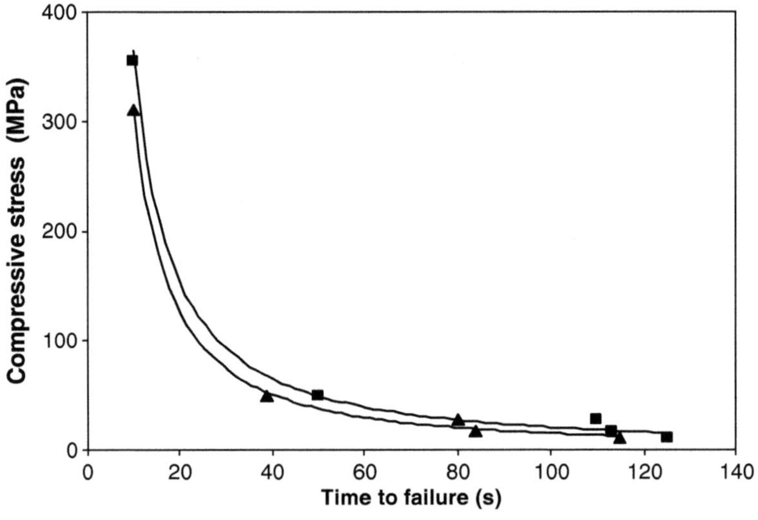

Fig. 17 Compressive stress rupture curves (stress vs time-to-failure) for glass vinyl ester laminates (squares) and glass polyester laminates (triangles) at a heat flux of $75\,kW\,m^{-2}$

time decreases when the applied stress increases. The failure times under tensile stress, however, are almost an order of magnitude larger than those for compressive stress. This reflects the fact that tensile failure is largely determined by the strength of the glass fibres, whereas compressive failure is resin dominated. The tension failure data for the two resin systems are rather close to one another, the vinyl ester system performing only slightly better than the polyester, again reflecting the fibre dominated nature of the failure process. The curves in this figure were fitted on the assumption of a power law relationship between tensile stress and time-to-failure. A point was added corresponding to the undamaged tensile strength, 440 MPa, which was assumed to correspond to an arbitrary failure time-to-failure of 10 s.

In the tensile case it is possible, over about 10 min (600 s) to support composite stresses of the order of 30 MPa. Neglecting any contribution from the matrix, this corresponds to fibre stresses in excess of 130 MPa in those plies that are under stress. While significant, this is quite low, compared to the room temperature strength of commercial glass fibres, which is about 1,800 MPa. It has been reported, however, that the strength of glass fibres declines significantly and irreversibly at high temperatures, even in the absence of stress [17], probably due to a corrosion process. The 130 MPa stress value at 600 s corresponds fairly closely to the diminished strength value at elevated temperature predicted by the model reported in [17, 18]. Figure 18 shows a sample after failure in tension. The black appearance indicates a significant amount of residual decomposed resin. Fibre failure was distributed throughout the gauge length of the sample, with multiple points of fracture. The specimen did not separate into two clear parts, because pumping to the hydraulic cylinder was terminated when the load fell to zero.

Figure 19 shows one of the samples that failed in compression. The compressive failure mode involved localised kinking of the plies, followed by more extensive crumpling, accompanied by delamination of some of the hot surface plies. The kinking effect is similar to that described by Budiansky and Fleck [29] for isothermal compressive failure of unidirectional samples. This model assumes that some local region of material, in which the fibres are imperfectly oriented, initiates a shear-dominated local collapse. Despite the greater complexity of the present case, which

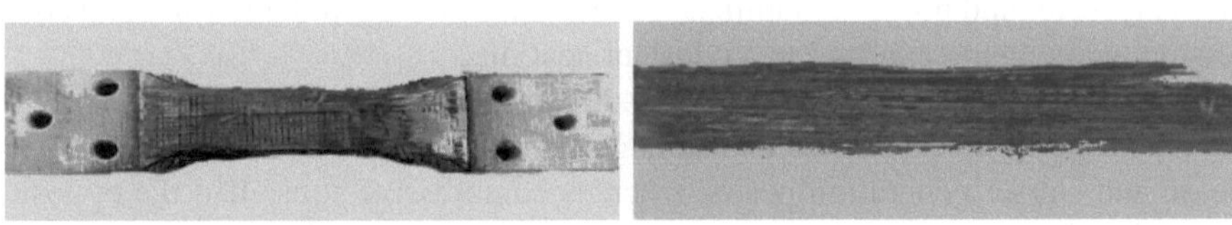

Fig. 18 Vinyl ester laminate failed under a tensile stress of 64 MPa and a heat flux of 75 kW m^{-2}. *Left*: General view of specimen. *Right*: Side view

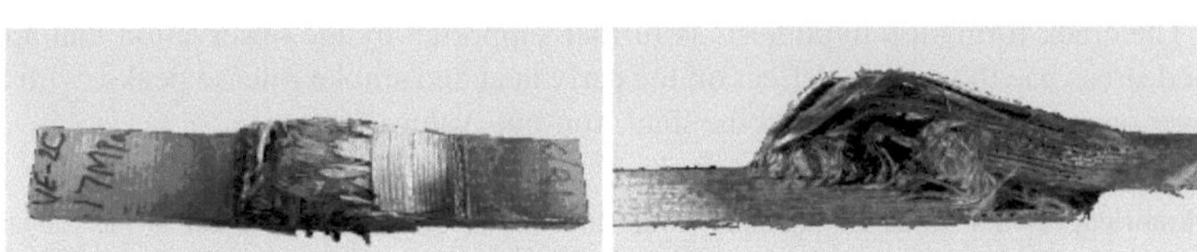

Fig. 19 Vinyl ester laminate failed under a compressive stress of 17 MPa and a heat flux of 75 kW m^{-2}. *Left*: General view of specimen. *Right*: Side view (hot side uppermost) showing sheared material near to the cold face and delaminated plies at the hot face

is non-isothermal and which does not involve a unidirectional composite, the mechanism is clearly rather similar to this. The resin shear-dominated nature of the failure accounts for why it occurs at such a short time, or low average laminate temperature. Again a power law relationship was assumed when plotting the curves. Undamaged compressive strength values were 355 and 310 MPa for vinyl ester and polyester, respectively. Again these were assumed to correspond to a time-to-failure of 10 s.

The longer-term stress that could be supported, at less than 10 MPa, is certainly lower than that observed in the tension case. It reflects the substantial softening of the resin at its glass transition temperature. The most pronounced kinking of the sample is near to the rear (cold) face. Due to the non-isothermal regime of the cone calorimeter test this cool region will concentrate most of the compressive stress, due to the fall in Young's modulus of the layers nearer to the laminate hot surface. Compressive failure probably initiates in this region.

The times to failure in tension and compression are somewhat shorter than those reported previously for similar laminates [1, 8, 10, 11]. The reason is probably the smaller sample size used here and the greater edge area. Although the edge region was well-insulated during the tests it is known that there is an 'edge' effect, due mainly to the greater ease with which the volatiles can escape from this region. The shorter times to failure in the present work do not affect the main conclusions with regard to the effect of stress. Indeed it is possible the specimen edges reduced the apparent stress effect.

4 Conclusions

A loading frame fitted to the cone calorimeter allowed 100×30 mm rectangular glass vinyl ester and polyester laminate samples to be tested under tension and compression, while being exposed to a constant heat flux.

It was found that there was a small but significant effect of stress on the fire reaction properties, heat release and smoke generation. Tensile stress increased heat release and smoke generation peaks, whereas compressive stress had the opposite effect. Tensile stress also shortened the time-to-ignition, with compressive stress having the opposite effect. The effect was attributed to an increased tendency for cracks to form when the laminate was subjected to tension, the opposite being true for compression.

The crack formation hypothesis is further supported by the observation that applied stress has the greatest effect on the early heat and smoke release peaks, with a lower or even negative effect on the final 'run-out' values.

Acknowledgements Awad Elmughrabi, would like to acknowledge the support of the Libyan Embassy during this study, which forms part of his Ph.D research at Newcastle University. Geoff Gibson would like to acknowledge the support of the US Office of Naval Research, NICOP program, as well as the collaboration and advice of Professor A.P. Mouritz and his research group at RMIT Melbourne, Australia.

References

1. Mouritz AP, Gibson AG. Fire Properties of Polymer Matrix Composites. Springer, Dordrecht, 2006.
2. Lyon RE, Demario J, Walters RN, Crowley S. Flammability of glass fibre-reinforced polymer composites. In: Proceedings of the 4th International Conference on Composites in Fire, Newcastle upon Tyne, England, 2005.
3. Mouritz AP, Mathys Z, Gibson AG. Heat release of polymer composites in fire. Composites, Part A 2006; 37:1040–1054.
4. Koo JH, Muskop F, Venumbaka R, Van Dine R, Spencer B, Sorathia U. Flammability properties of polymer composites for marine applications. In: Proceedings of the 32nd International SAMPE Technical Conference, 5–9 November 2000.
5. Brown JR, Mathys Z, Reinforcement and matrix effects on the combustion properties of glass reinforced polymer composites. Composites, Part A 1997; 28(7):675–681.
6. Beeson HD, Hshieh F-Y. Flammability testing of flame-retarded epoxy composites and phenolic composites. Fire and Materials 1997; 21(1):41–49.
7. Lyon RE. Plastics and rubber. In: C.A. Harper, editor. Chapter 3 in Handbook of Building Materials for Fire Protection.McGraw-Hill, New York, 2004.
8. Mouritz AP, Mathys Z. Post-fire mechanical properties of marine polymer composites. Composite Structures 1999; 47: 643–653.
9. Mouritz AP, Mathys Z. Post-fire mechanical properties of glass-reinforced polyester composites. Composites Science and Technology 2001; 61(4):475–490.
10. Gibson AG, Wright PNH, Wu Y-S, Mouritz AP, Mathys Z, Gardiner CP. Modelling residual mechanical properties of polymer composites after fire. Plastics, Rubber and Composites 2003; 32(2):81–90.
11. Gibson AG, Wright PNH, Wu Y-S, Mouritz AP, Mathys Z, Gardiner CP. The Integrity of polymer composites during and after fire. Journal of Composite Materials 2004; 38(15):1283–1307.
12. Bausano J, Boyd S, Lesko J, Case S. Composite lifetime during combined compressive loading and one-sided simulated fire exposure. Composites, Part A: 2005; 37(7):1092–1100.
13. Easby RC, Feih S, Konstantis C, La Delfa G, Urso Miano V, Elmughrabi A, Mouritz AP, Gibson AG. Failure model for phenolic and polyester pultrusions under load in fire. Plastics, Rubber and Composites 2007; 36(9):379–388.
14. Feih S, Mathys Z, Mathys G, Gibson AG, Robinson M, Mouritz AP. Influence of water content on failure of phenolic composites in fire. Polymer Degradation and Stability 2008; 93(2):376–382.
15. Gibson AG, Wright PNH, Wu Y-S, Evans JT. Laminate theory analysis of composites under load in fire. Journal of Composite Materials 2006; 40(7):639–658.
16. Feih S, Mathys Z, Gibson AG, Mouritz AP. Modelling the compression strength of polymer laminates in fire. Composites, Part A 2007; 38(11):2354–2365.
17. Feih S, Mouritz AP, Mathys Z, Gibson AG. Tensile strength modelling of glass fiber-polymer composites in fire. Journal of Composite Materials 2007; 41(19):2387–2410.
18. Feih S, Mathys Z, Gibson AG, Mouritz AP. Modelling the tension and compression strengths of polymer laminates in fire. Composites Science and Technology 2007; 67(3–4):551–564.
19. Babrauskas V. Development of the Cone Calorimeter: A Bench Scale Heat Release Rate Apparatus Based on Oxygen Consumption (NBSIR 82-2611), USA. National Bureau of Standards, 1982.
20. Babrauskas V, Peacock RD. Heat release rate: the single most important variable in fire hazard. Fire Safety Journal 1992; 18:255–272.
21. Huggett C Estimation of rate of heat release by means of oxygen consumption measurements. Fire and Materials 1980; 4:61–65.
22. Henderson JB, Wiecek TE. A mathematical model to predict the thermal response of decomposing expanding polymer composites. Journal of Composite Materials 1987; 21:373–393.
23. Henderson JB, Wiebelt JA, Tant MR. A model for the thermal response of polymer composite materials with experimental verification. Journal of Composite Materials 1995; 19:579–594.

24. Gibson AG, Wu Y-S, Chandler HW, Wilcox JAD, Bettess P. A model for the thermal performance of thick composite laminates in hydrocarbon fires. Revue de l'Institut Français du Pétrole 1995; 50(1):69–74.
25. Dodds N, Gibson AG, Dewhurst D, Davies JM. Fire behaviour of composite laminates. Composites, Part A 2000; 31(7):689–702.
26. Lua J, O'Brien J, Key C, Wu Y-S, Lattimer B. A temperature and mass dependent thermal model for fire response prediction of marine composites. Composites, Part A 2006; 37(7):1024–1039.
27. Liu L, Kardomateas GA, Simitses GJ, Li R. Response of a sandwich panel subject to fire or elevated temperature on one of the surfaces. Composites, Part A 2006; 37(7):981–988.
28. Liu L, Kardomateas GA, Birman V, Holmes JW, Simitses GJ. Thermal buckling/bending of a heat-exposed, axially restrained composite column. Composites, Part A 2006; 37(7):981–988.
29. Budiansky B, Fleck N. Compressive failure of fiber composites, Journal of the Mechanics and Physics of Solids, 1993; 41:183–211.

Investigation of Failure and Failure Progression in Stiffened Composite Structures

Hans-G. Reimerdes and Sebastian Büsing

Abstract The reliable simulation of the mechanical behaviour of structures made of composite materials has become more and more important. For the prediction of the behaviour up to collapse, initiation and propagation of failure have to be taken into account. In stiffened laminated fibre-reinforced composite structures, a common failure type is delamination. As it is critical for the load-carrying capacity of the structure, especially delaminations between stringer foot and skin in thin-walled stiffened structures have to be taken into account.

In this paper, the results of Pull-off tests on specimens representing a strip of a stiffened composite panel are presented. They were carried out in order to investigate failure load and failure mode of skin-stiffener separation. Furthermore a tool is presented, that uses an element consisting of two shells connected by an adhesive layer. A degradation model is implemented into the adhesive layer that accounts for failure by introducing a damage parameter, that reduces the stiffness of the adhesive layer gradually up to complete decohesion. This tool is applied to simulate the Pull-off tests, and the results are compared to those of the experiments.

Keywords Composite materials · Adhesive layer · Linear softening · Delaminations · Pull-off test

1 Introduction

Thin-walled stiffened composite structures are commonly used in aerospace structures. Due to the complex properties of laminated fibre-reinforced composites, the reliable simulation of the mechanical behaviour has become more and more important, in order to reduce costs during design process by minimizing the testing effort [1]. For the prediction of the mechanical behaviour up to collapse, it is inevitable to take

H.-G. Reimerdes (✉) and S. Büsing
Institut für Leichtbau, RWTH Aachen University, Germany
e-mail: hg_reim@ilb.rwth-aachen.de

S. Pantelakis, C. Rodopoulos (eds.), *Engineering Against Fracture: Proceedings of the 1st Conference*,

failure and failure progression into account. In laminated fibre-reinforced composites, a common failure type is delamination, due to the low strength perpendicular to the layers of the laminate.

To investigate failure load and failure mode of the skin-stiffener connection, Pull-off tests on specimens representing a strip of a stiffened composite panel were carried out. Two sets of specimens were investigated, that differ in material, layup and stiffener type: a symmetric T-shaped stiffener and an L-shaped stiffener. Both sets of specimens were tested with two different span lengths.

The focus of these tests is to investigate the strength of the skin stringer connection depending on the configuration, and to provide data that allow to demonstrate the capability of numerical simulations.

2 Analysis Tool

To determine the stresses in the connection between two composite shells in general, or between the skin and the stiffener as a special case, an element consisting of two shell layers connected by an adhesive layer has been developed. In connection with a failure model for the interface layer, the onset of delamination and complete failure of this interface can be computed. In a first step, a two dimensional model was developed, where the shells are reduced to beams.

2.1 Theoretical Model

The formulation of the element is based on a three-layer model (see Fig. 1) as it is used for the investigation of adhesive bondings [2], or of sandwich structures [3]. For the intermediate layer the in-plane stiffnesses are neglected, thus only out of plane normal and shear stresses are computed in this layer. The connection between the shells (beams) and the adhesive layer is assumed to be perfect.

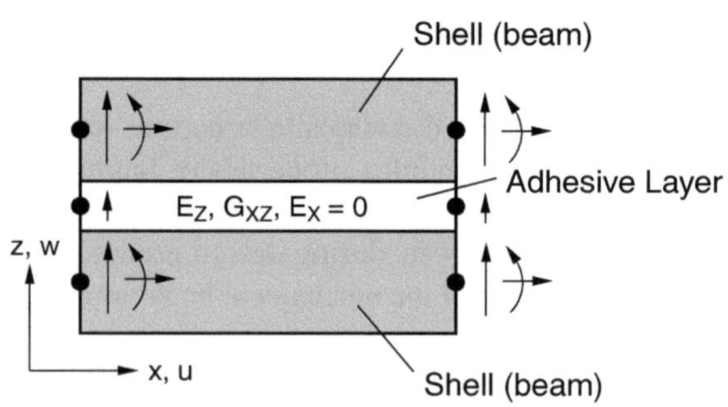

Fig. 1 Element with adhesive layer

2.2 Element Description

A finite element is derived based on the principle of virtual displacements [4]:

$$\delta U - \delta W = \int \vec{\sigma}^T \delta \vec{\varepsilon}_l dV = \delta \vec{u}^T \vec{F} \tag{1}$$

The integral in Eq. (1) is solved seperately for the individual components (shells and adhesive layer) [5]:

$$\int \vec{\sigma}^T \delta \vec{\varepsilon}_l dV = \delta \vec{u}^T \left(\sum_{i=1}^{2} K_{f_i} + K_c \right) \vec{u} \tag{2}$$

$\boldsymbol{K}_{fi}$ is the stiffness matrix of the shell (beam) fi and $\boldsymbol{K}_c$ the stiffness matrix of the adhesive layer. For the shells (beams), geometrical nonlinearity is considered in the v. Kármán sense.

2.3 Failure Model

The linear softening model [6, 7] is used to describe the degradation of the adhesive layer. The initial stiffness of the adhesive layer holds the shells together in the linear-elastic range. Exceeding the effective strain at delamination onset, the initial stiffness is gradually reduced to zero, i.e. complete decohesion.

The damage behaviour is defined by a triangular stress-displacement curve, as shown in Fig. 2. It is described by the effective strain and the respective stress at onset of degradation ε_m^0 and σ^0, and the strain at decohesion ε_m^f. The area under this stress-displacement curve is the respective fracture energy.

The load case is defined by the normal and shear strain ε_n and γ, the Mixed Mode ratio β and the mean strain ε_m:

$$\beta = \frac{\gamma}{\varepsilon_n} \tag{3}$$

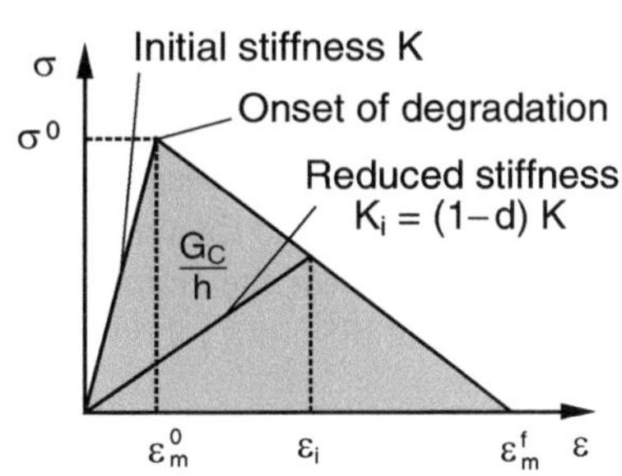

Fig. 2 Linear softening model

$$\varepsilon_m = \begin{cases} \sqrt{\varepsilon_n^2 + \gamma^2} & \varepsilon_n > 0 \\ \gamma & \varepsilon_n \leq 0 \end{cases} \tag{4}$$

The effective strains at delamination onset and at decohesion are defined by the ultimate strength σ_n^0 and τ^0, the respective strain, and the Mixed Mode ratio:

$$\varepsilon_n^0 = \frac{\sigma_n^0}{E} \qquad \gamma^0 = \frac{\tau^0}{G} \tag{5}$$

$$\varepsilon_m^0 = \begin{cases} \varepsilon_n^0 \gamma^0 \sqrt{\dfrac{1+\beta^2}{(\gamma^0)^2 + (\beta \varepsilon_n^0)^2}} & \varepsilon_n > 0 \\ \gamma^0 & \varepsilon_n \leq 0 \end{cases} \tag{6}$$

$$\varepsilon_m^f = \begin{cases} \dfrac{2G_C}{E\varepsilon_m^0 h} & \varepsilon_n > 0 \\ \dfrac{2G_{IIC}}{G_\tau{}^0 h} & \varepsilon_n \leq 0 \end{cases} \tag{7}$$

To determine the total fracture toughness G_{tC} for the present mode ratio, the criterion proposed by Benzeggagh and Kenane (BK-Criterion) [8] is used:

$$G_{tC} = G_{IC} + (G_{IIC} - G_{IC}) \left(\frac{\beta^2}{1+\beta^2} \right)^{\alpha} \tag{8}$$

The damage parameter d is defined with the effective strains at delamination onset, at decohesion and the maximum mean strain ε_m^{max} from the present and the previous load cases:

$$d = \frac{\varepsilon_m^f \left(\varepsilon_m^{max} - \varepsilon_m^0\right)}{\varepsilon_m^{max} \left(\varepsilon_m^f - \varepsilon_m^0\right)} \qquad 0 < d < 1 \tag{9}$$

Between effective strain at delamination onset and at decohesion, both Young's modulus and shear modulus of the adhesive layer are reduced by $(1-d)$. If the adhesive layer is loaded in compression ($\varepsilon_n \leq 0$), the initial stiffness is reapplied to avoid interpenetration of the shells.

The properties required for the linear softening model are the initial values of Young's modulus and shear modulus of the adhesive layer, the Mode I and Mode II fracture energies and the corresponding tensile and shear strengths. The exponent α that is required for the Benzeggagh–Kenane criterion can be obtained from Mixed Mode tests at different mode ratios.

3 Experimental Investigation

Tests were performed to determine the interlaminar fracture toughness, which is the most important material property related to delamination. Furthermore, pull-off tests were performed to investigate initiation and progression of delaminations between stiffener and skin.

3.1 *Interlaminar Fracture Toughness Tests*

The interlaminar fracture toughness is expressed in terms of the critical energy release rate G_C. This is the energy consumed by the material as the delamination front advances through a unit area. Tests were performed to determine the interlaminar fracture toughness for Mode I (opening), Mode II (shear) and combination of these modes.

The following tests were performed:

- **Mode I:** Double Cantilever Beam (DCB) test defined by ASTM Standard D6671 [9]
- **Mode II:** End Notched Flexure (ENF) test defined by Standard DIN EN 6034 [10]
- **Mixed Mode:** Mixed Mode Bending (MMB) tests defined by ASTM Standard D5528 [11]

During the tests, load and crosshead displacement were recorded. From this data, the critical energy release rate G_C was calculated, the mean values are listed in Table 1. To determine the critical energy release rate for Mixed Mode ratios in the failure model, the BK-Criterion (see Eq. 8) is used.

3.2 *Pull-Off Tests*

Experiments were performed in order to investigate skin-stiffener separation. The type of test was a Pull-off test, for which strips were cut out of stiffened plates.

Table 1 Interlaminar fracture toughness of 950-GF3-5H-1000

Pure mode I/mode II		Mixed mode	
G_I	$139\,\mathrm{J/m^2}$	$G_{28.0\%}$	$215\,\mathrm{J/m^2}$
G_{II}	$695\,\mathrm{J/m^2}$	$G_{52.0\%}$	$334\,\mathrm{J/m^2}$
		$G_{74.5\%}$	$398\,\mathrm{J/m^2}$
Exponent for BK-Criterion		$\alpha = 1.913$	

Fig. 3 Test setup with loaded specimen

The skin of those strip specimens was pinned down on the base plate by one bar on each side, which is adhesivly bonded to the specimen and bolted to the base plate. The web of the stiffener was clamped in the crosshead of the testing machine (see Fig. 3). It was loaded controlled in displacement until complete separation of stiffener flange and skin. Two sets of specimens were tested, that had a different layup and a different stiffener type. Furthermore, specimens of both type were tested at two different span lengths between the clamps fixing the skin: 100 and 150 mm.

During the tests, the deformation of the specimens was measured using an ARAMIS measuring system. Furthermore, crosshead displacement and load were recorded.

3.2.1 Specimen Geometry

The first type was a symmetrical T-shaped stiffener, that was co-cured with the skin. Both skin and stiffener consist of four plies each, one additional ply is placed between skin and stiffener. The geometry of the specimens is given in Fig. 4, the material is the woven fabric prepreg 950-GF3-5H-1000.

The second type was an L-shaped stiffener, that was attached to the skin by secondary bonding (Fig. 5). The skin consists of woven fabric prepreg GR/EP 199-45-005 FB, the stiffener is made of woven fabric and unidirectional prepreg (GR/EP 199-45-005 FB and GR/EP 199-45-002 TP).

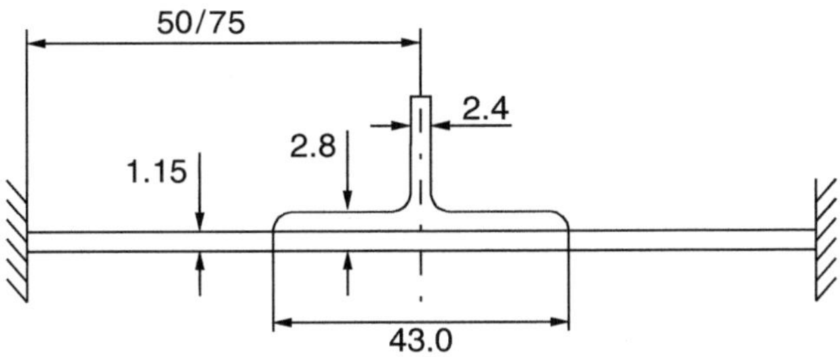

Fig. 4 Symmetrical specimen geometry and boundary conditions

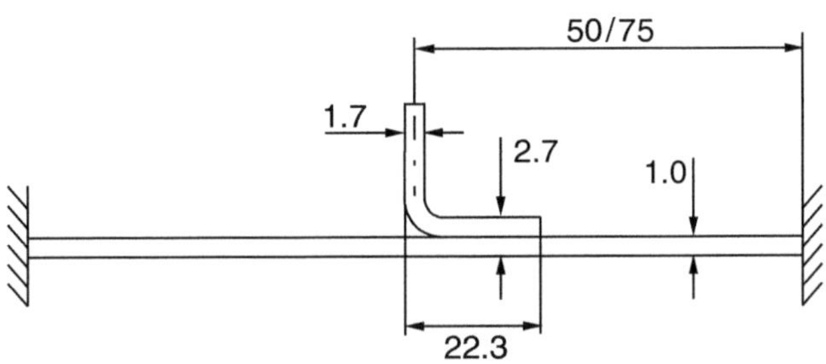

Fig. 5 Asymmetrical specimen geometry and boundary conditions

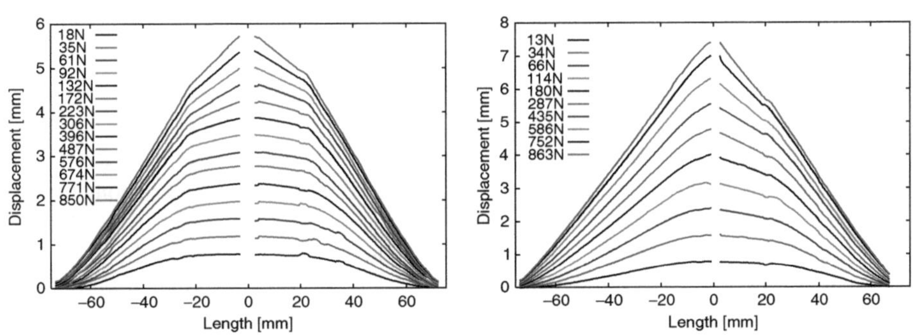

Fig. 6 Deformation of symmetrical and asymmetrical specimens at different load steps (span width 150 mm)

3.2.2 Results

The scope of the experiments was the determination of the curve of load versus load point displacement and the displacement of skin and stiffener flange. The deformation of the specimens at different load steps is shown exemplarily for one symmetrical and one asymmetrical specimen in Fig. 6. Kinks in the displacement curves of the specimens near ± 20 mm result from inaccuracies of the optical measurement at the change in thickness at the end of the stiffener flange.

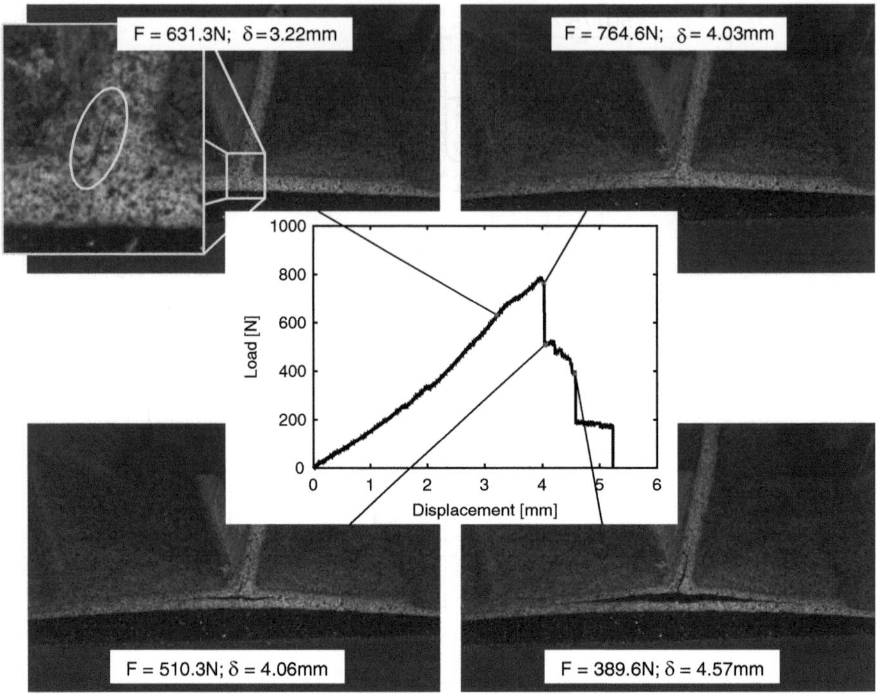

Fig. 7 Crack initiation and propagation in specimen T13

In the symmetric, T-shaped specimens, first failure was delamination in the stiffener web, while a small decrease in the slope of load-displacement curve could be noticed. The crack leading to total failure between skin and stiffener initiated below the web, between the additional 0° ply and the stiffener flange, and propagated to the free ends of both flanges (see Fig. 7).

As for the symmetric specimens, a crack initiated below the web of the L-shaped stiffener and propagated to the free end of the flange, leading to sudden total failure. Additionally, delaminations occurred in the radius between stiffener flange and web. All specimens of each type showed the same failures typical for the respective type.

The measured load-displacement curves for all tested samples are presented in the next section.

4 Numerical Simulation

4.1 Mode I, Mode II and Mixed Mode Tests

In order to validate the element and the degradation model, DCB, ENF and Mixed Mode tests were simulated. The specimens were modelled with 300 elements, the material properties used for the interface layer are listed in Table 2. The initial

Table 2 Properties of the interface layer used for simulations

	Interface layer
Thickness h	0.05 mm
E_z	$3000\,\text{N/mm}^2$
G_{xz}	$1200\,\text{N/mm}^2$
σ_n^0	$45\,\text{N/mm}^2$
τ^0	$80\,\text{N/mm}^2$
G_{IC}	$139\,\text{J/m}^2$
G_{IIC}	$695\,\text{J/m}^2$
α	1.913

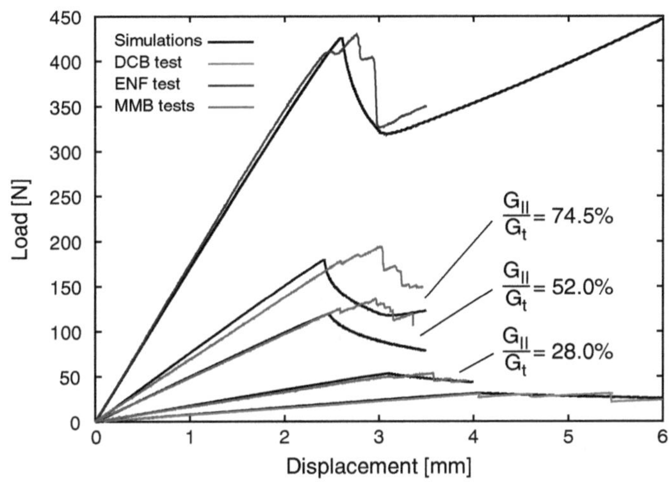

Fig. 8 Comparison of tests (Mode I, Mode II and Mixed Mode) and simulations

delamination of the specimens was taken into account by assuming total decohesion ($d = 1.0$) of the interface layer in the respective elements.

The comparison of the curves of load vs. displacement of simulations and tests are shown in Fig. 8. For pure Mode I and Mode II loading as well as for the Mixed Mode ratio of 28.0% the simulations show a very good correlation to the experiments. For the Mixed Mode ratios of 52.0% and 74.5% the maximum load of the simulations is lower than those of the tests, while the load at onset of degradation is on the same level.

4.2 Pull-Off Tests

For the simulation of the pull-off tests, simple models were established. The web of the stringer profile was not idealised (see Figs. 9 and 10). The deformation was applied to the flange as shown in Figs. 9 and 10. An adhesive layer between stringer

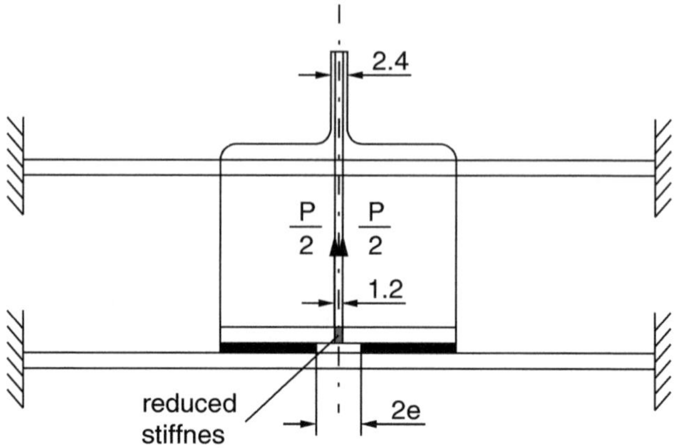

Fig. 9 Model of the symmetrical specimen

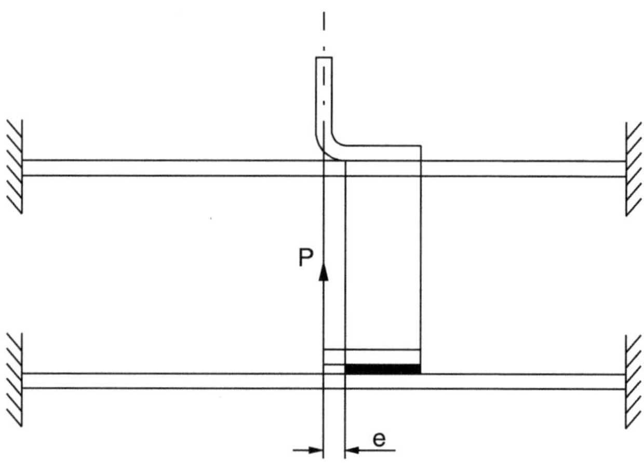

Fig. 10 Model of the asymmetrical specimen

and skin was not idealised within the Bermuda-triangle below the web of the stiffener. In case of the T-stiffener the stiffness of the flange was reduced between the load introduction points, in order to idealise the low stiffness of the web normal to its mid plane and of the Bermuda-triangle. In order to idealise the stiffness of the test fixture, rotation of the load introduction points in the flange was constrained. In case of the L-stiffener, the horizontal deformation of the load introduction point was also set to zero. For both types of specimen the properties given in Table 2 are used for the interface layer.

For each of the four investigated configurations two types of simulation were performed: (a) no degradation within the interface layer and (b) degradation within the interface layer. As expected, the computed vertical deformations of the load introduction points are smaller than in case of the experiments, which is due to the idealisation of clamped boundary conditions which could not be realised in the experiments.

4.3 T-Stiffener

The results for the T-Stiffener are presented in Fig. 11. The experiments show above about 300 N degradation of the stiffness resulting from first damages in the adhesive layer. The failure load is independent on the length of the specimens at about 750–800 N. The results from the numerical simulations show the degradation of the interface at the same load level observed in the experiments, while the predicted failure load is higher than that achieved in the experiments. It is expected, that the simulation results will improve, if the model includes the stringer web as well. This will be the next development step for the simulation tool. Nevertheless the main result is the prediction of the load leading to the first interface degradation, which was already predicted by this very simple model.

4.4 L-Stiffener

The results for the L-Stiffener are presented in Fig. 12. The experiments show nearly no degradation of the stiffness up to the failure load and sudden failure. The results from the numerical simulations show also no degradation of the interface and sudden failure. The predited failure load is close to that observed in the experiments. The simulation results will aigain improve, if the model includes the stringer web as well.

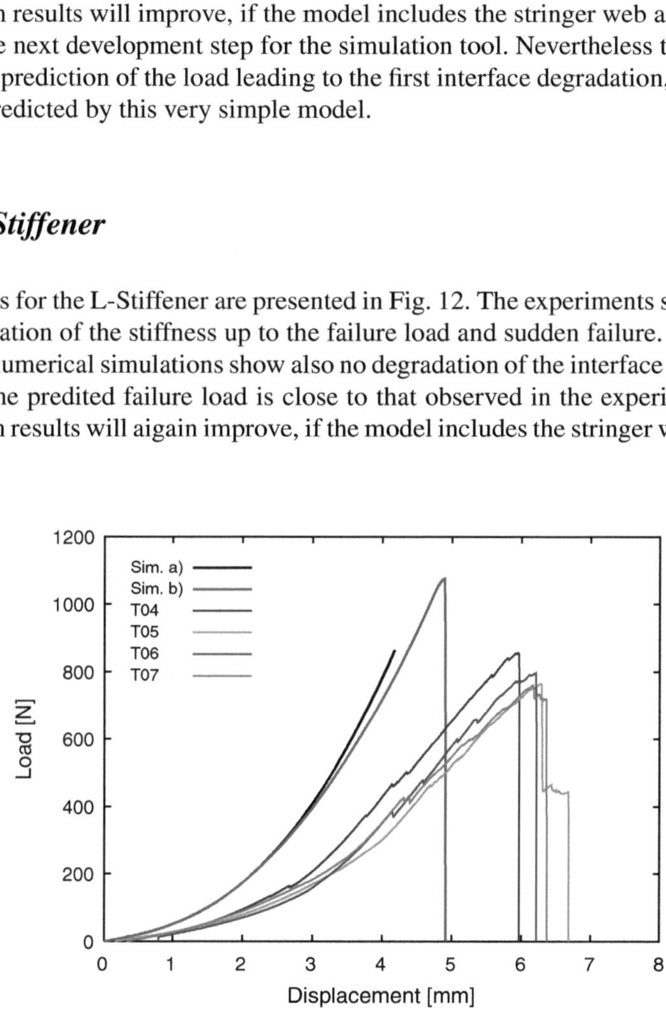

Fig. 11 Comparison of load-displacement curves of experiments and simulations (symmetrical specimen, span width 150 mm)

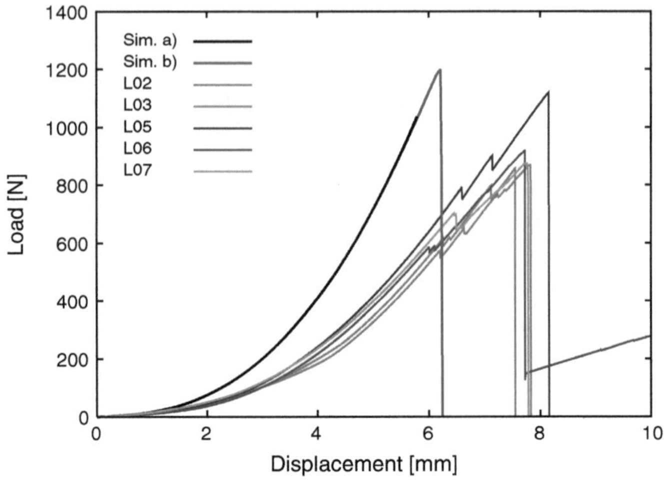

Fig. 12 Comparison of load-displacement curves of experiments and simulations (asymmetrical specimen, span width 150 mm)

5 Conclusions

The results of DCB, ENF and Mixed Mode tests as well as of Pull-off tests are presented and compared to numerical simulations. These numerical simulations were performed using a fast tool, based on the idealisation of the connection between skin and stringer by an adhesive layer. For this layer a linear softening model is used to describe its degradation. The comparison between experimental results and numerical simulations showed a good agreement for the DCB, ENF and Mixed Mode tests. For the pull-off tests the deformations observed within the experiments were higher than in the computations, as clamped boundary conditions could not be realised within the tests. The load leading to onset of degradation was predicted very good. In the next development step, the numerical model will be improved by idealising the stringer web as well. This is especially needed for the correct prediction of the failure load of the T-stiffener specimens.

Acknowledgement This work is supported by the European Commission, Priority Aeronautics and Space, Contract AST3-CT-2003-502723. The information in this paper is provided as is and no guarantee or warranty is given that the information is fit for any particular purpose. The user thereof uses the information at its sole risk and liability.

References

1. Degenhardt, R.; Rolfes, R.; Zimmermann, R.; Rohwer, K.: COCOMAT - improved material exploitation of composite airframe structures by accurate simulation of postbuckling and collapse. In: *Composite Structures* 73 (2006), pp. 175–178

2. Bansemir, H.: Krafteinleitung in versteifte orthotrope Scheiben. In: *Ingenieur Archiv* 42 (1973), pp. 127–140
3. Zhong, A.; Reimerdes, H.-G.: Stability behavior of cylindrical and conical sandwich shells with flexible core. In: *Journal of Sandwich Structures and Materials* 9 (2007), pp. 143–166
4. Bathe, K.-J.: *Finite-Elemente-Methoden* 2. Edition (2002), Springer, Berlin
5. Linke, M.; Wohlers, W.; Reimerdes, H.-G.: Finite element for the static and stability analysis of sandwich plates. In: *Journal of Sandwich Structures and Materials* 9 (2007), pp. 123–142
6. Camanho, P. P.; Dávila, C. G.: Mixed-Mode Decohesion Finite Elements for the Simulation of Delamination in Composite Materials. NASA/TM-2002-211737 (2002)
7. Turon, A.; Camanho, P. P.; Costa, J.; Dávila, C. G.: An Interface Damage Model for the Simulation of Delamination Under Variable-Mode Ratio in Composite Materials. NASA/TM-2004-213277 (2004)
8. Benzeggagh, M. L.; Kenane, M.: Measurement of mixed-mode delamination fracture toughness of unidirectional glass/epoxy composites with mixed-mode bending apparatus. In: *Composites Science and Technology* 56 (1996), pp. 439–449
9. Standard D6671: Standard Test Method for Mode I Interlaminar Fracture Toughness of Unidirectional Fiber-Reinforced Polymer Matrix Composites (1994)
10. Standard DIN EN 6034: Aerospace series - Carbon fibre reinforced plastics - Test method - Determination of interlaminar fracture toughness energy Mode II - G_{IIC} (1995)
11. Standard D5528: Standard Test Method for Mixed Mode I - Mode II Interlaminar Fracture Toughness of Unidirectional Fiber Reinforced Polymer Matrix Composites (2001)

Next Generation Composite Aircraft Fuselage Materials under Post-crash Fire Conditions

Gaetano La Delfa, Johannes Luinge, and Arthur Geoff Gibson

Abstract This paper summarizes a series of small-scale tests carried out to evaluate and model the post-crash fire integrity of composite aircraft fuselage structures.

The US Federal Aviation Administration regulations for the penetration of an external fuel fire into an aircraft cabin after crash require a burn-through period of 4 min (FAA § 25.856 Appendix F, Part VII). Different candidate structures for the next generation of composite aircraft fuselage, provided by Airbus, were investigated, including CFRP monolithic laminate and a folded-core CFRP sandwich. Those materials were subjected to constant heat flux from a propane gas burner, while being held under compressive load in a small, specially designed compression test rig with anti-buckling guides. The propane burner was calibrated to produce a constant heat flux up to $182\,kW/m^2$. The sample time-to-failure was measured, along with the temperatures at various points through the thickness.

Modelling the thermal and structural behaviour under load required the use of a modified version of the Henderson Equation, which describes heat transfer through composites under ablative fire conditions. This has been incorporated into the Com-Fire software model. Kinetic parameters for the resin decomposition reaction were determined from thermo-gravimetric data and other thermal parameters, conductivity and diffusivity were measured experimentally. The paper will compare the behaviour of single and double-skinned structures and will examine measured and modelled behaviour.

Keywords Composite structures · Aerospace · Fire resistance · Mechanical property

G.L. Delfa (✉) and A.G. Gibson
NewRail, Newcastle University, Stephenson Building, Claremont road, Newcastle upon Tyne, NE1 7RU, UK
e-mail: Gaetano.La-Delfa@ncl.ac.uk

J. Luinge
EADS Innovation Works, TCC1/Department IW-CT, 81663, Munich, Germany

S. Pantelakis, C. Rodopoulos (eds.), *Engineering Against Fracture: Proceedings of the 1st Conference*,

1 Introduction

This paper describes small-scale tests and a model to evaluate the integrity of composite aircraft materials exposed to fire conditions [1]. In the past decade there has been rapid development of polymer composite materials in aircraft due to the strength, stiffness and lightweight advantages. This trend is expected to continue in the future as these materials are replacing aluminium and other aerospace alloys in the primary structure and control surface [2]. One question that needs to be asked, however, is whether the growing use of polymer composite materials has increased the fire hazard due to the flammable nature of the organic matrix. Hence a serious situation known as "Post crash fire condition" has to be considered since this has important consequence during many accidents. As shown in Table 1 the Civil Aviation Authority (CAA) reported that "Collision with terrain/water/obstacle" was a consequence in 47% of accidents, whilst "Controlled flight into terrain (CFIT)" was a consequence in 35% of accidents. "Post crash fire" involves 22% of the total accidents [3].

Two important factors are considered in this paper: burn-through resistance and structural integrity. The fuselage burn-through refers to the penetration of an external post-crash fuel fire into an aircraft cabin. In survivable aircraft accidents the time to burn-through is critical because of the hazard associated with burning cabin materials which have been ignited by external fuel fire. This may incapacitate passengers due to smoke inhalation and lack of oxygen before they are able to escape from the aeroplane. The goals set by the Federal Aviation Administrative (FAA) state that during passenger evacuation the flame penetration should not reach the cabin interior 4–5 min after a crash. Also, the fuselage structure should have a low heat release rate and sufficient structural integrity. The fire response requirements by the FAA are that the maximum heat release rate during a 5 min test will not exceed $65\,\mathrm{kW/m^2}$ and the total heat released during the first 2 min will not exceed $65\,\mathrm{kW\,min/m^2}$ [4].

Table 1 Consequences of fatal civil aircraft accidents

Type of incident	Fatalities
Collision with terrain/water/obstacle	289
Controlled flight into terrain	219
Loss of control in flight	178
Post-crash fire	134
Overrun	55
Undershoot	53
Collision on ground	39
Forced landing	30
Structural failure	27
Fire/smoke during operation	24

Consequences are not mutually exclusive, as each accident generally involved more than one type of incident.

2 Materials and Experimental

2.1 *Materials*

Two types of material construction were investigated. The first was a Carbon Fibre Reinforced Polymer CFRP not-isotropic monolithic laminate (Fig. 1). The stacking sequence of a single laminate was $[+45°/0°/0°/-45°/90°]$ and the fibre volume content was 66%. Two separate laminate thicknesses were considered (2.1 and 1.7 mm thickness). The second material construction considered a sandwich panel (Fig. 2). The skins of which were made from the same CFRP construction as described previously. The core which was adhesively bonded to the skins was made from "folded mechanical paper and phenolic resins". The laminate skins have good mechanical characteristics compared with the weak core. In addition, the core gives good thermal insulation properties. The main drawbacks are susceptibility to buckling and lack of resistance to penetration of sharp objects.

2.2 *Thermo-gravimetric Analysis (TGA)*

The thermal model required certain kinetic parameters in order to calculate the rate of decomposition of the composite resin. Thermo-gravimetric analysis was used in this work to study the degradation kinetics of the epoxy matrix. The TGA was performed in a nitrogen atmosphere at three heating rates of $5\,°C/min$, $10\,°C/min$

Fig. 1 CFRP carbon epoxy laminate

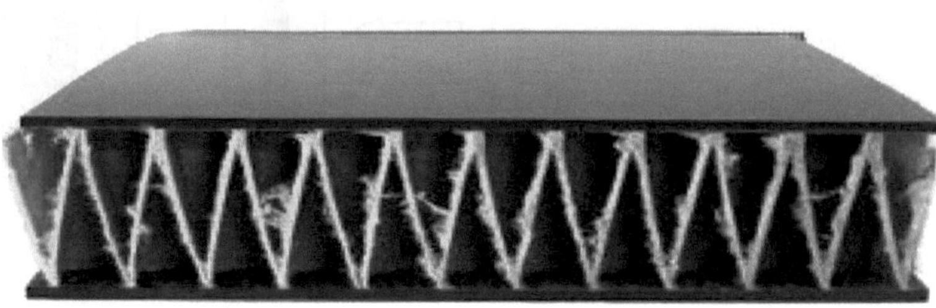

Fig. 2 Folded core CFRP sandwich panel

and 20 °C/min. The TGA measured the mass loss of the materials with increasing temperature, this could then be correlated to the amount of decomposition. After that the decomposition reaction rate constants A, E and n are determined. The procedure is explained in more detail in Sect. 4.1.

2.3 Fire Under Compression Load Test

Compression performance of advanced composites is a limiting design parameter. While tensile performance of carbon fibres has continued to increase, the compression performance has remained relatively stagnant for a wide variety of fibre types and diameters. For this reason the specimens were tested in compression. Fire testing under load were carried out on small scale samples. A propane burner was used to produce a constant heat flux of $75\,\mathrm{kW/m^2}$ and also $182\,\mathrm{kW/m^2}$ to the front face. As well as measuring the time to failure, the temperatures at different points through the thickness also was recorded. These results were used to validate the CFRP laminate compressive strength model predictions [5–8]. Figure 3 shows the test method. Anti-buckling guides were set up to avoid the column failure mode. This type of system is comparable with the well-known Boeing compression test [9]. The lateral constraints raised the load required for buckling failure, this enabled specimens of reasonable size to be placed under significant compressive stress during the fire tests. The material outside the heated region and the back surface were not insulated with a thermal blanket to simulate a real circumstance.

Two important parameters were needed in order to maintain a constant heat flux: the distance from the burner to the sample and the pressure of gas to the burner was held constant.

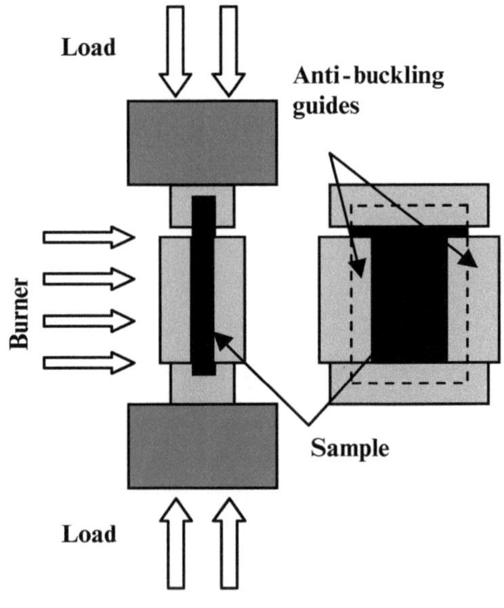

Fig. 3 Propane burner test under compressive load

2.4 Thermal Diffusivity Measurements

The thermal model required calculation of same thermal parameters to give accurate prediction of the fire response of the polymer composite. In particular, the thermal conductivity of the materials was calculated with the following relation:

$$k = \rho C_p \alpha \tag{1}$$

where ρ is density, C_p is specific heat and α is thermal diffusivity.

The density was taken to be 1,310 kg/m^3 and the specific heat as 1,278 J/kg K. Laplace's equation can be used to form a model for unsteady state heat flow:

$$\frac{dT}{dt} = \alpha \cdot \frac{d^2T}{dx^2} \tag{2}$$

The model considered the temperature of the centre of a slab at the point the sample was removed from the oven, T0, and also while the sample cooled to a lower temperature T1. Equation (2) can be expanded to describe the behaviour of the heat flow as follows:

$$\frac{T - T_0}{T_1 - T_0} = 1 - \frac{4}{\pi} \exp\left(\frac{-\pi^2 F_0}{4}\right) \tag{3}$$

$$F_0 = \frac{\alpha \cdot t}{b^2} \tag{4}$$

F0 is Fourier number, t is the time in (s) and b is the thickness of the specimen (mm). The cooling was measured experimentally, submerging the sample in water and recording the variation of temperature from 50°C to 15°C. Once the experimental cooling curve was obtained, this was then super imposed over the curve obtained from the model. The value of α thermal diffusivity was then manually altered in order to best fit the actual results. Finally the thermal conductivity was calculated using Eq. (1).

3 Results and Discussion

3.1 Fire Test Under Compressive Load

Fire tests under compressive load have been carried out on the CFRP NI (not isotropic) laminate specimens and on the folded-core sandwich with different skin thicknesses. The specimens were subjected to a constant compressive load with one side exposed to fire using a propane burner. The loads were applied to the specimens

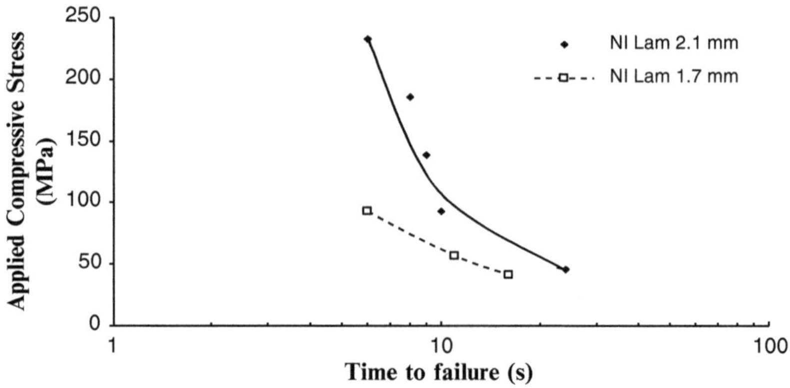

Fig. 4 Failure times for CFRP carbon epoxy laminates at different thickness exposed to an heat flux of $182\,\mathrm{kW/m^2}$

using a 295 kN hydraulic machine. The specimens, in the vertical position, were exposed to a calibrated high heat flux of $182\,\mathrm{kW/m^2}$. The distance of the propane torch to the surface of the panel was 35 cm.

The time to failure was recorded. This was defined as the time needed for the exposed face to fail. In preliminary experiments CFRP NI laminates have been studied. The results of this study indicate that the failure time for the laminates are very short. Figure 4 shows that this ranged from 8 to 15 s. It is apparent from the graph below that the time to failure increased with the laminate thickness. These results are consistent with those of other studies [10]. The most likely form of failure found seems to be delamination between the plies nearest the hot surface.

The burn-through resistance was considered on the basic that no flame should penetrate on the back face before 4–5 min. Laminates reasonably withstand high temperatures; however, the burn-through resistance gives better performance with sandwich panels. Therefore more consideration will be given to the burn through thickness resistance of the sandwich panel.

The constant compressive stress was applied on both skins until failure of the samples occurred. Different failure modes arose depending on the size of the applied compressive stress. When the stress was high the time to failure was short as shown in Fig. 5.

Mainly the failure mode involved collapse of the front skins with little delaminations and kinking of the ply [11, 12]. This can be observed in Fig. 6 below. The fibres were broken at each kink band boundary. Failures predominantly occurred in the centre of the specimens. At the boundary the main failure mode was local buckling.

A different mode of failure was observed when the stress was very low. Failure occurred over a longer period of time. The main failure mode noted was debonding of the front skins from the core materials for both thickness of the FS NI sandwich panel. A possible explanation for this might be that since the load was very low, the skins were unaffected and the thin adhesive subjected at high temperature and load

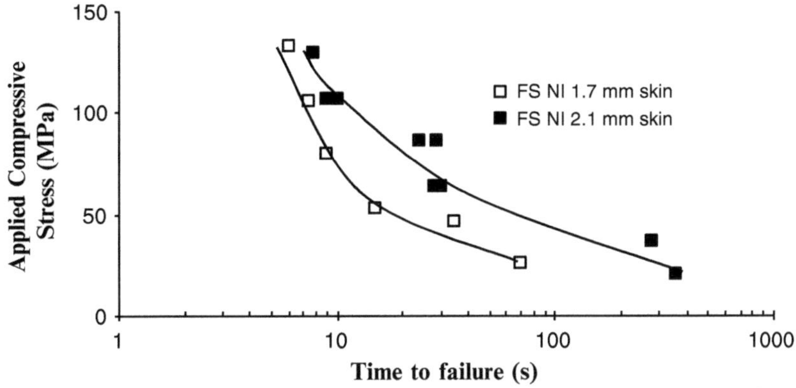

Fig. 5 Failure times for sandwich composite with CFRP carbon epoxy skins at different thickness exposed to a heat flux of $182\,kW/m^2$. The compressive load was applied in both skins

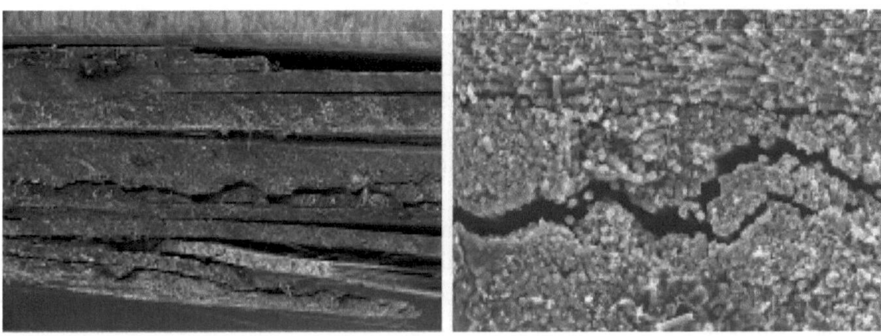

Fig. 6 Scanning Electron Microscopy of CFRP carbon epoxy skins exposed to a heat flux of $182\,kW/m^2$ and compressive load. Formation of delaminations due to the accumulation of volatile gas. Enlarged view of the carbon epoxy delaminations on the right side picture

failed. The time to failure of the sandwich panels with thicker skins response was three time more than the thinner ones at medium-low load. The burn-through resistance was successful as the sandwich was able to withstand the flame for 4–5 min as FAA regulations required.

The experiment for the sandwich panels was repeated under different conditions in which the specimens were reduced in length by 2 mm per sides. The purpose of this modification was that the skin unexposed to fire should carry out all the stress. This implies that the front skin was subjected only to fire. The time to failure was recorded when the back skins failed. The results of this part of analysis led to the conclusion that the panels lasted longer time than the previous method as shown in Fig. 7. The mode of failure recognized a kink band broadening without fibre fracture on the internal skins with debonding of the core from the exposed skins. This was accompanied with evidence of structural buckling and delaminations. Further research should be done to investigate the failure of the unexposed skin.

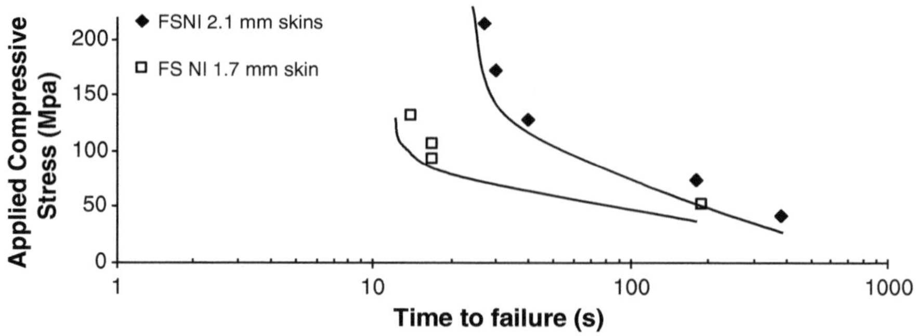

Fig. 7 Failure times for sandwich composite with CFRP carbon epoxy skins at different thickness exposed to a heat flux of $182\,\mathrm{kW/m^2}$. The compressive load was applied only on the skin not exposed to fire

3.2 Thermal Diffusivity Measurement

The thermal diffusivity was calculated recording the variation of temperature from 50 °C to 15 °C which was the water temperature. A thermocouple was embedded in the middle between two faces of the specimens. The setup procedure was the same for a Macor glass ceramic, CFRP NI laminate and sandwich panels. For the latter, two thermocouples were used.

The materials were surrounded with adhesive tape to prevent water intrusion. To achieve the required temperature, testing was performed using an oven. After thermal equilibrium was achieved, the specimens were dipped in water and cooling rate was recorded.

Validation of the test method was carried out using Macor glass ceramic. The thermal conductivity of which was given as 1.46 W/m K. The dimensions of the specimens were: length 50 mm, width 50 mm and 20 mm thickness. The results obtained from the analysis for Macor glass ceramic shown a significant positive correlation between the predictive and experimental curves. The response showed that the thermal conductivity, K, was 1.43 W/m K at 50 °C. This was less than 3% lower than the one given from the supplier.

Tests were conducted on the laminate and on the sandwich panels with 2.1 mm skins thickness. The dimensions of the specimens were standard as previous described. The laminate thickness was 2.1, and 34 mm for the sandwich panels. Good agreement exists between the experimental and the calculated temperatures variation for the CFRP single skin laminate and the sandwich panels as shown in Fig. 8. The response was that the thermal conductivities K obtained was 0.21 W/m K at 50 °C for the 977 laminate and 0.05 W/m K for the sandwich panels. The most important limitation lies in the fact that this method cannot be used at higher temperature higher than 50 °C because comparison between experimental and prediction was not well correlated.

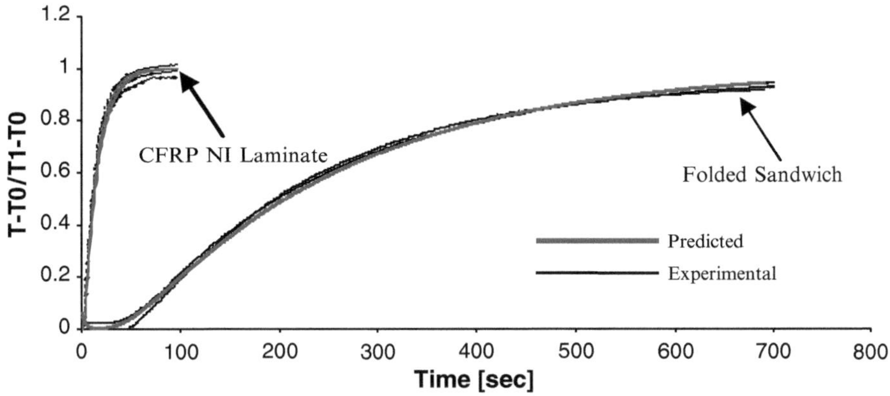

Fig. 8 Variations of temperature with time for the CFRP NI epoxy laminate and the sandwich panel

4 Modelling and Results

4.1 Analysis TGA

The degradation of linear polymers is a complex phenomenon and although many authors have studied it, many aspects are still unclear [13–15]. Analysis and comparison of the TGA curves allowed the calculation of the reaction rate factor (A) and activation energy (E) for each material. The general form of the kinetic expression to analyse the dynamic TG data was based on the nth order of the reaction mechanism and the temperature dependence could be sufficiently described by the Arrhenius equation which is written in the following form:

$$\frac{d\alpha}{dt} = -A \cdot e^{-\frac{B}{T}} \left(\frac{m - m_f}{m_o - m_f} \right)^n \tag{5}$$

where: dα./dt is the rate of change of mass (kg/s), A is the rate constant (1/s), mo, m and mf are the original, instantaneous and final mass of the polymer during the decomposition process (kg), B = E/R where E is the activation energy (J/mol), R is the universal gas constant (8.314 J/mol K), T is the temperature (K), n is the order of the reaction.

From the results two stages of reaction are evident. The first was a volatile process occurred over the temperature interval of 350–450 °C. The second was a decomposition of the non-volatile residues formed in the first stage occurred from 450 °C to 520 °C. The maximum rates of weight loss were 0.08/min, 0.15/min and 0.29/min respectively for 5 °C/min, 10 °C/min and 20 °C/min heating rates. Both mass loss curves had the same shape, however the different heating rates produced an effective shift in the temperature around which most of the decomposition occurred.

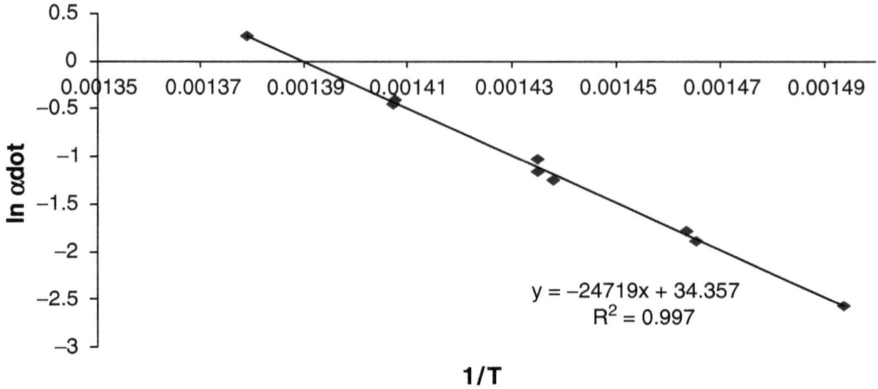

Fig. 9 Plot of ln $\dot{\alpha}$ versus 1/T for three different values of α and the slope provides a value of B = −E/R

Table 2 Activation energy values as function of α

α	Ea (kJ/mol)	R^2
0.4	212	0.9971
0.6	197	0.9971
0.8	195	0.999

A linear first order decomposition reaction type was observed, which accounted for the major degree of decomposition region between .0.26 to 0.85. From a preliminary analysis a value of B was found from the slope of ln $\dot{\alpha}$ versus 1/T (K) (Fig. 9). A dozen values of α. were selected ranging from 0.4 to 0.8. Values of $\dot{\alpha}$ (rate of reaction) and T were determined from each heating rates. As shown in Table 2, the correlation coefficient is always greater than 0.997 when $0.4 < \alpha < 0.8$. The evolution of the activation energy was noted by a lower value at the beginning of the degradation. Further analysis showed that an initial Ea value was about 93 kJ/mol. In addition to the resin degradation parameters, the heat of combustion and order of reaction, n, were required for each material. The heat of combustion was assumed to be 226,400 kJ and n was assumed to be 1 for all polymeric materials, based on the work of Friedman [13] and Henderson [16].

Modelled mass loss curves were constructed using the measured mass loss rate and the derived decomposition rate parameters. Figure 10 shows a comparison between the measured and modelled mass loss curves for the CFRP NI epoxy laminate. The main but small deviation between experimental and theoretical analysis occurred during early degradation, where the calculated loss exceeded the actual weight loss. In conclusion the decomposition parameters are a reaction rate factor A equal to 9.00× 1,014 1/s and activation energy E equal to 2.05 × 105 J/mol. These were sufficiently precise to be used in the thermal model.

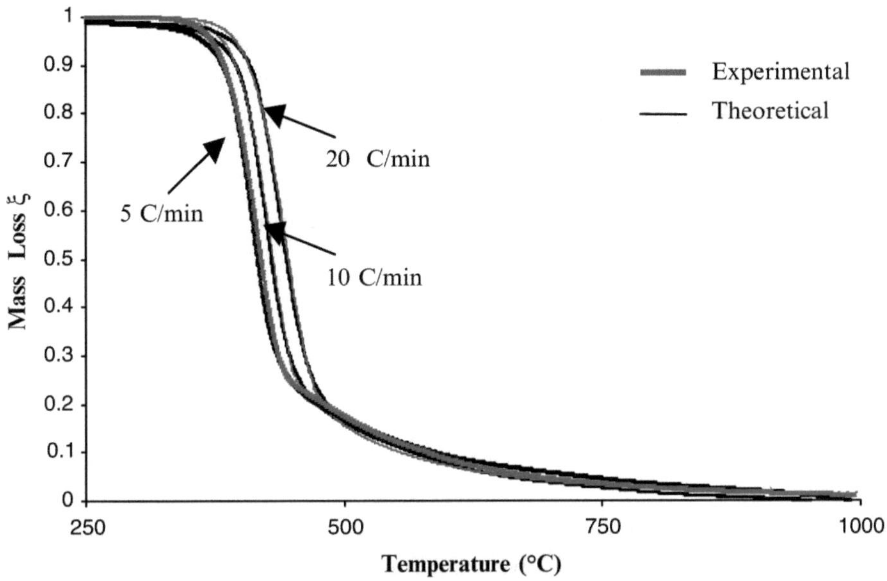

Fig. 10 Measured and modelled mass loss curves for CFRP epoxy laminate exposed to a 5 °C/min, 10 °C/min and 20 °C/min heating rates. The main decomposition region (350–480 °C) is shown

4.2 The Henderson Model

The Henderson equation has been adapted by Gibson et al. [8, 17] and Dodds [7] to predict the fire performance of glass reinforced plastic laminates at different levels of heat flux:

$$\rho C_p \frac{\partial T}{\partial t} = \frac{\partial}{\partial x}\left(k\frac{\partial T}{\partial x}\right) - \dot{M}_G \frac{\partial}{\partial x} h_G - \rho A \left[\frac{m - m_f}{m_o}\right]^n e^{\frac{-E}{RT}} (Q + h_C + h_G) \quad (6)$$

Equation (6) shown the adapted model which has been simplified in a number of ways. The thermal conductivity and specific heat properties were assumed to remain constant with the increase in temperature. Thermal and gas transport properties were considered to be constant during the decomposition process. The three main processes involved in energy transfer described in Eq. (6) are: heat conduction through the material, the convective mass flow of volatile products and the endothermic reaction of resin decomposition. The model was used to predict the temperature profile versus time using finite difference techniques.

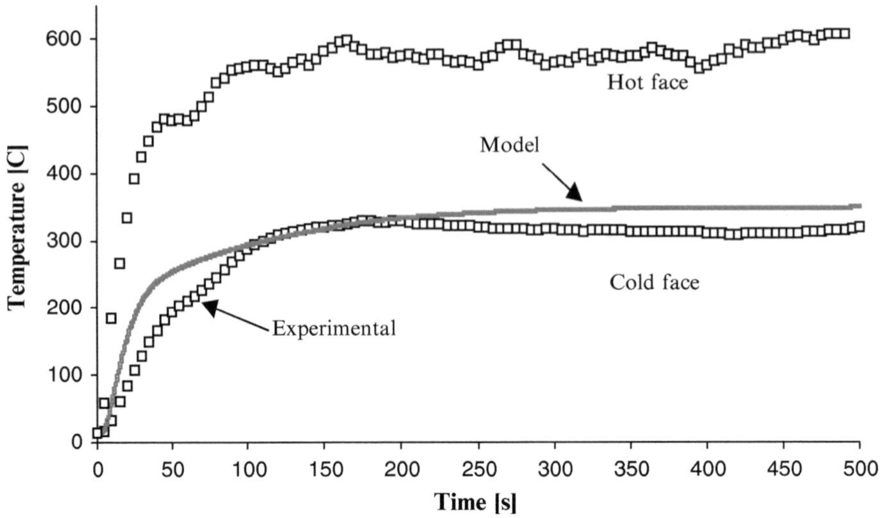

Fig. 11 Predicted and measured temperature profiles for a 2 mm carbon epoxy laminate exposed to a $75\,kW/m^2$ heat flux with natural convection and radiation on the cold face

4.3 Thermal Model Temperature Profile

Figure 11 compares the modelled and experimental rear face temperature profile for CFRP carbon epoxy laminate in $75\,kW/m^2$ fire. This material was modelled reasonably well and the full temperature will be used to provide the necessary input for modelling the structural response of laminates in fire.

5 Conclusions

A number of conclusions can be drawn from the present study. A thermal model was based on TGA and thermal analysis. This enabled the temperature profile to be predicted for the carbon epoxy laminate as shown in Fig. 11. The response of the laminate and sandwich panels under fire and load was performed. The results showed that the unloaded laminate and sandwich panels met the (4–5 min) criteria required by FAA with respect to the burn-through thickness. However, when a high load was applied, the time to failure was significantly reduced. Although, this is a small scale test, it does nonetheless more accurately simulate the integrity of the material during a crash were fire and load are likely to be present. It would be interesting to assess the same effects on a larger scale test. Considerably more work will need to be carried out in order to better predict the temperature profile for the sandwich panels due to the complexity of the sandwich panel construction. Future studies will focus on a structural model which will provide the time to failure of the materials tested.

Acknowledgement Research on integrity of composite materials in fire Newcastle University is funded by the Marie Curie Momentum Research Training Network.

References

1. Gibson, A.G., Wright, P.N.H., Wu, Y-S., Mouritz, A.P., Mathys, Z. and Gardiner, C.P. (2004). The integrity of polymer composites during and after fire. *Journal of Composite Materials* 38:15, 1283–1308.
2. Mouritz, A.P. (2006) Fire Safety of Advance Composite for Aircraft, Aviation Safety Research Grant B2004/0046, School of Aerospace, Mechanical & Manufacturing Engineering RMIT University.
3. CAP681 Global Fatal Accident Review 1980–1996. www.caa.co.uk
4. Committee on fire- and smoke-resistant amterials for commercial aircraft interiors (1995) Fire and Smoke-Resistant Interiors Materials for Commercial Transport Aircraft, The National Academy of Sciences, Washington, DC.
5. Dao, M. and Asaro, R.J. (1999) A study on failure prediction and design criteria for fiber composites under fire degradation. *Composites* 30A:123–131.
6. Mouritz, A.P. and Mathys, Z. (2001) Post-fire mechanical properties of glass-reinforced polyester composites. *Composites Science and Technology* 61:475–490.
7. Dodds, N. et al. (2000) Fire behaviour of composite laminates. *Composites: Part A* 31(7): 689–702.
8. Gibson, A.G. et al. (2003) A Low Cost Burner Technique for the Development and Modelling of Laminates in Fire. In Composites in Fire 3: 3rd International Conference on the Response of Composite Materials to Fire, Newcastle upon Tyne, England.
9. Greene, E. (1993) Fire Performance of Composite Materials for Naval Applications, US Navy Contract N61533-91-C-0017, Structural Composites, Melbourne, FL.
10. Gibson, A.G. et al. (2006) Laminate theory analysis of composites under load in fire. *Journal of Composite Materials* 40(7):639–658.
11. Budiansky, B. and Fleck, N.A. (1993) Compressive failure of fibre composite. *Journal of the Mechanics and Physics of Solids* 41/1:183–211.
12. Schultheisz, R. and Waas, A.M. (1996) Compressive failure of composite, Part I: Testing and micromechanical theories. *Progress in Aerospace Science* 32:1–42.
13. Friedman, H.L. (1959) Kinetics of thermal degradation of char-forming plastics from thermogravimetry: Application to a phenolic plastic. In Proceedings of the 136th American Chemical Society meeting, Atlantic City, NJ.
14. Cai, X.E., Shen, H., Zhang, C.H., Wang, Y.X. and Kong Z. (2000) Application of constant reaction rate TG to the determination of kinetic parameters by Hi-Res TG. *Journal of Thermal Analysis and Calorimetry* 60:623–628.
15. Opfermann, J. (2000) Kinetic analysis using multivariate non-linear regression. *Journal of Thermal Analysis and Calorimetry* 60:641–658.
16. Henderson, J.B., Wiebelt, J.A. and Tant, M.R. (1985) A model for the thermal response of polymer composite materials with experimental verification. *Journal of Composite Materials* 19:579–594.
17. Gibson, A.G. et al. (1995) A model for the thermal performance of thick composite laminates in hydrocarbon fires. *Revue de l'Institut Français du Pétrole* 50(1):69–74 (special issue).

Progressive Fracture Analysis of Planar Lattices and Shape-Morphing Kagome Structure

Konstantinos I. Tserpes

Abstract The fracture behaviors of three defected planar lattices loaded in axial tension and the 3D shape-morphing Kagome structure loaded as a cantilever beam are explored by using finite element-based progressive fracture analysis. The assumed defects are in the form of symmetrical notches introduced in the lattices by removing the struts in single rows. Numerical results reveal that the presence of the notches significantly reduces the tensile strength of the lattices. On the other hand, with increasing the load in the Kagome structure, yielding and buckling of the struts in the core and yielding of the face-sheet appear consecutively inducing degradation of structure's bending stiffness and large dips of the loaded end.

Keywords Lattice structures · Shape-morphing · Kagome-structure · Finite element analysis

1 Introduction

Lattice solids are interconnected networks of cells stacked in two or three dimensions. They belong to the family of cellular solids and are being also referred as honeycombs. The enormous range of the density, the Young's modulus the thermal conductivity and the compressive strength of lattices creates structural applications which cannot easily be filled by dense materials and offers potential for engineering ingenuity [1]. For example, the low density permits the design of light, stiff components such as sandwich panels and the low strengths and large compressive strains make cellular solids attractive for energy absorbing applications. Perhaps the most challenging application of all, emanating from the ability of lattice solids to sustain large external loads and at the same time undergo large changes of shape, is shape-morphing aircraft structures [2].

K.I. Tserpes
Laboratory of Technology & Strength of Materials, Department of Mechanical Engineering & Aeronautics, University of Patras, Patras 26500, Greece
e-mail: lkit2005@mech.upatras.gr

S. Pantelakis, C. Rodopoulos (eds.), *Engineering Against Fracture: Proceedings of the 1st Conference*,

In view of the prospective use of lattice solids in critical applications, their fracture behavior and defect tolerance must be understood. In the present work, these two issues are investigated for three basic planar lattices as well as for the 3D shape-morphing Kagome structure. To this effect, finite element-based progressive fracture analysis is used.

2 Problem Definition

2.1 Planar Lattices

According to the geometrical shape of the cells, different honeycombs may appear. In this work, three basic honeycombs; namely, the triangular (T), the hexagonal (H) and the Kagome (K) honeycomb are considered. The honeycombs are schematically shown in Fig. 1.

In all lattices, the struts are of uniform length L. They are made of steel with a Young's modulus of 200 GPa, Poisson ratio of 0.3 and fracture strength of 3.6 GPa. The material of the struts is considered to be homogeneous, isotropic, linear-elastic and brittle. The relative density of the honeycombs $\bar{\rho}$, as defined by the ratio of the density of the lattice material to the density of the solid [3], can be expressed in terms of the ratio t/L as follows

$$\bar{\rho}_{\mathrm{T}} = \frac{t_T 2\sqrt{3}}{L}\left(1 - \frac{t_T\sqrt{3}}{2L}\right) \tag{1}$$

$$\bar{\rho}_{\mathrm{H}} = \frac{2t_H}{L\sqrt{3}}\left(1 - \frac{t_H}{2\sqrt{3}L}\right) \tag{2}$$

$$\bar{\rho}_{\mathrm{K}} = \frac{t_K}{L}\left(\sqrt{3} - \frac{t_K}{L}\right) \tag{3}$$

where t_i (i = T, H, K) is the thickness of the struts in each lattice. For all analyses conducted within the frame of the present work, the value of 17% was

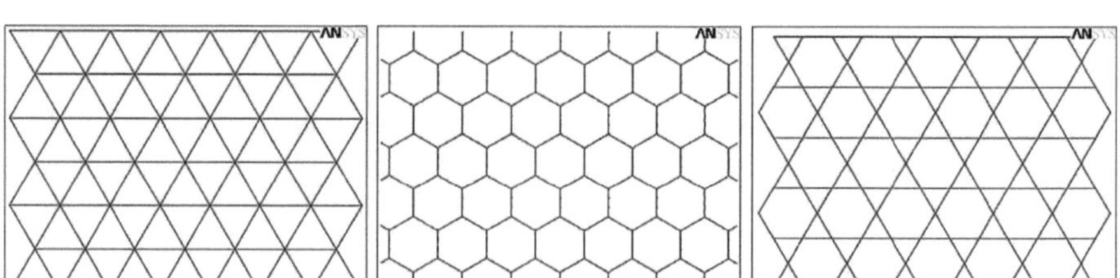

(T) (H) (K)

Fig. 1 Schematic representations of the honeycombs

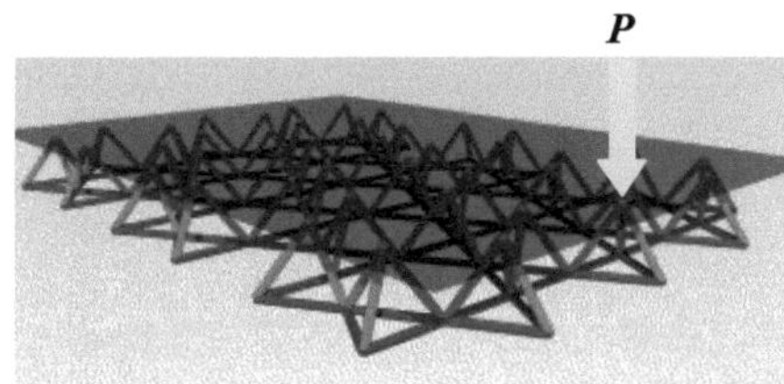

Fig. 2 Schematic representation of Kagome-structure, taken from [2]. The face-sheet is shown in blue, the tetrahedral core in green and the Kagome back-plane in red

adopted for the relative density. Using L = 1 mm, from Eqs. (1)–(3) we obtain tT = 0.05 mm, tH = 0.15 mm and tK = 0.1 mm, respectively.

It is desired to assess the effect of notch's presence in the tensile behavior of the lattices.

2.2 Kagome Structure

The Kagome structure is a very attractive structure for application as a shape-morphing material because of its unique combination of low actuation resistance and high passive strength. Hence, it can be easily actuated into relatively intricate surface shapes. As shown in the schematic of Fig. 2, it consists of a solid face-sheet with a Kagome, active back-plane and a tetrahedral core. Shape change of the face-sheet is achieved by replacing various struts in the back-plane with linear actuators. During service, the structure is loaded simultaneously by passive loads and actuation loads. The length of the struts in both the back-plane and the core is 51 mm while their diameters are 3.0 and 1.4 mm, respectively. All members of the structure are assumed to be made of steel with a Young's modulus of 200 GPa, Poisson ratio of 0.3 and yielding stress of 200 MPa.

It is desired to simulate the behavior of the Kagome structure loaded as a cantilever beam by the passive load P.

3 FE Analysis

3.1 Planar Lattices

FE models of the lattices have been built using the ANSYS FE code [4].The struts are represented by 3D elastic beam elements. A total of ten elements have been used per strut-member in order to accurately capture bending.

Notch-shaped defects have been introduced in the lattices by removing the struts in a single row. In Fig. 3, the FE meshes of the defected lattices are depicted.

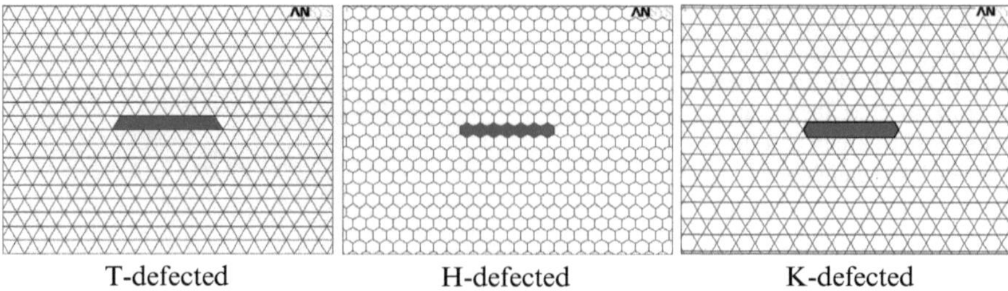

Fig. 3 FE meshes of the defected planar lattices

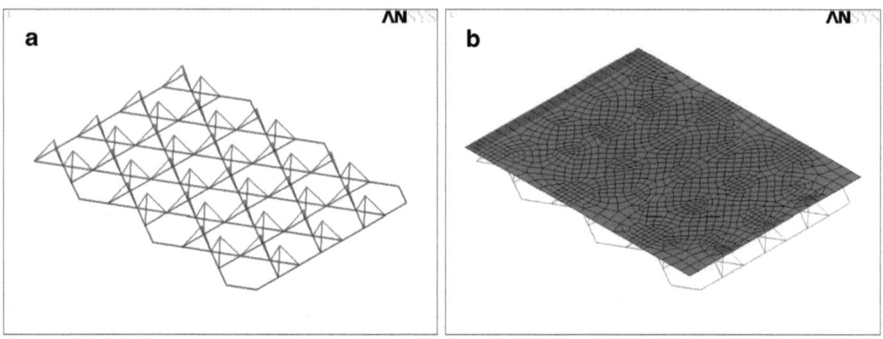

Fig. 4 FE mesh of **a** the Kagome back-face and grid, **b** the complete Kagome structure

The ratio of the notch length to the lattice width is similar for all three lattices. Tensile loading is simulated in the honeycombs through the application of an incremental axial displacement at both of their ends.

3.2 *Kagome Structure*

A 3D FE model of the Kagome structure has been also built. The struts are modeled as in the planar lattices. The face-sheet is modeled using elastic shell elements. FE meshes of the back-face and core are shown in Fig. 4a while the FE mesh of the complete Kagome structure in Fig. 4b.

To simulate cantilever bending in the Kagome structure, one end is fully clamped and an incremental normal force is applied at the central node of the opposite end. Three different failure modes have been considered: yielding and buckling of the struts as well as yielding of the face-sheet. Since the back-face and core are mainly in compression, fracture of the struts is not expected to occur before the structure fails in yielding and buckling. The stresses computed at each element are compared with the yielding stress and each time yielding is found the Young's modulus of the specific member is degraded. Buckling is predicted using the Euler formula for hinged bars

$$P_{cr} = \frac{\pi^2 EI}{L^2} \tag{4}$$

where P_{cr} is the critical buckling force, E the Young's modulus and I the secondary moment of inertia of the member. The values of P_{cr} derived from Eq. (4) for the struts of the back-face and core are 0.88 and 0.19 kN, respectively.

4 Numerical Results

4.1 Planar Lattices

Figure 5 compares the tensile stress–strain curves of the pristine and defected honeycombs. Strain has been evaluated as the ratio between the applied displacement and honeycomb's length at the stretched direction and stress as the ratio between the total reaction force at one end and the effective stressed area defined as the product of honeycomb's thickness (strut's diameter) and width. In all three cases, the presence of the notch reduces significantly the tensile strength, which corresponds to the maximum sustainable stress. The reduction is 31% for the T lattice, 35.8% for the H lattice and 26.47% for the K one. The effect on stiffness of the honeycombs is less significant: there is a small decrease at the initial loading stages which increases as soon as fracture initiates. Of course, the influence of the notches to the response of the honeycombs is expected to be analogous to the ratio of notch's length and honeycomb's width. Between the three different honeycombs,

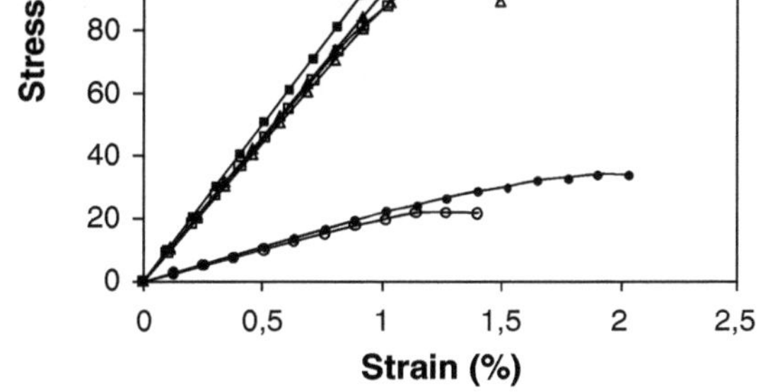

Fig. 5 Predicted tensile stress–strain curves for the pristine and defected honeycombs

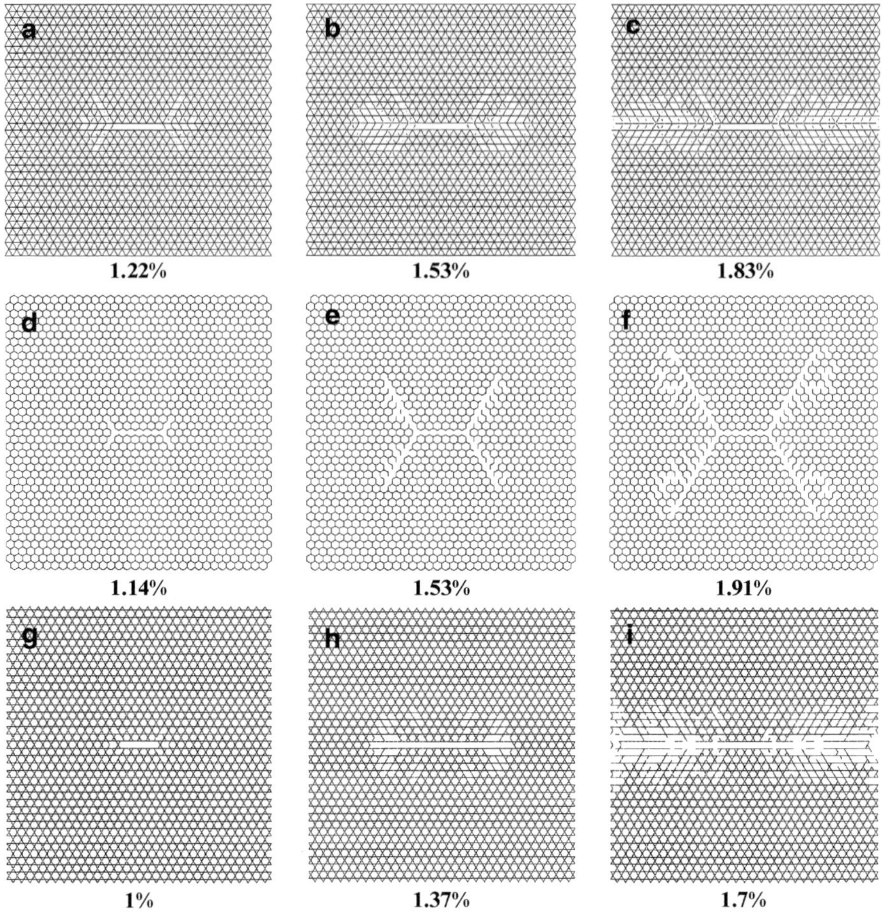

Fig. 6 Progression of fracture in the defected T **a–c**, H **d–f** and K **g–i** lattices as a function of applied strain

the T and K ones are much stiffer and stronger than the H. Furthermore, based on the obtained percentages of strength reduction, it is concluded that the K lattice shows a greater notch tolerance.

Figure 6 illustrates the progression of fracture in the three defected lattices as a function of applied strain. As expected, fracture initiated at the edges of the notches and propagated outwards. Some early fractures at the edges of the lattices, owing to the boundary conditions, are also observed. In all cases, fracture initiated at around 1% strain. In T and K lattices, fracture propagated vertical to the loading direction and finally separated the lattice. On the other hand, in the H lattice, fracture propagated in 60° and -60° with the regard to the loading direction following the members of the hexagonal cells.

4.2 Kagome Structure

Figure 7 plots the predicted relation between the applied force and deflection of the loaded node for the Kagome structure. The relation is linear up to the force of

Fig. 7 Predicted force–deflection curve for the Kagome structure

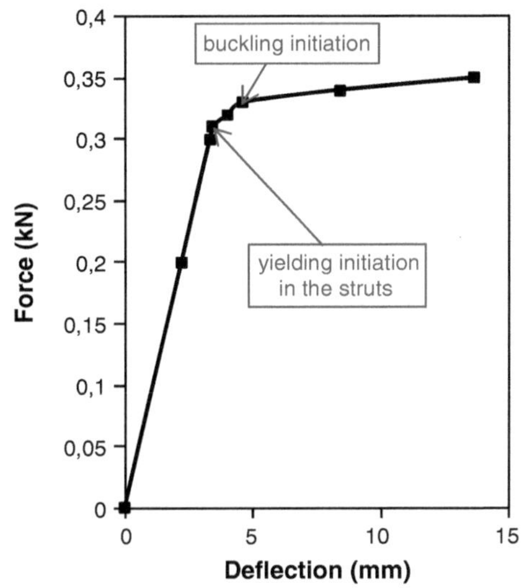

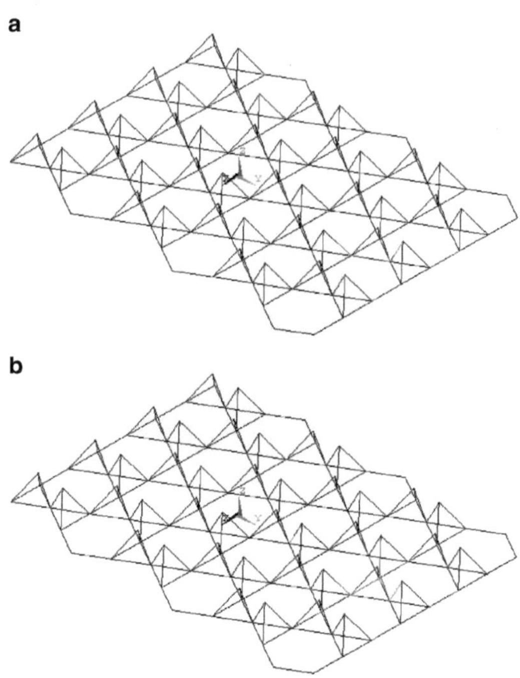

Fig. 8 Yielded (red colored) and buckled struts (green colored) at **a** 0.33 kN and **b** 0.36 kN

0.31 kN where yielding initiated in the struts of the core. At this point, the bending stiffness of the structure is significantly degraded. By increasing loading, buckling initiated at 0.33 kN in one of the three struts that hold the loaded node as can be seen in Fig. 8a. Consequently, a sharp increase in the deflection of the node occurs. By further increasing of the load, yielding in the core's struts is accumulated increasing the deflection. At 0.36 kN, a second strut buckles (see Fig. 8b) and large deformation of the loaded end of the structure takes place as shown in Fig. 9. Just after buckling initiation, yielding is initiated in the material of the face-sheet at the areas close to the clamp and around the loaded node as can be seen in Fig. 10.

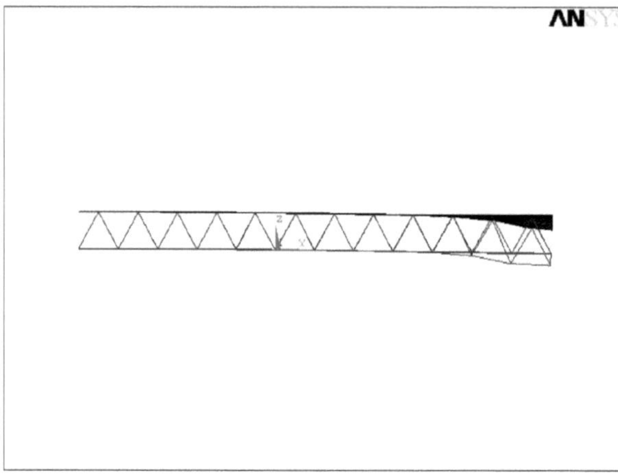

Fig. 9 Deformed shape of the Kagome structure at 0.34 kN

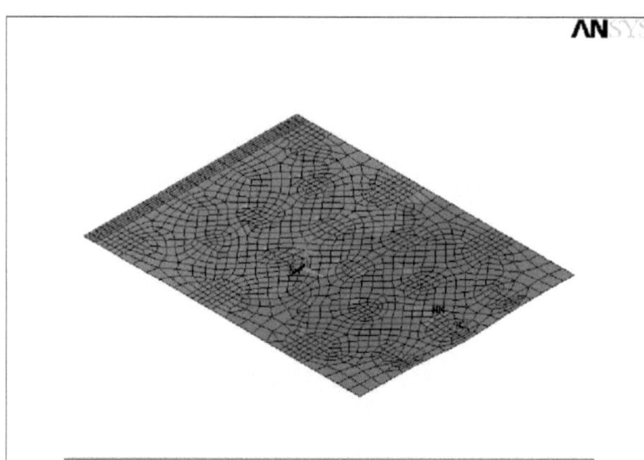

Fig. 10 Yielding initiation in the face-sheet at 0.34 kN

5 Conclusions

The conclusions of the paper are summarized as follows:

- The fracture behaviors of three defected planar lattices loaded in axial tension and the 3D shape-morphing Kagome structure loaded as a cantilever beam have been explored by using finite element-based progressive fracture analysis.
- The presence of the notch significantly reduces the tensile strength of the lattices.
- With increasing the load in the Kagome structure, yielding and buckling of the struts and yielding of the face-sheet appear consecutively inducing degradation of structure's bending stiffness and large dips of the loaded end.

References

1. Gibson LJ, Ashby MF (1997) Cellular solids, Cambridge University Press, Cambridge, UK.
2. dos Santos e Lucato SL, Wang J, Maxwell P, McKeening RM, Evans AG (2004) Design and demonstration of a high authority shape morphing structure. International Journal of Solids and Structures 41: 3521–3543.
3. Lipperman F, Ryvkin M, Fuchs MB (2007) Nucleation of cracks in two-dimensional periodic cellular solids. Computational Mechanics 39: 127–139.
4. ANSYS User's Manual (2007). Version 11.0. Swanson Analysis Systems, Inc.

Session IV
Fatigue Damage (Experimental)

Fatigue Behavior of Non-crimp Fabrics

Peter Horst and Stephan Adden

Abstract The paper presents experimental data on the fatigue of NCFs, with special emphasis on the development of damage, i.e. the crack density, stiffness degradation and fatigue life under multi-axial loading. Non-crimp–fabrics (NCF) consist of several sub-layers with different orientations, all stitched together by a stitching yarn. Its lay-up can be tailored to specific needs. It offers cost-savings in production processes because instead of laying several single layers under different orientations into a mould it is possible to lay just one NCF that includes all sub-layers. One disadvantage is the stitching that causes local inhomogenities and fiber misalignment in the material.

Keywords Composites · Non-crimp fabrics · Fatigue

1 Introduction

NCFs have been subject of several research projects; Hogg et al. [1] compared NCFs to woven fabrics by performing experimental tests (tension and flexure tests) and showed that the NCFs have superior mechanical properties. Wang et al. [2] performed tension, bending and short beam shear tests. Their results were compared with the Classical Laminate Theory with good agreements. They came to the conclusion that NCF composite laminates behaved like prepreg composite laminates. Bibo et al. [3] performed tension and compression tests with glass-fiber and carbon-fiber NCF and showed that the stitching yarn has an influence on the compression strength, while the influence on the tension strength is very low. However, the work performed by Asp et al. [4] showed a completely different effect. Asp et al. analysed

P. Horst (✉) and S. Adden
Institute of Aircraft Design and Lightweight Structures, TU Braunschweig,
Hermann-Blenk-Str. 35, D-38108 Braunschweig, Germany
e-mail: p.horst@tu-bs.de

S. Pantelakis, C. Rodopoulos (eds.), *Engineering Against Fracture: Proceedings of the 1st Conference*,

different aircraft-grade NCF carbon fiber composites with different stitching parameters and did not find an influence on the compression and tension strength. Strangely Asp et al. did not compare their NCF-results to non-stitched materials, so the comparison of both results is difficult. It may be that any stitching pattern reduces the compression strength of the material, compared to the non-stitched material. Which stitching pattern is actually used might be of less importance. Vettori et al. [5] performed tension tests on plates made of different NCFs and used several NDT-techniques (such as acoustic emission, ultrasonic scanning and radiography) to characterize the damage evolution. They showed that the stitching pattern influences the periodicity of the damages seen in the NCFs. Carvelli et al. [6, 7] created a finite element unit cell, which included the stitching yarn and correlated the strength and the stiffness with the results from experiments. They showed that their finite element-unit cell without stitching overestimated and the unit cell with stitching underestimated the mechanical properties. Important for the research presented in this paper is the fact that NCFs can widely be treated like non-stitched laminates, when looking at their behavior. This will be important when looking at the modeling part. The paper is focussed on the multi-axial fatigue behaviour. Therefore special specimens have to be used. In this case, these are tube specimens.

2 Specimens

The specimens are made of E-Glass-fiber non-crimped-fabric and a resin material using resin transfer moulding (RTM). The NCF has been provided by SAERTEX^R and consists of four sub-layers, which lay-up is given in Table 1. The sub-layers are tied together using a tricot-type of stitching. The material of the stitching yarn is PES, with a mass per unit area of about $12\,\mathrm{g/m^2}$. The roving used during manufacturing of the NCF is OC111A (Owens-Corning). The resin/hardener-combination used for manufacturing the specimens is L135i/H137/H134 (MGS), which has been for use in wind-energy industry. The mechanical properties according to the company are given in Table 2. The tube specimen consists of two layers of the above mentioned NCF, which are wrapped around a mandrel in a way, that the plies orientated in 0° direction are the inner and outer ply. The tube specimens therefore have a slightly anti-symmetric lay-up. Due to the fact that the refraction index of fibre and

Table 1 Lay-up of the NCF

SAERTEX V92402-01310-01270-000000	
Orientation (°)	Mass per unit area ($\mathrm{g/m^2}$)
0	638
−45	301
90	63
45	301

Table 2 Mechanical properties of the gfrp-fibre and resin

Fibre OC111A	
Young's-modulus	80.7 GPa
Tensile strength	2.56 Gpa
L135i/H137i	
Tensile modulus	3.2–3.5 GPa
Tensile strength	70–80 Mpa
Compression strength	120–140 Mpa
Elongation at rupture	5.0–6.5%
Density	1.18–1.20 g/cm^3

resin are nearly similar, the material is, apart from the stitching yarn, transparent, which allows monitoring the crack density by means of a light microscope. This is used to calculate the damage directly.

The ends of the tube specimens are enforced by a doubler at a length of 70 mm and strengthened by a bonded steel insert, in order to prevent failure caused by the fixing pressure of the tension-torsion machine. The 0° direction of the NCF is identical to the axial direction of the tube specimen. Therefore, the expression normal or tensional loading indicates an axial loading. Torsion stands for turning the upper end of the specimens against the lower end. This kind of specimen is used because of two reasons: it is possible to apply bi-axial loads (tension/compression and torsion) with arbitrary load ratios and the so called free-edge-effect, which occurs under fatigue-loading by using coupon specimens, can be minimized.

The lay-up of the material promises high intra- and inter-laminar stresses, because (depending on the loading condition) a significant percentage of the fibers in the different layers will not be orientated in the loading-direction. This leads to the development of cracks, which in turn decreases the stiffness of the material. Both, the experimental as well as the theoretical investigations are trying to link the three parameters: loading cycles, crack density and stiffness. This paper shows a modeling approach which is the basis for further works and links the crack density to the stiffness.

3 Testing

The experimental work was performed with a servo-hydraulic testing machine MTS 319.25, which allows to apply tensional as well as torsional loading. The maximum tensional force is ±250 kN and a torsional moment of ±2,200 Nm. All fatigue tests reported here were performed in a force-controlled manner under $R = -1$. Different bi-axial loading ratios were introduced by using different ratios of tension force

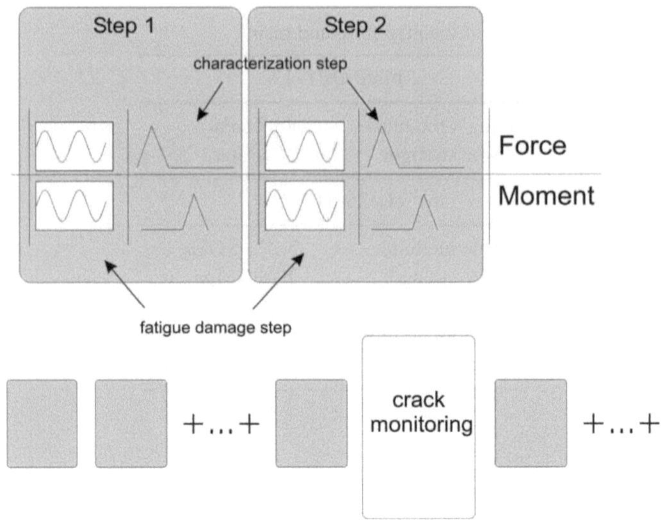

Fig. 1 Fatigue testing of tube specimen

(introducing normal stress σ) and torsional moments (introducing shear stress τ). During the mechanical test a so-called characterisation step was followed by a fatigue step (see Fig. 1). The characterization step was used to calculate the Young's-modulus and the shear modulus of the tube specimen, by applying loads of 15 kN and 100 Nm, which are considerably smaller than the applied load in the fatigue steps, but allows to calculate the initial stiffness. The damage steps were varied according to the biaxial loading conditions.

The loads have been chosen after having performed static tests of the specimens, which can be found in [8]. In order to distinguish the different loading directions of the tube specimen a parameter β is introduced, which is the angle calculated by

$$\beta = atan(\tau/\sigma)$$

i.e. $\beta = 0°$ means pure tension/compression and $\beta = 90°$ pure shear. Tests were performed for the load ratios $\beta = 0°$, $30°$, $60°$ and $90°$. For all load ratios three different load amplitudes were chosen on the basis of static test results. Three specimens were tested per load-amplitude.

As mentioned above, the crack density is one of the most interesting parameters, because it may be interpreted as a true internal state variable, which helps to describe the material behaviour. The fatigue tests were stopped after several steps; the specimen was taken out of the testing machine and observed per light microscope. Photos were taken over up to 15 different areas of the specimen which were used for the crack counting. All conclusions presented here were derived after analysing each area. Results are therefore average values, sometimes with additional standard deviation, from the different areas.

During the tests experiments with plate acoustic waves were conducted in cooperation with Busse and Solodov from the IKT in Stuttgart. Experiments and crack densities were measured and compared with the resulting plate acoustic wave, which is influenced by the crack density. A dependence, which could be fitted by a linear approach, was found [9]. Gagel et al. [10] worked on the same material in cooperation with the authors.

4 Results

This section shows a few results on the fatigue behaviour of the tube specimens. In Fig. 2 the development of cracks in one tube specimen under pure shear loading is presented. It is clearly visible that cracks under 45° develop first in a considerable number, while cracks of other orientations are developing later. The axial, i.e. the 0°-direction is the horizontal direction in the photographs. It is easily possible to identify that certain cracks develop quite fast and do not extend over a certain size, while others seem to propagate as far as possible almost immediately. Although it is not possible to detect this in the almost transparent continuum, it is clear that the cracks occur in the layers of the same fibre orientation along or within the rovings.

As stated already, it is now possible to count the number of cracks of different orientation, in order to define a damage parameter for each layer. In Fig. 3 this is done without distinguishing between the sizes of the cracks. The crack densities are plotted versus the normalized life of the specimen. Just for comparison, Fig. 4 shows the crack densities plotted versus the normalized life of a specimen under pure tension and Fig. 5 shows the same for a specimen under combined tension/torsion load ($\beta = 60°$). While Fig. 4 shows a quite usual behaviour, i.e. a very rapid increase

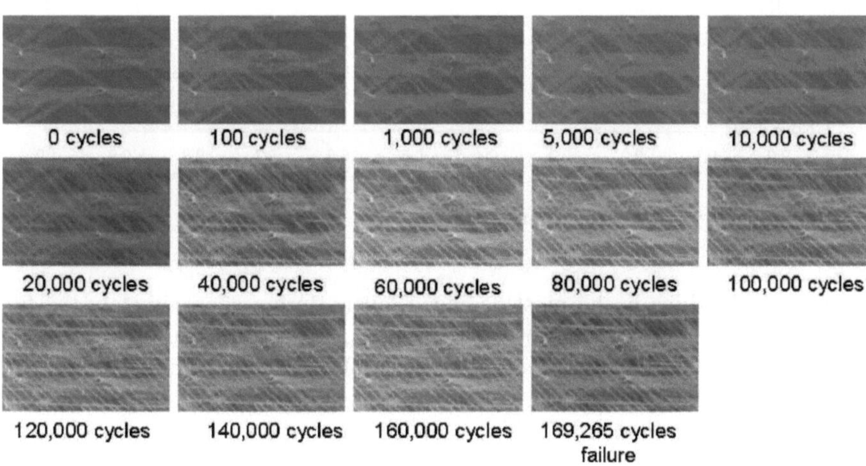

Fig. 2 Cracks developing under pure shear fatigue loading

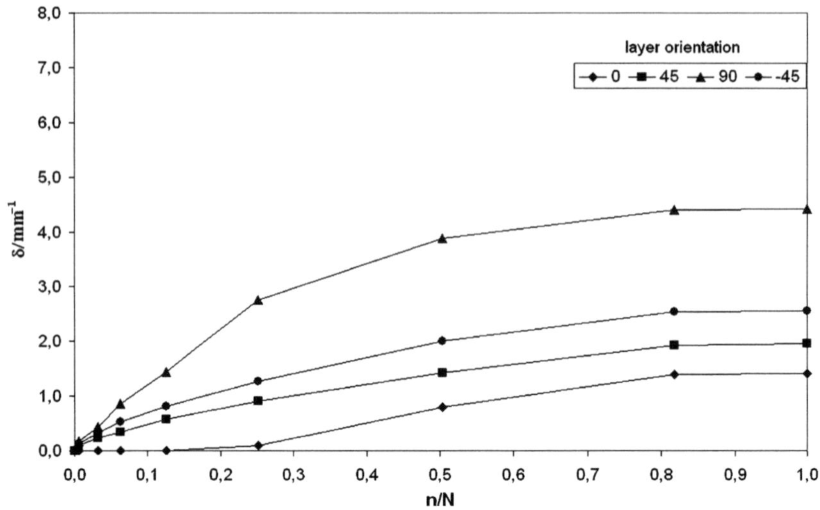

Fig. 3 Crack density under pure shear loading

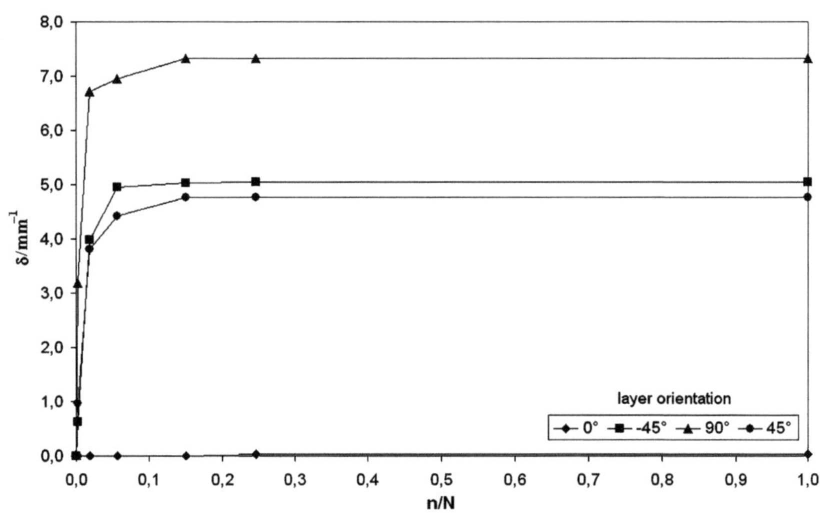

Fig. 4 Crack density under pure tension

of the crack densities in the most damaged layers during approximately the first 10–15% of the life, this is not the case in Fig. 3. The main reason for this is the normalization of the life with respect to the total life. Tube specimens made out of NCFs have to have an overlap, which under shear loads leads to a delamination problem. This delamination leads to a reduced life, but does not influence the development of the crack densities in the areas far away from the overlap. Far more interesting in both figures is the fact that the crack densities behave different under different loads. Under shear loads one observes a significant amount of cracks in

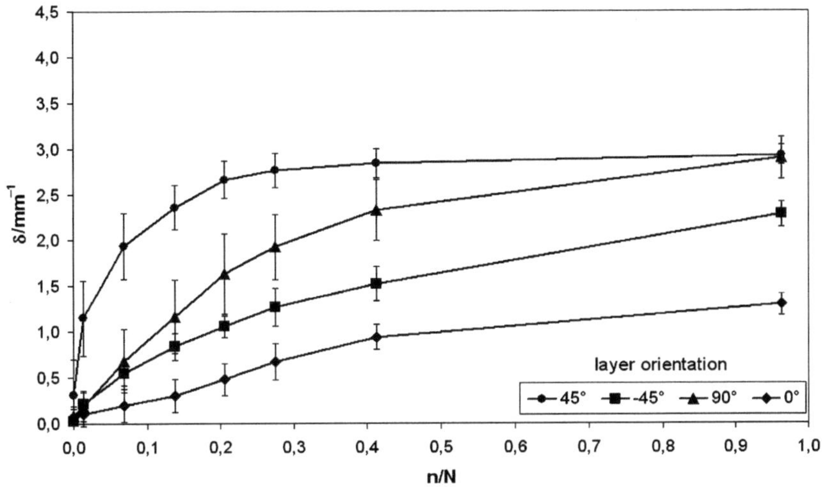

Fig. 5 Crack density under combined tension/torsion load

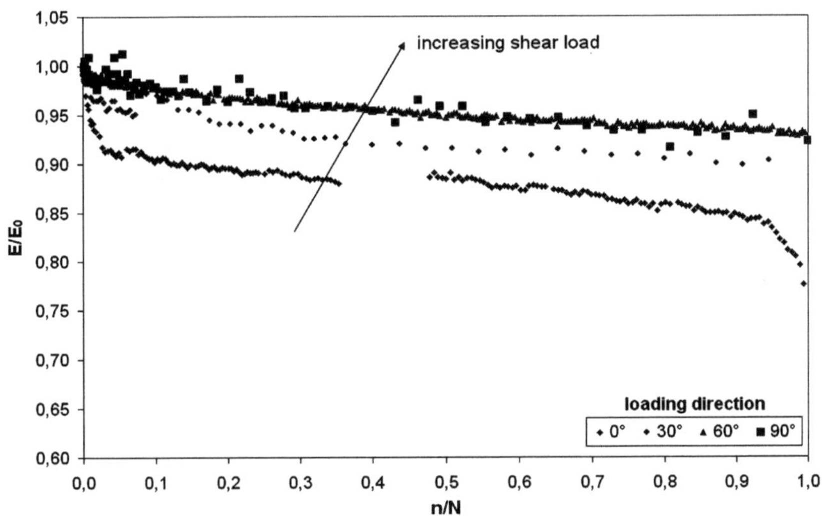

Fig. 6 Young's modulus versus normalized life

0°-direction, which result from the intralaminar loading of this layer. Under pure tension these cracks are not visible. Another point is that the absolute value of the crack densities is not the same, too. Naturally we would expect to see a similar behaviour for the stiffness degradation.

The stiffness degradation with respect to the Young's modulus is shown in Fig. 6 for several loading directions β. The $\beta = 0°$ curve shows the typical behaviour of a rapid decrease of the Young's modulus in the first 10–15% again and, after something like 80–90% of the life, a further rapid decrease, which ends in the final

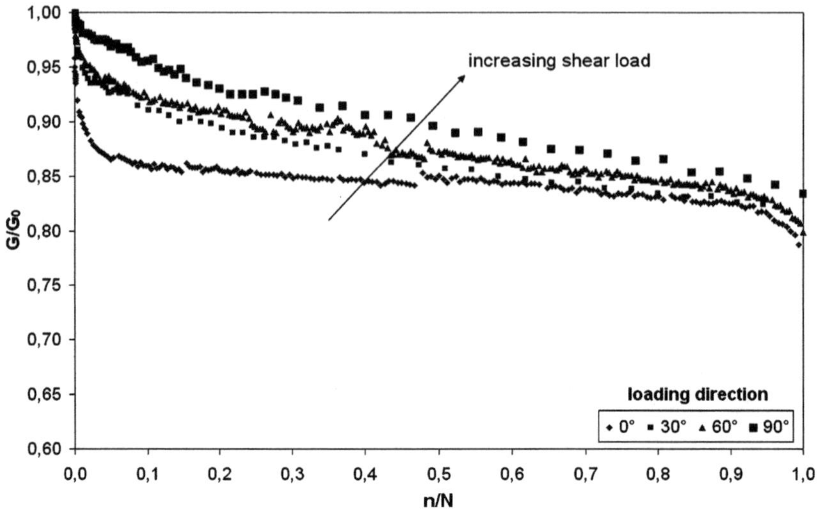

Fig. 7 Shear modulus versus normalized life

failure. Between these two phases, there is a phase with a slow, almost linear decrease. A similar behaviour is visible in Fig. 7 for the decrease of the shear modulus. The similarities between Figs. 3–7 suggest to look for a link between stiffness decrease and crack density as a first step, before trying to predict fatigue life itself. Finally, one can see that the increase of the crack density and the decrease of the stiffness are different for different levels of bi-axiality.

5 Conclusions

The fatigue behavior of NCF tube-specimens under multi-axial loading is presented, using the number of cracks counted under an optical microscope. The influence of bi-axiality is clearly visible. This means that a purely one-dimensional testing of such materials is a simplification.

Acknowledgements The authors wish to thank the German Research Foundation (DFG) for funding this research in the frame of the Focus Project SPP1123 "Textile Composites".

References

1. P. J. Hogg, A. Ahmadnia, F. J. Guild "The mechanical properties of non-crimped fabric-based composites", Composites, Vol. 24, No. 5, pp. 423–432, 1995
2. Y. Wang, J. Li, P. B. Do, "Properties of composite laminates reinforced with e-glass multiaxial non-crimp fabrics", Journal of Composite Materials, Vol. 29, No. 17, pp. 2317–2333, 1995

3. G. A. Bibo, P. J. Hogg, M. Kemp, "Mechanical characterisation of glass- and carbon-fibre-reinforced composites made with non-crimp fabrics", Composite Science and Technology, Vol. 57, pp. 1221–1241, 1997
4. L. E. Asp, F. Edgren, A. Sjögren, "Effects of stitch pattern on the mechanical properties of non-crimp fabric composites", Proceeding of the 11 ECCM, 31.05–03.06.04, Rhodes, Greece
5. M. Vettori, T. Chi, S. Lomov, I. Verpoest, "Progressive damage characterization of stitched, biaxial, multi-ply carbon fabrics composites", Proceeding of the 11 ECCM, 31.05–03.06.04, Rhodes, Greece
6. V. Carvelli, A. Corigliano, "Transversal resistance of long-fibre composites: Influence of the fibre-matrix interface", Proceeding of the 11 ECCM, 31.05–03.06.04, Rhodes, Greece
7. V. Carvelli, T. Chi, M. Larosa, S. Lomov, C. Poggi, D. Angulo, I. Verpoest, "Experimental and numerical determination of the mechanical properties of multi-axial multi-ply composites", Proceeding of the 11 ECCM, 31.05–03.06.04, Rhodes, Greece
8. S. Adden, P. Horst, "Damage propagation in non-crimp fabrics under bi-axial static and fatigue loading", Composite Science and Technology, Vol. 66, pp. 626–633, 2006
9. S. Adden, K. Pfleiderer, I. Solodov, P. Horst, G. Busse, "Characterization of stiffness degradation caused by fatigue damage using circumferential plate acoustic waves", accepted: Composite Science and Technology, 2008
10. A. Gagel, D. Lange, K. Schulte, "On the relation between crack-densities, stiffness degradation and surface temperature distribution of tensile fatigue loaded glass-fibre non-crimp-fabric reinforced epoxy", Composites Part A: Applied Science and Manufacturing, Vol. 37(2), pp. 222–228, 2006

Fatigue Crack Growth Assessment of Corroded Aluminum Alloys

Alexis Th. Kermanidis and Spiros G. Pantelakis

Abstract The fatigue crack growth behaviour of pre-corroded aluminum alloy has been investigated analytically. For this purpose the fatigue crack growth LTSM-F model has been implemented and upgraded to account for the local material properties of the pre-corroded material ahead of the crack tip. In the model the local fracture behaviour at the crack tip of the corroded alloy is governed by the local fracture toughness of the material. The analytical results obtained show good correlation with fatigue crack growth experiments under irregular aircraft service spectra from the literature and demonstrate the significance to consider the degradation of local material properties of structural members including corroded areas.

Keywords Fatigue crack growth · Modeling · Corrosion · Aluminum · Local fracture toughness · Aircraft service spectra

1 Introduction

Corrosion of metallic airframes is correlated to the degradation mechanisms due to the complex process of oxidation that affect the structural integrity and hence the safety of aircraft. Several investigators have experimentally studied the effect of corrosion on the fatigue performance of aluminum alloys [1–6]. Bray et al. [3] have demonstrated the influence of prior corrosion pitting on the fatigue strength

A. Th. Kermanidis (✉)
Department of Mechanical and Industrial Engineering, University of Thessaly 38334, Volos, Greece
e-mail: akermanidis@mie.uth.gr

S.G. Pantelakis
Laboratory of Technology and Strength of Materials (LTSM), Department of Mechanical Engineering & Aeronautics, University of Patras, 26500 - Patras, Greece
e-mail: pantelak@mech.upatras.gr

S. Pantelakis, C. Rodopoulos (eds.), *Engineering Against Fracture: Proceedings of the 1st Conference*,

of 2024-T3 and 2524-T3 aluminum alloys. Furthermore in [5, 6] the effects of pre-existing localized surface pitting corrosion on the fatigue lives of 7075 alloy have been investigated. Concerning the influence of existing corrosion on damage tolerance, it was found in [7] that there is no significant effect of prior exfoliation corrosion on the fatigue crack growth rate of 2024-T351 specimens. On the other hand, a correlation between existing corrosion damage and reduction of critical crack length to failure has been demonstrated in [8].

Despite the efforts made and although essential for damage tolerance design, the effect of corrosion on damage tolerance characteristics has not yet been sufficiently understood. Moreover, limited investigations and experimental data exist on the influence of pre-existing corrosion on the damage tolerance performance of aluminum alloys. Corrosion repair and maintenance procedures are typically over-conservative due to the lack of reliable methodologies to predict the future effects of corrosion. Therefore, an increasing need exists for the development of analytical tools to evaluate the effects of corrosion on structural behaviour of aircraft components [9]. The considerable interest in keeping aging airframes in service as long as possible is facilitated through damage tolerance design concepts with the requirement of periodic maintenance to assess the state of corrosion and fatigue damage in the airframe. Since fatigue cracks will initiate preferentially from any surface pits, including corrosion pits, the presence of corrosion reduces the fatigue life of the component. Hence, it is desirable to have analytical tools that will allow an assessment of the effect of corrosion damage on the remaining fatigue life of the component. These tools can be used to identify corrosion damage that needs to be removed and to establish appropriate inspection intervals.

In the present work an analytical methodology is presented for the fatigue crack growth assessment of corroded aircraft aluminum alloys. The methodology takes into account the degradation due to corrosion of local material properties ahead of the crack tip in the corrosion affected areas. For the assessment of properties in local scale, a synergetic effect of corrosion and corrosion-induced hydrogen embrittlement (e.g. 10–12) is used. The analytical results are compared against fatigue crack growth tests under aircraft service spectra provided by ASTM [13].

2 Corrosion-Dependent Local Fracture Toughness

In recent years experimental efforts on a series of aircraft aluminum alloys have provided evidence that corrosion is the cause for diffusion controlled material hydrogen embrittlement (e.g. 10–12). The significant information gained on the issue of the corrosion induced hydrogen embrittlement of 2024 alloy has been used in [14] for the suggestion of a simplified corrosion and hydrogen embrittlement process, for evaluating the corrosion induced local fracture toughness of the material.

Accrding to the work in [14] the surrounding area of the crack tip in a corroded fracture toughness specimen consists of a sequence of prospective material

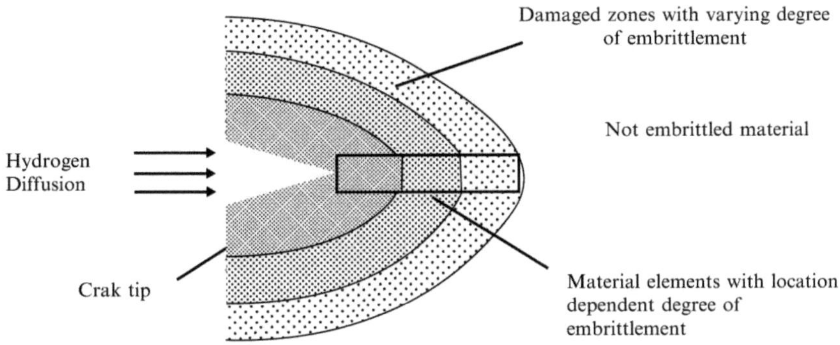

Fig. 1 Location-dependent degree of embrittlement at the crack tip

elements with varying degree of damage and hence varying degree of embrittlement (Fig. 1). Since the underlying damage processes are diffusion controlled, the severity of the corrosion and hydrogen-induced damage in each of these elements depends on location and time.

In [14], the local fracture toughness of the material ahead of the crack tip has been evaluated according to the equation:

$$K_{cr}{}^{cor} = K_{cr}\sqrt{\lambda} \tag{1}$$

Where Kcr is the fracture toughness of the uncorroded material and the corrosion factor μ can be calculated from equation:

$$\mu = [1 - \bar{\varepsilon_f}(1 - e^{-\beta t})](1 - Pt^m) \tag{2}$$

When the case specific constants P, m, $\bar{\varepsilon}_f$, β are known [14]. The corrosion factor values of the material range between μ = 1 for the uncorroded material and a minimum corrosion factor microminutes referring to the material which has reached maximum embrittlement for the case under consideration.

3 Fatigue Crack Growth Code

For the calculation of fatigue crack growth under irregular loading conditions the LTSM-F code is implemented. It consists of two modules, the first of which supports the simulation of irregular service spectra by equivalent stress cycles, while the second serves as a basis for the calculation of fatigue crack growth, accounting also for overload and overload interaction effects. The fatigue crack growth retardation model has been first introduced in [15] and further upgraded in [16].

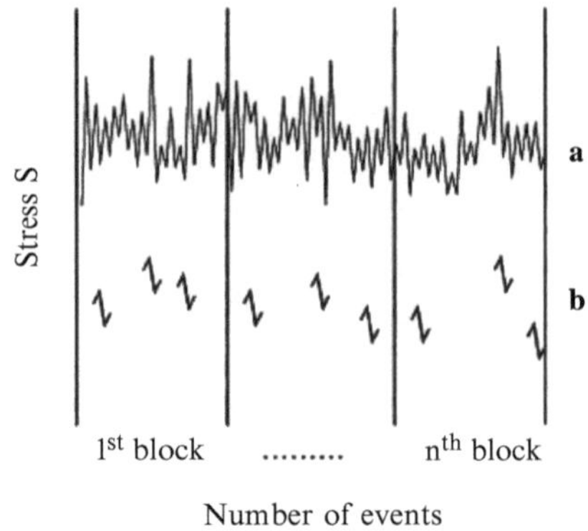

Fig. 2 **a** Typical aircraft stress spectrum discretized into n loading blocks. **b** Equivalent spectrum consisting of full, distinguished cycles

4 Stress Spectrum Simulation

Implementation of the model presented for the assessment of fatigue damage of the material subjected to service spectra requires the transformation of the loading spectrum from Fig. 2a to a sequence of distinguished reversed stress cycles as they are shown in Fig. 2b. Plastic deformation is considered as an accumulating damage after each load event, while additional damage effect due to unloading or compressive events is considered as negligible. The implementation of the rainflow counting method can satisfy the above requirements. However, classical rainflow algorithms do not account for the relative position of each event within the service fatigue spectrum. This may be interpreted as the a priori assumption of linear damage accumulation during fatigue. In the LTSM-F code the rainflow algorithm of [17] has been modified; the service loading spectrum is discretized into n loading blocks. Each block consists of a small number of events as it is shown in Fig. 2. Then, the rainflow algorithm is implemented for each of these blocks separately. The introduced modification limits the assumption of linear fatigue damage accumulation within each block. Mathematically it refers to the approximation of a power function by a sufficient number of prospective, short straight lines with increasing slopes. In addition, this modification allows the consideration of the relative position of the overloads within the entire service spectrum.

5 Fatigue Crack Growth Calculation

In the fatigue crack growth analysis code [16] the crack growth rate is described by the equation:

$$\frac{d\alpha}{dN} = \left(\frac{\sqrt{A}}{\sqrt{2}K_{cr}}\right)^{2/m} \left(\frac{\pi}{32S_y{}^2}\right)^{1-\frac{\beta-1}{m}} \Delta\mathrm{K}^{2(1-\frac{\beta-2}{m})} \tag{3}$$

where the quantities

$$C = \left(\frac{\sqrt{A}}{\sqrt{2}K_{cr}}\right)^{2/m} \left(\frac{\pi}{32{S_y}^2}\right)^{1-\frac{\beta-1}{m}} \tag{4}$$

and

$$\mathrm{n} = 2\left(1 - \frac{\beta - 2}{\mathrm{m}}\right) \tag{5}$$

are constant and depend on the material and the stress ratio. Equation (3) is of the same form as the Paris law and may be applied for fitting constant stress amplitude fatigue crack growth curves. Fitting parameters are the quantities A and β. For all metals the parameter m ranges between 0.5–0.7. For the assessment of the crack growth behavior after an overload a retardation factor λ is defined as:

$$\lambda = \frac{(d\alpha/dN)_{ret}}{(d\alpha/dN)_{nom}} \tag{6}$$

where (dα/dN)ret is the crack propagation rate following the overload, while (dα/dN)nom corresponds to the rate without the overload. Using Eq. (3) for calculating both, (dα/dN)ret and (dα/dN)nom the retardation factor can be derived as

$$\lambda = \left(\frac{S_{y0}}{S_y}\right)^2 \tag{7}$$

where Sy* and Sy are respectively the cyclic yield stress without and with overload plasticity. Sy decreases with the propagation of the crack through the overload plastic zone up to the value $\lambda = 1$ (Fig. 3). A linear variation from Sy* at the crack front to Sy0 at the region where the plastic zone created by an overload meets the virgin material is assumed as shown in Fig. 3. The linear approximation of Sy inside the plastic zone is given by:

$$S_y = {S_y}^* - \frac{\alpha + (\omega_s/2) - \alpha_0}{\omega_0}({S_y}^* - S_{y0}) \tag{8}$$

The calculation of the actual crack length α after application of i fatigue blocks of variable stress amplitudes is based on the relation:

$$\alpha = \alpha_i + \sum_{i=1}^{k} \lambda_i \left(\frac{d\alpha}{dN}\right)_{nom} \tag{9}$$

where αi is the crack length before application of the ith block, k the number of cycles involved in the block i and (dα/dN)nom can be estimated from the constant amplitude crack growth rule.

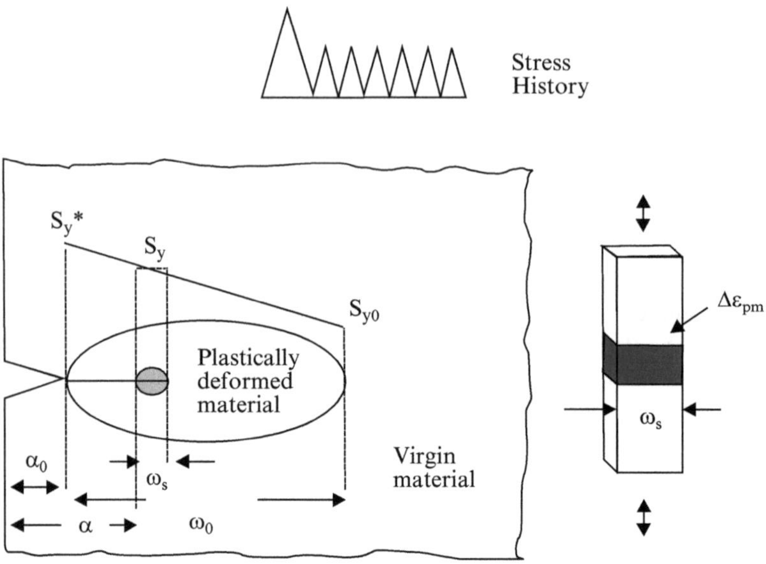

Fig. 3 Basic considerations for the retardation model

6 Implementation of Crack Growth Model for Corroded Material

Based on the described model an assessment of the influence of corrosion damage on the crack growth behaviour of the material can be performed. For this, the local fracture toughness value determined by Eq. (1) and the yield stress value of the corroded material is implemented. Accordingly, the crack growth rate in the corrosion damaged material may be assessed by using Eq. (3) in the form:

$$\left(\frac{d\alpha}{dN}\right)^{cor} = \left(\frac{\sqrt{A}}{\sqrt{2}K_{cr}^{cor}}\right)^{2/m} \left(\frac{\pi}{32\sigma_F^{cor^2}}\right)^{1-\frac{\beta-1}{m}} \Delta K^{2\left(1-\frac{\beta-2}{m}\right)} \tag{10}$$

For the assessment of crack growth rate following the application of an overload the retardation factor λ is implemented in the form:

$$\lambda = \left(\frac{\sigma_{F0}^{cor}}{\sigma_F}\right)^2 \tag{11}$$

Finally calculation of crack growth is performed via the formula:

$$\alpha = \alpha_i + \sum_{i=1}^{k} \lambda_i \left(\frac{d\alpha}{dN}\right)_{nom}^{cor} \tag{12}$$

For the assessment of fatigue crack growth in pre-corroded material the LTSM-F code has been implemented in the case of 2024 T351 aluminum alloy. Experimental results obtained in [8] of yield strength and fracture toughness for 2024 aluminum alloy which has been exposed in exfoliation corrosion solution for 36 h have been used for this purpose.

The crack growth rate in the case of the corroded aluminium alloy is calculated using Eq. (10). The material parameters A, β were determined using the crack growth curves of the uncorroded material [8]. For the low-cycle fatigue parameter m the value 0.7 has been used and was taken from [18]. The failure criterion for the calculation of fatigue life of the corroded material used was $K^{corcr} \geq$ Kcr local, where Kcr local is the local fracture toughness value of the corroded material as described in the previous paragraph. For the simulation of fatigue crack growth the aircraft service spectra provided in [13] have been used and correspond to Air to Air, Air to Ground and Instrumentation-Navigation flight phases (Fig. 4b). They are given as percentage values of a reference limit stress. The stress spectrum is generated by multiplication of the spectrum percentage factor with the limit stress.

The local fracture toughness value as described in the previous paragraph, was calculated for the specimen of Fig. 4a exposed for 36 h in exfoliation solution environment as $K_{crlocal} = 55\,\mathrm{MPa}\sqrt{m}(\mu = 0.155)$ [19]. With this value, the fracture toughness of the corroded specimen Kcor cr was estimated in [19] as $K_{cr}^{cor} = 110\,\mathrm{MPa}\sqrt{m}$. The yield strength of the corroded alloy exposed in the same environment was determined from experimental results in [8] as 336 MPa.

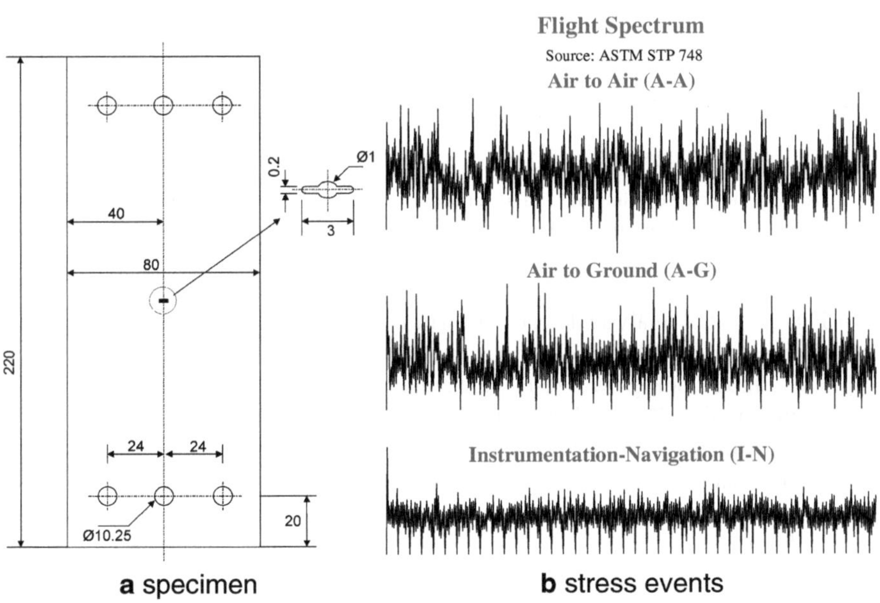

Fig. 4 **a** Fatigue crack growth specimen used for simulation and **b** aircraft service spectra according to ASTM

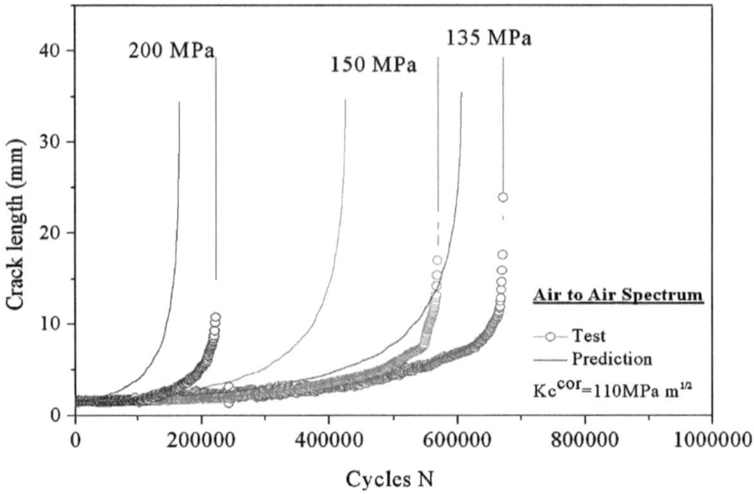

Fig. 5 Analytical and experimental fatigue crack growth curves for Air to Air spectrum

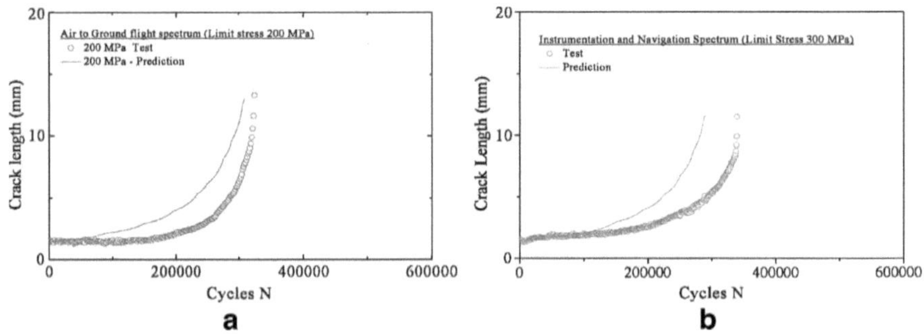

Fig. 6 Analytical and experimental fatigue crack growth curves for **a** Air to Ground spectrum **b** Instrumentation-Navigation spectrum

In Figs. 5–6 the analytical crack growth curves obtained by implementing the methodology described, are compared against experimental results performed in [19]. The fatigue crack growth tests were conducted for the Air to Air, Air to Ground and Instrumentation-Navigation flight phases for specific values of limit stress. In all cases examined a good correlation between analytical results and experiment exist with the model leading to a more conservative prediction in terms of fatigue life. Herein we should mention that in the crack growth model the compressive residual stresses due to load interaction effects have not been taken into account. Furthermore, the effect of existing corrosion notches on the fatigue crack growth behaviour has been neglected in the analysis. Finally, in the evaluation of analytical results the difficulty in assessing accurately the degree of embrittlement of the corroded alloy and hence the respective corrosion factor μ should be considered.

7 Summary

In the present paper the effect of corrosion on the fatigue crack growth behaviour of pre-corroded aluminum alloy has been investigated. For this purpose the LTSM-F fatigue crack growth model has been upgraded to account for local material property degradation ahead of the crack tip for the corroded alloy. The fatigue crack growth model presented uses the local fracture toughness value of the corroded material. The analytical results showed good correlation with experimental fatigue crack growth tests under irregular aircraft service loadings and demonstrate the capability to account for the influence of pre-existing corrosion on the material's properties for the reliable fatigue and damage tolerance analyses of components involving corroded areas.

References

1. Chen GS, Liao CM, Wan KCC, Gao M, Wei RP. Pitting corrosion and fatigue crack nucleation. In: Van Der Sluys WA, Piascik RS, Zawieruca R, editors. Effects of environment on the initiation of crack growth, ASTM STP 1298. West Conshohocken, PA: American Society for Testing and Materials, 1997, pp. 18–33.
2. Sankaran KK, Perez R, Jata KV. Effects of pitting corrosion on the fatigue behavior of aluminum alloy 7075-T6: modeling and experimental studies. Mater Sci Eng A 2001;A297: 223–229.
3. Bray GH, Bucci RJ, Colvin EL, Kulak M. Effect of prior corrosion on the S/N fatigue performance of aluminum sheet alloys 2024-T3 and 2524-T3. In: Van Der Sluys WA, Piascik RS, Zawieruca R, editors. Effects of environment on the initiation of crack growth, ASTM STP 1298. West Conshohocken, PA: American Society for Testing and Materials, 1997, pp. 89–101.
4. Rokhlin SI, Kim JY, Nagy H, Zoofan B. Effect of pitting corrosion on fatigue crack initiation and fatigue life. Eng Fract Mech 1999;62(4–5):425–444.
5. Pao PS, Feng CR, Gill SJ. Corrosion fatigue crack initiation in aluminum alloys 7075 and 7050. Corrosion 2000;56(10):1022–1031.
6. Pao PS, Gill PS, Feng CR. On fatigue crack initiation from corrosion pits in 7075-T7351 aluminum alloy. Scr Mater 2000;43:391–396.
7. Chubb JP, Morad TA, Hockenhull BS, Bristow LW. The effect of exfoliation corrosion on the fatigue behaviour of structural aluminium alloys, structural integrity of ageing airplanes, Berlin: Springer, 1991, p. 87.
8. Kermanidis AlTh, Petroyiannis PV, Pantelakis SpG. Fatigue and damage tolerance behaviour of corroded 2024 T351 aircraft aluminum alloy. Theoret Appl Fract Mech 2005;43:121–132.
9. Aging of US Air Force Aging Aircraft. National Materials Advisory Board Publication NMAB-488–2, Washington, DC: National Academy Press, 1997, pp. 54.
10. Pantelakis SpG, Haidemenopoulos GN. Corrosion and hydrogen embrittlement of aircraft aluminum alloys. In: Proceedings of the 4th International Conference on New Challenges in Mesomechanics, Aalborg University, Denmark, 2002.
11. Petroyiannis PV, Kermanidis AlTh, Kamoutsi E, Pantelakis, SpG, Bontozoglou V, Haidemenopoulos GN. Evidence on the corrosion-induced hydrogen embrittlement of the 2024 aluminum alloy. Fatigue Fract Eng Mater Struct 2005;28:565–574.
12. Petroyiannis PV, Kermanidis AlTh, Papanikos P, Pantelakis SpG. Corrosion-induced hydrogen embrittlement of 2024 and 6013 aluminum alloys. Theoret Appl Fract Mech 2004;41:173–183.

13. Chang JB. Round robin crack growth predictions on center-cracked tension specimens under random spectrum loading, methods and models for predicting fatigue crack growth under random loading. ASTM STP 1981;748:3–40
14. Kermanidis AlTh, Pantelakis SpG. A fracture mechanical model for the prediction of the corrosion dependent local fracture toughness of 2024 Alloy. J Strain Anal Eng Design 2006;43(3):253–261.
15. Pantelakis Sp, Kermanidis Th, Pavlou D. Fatigue crack growth retardation assessment of 2024-T3 and 6061-T6 aluminium specimens. Theoret Appl Fract Mech 1995;22:35–42.
16. Kermanidis AlTh, Pantelakis SpG. Fatigue crack growth analysis of 2024 T3 aluminum specimens under aircraft service spectra. Fatigue Fract Eng Mater. Struct 2001;24:699–710.
17. Matsuishi M, Endo T. Fatigue of metals subjected to varying stress. In: Proceedings of the Kyushu Branch of Japan Society of Mechanical Engineering, Fukuoka, Japan (in Japanese), 1968, pp. 37–40.
18. Dowling NE. Mechanical behavior of materials, Upper Saddle River, NJ: Prentice-Hall, 1993.
19. Kermanidis, AlTh. Effect of Corrosion on the Structural Integrity of Lightweight Aircraft Structures, Ph.D. thesis, University of Patras, Department of Mechanical and Aeronautical Engineering, Patras, Greece, 2003.

Fatigue Crack Growth Behavior under Spectrum Loading

Daniel Kujawski and Stoyan Stoychev

Abstract The purpose of this study is to investigate experimentally the fatigue crack growth behavior under typical service loading spectrum recorded from military P-3 cargo airplane. The material is 2324 – T39 Al alloy widely used in the aircraft industry. Experiments were conducted using the full spectrum and modified versions of it such as only 'positive' (no negative loads) or with reduced (clipped) high peaks. The results show that the compressive loads decrease fatigue life of the specimen by ~300%. Furthermore, by running tests with clipped peaks it was found that the fatigue life was shorten significantly due to reduction of crack growth retardation caused by highest tensile peaks. Multiple tests were conducted in order to establish a scatter in the experimental data under spectrum loads.

Keywords Spectrum loading · Spectrum modification · Fatigue crack growth

1 Introduction

Characterization of fatigue crack growth (FCG) under constant amplitude loading is fairly well established using fracture mechanics framework [1–3]. However, most of the structures in service are subjected to cyclic loads with randomly varying amplitudes that introduce additional complexities in FCG description related to load interaction effects. In the past, most of the experimental studies have been done using simplified load interaction effects associated with a single overload, underload or combinations of both.

D. Kujawski (✉)
Department of Mechanical and Aeronautical Engineering, College of Engineering and Applied Sciences, Western Michigan University, Kalamazoo, MI 49008–5343, USA
e-mail: daniel.kujawski@wmich.edu

S. Stoychev
Dana Corporation, 6938 Elm Valley Dr., Kalamazoo, MI 49009, USA

S. Pantelakis, C. Rodopoulos (eds.), *Engineering Against Fracture: Proceedings of the 1st Conference*,

The purpose of this paper is to investigate experimentally the fatigue crack growth behavior under typical service loading spectrum, recorded from military P-3 cargo airplane. In particular, the effects of compression loads and the redaction in magnitude (clipping) of the highest peaks in the spectrum is systematically investigated and analyzed. Simulations and analytical predictions of these results using a two-parameter fatigue crack driving force approach [4] will be reported in separate paper.

2 Experimental Setup

The material used in is 2324-T39 Al alloy (yield strength 370 MPa, ultimate strength 475 MPa, and elongation at break 8%). The specimen is a flat plate with a notch machined at one of the edges (Fig. 1). The actual dimensions of all specimens used in the tests are listed in Table 1. It can be seen that there is some variation in the thickness and an initial crack length of the specimens which can partially explain the scatter observed in the experimental data.

This geometry is chosen due to compressive loads present in the spectrum. Also a single edge geometry provides a large crack propagation interval which allows applying multiple passes of the spectrum on the same sample. The gage length of the specimen is shorter in comparison to the standard requirements [5, 6]. This allows

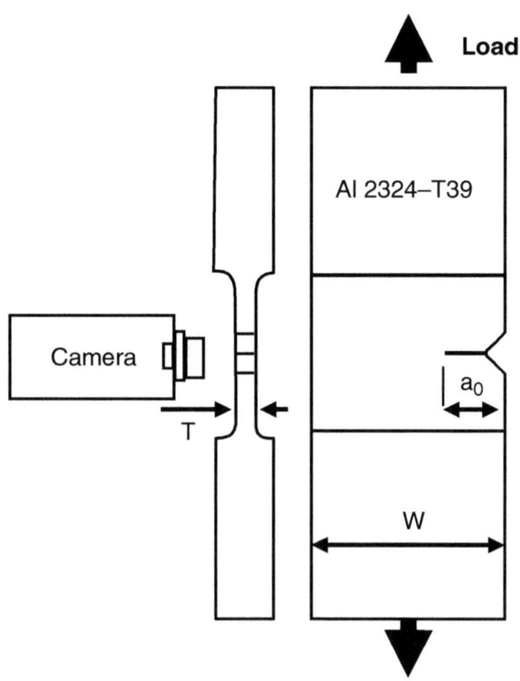

Fig. 1 Specimen geometry

Table 1 Specimens dimensions

Specimen name	W (mm)	T (mm)	a_0 (mm)
N3	44.80	2.64	13.79
N5	44.84	2.55	8.28
N7	44.83	2.80	8.24
S1	44.28	2.87	8.35
S2	44.61	2.81	7.79
S3	44.58	2.59	8.98
S4	44.55	2.79	7.70
S5	44.22	3.02	8.00
S6	44.54	3.01	7.32
S7	44.38	2.92	6.71
S8	44.42	3.10	6.97
S9	44.13	2.79	7.54
S10	44.27	2.99	8.11
S11	44.23	2.66	8.69
S12	44.74	3.00	9.26
S13	44.20	2.92	9.83

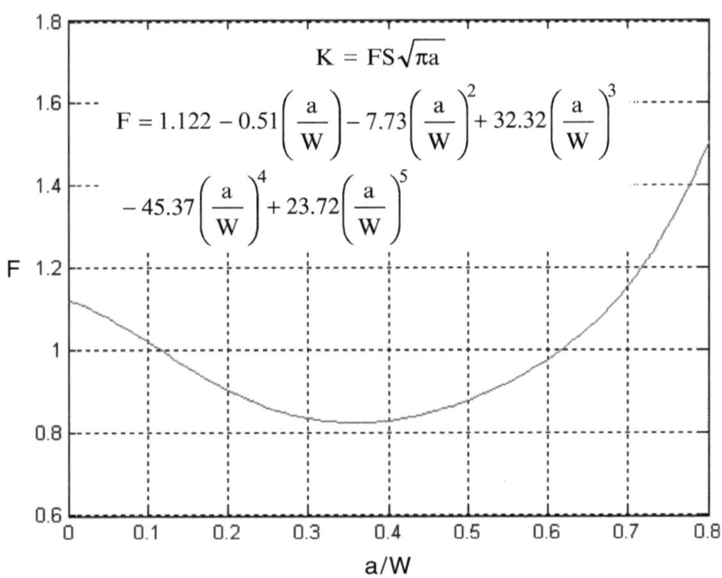

Fig. 2 Custom function for calculation of the stress intensity factor [7]

for more efficient use of the material, but requires a custom solution for the stress intensity factor as a function of the crack length (Fig. 2). This solution was obtained through finite element analysis of the stress field around the crack tip at different crack lengths [7].

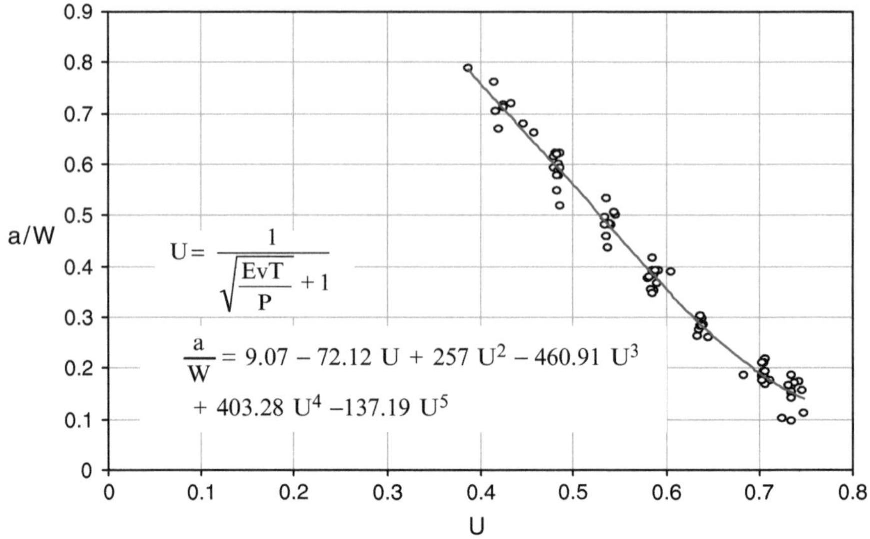

Fig. 3 Function for calculation of the crack length from compliance

The crack length extension was recorded every 0.04 mm from the compliance measurement of the specimen. In addition to that, the crack growth was monitored periodically (every 5 mm) using optical system consisting of microscope and camera. Figure 3 shows the calibration procedure for the compliance measurements of the crack length. The crack opening displacement, v, is measured automatically with a clip gage, mounted at the notched edge of the specimen (crack mouth). Only the upper linear part of the load-displacement records (fully open crack) was used for calculation of the compliance, v/P (mm/N). The thickness of the specimen, T, is in (mm), the elastic modulus, E, is measured in (GPa) and the load, P, in (N). The open circles in Fig. 3 show experimental data from all tests.

All specimens were pre-cracked using two-step procedure. First, the crack was initiated and propagated under fully compressive loading (Pmin = −19 kN and Pmax = −0.15 kN) for 200,000 cycles. This procedure generates self arrested crack of about 0.5 mm in length. Then, the crack was propagated under zero-to-tension loading with Pmax = ΔP = 6 kN until it reached ∼8 mm (including the notch).

2.1 Load Spectrum

The load spectrum for the tests is a typical service loading recorded from military P-3 cargo airplane. The raw spectrum (provided by the US Navy) was filtered an analyzed using the dedicated 'Spectrum Explorer' software developed specifically for this project. Figure 4 shows the normalized load spectrum along with the results from the procedure for filtering of the redundant (no peaks) data points. The max-

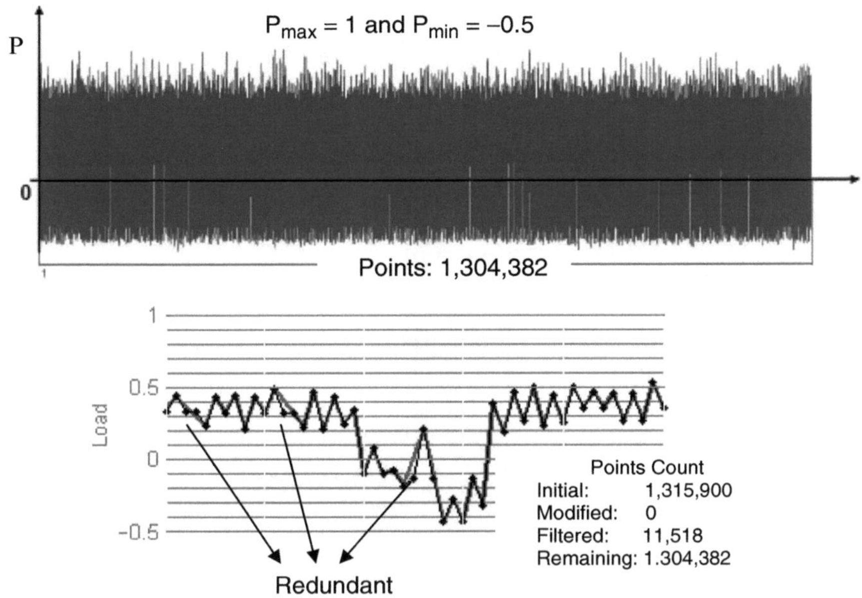

Fig. 4 Normalized full spectrum

imum load, Pmax = 18 kN (or so called spectrum load multiplier), is kept constant for most of the tests and is chosen low enough to allow at least a few repetitions of the spectrum to be applied before specimen failure. The insert in Fig. 4 illustrates filtering of the redundant (no peaks) data points.

The above load spectrum can be represented more concisely using the well known rainflow cycle-counting algorithm as it is shown in Fig. 5.

At constant amplitude loading the negative loads are often ignored in the fatigue crack growth description as damaging. There is however, experimental evidence [1–3, 8] that a negative load applied after overload can significantly decrease the effect of the overload on the fatigue crack growth. Therefore, in order to estimate the effect of the negative loads, the full spectrum shown in Fig. 4 is modified by removing the compressive loading.

Figure 6 illustrates how this modification is done. First, all negative peaks are moved up to zero load. Those are the 'modified' points. This creates some additional redundant points which are 'filtered' during the second pass with the filter. These steps are repeated until there are no more filtered or modified points.

The procedure illustrated in Fig. 6 effectively eliminates any compressive loads from the spectrum which is called 'positive spectrum'. Figure 6 shows also the clipped load levels corresponding to 95%, 85% and 65% of the based load, Pmax = 18 kN, used in this investigation. The positive spectrum is represented in Fig. 7 by rainflow cyclic-counting procedure. All spectra used in this analysis can be found in [9].

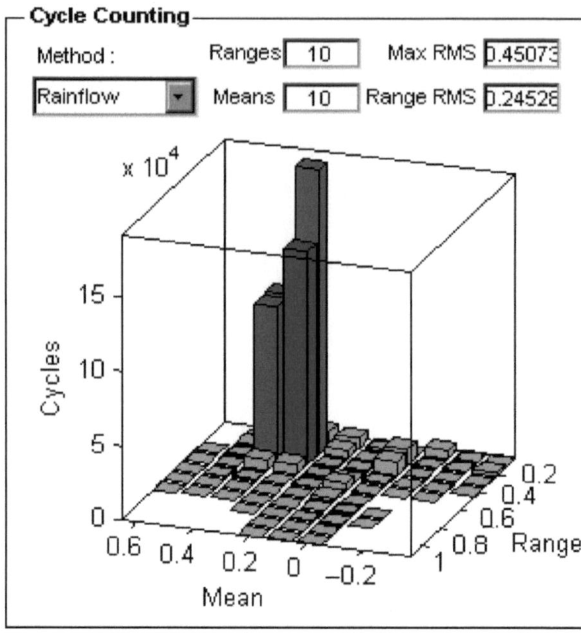

Fig. 5 Results from the rainflow cycle-counting procedure

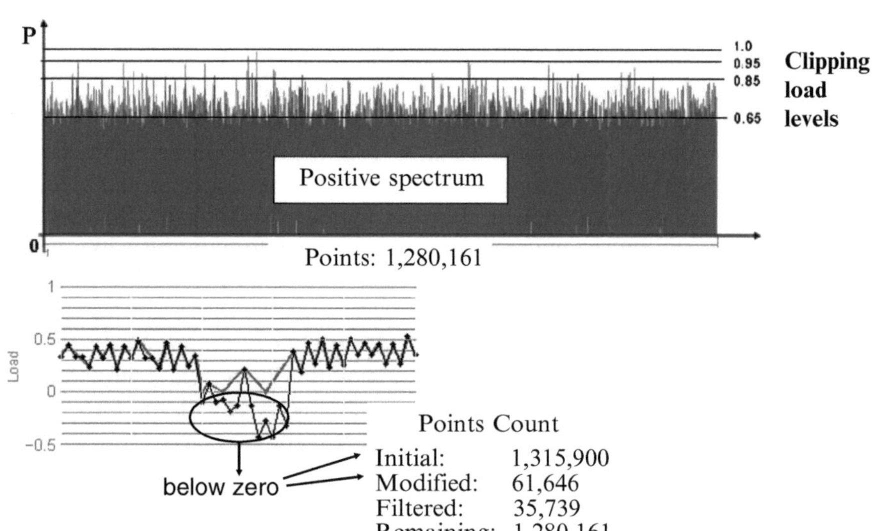

Fig. 6 Normalized positive spectrum with clipping load levels and the procedure for elimination of the compressive loads from the spectrum

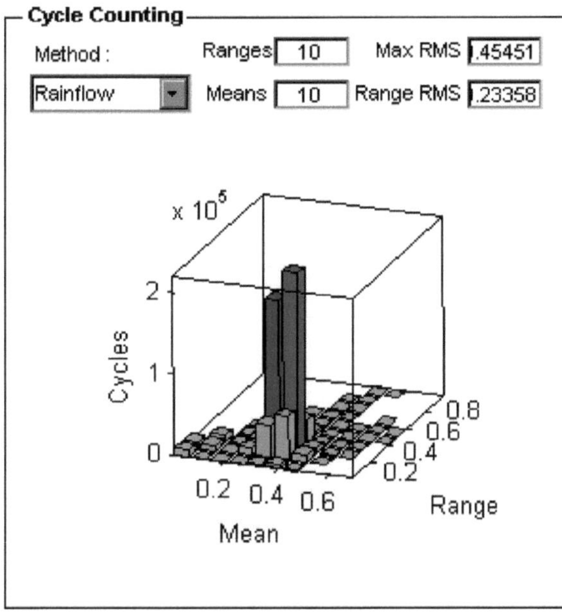

Fig. 7 Positive load spectrum representation using rainflow counting procedure

3 Results

Figures 8–11 depict the raw experimental results of the whole test program in terms of crack length, a, versus number of cycles, N. There are three aspects of the load spectrum modification that were studied, namely the effects associated with:

(a) The removal (or truncation) of low magnitude cycles from the spectrum
(b) The reduction (or clipping) in magnitude of the high peaks in the spectrum (see Fig. 6) and
(c) The spectrum load scaling

Figures 8–10 show the effect of each one of those factors on the fatigue crack growth, while the other two are kept constant. Figure 11 is showing the combined effect of multiple factors. The dashed, vertical lines mark the ends of each spectrum repetition.

Figure 8 shows the effect of the compressive load truncation on the fatigue crack growth. The original full spectrum has a minimum limit Pmin $= -0.5$Pmax. This minimum was limited to Pmin $= 0$ and later to Pmin $= -0.2$Pmax, using the procedure from Fig. 6.

The results for full and the positive load spectra are used as baselines for comparisons later in the study. Therefore, the tests with both full and positive spectra were repeated four times each in order to establish a scatter in the experimental data. The average curves shown in Fig. 8 were determined by averaging the number of cycles for a particular crack length.

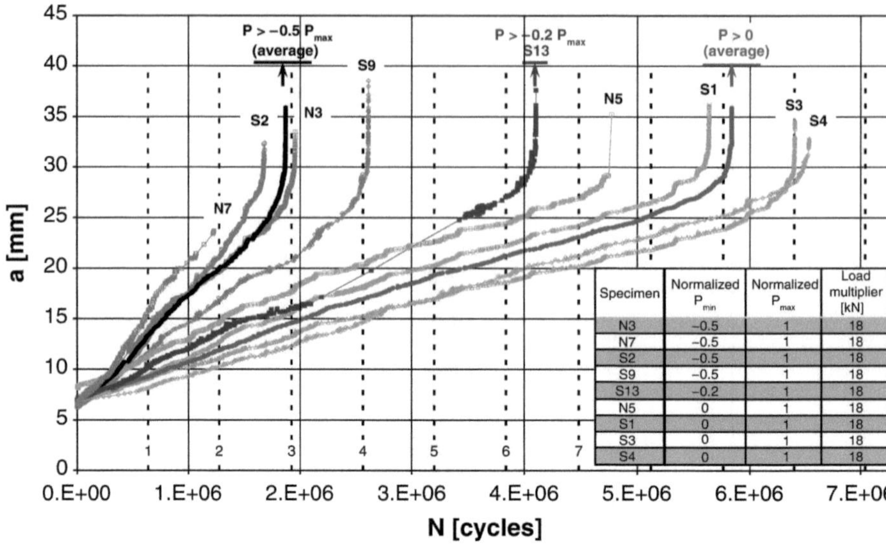

Fig. 8 Effect of the minimum peaks truncation on the fatigue crack growth

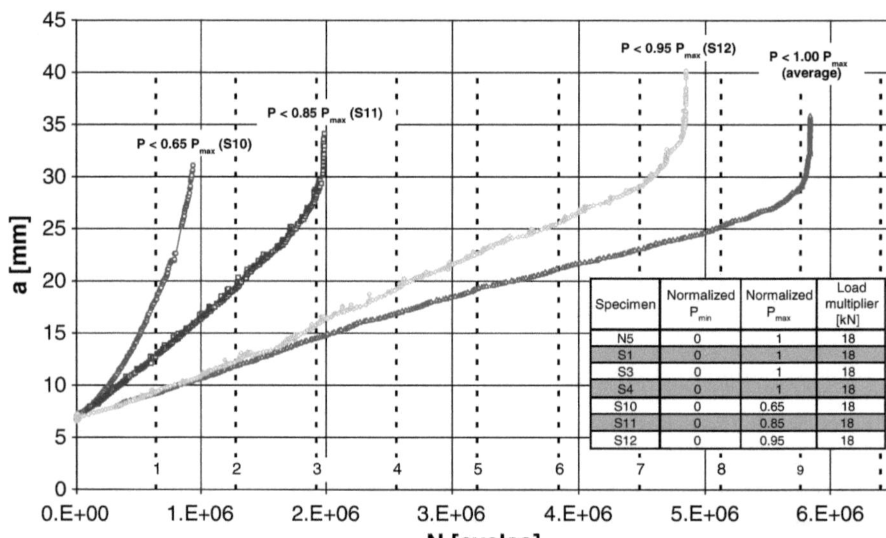

Fig. 9 Effect of the maximum load clipping on the fatigue crack growth

Figure 9 shows the effect of the maximum load clipping on the fatigue crack growth. All tests are done using the positive load spectrum which was modified further to accommodate the maximum loading peaks of interest. The spectrum was modified using a variation of the procedure in Fig. 6, where the loads above a chosen

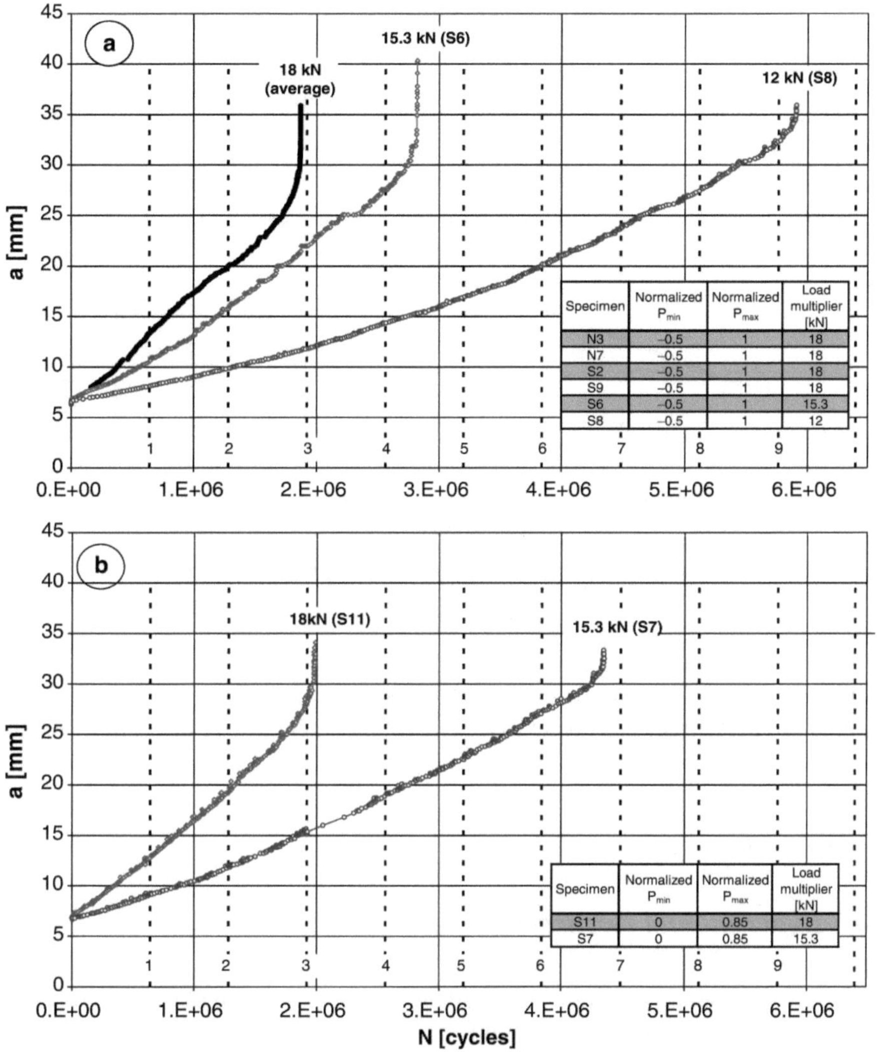

Fig. 10 Effect of the load scaling on the fatigue crack growth

limit were clipped to the limit level. In the rare occasions when both the peak and the valley of a given cycle are modified, one of the points is filtered as redundant according to the procedure in Fig. 4.

Figure 10 shows the effect of the load scaling on the fatigue crack growth. The load scaling is achieved simply by multiplying the normalized spectrum peaks with a different load multiplier. This effect is investigated both using the full spectrum (Fig. 10a), and the positive spectrum clipped to maximum limit of 0.85Pmax (Fig. 10b).

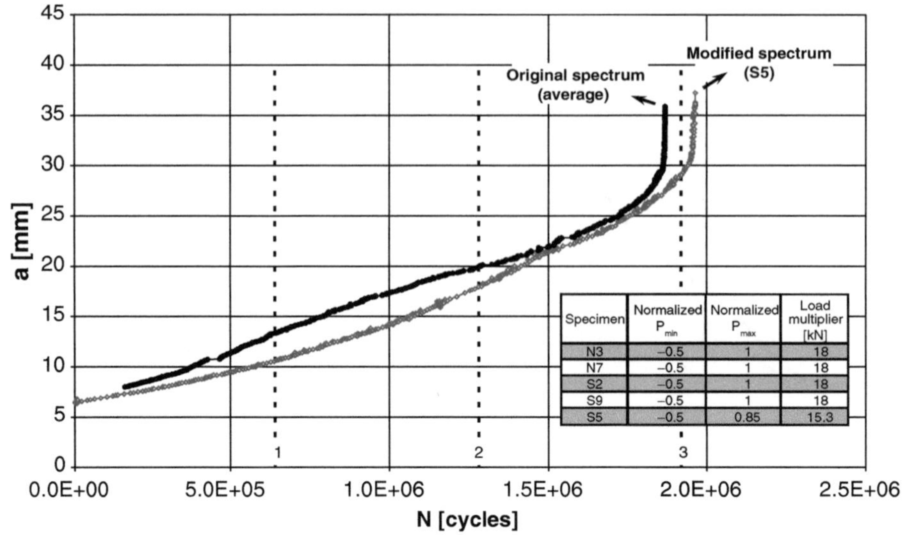

Fig. 11 Combined effects of clipping and scaling on the fatigue crack growth

Figure 11 shows the combined effect of the maximum load clipping and the load scaling on the full load spectrum. This test was designed as a benchmark for examining of the performance of the existing and future crack growth prediction models.

4 Discussion

4.1 Experimental Accuracy

The spectrum loading tests are very challenging both for the load frame and for the data acquisition system (DAQ). Therefore, in this study the current state of the art equipment is used. The load frame is MTS 810, the controller and dedicated DAQ systems (provided by Fracture Technology Associates). The adaptive load compensation option allowed the tests to be conducted fast and efficient. Constant load rate (38.3 V/s at 8.896 kN/V) is used for all tests. The loading frequency is adjusted automatically to achieve the desired load rate. This allows the cycles that have small amplitudes to be executed much faster than the large ones. Additionally, the loading frequency is kept below the load frame limit of 50 Hz and above 20 Hz (just to make the tests faster).

When the test is run so fast, there are variable phase lags between the load command and the feedback. Additionally, the DAQ system is challenged to acquire 400 load/displacement pairs and calculate the compliance each cycle. Finally, as the crack grows, so do the displacements and there is a possibility that the system can

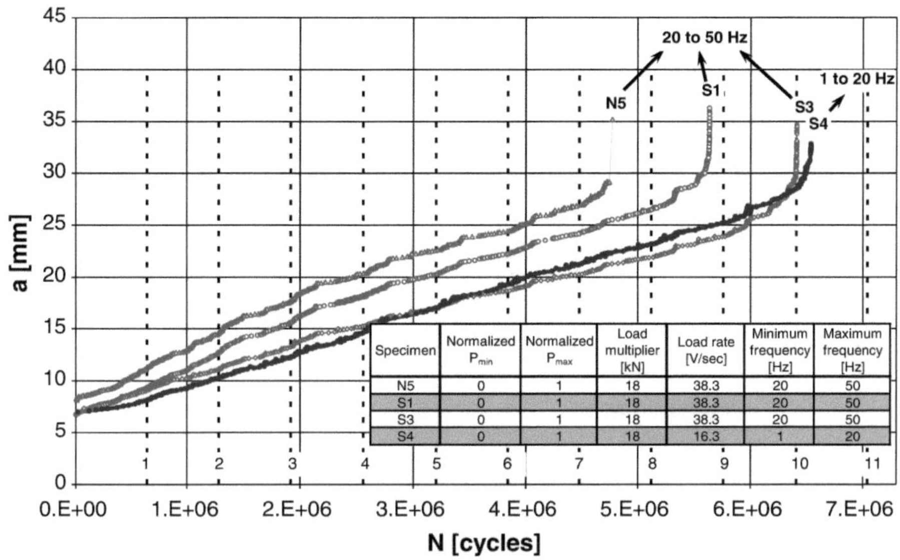

Specimen	Normalized P_{min}	Normalized P_{max}	Load multiplier [kN]	Load rate [V/sec]	Minimum frequency [Hz]	Maximum frequency [Hz]
N5	0	1	18	38.3	20	50
S1	0	1	18	38.3	20	50
S3	0	1	18	38.3	20	50
S4	0	1	18	16.3	1	20

Fig. 12 Loading rate effect on the fatigue crack growth

not reach the maximum load towards the end of the test. Those problems, if they present might erode the confidence in the validity of the experimental results. In this study, special care is taken to avoid them. The error between the target load and the feedback signal is monitored continuously and the test would stop if it is bigger than 2% of Pmax. Also, one of the positive spectrum tests, S4, was conducted more than two times slower (Fig. 12). It can be seen that the crack growth rate remains approximately the same, irrespective of the loading rate. Therefore, it can be concluded that the load was always applied properly and to the full extent of Pmax.

The remaining scatter in the experimental results can be attributed partially to variability in the specimen geometry as shown in Fig. 13. It can be seen that, in general, there is some correlation between the life of the specimen and the variation in thickness 'T' and the initial crack size 'a0'. This is expected, since the decrease in area with decreasing thickness leads to the higher stresses for the same applied load. However, there are two specimens that do not follow this trend, S3 and N3. It can be speculated, that this is due to the fact that the humidity in the laboratory is not maintained constant. Typically, it is changing between 20% in winter to about 80% in summer. Unfortunately, this fact was discovered too late in the test program and the humidity was not recorded for most of the specimens.

Figure 13 shows that the amount of scatter in the fatigue life remains the same for both positive and full spectrums. Therefore the relative error in the results will be smaller for the specimens with longer fatigue life. This makes the positive spectrum, a logical choice for further experimental investigations.

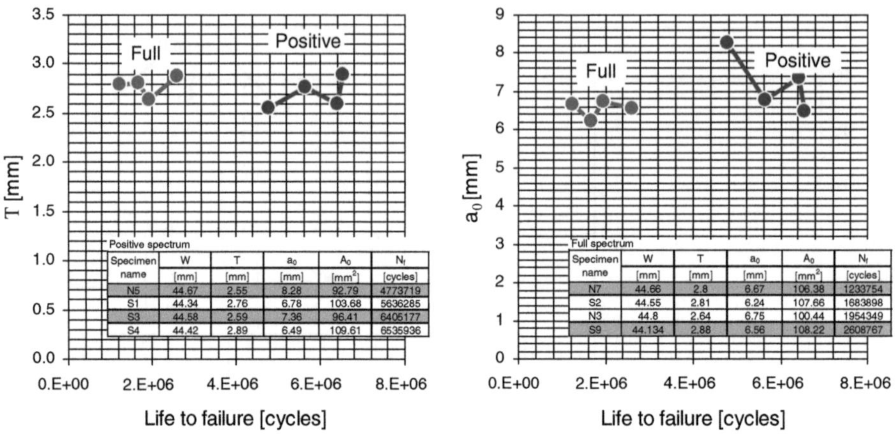

Fig. 13 Effect of the specimen geometry on the fatigue life

4.2 Compressive Loads

In the analysis of the fatigue crack growth under constant R-ratio loading, the influence of the compressive loads is generally neglected. For example, the ASTM standard [5] stipulates that ΔK is assumed to be equal to Kmax for negative load ratios. This basically comes from the notion that if the crack is closed under compressive load (or stress), it does not have a stress intensity factor associated with it. Under constant loading this assumption is mostly valid and does not cause any significant errors in the fatigue crack growth predictions [1–3]. In the case of more involved load histories the compressive cycles, although "not damaging" for steady state condition (constant R-ratio test), can modify significantly the stress field ahead of the crack for spectrum loading situation. As a result, the effect of the subsequent tension cycles on the fatigue crack growth could be modified too. This was observed experimentally in earlier investigation by using sequences of overloads and underloads [1–3]. Such simple over- and under-load histories are instrumental in the development of crack growth prediction models. In the reality however, the load interactions in spectrum are much more complex. The reliable prediction of the actual crack growth acceleration/retardation due to the compressive part of the spectrum loading still remains a challenge.

In the present study the effect of the compressive cycles was studied experimentally by imposing a limit on the negative peaks as it is depicted in Fig. 8. The results are summarized in Fig. 14.

It can be seen that the life of the specimen increases with reduction (truncation) of the negative loading peaks. Furthermore, the presence of compressive peaks which are only half of the magnitude of the maximum positive peak reduces the life of the specimen by 300%. Therefore, the negative loads in spectrum loadings should be considered always in the predictions of the crack growth.

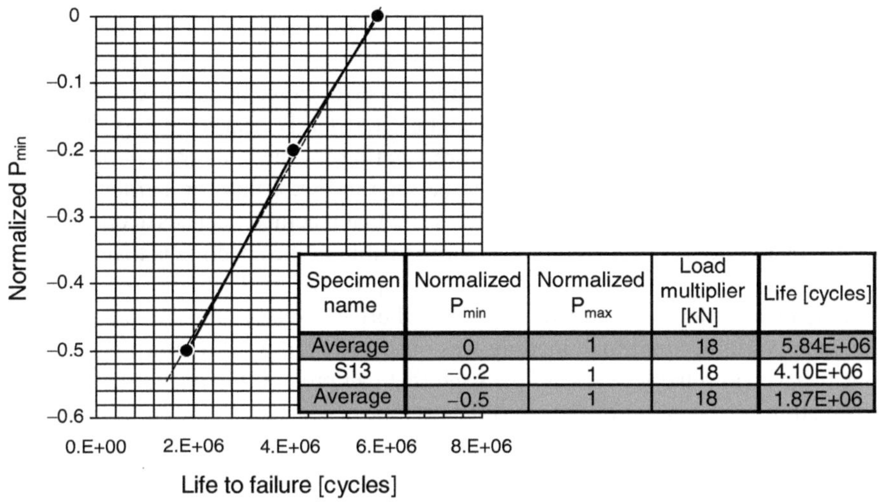

Specimen name	Normalized P_{min}	Normalized P_{max}	Load multiplier [kN]	Life [cycles]
Average	0	1	18	5.84E+06
S13	−0.2	1	18	4.10E+06
Average	−0.5	1	18	1.87E+06

Fig. 14 Summary of the compressive loads investigation

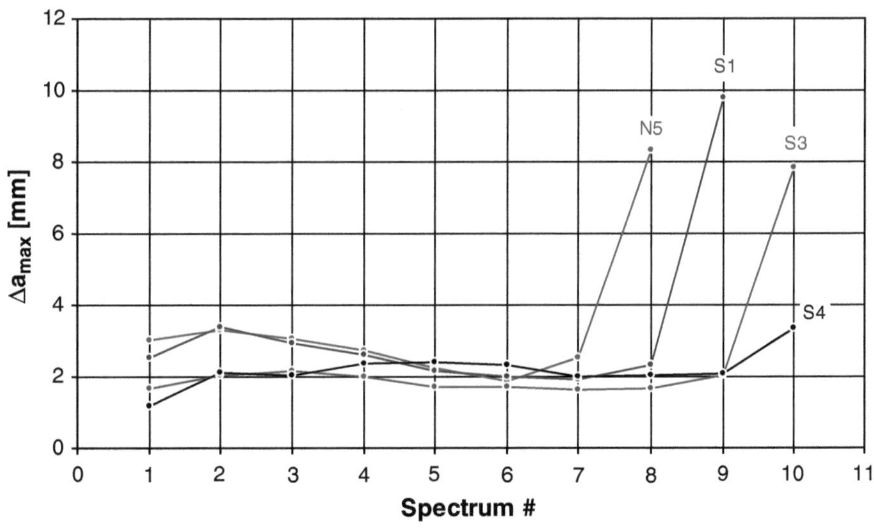

Fig. 15 Amount of crack growth per spectrum

4.3 Overloads

Figure 15 shows that the amount of crack extension for positive spectrum remains essentially the same for each spectrum repetition, irrespective of the size of the crack. As a result, the overall crack growth rate remains approximately constant for most of the specimen life. This can be explained by taking into account the existence of two competing mechanisms that alter the crack growth rate. First, it is the increase of the crack length which induces gradual increase of the crack driving force in terms of ΔK and Kmax as the test progresses.

And second, as a crack extends the highest peaks cause bigger overloads which induce larger retardation effects. This can be seen clearly near the end of the specimen life, where the crack almost stops propagating for the later spectra passes as it is shown in Fig. 15. Furthermore, it can be seen in Fig. 8 that for a crack length at about 20 mm the overload effect dominates and that it actually causes decrease in the crack growth rate. As expected, this is more visible in the specimens that are tested under the positive spectrum, because the compressive loads in full spectrum can partially or completely eliminate the overload effect. Nearly constant crack growth rate under variable amplitude loading was observed before [8]. Unfortunately, until now, there is no reliable software package that can predict this phenomenon. More information on the effect of the overloads can be obtained by plotting the observed crack growth rate '$\Delta a/\Delta N$' together with the applied load (Fig. 16). I can be seen that even though there are only ten peaks bigger than 0.9 Pmax, they are causing all the major retardations in the crack growth rate in the spectrum. The rate $\Delta a/\Delta N$ in Fig. 16 is an average for all spectrums tested at positive spectrum (four specimens and approximately ten spectrums per specimen to failure). The trends shown in Fig. 16 are essentially the same for each individual spectrum.

Figure 17 shows the number of peaks exceeding particular load level. The numbers that are circled correspond to the clipped load levels that are investigated in this study, namely 95%, 85% and 65% (see Fig. 4). It can be seen that even at 65% clipping only ~2,000 peaks were modified, which is negligible compared to the total number of peaks in the load spectrum (~1.2 million). Finally, Fig. 18 shows the test results from the clipping of the maximum load limit investigations.

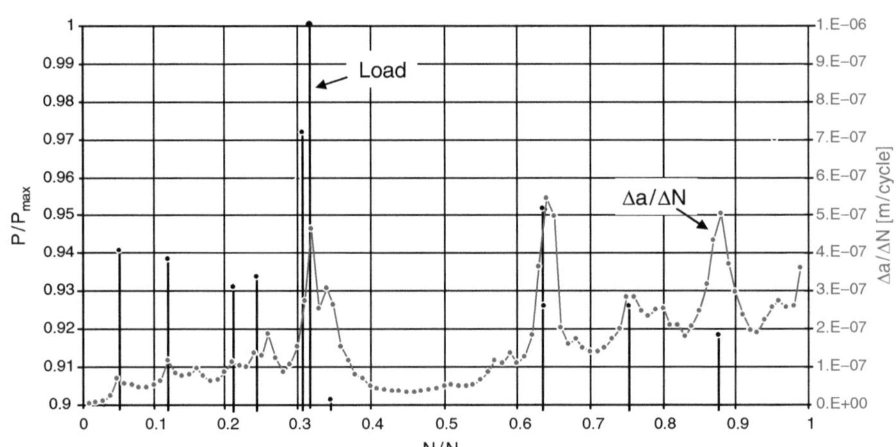

Fig. 16 Overload locations and effect on the crack growth under positive spectrum loading

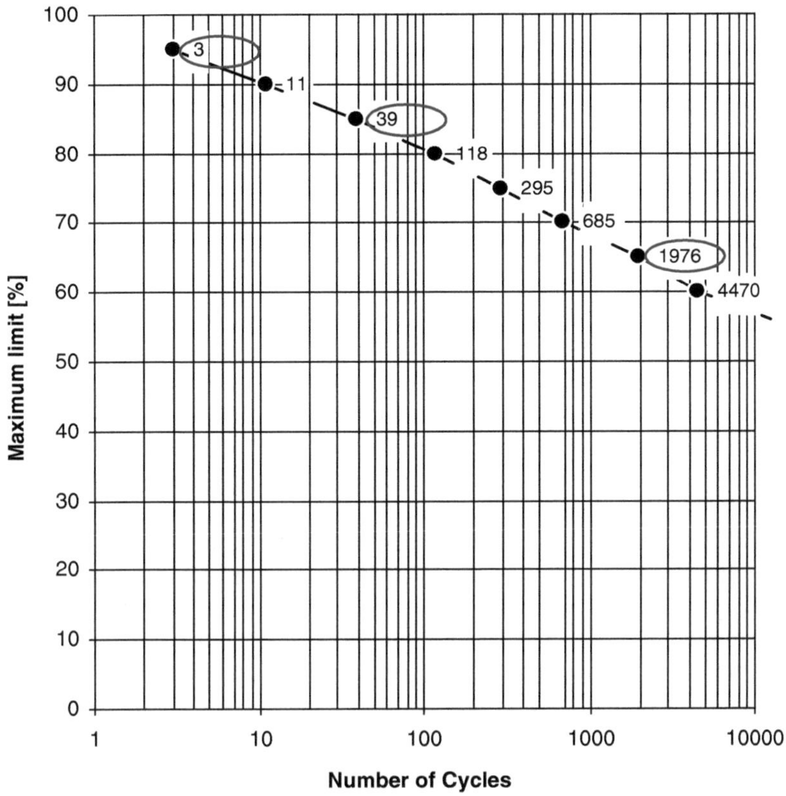

Fig. 17 Number of cycles exceeding particular load limit

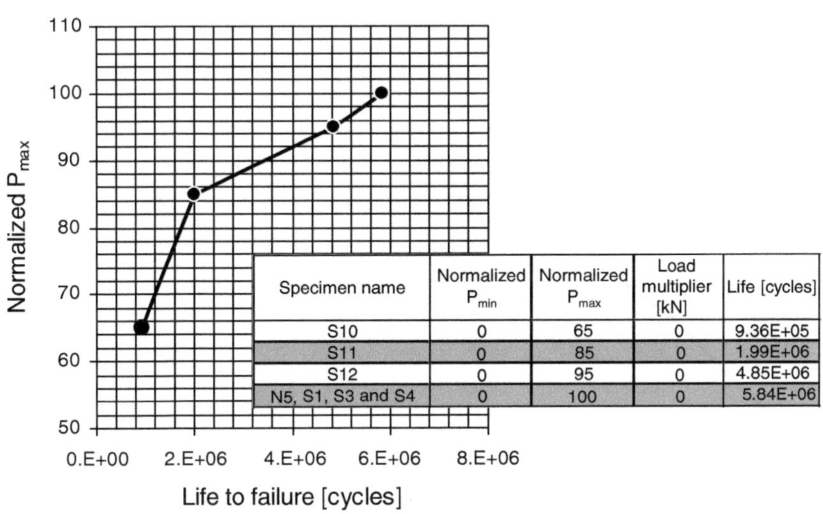

Specimen name	Normalized P_{min}	Normalized P_{max}	Load multiplier [kN]	Life [cycles]
S10	0	65	0	9.36E+05
S11	0	85	0	1.99E+06
S12	0	95	0	4.85E+06
N5, S1, S3 and S4	0	100	0	5.84E+06

Fig. 18 Summary of the maximum load clipping experiments

The fatigue life is increasing with the increased clipping of the maximum load limit. In other words, the fatigue life is increased by 600% (from one million for 0.65 Pmax to six million cycles for unmodified positive spectrum) due to the crack growth retardation effects associated with the tensile peaks (overloads) above 65% of Pmax.

4.4 Load Scaling

It has to be noted that all tests start with a load spectrum which is normalized between −0.5 and 1.0 as shown in Fig. 4. Then the each normalized peak is multiplied by a given Pmax called a spectrum load multiplier. Therefore, changing Pmax scales each peak and valley proportionally, instead of changing only the peaks or valleys. The experimental results shown in Fig. 19 indicate that the life of specimen S6 does falls just outside of the scatter band for the tests conducted under full load (specimens N7, S2, N3 and S9 shown in Fig. 8). Therefore, 15% reduction of overall load (or load multiplier) does not affect the life of the specimen significantly.

In the S8 experiment, Pmax was reduced even further to Pmax = 12 kN (or 66% load scaling). This finally yielded result comparable to the positive load spectrum tests (specimens S1, S3, N5 and S4 see Fig. 8). Therefore, the effect of the negative loads in the investigated spectrum is approximately the same as the effect caused by a 33% load reduction, (18–12)/18 kN.

The life of S8 specimen falls in the same scatter band as the positive spectrum specimens. There are, however, two notable differences. Firstly, the changes in periodicity of the crack growth within each spectrum repetition are not apparent

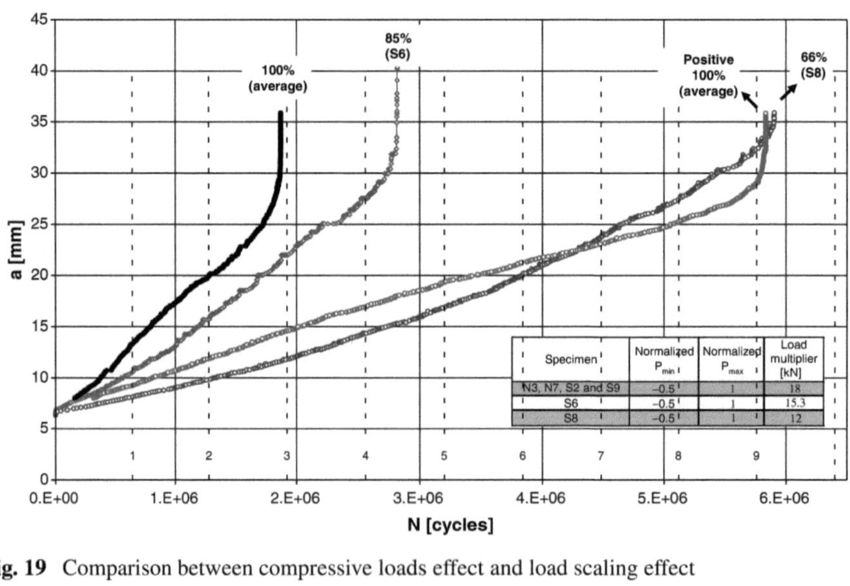

Fig. 19 Comparison between compressive loads effect and load scaling effect

in the S8 test. Secondly, the overall crack growth rate increases gradually with the length of the crack in the S8 case as opposed to the positive spectrum tests where it stays approximately constant or even decreases. These observations can be rationalized using an overload effects argument. The overload effect is much smaller in the S8 experiment because Pmax is smaller (only 12 kN instead of 18 kN), and also because of the presence of negative loads that reduces the actual retardation caused by the overloads.

Figure 20 shows the effect of the load scaling on the failure life. It can be seen that the failure life increases, as expected, when the load multiplier is reduced.

Finally, Fig. 21 compares the effects on fatigue life between (a) the load clipping (or overloads, OL) versus (b) the truncation of compressive loads (or underloads,

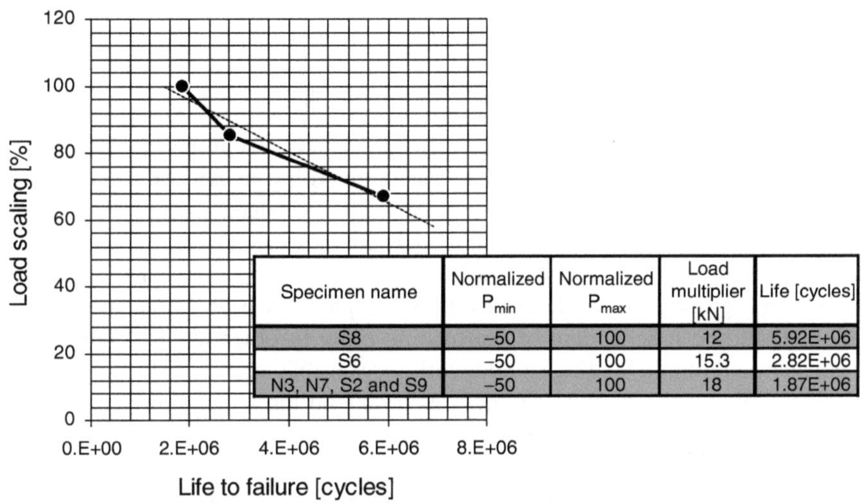

Specimen name	Normalized P_{min}	Normalized P_{max}	Load multiplier [kN]	Life [cycles]
S8	−50	100	12	5.92E+06
S6	−50	100	15.3	2.82E+06
N3, N7, S2 and S9	−50	100	18	1.87E+06

Fig. 20 Effect of the load scaling on the failure life

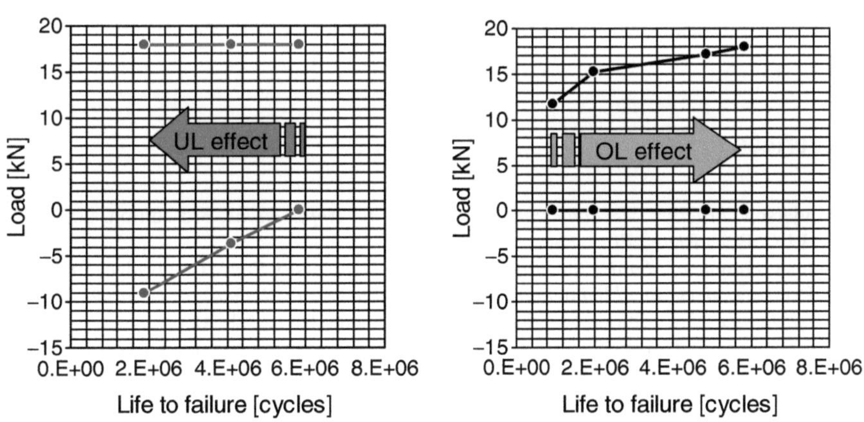

Fig. 21 Comparison between maximum load clipping (or overloads, OL) and minimum load truncation (or underloads, UL) on fatigue life

UL). An examination of Fig. 21 indicates that both the minimum and the maximum load limits have a significant effect on the observed fatigue life.

5 Conclusions

- The fatigue life for the positive spectrum without clipping is six times longer than for the positive spectrum clipped at 65% of Pmax = 18 kN where overload effects are largely eliminated.
- Negative loads, between 0 and −0.5 Pmax, reduce the amount of the crack growth retardation caused by the positive peaks (overloads) and cause the specimen to fail three times faster.
- The compressive loads are just as important as the overloads (see Fig. 21).
- The effect of the compressive loads in the investigated load spectrum on the fatigue life is equivalent to a 34% increase in the applied load.

This results clearly indicate that the negative loads are very important in the spectrum loading cases and should not be treated according to ASTM stipulation for constant amplitude loading, where for $R < 0$, $\Delta K = Kmax$. Usually the parameters in fatigue crack growth prediction models are calibrated twice i.e. for constant amplitude and separately for variable or spectrum loading situations. Hence, the fatigue crack growth model describes the same physical process the models calibration should be independent of the loading type. In other words – what works for constant amplitude loading should work for variable amplitude loading as well. This is clearly not the case when one tries to extend the stipulation of the ASTM standard for constant amplitude loading to spectrum loading with $R < 0$. Therefore, the customary treatments of the compressive loads and the underlying crack driving force assumptions have to be critically reexamined for spectrum loading situations.

Acknowledgements This study is supported by the Office of Naval Research under grand N00014-04-1-0718 and N00014-05-1-0872 (Program Director Dr. A.K. Vasudevan).

References

1. L. Anderson, *Fracture Mechanics: Fundamentals and Applications*, 3rd edition, 2006, CRC Press LLC, Boca Raton, FL.
2. S. Suresh, *Fatigue of Metals*, 2nd edition, Cambridge University Press, Cambridge, 1998.
3. N.E. Dowling, *Mechanical Behavior of Materials: Engineering Methods for Deformation, Fracture and Fatigue*, 3rd edition,, Prentice-Hall, Upper Saddle River, NJ, 2006.
4. S. Stoychev and D. Kujawski, Analysis of crack propagation using ΔK and Kmax, International Journal of Fatigue, Vol. 27, 2005, pp. 1425–1431.
5. ASTM E 647–05 "Standard Test Method for Measurement of Fatigue Crack Growth", Annual Book of ASTM Standards, Vol. 03.01, 2006, American Society For Testing and Materials, West Conshohocken, PA.

6. H. Tada, P.C. Paris and G.R. Irwin, *The Stress Analysis for Cracks Handbook*, 2nd edition, Paris Productions, St. Louis, MO, 1985.
7. J. Newman, Jr., Private communication, 2002.
8. J. Schijve and D. Broek, Crack propagation tests based on a Gust spectrum with variable amplitude loading, Aircraft Engineering, Vol. 34, 1962, pp. 314–316.
9. http://www.mae.wmich.edu/labs/Solid_Mechanics/experimental/spectrum_data/spectrum_data.html

Small Crack in a Simulated Columnar Polycrystalline Aggregate with Random 2D and 3D Lattice Orientations

Igor Simonovski and Leol Cizelj

Abstract The behaviour of microstructurally small cracks is heavily dependent upon the microstructural features in its immediate surroundings such as grain boundaries, crystallographic orientations, phases, etc. Crystallographic orientations of grains is one such feature and it's influence on the crack tip opening displacements (CTOD) of a small crack is investigated in presented work. The principal objective is to ascertain the difference in influence of crystallographic orientations between 2D lattice rotations (2D models) and full 3D lattice rotations (3D models). A finite-element based model that accounts for randomly shaped and sized grains is used. A small, inclined surface crack is introduced in a selected surface grain. Since a Stage I crack is assumed, the crack is always placed in a slip plane. Models are then loaded monotonically in uniaxial tension up to a maximum load of 1.12 yield stress. The influence that a random grain structure imposes on a Stage I crack is then assessed by calculating the CTOD values for single crystal and polycrystal models. It is shown that crystallographic orientations significantly impact the crack tip displacements for both 2D and 3D models. In some cases this effect is more pronounced for the 3D models.

Keywords Crack tip displacements · Crystal plasticity · Polycrystals · Small cracks

1 Introduction

Our understanding of microstructurally small cracks is still not complete. Their behaviour is strongly influenced by local microstructural features such as grain boundaries, crystallographic orientations, inclusions, voids and material phases, etc. [1–3]. Crack growth rate can be decreased when the crack approaches the grain boundary [3–5]. Cracks can change the slip plane after the crack has passed the

---

I. Simonovski (✉) and L. Cizelj
Jožef Stefan Institute, Reactor Engineering Division, Jamova cesta 39, SI-1000 Ljubljana, Slovenia
e-mail: Igor.Simonovski@ijs.si, Leon.Cizelj@ijs.si

S. Pantelakis, C. Rodopoulos (eds.), *Engineering Against Fracture: Proceedings of the 1st Conference*,

boundary [6], often resulting in zigzag patterns [7]. Different crystallographic orientations of the grains may also increase, decrease or arrest the crack growth [8, 9]. The crack tip loading is therefore generally mixed-mode with strong shear component. The plastic zone size for these cracks is comparable to the crack length even at relatively small loads, therefore severely thwarting the applicability of linear elastic fracture mechanics.

To expand our knowledge of these cracks further experimental work and modeling should be performed. The present study proposes a short crack model with incorporated basic microstructural features-random grain shapes and orientations. Crystal plasticity is combined with random grain structure. A small, Stage I crack, is embedded in a surface grain. Crack tip opening displacements (CTOD) values are calculated to estimate the spread of the CTOD values due to the surrounding crystallographic orientations for a 2D and 3D configurations under monotonic uniaxial tensile load.

The outline of the paper is as follows. First, a description of the finite element model with respect to kinematic constraints, boundary conditions and meshing is given. Basics of the crystal plasticity are also presented. Finally, results for different crystallographic orientations are presented.

2 Model description

The initial structural model includes a planar rectangular aggregate with 212 randomly sized and shaped grains. The grain structure is a planar Voronoi tessellation generated using code VorTESS [10]. To avoid violations of finite element shape constraints, only a subset of "meshable" Voronoi tessellations has been considered in the analysis. The "meshable" tessellations basically assume reasonably small aspect ratios of cord lengths. Further details are available in [11]. Such approach essentially prevents use of tessellations with very small grains. In this work we deal with 316L stainless steel that is for example used in certain nuclear power plants piping systems. The average grain size of 316L steel is between 50 and 80 μm. In this study a value of 52.9 μm is used.

Two-dimensional and 3D finite element models of grain structure are constructed. The 2D finite element model of the grain structure with a crack is presented in Fig. 1. Each grain is subdivided into eight-noded, reduced-integration, plane strain finite elements. Extensive mesh sensitivity study was performed in [12] resulting in the optimal mesh shown in Fig. 1. The applied mesh is expected to underestimate CTOD and CTSD by about 4.6% and 8.4% [12], respectively, which is deemed sufficient for the presented analysis. To obtain a 3D configuration, a 2D model is extruded in the depth (out of plane) direction to obtain a depth of 3.3 μm, Fig. 2. Each grain is subdivided into 20-noded, reduced-integration, 3D stress finite elements.

Each grain is assumed to behave as a randomly oriented monocrystal governed by the anisotropic elasticity and crystal plasticity models as described in a later

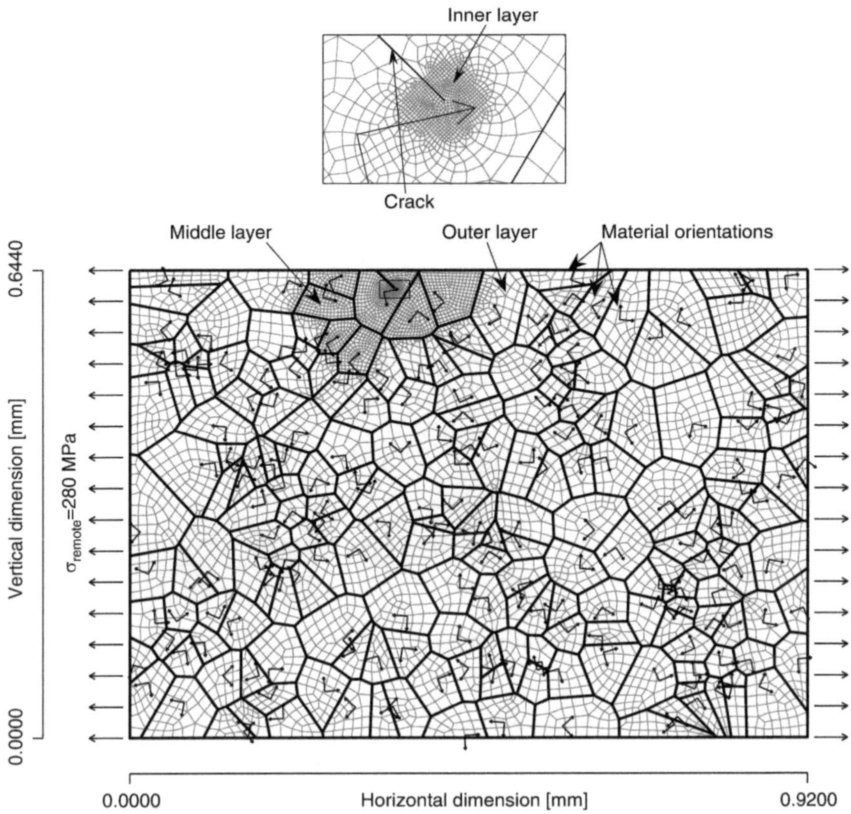

Fig. 1 The outline of the 2D finite element model. Details of the crack tip meshing are shown in the insert

section. Appropriate crystallographic orientations are defined by: (a) setting the angle between the crystallographic [100] direction (using Miller indices notation) and the macroscopic X axis of all crystals in the model to 135° as shown in Fig. 3 and (b) random rotation of the initial position about the global Z-axis. In a 2D configuration this results in a planar slip system model compatible with the planar macroscopic model. For a 3D configuration additional random rotation about the global X- or Y-axis is applied. Within the grain the initial crystallographic orientations are identical. The initial projections of the primary and conjugate slip planes are illustrated in Fig. 4.

A small surface crack, representative of a Stage I fatigue crack, is introduced in grain 38 in one of the two maximum shear planes, see Fig. 4. Crystallographic orientation of the crack-containing grain is set to $\alpha = 9.735°$ so that the crack is also placed in the slip plane P2. Crack tip opening (CTOD) displacements are calculated at a distance of 2.5% of the average grain size behind the crack tip (i.e. $0.025 \cdot 52.9 = 1.3\,\mu m$), see Fig. 5. This is consistent with examples found in the literature [13, 14].

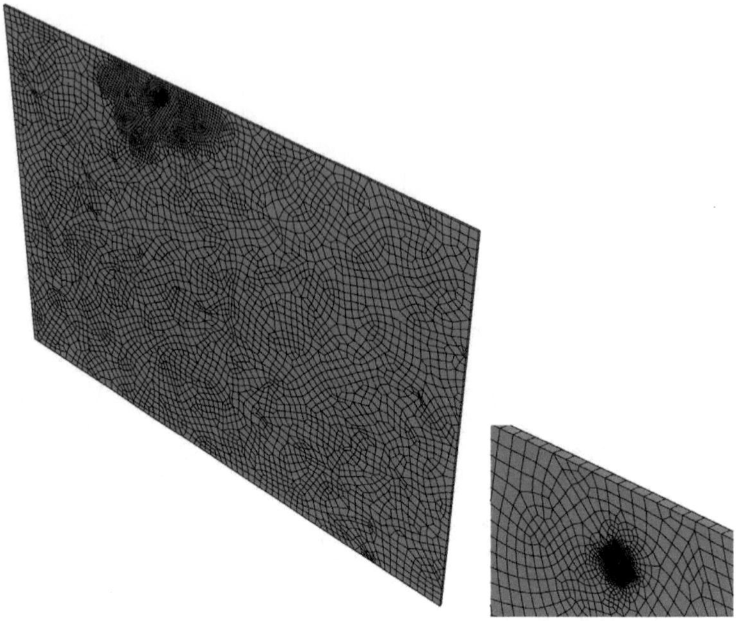

Fig. 2 The outline of the 3D finite element model. Details of the crack tip meshing are shown in the insert

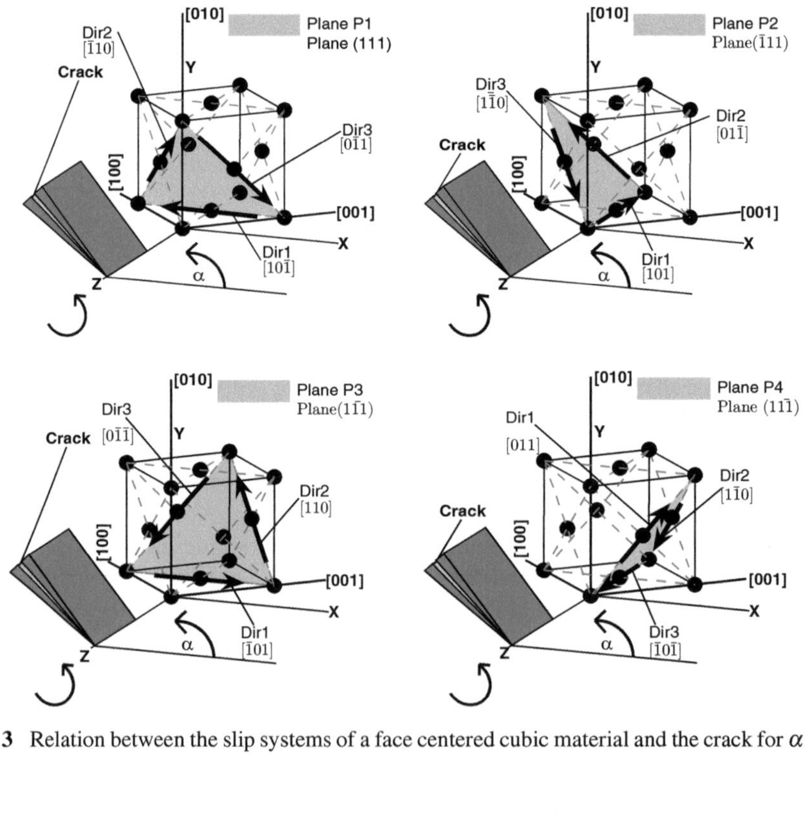

Fig. 3 Relation between the slip systems of a face centered cubic material and the crack for $\alpha = 0°$

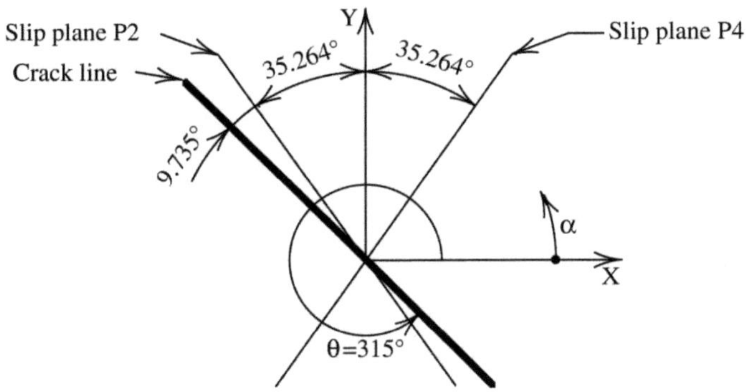

Fig. 4 Relation between the slip systems of a face centered cubic material and the crack for $\alpha = 0°$

Fig. 5 The definition of the CTOD

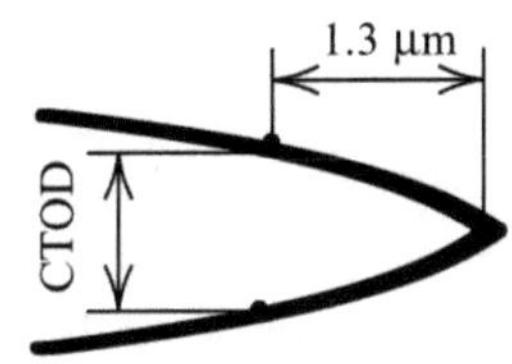

2.1 Constitutive Model

The elastic deformation at the monocrystal level is generally anisotropic and is governed by the generalized Hooke's law, $\sigma_{ij}^{e} = C_{ijkl}\varepsilon_{kl}^{e}$, where σ_{ij}^{e} represents the second-rank elastic stress tensor, C_{ijkl} the fourth-rank stiffness tensor and ε_{kl}^{e} the second-rank elastic strain tensor. The number of independent elastic constants for a cubic crystal system (body and face centered cubic) is 3.

Crystal plasticity theory is used [15, 16] to describe the material's plastic behaviour at the grain level. The plastic deformation in monocrystals is assumed to take place via a simple shear on a specific set of slip planes. Deformation by other mechanisms such as for example diffusion, twinning and grain boundary sliding is currently not taken into the account. The combination of a slip plane, denoted by its normal m_i^{α}, and a slip direction, s_i^{α}, is called a slip system, (α). The plastic velocity gradient, $\dot{u}_{i,j}^{p}$, due to a crystallographic slip can be written as [17],

$$\dot{u}_{i,j}^{p} = \sum_{\alpha} \dot{\gamma}^{(\alpha)} s_i^{(\alpha)} m_j^{(\alpha)}, \tag{1}$$

where the summation is performed over all active slip systems, (α), while $\dot{\gamma}^{(\alpha)}$ represents the shear rate. The cumulative slip, γ, is defined as,

$$\gamma = \sum_{\alpha} \int_0^t \left|\dot{\gamma}^{(\alpha)}\right| dt. \tag{2}$$

From the well-known relation for small strain $\varepsilon_{ij} = \frac{1}{2}(u_{i,j} + u_{j,i})$ one can obtain the plastic strain rate, Eq. (3). The constitutive relation of the elastic-plastic monocrystal is now given in terms of stress and strain rates as $\dot{\sigma}_{ij} = C_{ijkl}\left(\dot{\varepsilon}_{kl} - \dot{\varepsilon}_{kl}^{p}\right)$ [17].

$$\dot{\varepsilon}_{ij}^{p} = \sum_{\alpha} \frac{1}{2}\dot{\gamma}^{(\alpha)}\left(s_i^{(\alpha)} m_j^{(\alpha)} + s_j^{(\alpha)} m_i^{(\alpha)}\right) \tag{3}$$

It is assumed that the shear rate $\dot{\gamma}^{(\alpha)}$ depends on the stress only via the Schmid resolved shear stress, $\tau^{(\alpha)}$, Eq. (4). This is a reasonable approximation at room temperature and for ordinary strain rates and pressures [17]. Yielding is then assumed to take place when the Schmid resolved shear stress exceeds the critical shear stress τ_0.

$$\dot{\gamma}^{(\alpha)} = \dot{a}^{(\alpha)}\left(\frac{\tau^{(\alpha)}}{g^{(\alpha)}}\right)\left|\frac{\tau^{(\alpha)}}{g^{(\alpha)}}\right|^{n-1} \tag{4}$$

$$\tau^{(\alpha)} = s_i^{(\alpha)}\sigma_{ij} m_j^{(\alpha)} \tag{5}$$

$\dot{a}^{(\alpha)}$ represents the reference strain rate, n the strain-rate-sensitivity parameter and $g^{(\alpha)}$ the current strain-hardened state of the crystal. In the limit, as n approaches infinity, this power law approaches that of a rate-independent material. The current strain-hardened state $g^{(\alpha)}$ can be derived from,

$$\dot{g}^{(\alpha)} = \sum_{\beta} h_{\alpha\beta} \quad \dot{\gamma}^{(\beta)}, \tag{6}$$

where $h_{\alpha\beta}$ are the slip-hardening moduli defined by the adopted hardening law. In this work Peirce et al. hardening law is used [18], where self-hardening moduli $h_{\alpha\alpha}$ are defined by:

$$h_{\alpha\alpha} = \frac{h_0}{\cosh^2\left(\frac{h_0\gamma}{\tau_s - \tau_0}\right)}. \tag{7}$$

Here h_0 stands for the initial hardening modulus, τ_0 the yield stress (equal to the initial value of the current strength $g^{(\alpha)}(0)$) and τ_s a reference stress where large plastic flow initiates [19]. The latent-hardening moduli $h_{\alpha\beta}$ are given by,

$$h_{\alpha\beta} = q h_{\alpha\alpha}, \qquad (\alpha \neq \beta), \tag{8}$$

where q is a hardening factor. This model is implemented as a user-subroutine into the finite element code ABAQUS. Further details on its theory and implementation can be found in [19].

2.2 Loading and Boundary Conditions

The applied macroscopic loading and boundary conditions are illustrated in Fig. 1 and are identical for both 2D and 3D configurations. The left and right edges are loaded in macroscopic monotonic uniaxial tension up to a maximum load of 1.12 yield strength (280 MPa) with zero shear traction. This load is sufficient to trigger slip systems activity in all cases analyzed. The macroscopic yielding is achieved in most cases analyzed. The upper and lower edges are traction free. Prevention of rigid body movement is also imposed. For the 3D configuration the displacements in the depth direction are blocked on the front and back surface of the model to simulate the plain strain configuration which could than be compared with the 2D configuration.

Application of the macroscopic tension in the X-direction results in macroscopic maximum shear stress planes at $\pm 45^\circ$ to the X-axis. Now, coincidence of macroscopic shear planes and microscopic primary and conjugate slip planes is achieved when the crystals are rotated around the Z axis for: $\alpha = 9.735^\circ$, 80.265°, 99.735° or 170.265°.

2.3 Material Parameters

The following elastic constants for AISI 316L single crystal are used: $C_{iiii} = 163{,}680\,\mathrm{MPa}$, $C_{iijj} = 110{,}160\,\mathrm{MPa}$, $C_{ijij} = 100{,}960\,\mathrm{MPa}$ [20]. Crystal plasticity parameters have been optimized from the macroscopic plastic response of AISI 316L polycrystal [20]: $h_0 = 330\,\mathrm{MPa}$, $\tau_s = 270\,\mathrm{MPa}$, $\tau_0 = 90\,\mathrm{MPa}$, $n = 55$, $q = 1.0$ and $\dot{a}^{(\alpha)} = 0.001$. With these parameters the proposed plain strain model is deemed sufficient to provide a correct qualitative representation of the macroscopic response, see [20].

3 Results

The primary goal in this paper is to estimate the influence of a random grain structure on the CTOD of the inclined short surface crack. Investigation is focused on the randomness of orientations while keeping the grain shapes constant. Such approach is expected to reveal most of the variability contributed by the random grain structure [21].

In addition to the localized strain patterns, the orientations of the crack containing grain and grains in its immediate vicinity are expected to have dominant effect on the development of the CTOD. Two main cases with inclined stationary short surface crack have therefore been analyzed:

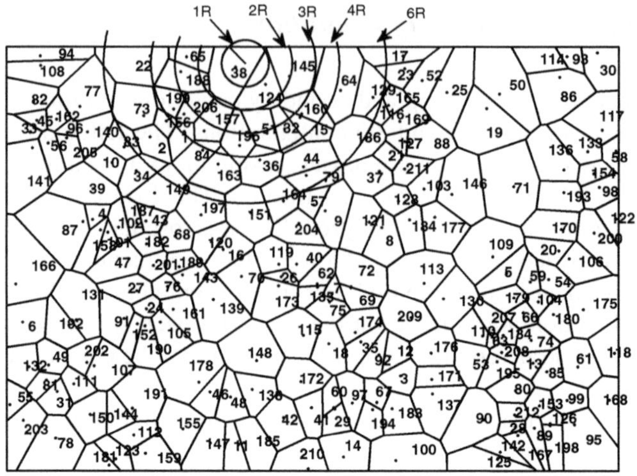

Fig. 6 Grain numbers and radii used

'$1(2,3,4,6)\times$R: 9.735, random' case. Increasing the size of the monocrystal containing the crack. This has been achieved by setting the orientation of the crack containing grain and all grains falling into radii of 1, 2, 3, 4 and 6$\times$ crack lengths to crystallographic orientation of $\alpha = 9.375°$. All remaining grains assumed random orientation (100 different realizations). A notation of '$3\times$R: 9.735, random' therefore means that all grains having their Poisson points (located in the center of a given grain) within radius of three times crack length ($3\times$R) are oriented at $\alpha = 9.735°$, while all other grains are oriented randomly. The radii are indicated in Fig. 6. Correspondingly, a notation '$1\times$R: 9.735, random' means that only the crack-containing grain is oriented at $\alpha = 9.735°$, while all other grains are oriented randomly.

'9.735, $1(2,3,4,6)\times$R: 135, random' case. Increasing the size of the bicrystal with crack. The orientation of the crack containing grain was set to $\alpha = 9.375°$, the orientation of the first neighbouring grain in the direction of the crack growth has been set to unfavourable $\alpha = 135°$. This orientation has been assessed from a 2D model. The size of the unfavourably oriented grain has then been increased by applying the $\alpha = 135°$ orientation to all grains falling into radii of 2, 3, 4 and 6$\times$ crack lengths (with the exception of the crack containing grain). All remaining grains assumed random orientation (100 different realizations). A notation of '9.735, $3\times$R: 135, random' therefore means that the orientation of the first grain is 9.735°, orientation of grains having their Poisson points within radius of three times crack length ($3\times$R) is 135°, while all other grains are oriented randomly.

For the '$1(2,3,4,6)\times$R: 9.735, random' cases we expect the results to tend towards the single crystal case, however, the calculations will show the range of scatter and are also useful for understanding more complex cases.

3.1 2D Configuration

Figure 7 shows the cumulative probability functions for the above cases in a 2D configuration. Min/max CTOD values for a limiting case of a single crystal are plotted for comparison. Also, the CTOD value for a bicrystal case (crack-containing grain: $\alpha = 9.375°$, all other grains: $\alpha = 135°$) is shown. One can see that for the '1 × R: 9.735, random' case the random crystallographic orientation can change the CTOD by a factor of 4. If we consider the values from all the analyzed cases, the ratio CTODmax/CTODmin increases to 11.8. Increasing the size of the crack-containing monocrystal to 2, 3, 4, 6 × R gradually increases the CTOD, but the spread of the values does not decrease significantly. It is interesting to note that the impact of the monocrystal size vanishes at large CTOD values, where cumulative densities of 2, 3, 4 and 6 × R overlap. This indicates that in this region the CTOD is dominated by the localized high strain areas (areas of grains with favourable orientation in the direction of maximum shear). Monocrystal size of 6 × R is the maximal size considered to avoid significant effects from the model size. Model with more grains would have to be employed to investigate this matter further in the future. It is also interesting to note that the max. CTOD can actually be higher than of a limiting case of a single crystal. We attribute this to the occurrence of localized high strain areas in the polycrystal case.

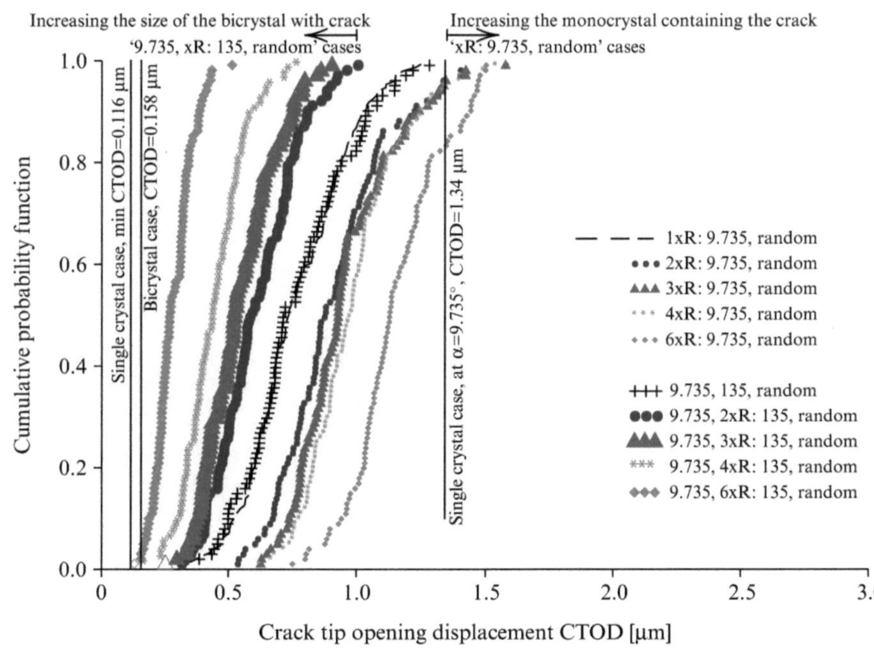

Fig. 7 Cumulative probability functions at a load of $1.12R_{p0.2}$. Two-dimensional models

As expected, increasing the size of the $\alpha = 135°$ layer ('9.735, $\times$R : 135, random' case) around the crack-containing grain decreases the CTOD values. The limit case in this situation is the bicrystal case. The spread of the CTOD values here gradually decreases and max/min CTOD ratio is still 3.8 at '9.7356, $6 \times$ R : 135, random' case. In this case the structure beyond $6 \times$ R significantly affects the CTOD values as well. It is also noted that the bicrystal configuration successfully minimizes the effects of the localized high strain areas due to its high stiffness.

3.2 3D Configuration

Figure 8 shows the cumulative probability functions for the above cases in a 3D configuration. One can immediately see that the CTOD values are higher for 3D configuration than for a 2D configuration. This is due to the fact that the crystallographic orientations in 3D configuration are fully 3D, while for a 2D configuration the crystallographic orientations change only due to the rotation about the global Z axis. The other significant difference is that for the 2D configuration the CTOD for the '9.735, $1 \times$ R : 135, random' cases was lower than for the '$1 \times$ R : 9.735, random' cases. This is not true for the 3D configuration and is slightly surprising. A possible explanation for this effect could be that the orientation 135° of the unfavorably oriented grain was estimated from a 2D model and is possibly not so

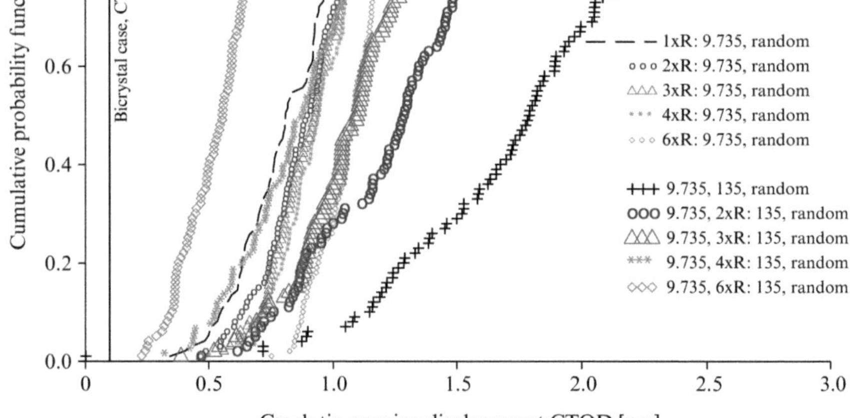

Fig. 8 Cumulative probability functions at a load of $1.12R_{p0.2}$. Three-dimensional models

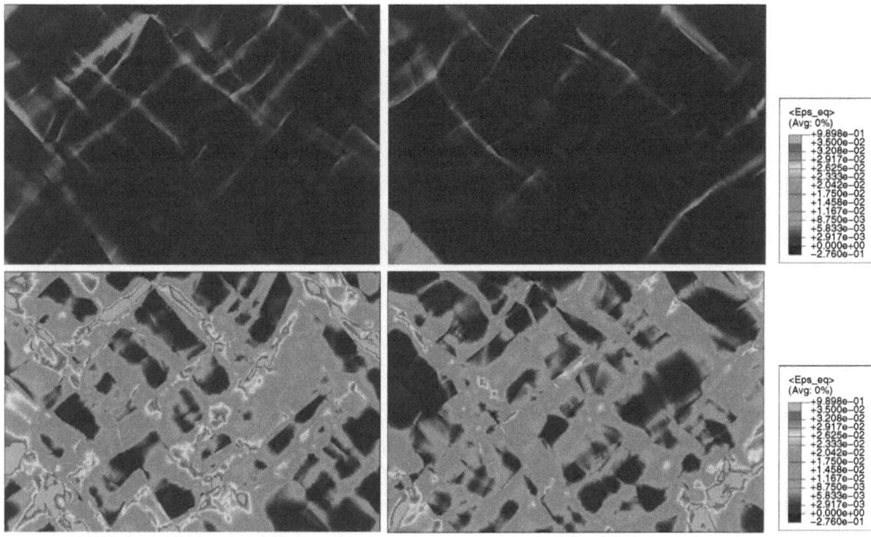

Fig. 9 Equivalent strain for 2D (*top*) and 3D (*bottom*) '9.735, $1 \times \mathrm{R}:$ 135, random' configurations at a load of $1.12R_{p0.2}$. *Left*, orientation set resulting in maximal CTOD, *right* orientation set resulting in minimal CTOD

unfavorable in 3D configuration. Also, there are more activated slip systems in a 3D configuration. As the size of this unfavorably oriented grain (135°) is increased, it increasingly blocks the deformation of the crack-containing grain, greatly reducing the CTOD values. This effect is the same as in a 2D configuration. The ratio between the Min/max CTOD values has remained approximately the same: 11.8 for the 2D configuration, compared to 12.2 for the 3D configuration.

Comparing 2D and 3D configurations, higher equivalent strain was observed for 3D configurations, see Fig. 9. This results from higher number of activated slip systems and agrees well with higher observed CTOD for 3D configurations. Higher CTOD values were also observed when a band of equivalent strain formed in the vicinity of the crack tip. It is interesting to note that even if a large, unfavorably oriented grain appears next to the crack-containing grain, a band of high equivalent strain can form across this unfavorably oriented grain towards the crack tip, increasing CTOD, see Fig. 10.

4 Summary

A study on the influence of the crystallographic orientations on the crack tip displacements of a microstructurally small crack is presented for both 2D and 3D configurations. We presume that the crack growth is linked to the crack tip displacements. Although the loading in this paper is monotonic, a number of key observations are still relevant for the fatigue case.

Fig. 10 Equivalent strain for 3D, '9.735, $6 \times \mathrm{R}: 135$, random' configuration resulting in a maximal (*left*) and minimal (*right*) CTOD. Load $1.12R_{p0.2}$

We observed that the size of the favorably oriented, crack-containing grains is linked with the CTOD. Larger CTOD values were observed for a larger grain for both 2D and 3D configurations. This effect is moderate. Larger difference was observed for increasingly large unfavorably oriented grain in the vicinity of the crack-containing grain. This grain decreases the CTOD values, however, for a 3D configurations this effect was significance more pronounced than for a 2D configuration.

We have shown that the crystallographic orientations significantly influence the CTOD, both for 2D and 3D configurations. Grains in the vicinity of the crack/containing grain affect the CTOD, however, larger patterns of bands of high equivalent strain can overcome this effect. If such a band of high strain forms in the vicinity of a crack tip, the CTOD will increase significantly. Even a large, unfavorably oriented grain can not prevent such a formation to take place, depending of course on the specific set of crystallographic orientations of grains.

Acknowledgements The authors acknowledge the financial support provided by the Slovenian Research Agency through the research project J2-9168-0106-06, Multiscale model of small crack initialization and propagation in pressure boundary components of a nuclear power plant, research contract 1000-07-219168.

References

1. K. J. Miller. The behaviour of short fatigue cracks and their initiation. Part II-A general summary. *Fatigue & Fracture of Engineering Materials & Structures*, 10(2):93–113, 1987.
2. K. Hussain. Short fatigue crack behaviour and analytical models: a review. *Engineering Fracture Mechanics*, 58(4):327–354, November 1997.
3. K. Hussain, E. R. de los Rios, and A. Navarro. A two-stage micromechanics model for short fatigue cracks. *Engineering Fracture Mechanics*, 44(3):425–436, February 1993.
4. W. L. Morris, O. Buck, and H. L. Marcus. Fatigue crack initiation and early propagation in Al 2219-T851. *Metallurgical Transactions A*, 7A:1161–1165, August 1976.
5. Y. H. Zhang and L. Edwards. On the blocking effect of grain boundaries on small crystallographic fatigue crack growth. *Materials Science and Engineering A*, 188(1–2):121–132, November 1994.

6. O. Dber, B. Knkler, U. Krupp, H.-J. Christ, and C.-P. Fritzen. Experimental characterization and two-dimensional simulation of short-crack propagation in an austenitic-ferritic duplex steel. *International Journal of Fatigue*, 28(9):983–992, September 2006.
7. S. Suresh. *Fatigue of materials*. Cambridge University Press, Cambridge, 1991.
8. V. Tvergaard, Y. Wei, and J. W. Hutchinson. Edge cracks in plastically deforming surface grains. *European Journal of Mechanics - A/Solids*, 20(5):731–738, September–October 2001.
9. T. Zhai, A. J. Wilkinson, and J. W. Martin. A crystallographic mechanism for fatigue crack propagation through grain boundaries. *Acta Materialia*, 48(20):4917–4927, December 2000.
10. H. Riesch-Oppermann. VorTess generation of 2–D random Poisson–Voronoi mosaics as framework for micromechanical modeling of polycrystalline materials-algorithm and subroutines description. Technical Report FZKA 6325, Forschungszentrum Karlsruhe, 1999.
11. S. Weyer, A. Frohlich, H. Riesch-Oppermann, L. Cizelj, and M. Kovač. Automatic finite element meshing of planar voronoi tessellations. *Engineering Fracture Mechanics*, 69(8):945–958, May 2002.
12. I. Simonovski. Mechanisms for thermal fatigue initiation and crack propagation in NPP components. 2nd mid term report. Technical Report, Jožef Stefan Institute, DG-JRC, Institute for Energy (accessible through http://www.rcp.ijs.si/isimonovski/Papers/Simonovski_2005_2.pdf), 2005.
13. G. P. Potirniche and S. R. Daniewicz. Analysis of crack tip plasticity for microstructurally small cracks using crystal plasticity theory. *Engineering Fracture Mechanics*, 70(13):1623–1643, September 2003.
14. V. P. Bennett and D. L. McDowell. Crack tip displacements of microstructurally small surface cracks in single phase ductile polycrystals. *Engineering Fracture Mechanics*, 70(2):185–207, January 2003.
15. R. Hill and J. R. Rice. Constitutive analysis of elastic-plastic crystals at arbitrary strain. *Journal of the Mechanics and Physics of Solids*, 20(6):401–413, 1972.
16. J. R. Rice. On the structure of stress-strain relations of time-dependent plastic deformation in metals. *Journal of Applied Mechanics*, 37:728–737, 1970.
17. A. Needleman. Computational mechanics at the mesoscale. *Acta Materialia*, 48(1):105–124, January 2000.
18. D. Peirce, R. J. Asaro, and A. Needleman. Material rate dependence and localized deformation in crystalline solids. *Acta Metallurgica*, 31(12):1951–1976, December 1983.
19. Yonggang Huang. A user-material subroutine incorporating single crystal plasticity in the ABAQUS finite element program. Technical Report, Division of Applied Sciences, Harvard University (accessible through http://www.columbia.edu/~jk2079/fem/umat_documentation.pdf), 1991.
20. I. Simonovski, K.-F. Nilsson, and L. Cizelj. Material properties calibration for 316L steel using polycrystalline model. In *13th International Conference on Nuclear Engineering*, May 16–20, 2005, Beijing, China, May 2005.
21. S. Weyer. *Experimentelle Untersuchung und Mikromechanische Modellierung des Schdigungsverhaltens von Aluminiumoxid unter Druckbeanspruchung*. Ph.D. thesis, Universitt Karlsruhe, 2001.

Thermo-Mechanical Methods for Improving Fatigue Performance of Wrought Magnesium Alloys

Muhammad Shahzad and Lothar Wagner

Abstract Wrought magnesium alloys AZ80 and ZK60 in as-cast condition were extruded at 300 °C with extrusion ratios of ER = 12 and 44. Resulting microstructures, crystallographic textures and mechanical properties were investigated. Extruding led to profound reduction in grain size in comparison to as-cast condition which drastically improved yield stress, tensile elongation and HCF performance. Strength differentials in ZK60 after extruding at ER = 12 were more pronounced than after extruding at ER = 44 whereas no such effect was observed in AZ80. Swaging after extruding further increased yield stress and endurance limit, while strength differential increased and ductility was lowered.

Keywords Magnesium alloys · Extrusion · Swaging · Fatigue

1 Introduction

Magnesium alloys have been in use since 1930s but despite the significant advantages of weight saving, magnesium alloys have not been able to get a wide spread usage. One of the reasons is the limited use of wrought alloys because of the difficulties in forming magnesium alloys. The close packed hexagonal lattice of magnesium alloys offers only limited slip systems at room temperature because the critical resolved shear stress of slip systems other than basal slip is quite high. However, non-basal slip as well as twinning becomes active at elevated temperatures resulting in reasonable ductility. Numerous studies are being pursued to investigate deformation mechanisms operating in magnesium alloys at elevated temperatures. Among the various forming processes, extrusion of magnesium alloys has developed a lot

M. Shahzad (✉) and L. Wagner
Institute of Materials Science and Engineering, Clausthal University of Technology, Agricolastrasse 6 38678, Clausthal-Zellerfeld, Germany
e-mail: muhammad.shahzad@tu-clausthal.de

S. Pantelakis, C. Rodopoulos (eds.), *Engineering Against Fracture: Proceedings of the 1st Conference*,

of interest and recently many studies have been carried out to investigate the effects of process parameters on mechanical properties [1–3]. The extrusion process has been shown to result in pronounced grain refinement which markedly improves the mechanical properties of the as-cast condition [4]. This paper studies the effect of extrusion ratio on mechanical properties particularly, under fatigue loading. In addition, swaging after extruding has been carried out to further improve the HCF performance.

2 Experimental

Direct chill-cast AZ80 and ZK60 alloys were received from Dead Sea Magnesium as Ø70 × 200 mm billets. The billets and the die were heated to the extrusion temperature of 300 °C for 1 h. MoS2 lubrication was applied before direct extruding the billets at a constant ram speed of 60 mm/min. Extrusion dies with inner exit diameters of 10.5 and 20 mm were used in this study. Extrusions were normal air cooled. Extrusion parameters for both operations are listed in Table 1. Twenty millimeters extrusions of both alloys were also progressively swaged at 300 °C to 17, 14, 12 and 10 mm diameter ($\varphi = 1.4$). Before swaging, materials were heated to 300 °C for 15 min. Between passes, materials were reheated for 5 min to 300 °C. After swaging, alloys were air cooled. Texture studies were performed using x-ray diffraction with cobalt x-ray source. Specimens for the various mechanical tests were machined with the load axis parallel to the longitudinal direction both for extruded and swaged materials. Quasi-static tensile and compression tests were conducted at initial strain rates of $10^{-3}\,s^{-1}$ in air. Some tensile tests were also conducted at elevated temperatures. Fatigue testing was performed on electrolytically polished hour-glass shaped specimens (6 mm gauge diameter except for swaged condition for which gauge diameter was 4 mm) in rotating beam loading ($R = -1$) in air at frequencies of 50 Hz.

3 Results and Discussion

The various microstructures of AZ80 and ZK60 are illustrated in Figs. 1 and 2, respectively. Due to relatively high cooling rates during DC casting, the average grain size in the as-received microstructure of AZ80 is only 100μm with fine eutectic components distributed along the grain boundaries (Fig. 1a).

Table 1 Comparison of extrusion parameters for the two dies used

	Extrusion ratio	Extrusion strain	Extrusion velocity (m/min)
Die (Ø10.5 mm)	44	3.8	2.6
Die (Ø20 mm)	12	2.5	0.7

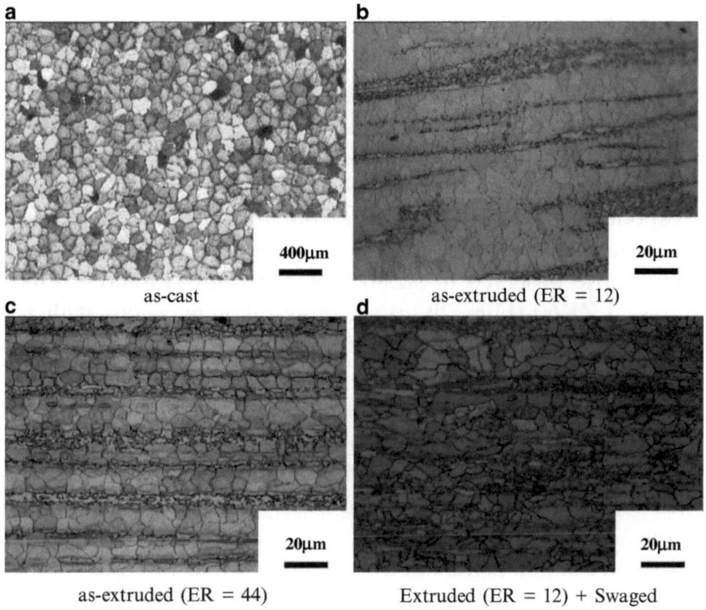

Fig. 1 Microstructures of AZ80

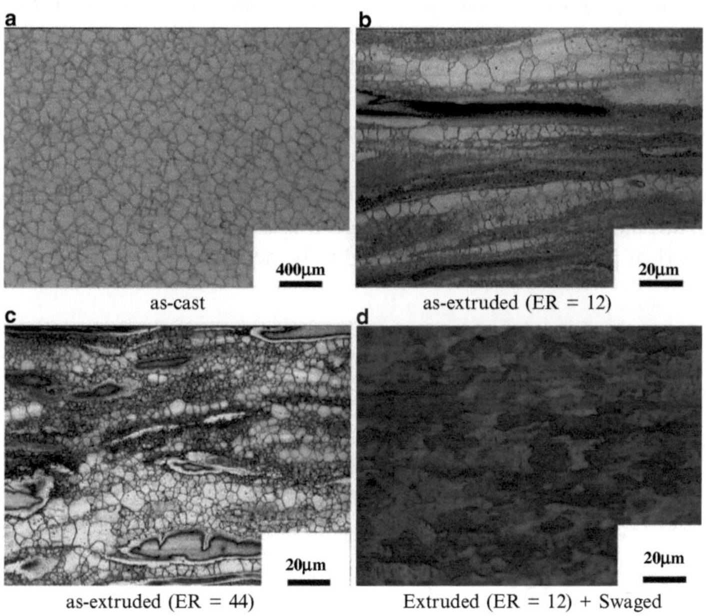

Fig. 2 Microstructures of ZK60

Extruding at ER = 12 (Fig. 1b) and ER = 44 (Fig. 1c) led to fully recrystallized equiaxed grains with sizes slightly below 10 μm. In addition, stringers with unrecrystallized grains parallel to the extrusion direction are visible. The post-extrusion swaged microstructure exhibits grains slightly elongated in the main deformation direction (Fig. 1d). No significant changes in grain size or volume fraction of unrecrystallized grains are seen (compare Figs. 1b, d).

The corresponding microstructures of ZK60 are shown in Fig. 2a–d. While the as-cast microstructure of ZK60 (Fig. 2a) shows equiaxed grains of about 100 μm very similar to those observed in AZ80 (compare Figs. 1a and 2a) the extruded microstructures of ZK60 (Figs. 2b, c) are quite inhomogeneous with at least two populations of grain sizes amounting to about 5 and 10 μm. In addition, unrecrystallized grains as found in AZ80 were also observed in ZK60, however, their size could be as large as the as-cast grain size. Increasing the extrusion ratio from ER = 12 to ER = 44 is seen to reduce the size of these unrecrystallized grains. After swaging previously observed large unrecrystallized grains seem to have been largely elongated and to some extent also broken but by large the microstructure is still inhomogeneous.

Both alloys had random texture in as cast condition as revealed by (0002) plane pole diagram in Fig. 3. After extrusion, it gets transformed to so called cylindrical texture in which basal planes are perpendicular to extrusion direction and have been reported in an earlier manuscript [5].

The recalculated (0002) pole figures for AZ80 and ZK60 alloys extruded with ratio 44 show the cylindrical arrangement of basal planes after extrusion (Fig. 4). This type of texture is very critical for mechanical properties as it causes so called texture strengthening, when tensile loaded parallel to extrusion direction by inhibiting basal slip owing to lowered Schmidt factor [6].

Relatively stronger texture has been observed for extruded ZK60 alloy in comparison to AZ80. This effect could be due to the relative lesser degree of recrystallization with the former. Presence of zirconium in Al alloys is known to hinder the recrystallization of the alloy due to the zener drag exerted by Al–Zr intermetallic [7]. A similar sort of phenomenon is suggested to be active in Mg alloys.

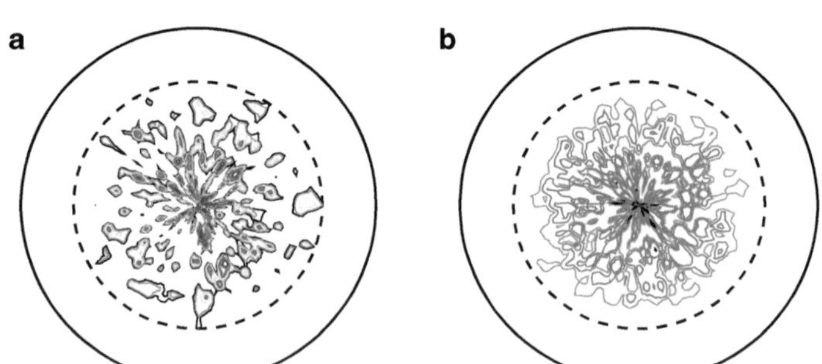

Fig. 3 (0002) pole diagram for **a** AZ80 as cast and **b** ZK60 as cast

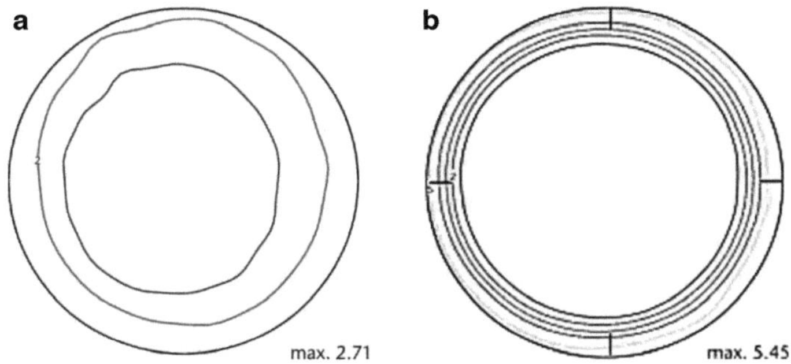

Fig. 4 Recalculated (0002) pole diagram for **a** AZ80 Ex (ER = 44) and **b** ZK60 Ex (ER = 44)

Table 2 Mechanical properties of the various alloys and conditions

(MPa)	YS (T) (MPa)	YS (C) (MPa)	SDF	UTS (MPa)	El (%)	$\varepsilon_f = \ln (A_0/A_F)$
AZ80 as-cast	105	95	1.10	150	2	0.03
AZ80 Ex (ER44)	225	205	1.08	325	16	0.30
AZ80 Ex (ER12)	225	210	1.06	295	14	0.20
AZ80 Ex (ER12) + Sw	300	235	1.28	335	6	0.13
ZK60 as-cast	150	120	1.25	210	6	0.09
ZK60 Ex (ER44)	280	250	1.12	340	19	0.55
ZK60 Ex (ER12)	290	210	1.39	335	16	0.50
ZK60 Ex (ER12) + Sw	350	240	1.46	390	12	0.30

Mechanical properties of the alloys are summarized in Table 2 where yield stresses in tension YS(T) and in compression YS(C) are listed along with strength differential factor SDF being the ratio of former over later.

The high values of YS, UTS and tensile elongation for extruded alloys in comparison to as cast condition manifest the positive effects of grain refinement attained during the extrusion process. On average, YS (T) and UTS values in ZK60 are roughly 50 MPa higher than in AZ80. Since YS (C) values in ZK60 are hardly higher than in AZ80, SDF values are markedly higher amounting to about 1.4 except of the material extruded at ER = 44. Strength differential is a common phenomenon in wrought magnesium alloys and is more pronounced in longitudinal/extrusion direction. It is mainly caused by texture and affected by microstructure. At room temperature deformation proceed mainly with (0001) <11–20> basal slip. When deformed in the longitudinal direction, the basal slip is inhibited because of unfavourable arrangement of the basal plane after extrusion and higher yield strength is observed. On the other hand the most easily activated twin system {10–12} <10–11> is a tensile twinning system, requiring tensile stress parallel to c axis and therefore is active in compression but not in tension [6]. Grain size also affects the deformation mechanism particularly under compression and lowering of yield strength has been observed by increasing the grain size [8] and thereby a higher SDF.

Extruded AZ80 alloys show very low SDF because of the fine grain size achieved after extrusion. The same hold for ZK60 alloy extruded with ER 44 but for ER 12, where there are large unrecrystallized grains, YS (C) is low whereas YS (T) is almost same thereby resulting in large SDF. Swaging after extruding resulted higher SDF for both alloys. This effect is mainly due to significant increase in YS(T) at the cost of ductility caused by strain hardening and stronger texture for swaged alloys.

Results from tensile testing at elevated temperatures are plotted for AZ80 and ZK60 in Figs. 5 and 6, respectively. Mechanical properties of both alloys in as cast condition show higher temperature stability than the extruded alloys. Given that there is no significant grain size increase up to 150°C, this rapid fall in YS

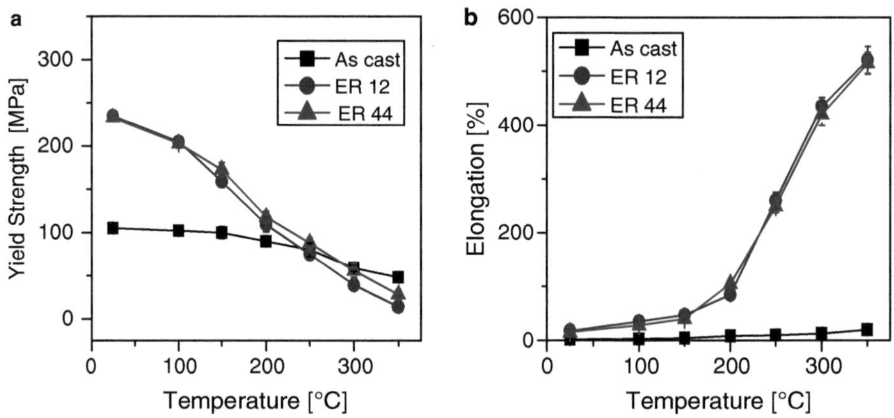

Fig. 5 Temperature dependence of **a** yield stress and **b** tensile elongation for AZ80 alloy

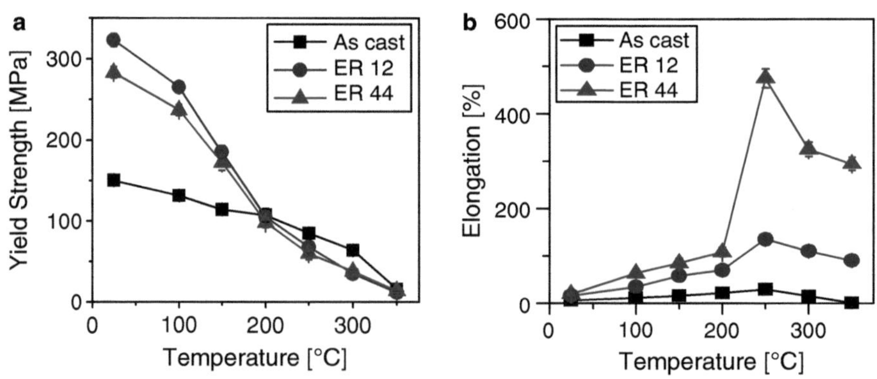

Fig. 6 Temperature dependence of **a** yield stress and **b** tensile elongation for ZK60 alloy

explains that the strengthening caused by extrusion process is not only due to the grain refinement, but so called texture strengthening and strain hardening caused by higher dislocation density is also a contributing factor.

The temperature dependence of the yield stress of both extrusion conditions in AZ80 is very similar and much more pronounced than that of the as-cast condition. At about 250°C, there is a cross-over in yield stress dependencies. Thus, the as-cast condition is superior to the extruded conditions for temperatures beyond 250°C. The yield strength dependence on temperature of the various conditions in ZK60 (Fig. 6) is very similar to that in AZ80. Again, the yield stress of the as-cast condition is superior to the extruded conditions at higher temperatures. At temperature over 250°C as cast alloy show higher strength because of its higher resistance to creep owing to its coarser grain size, but at 350°C which is just below the eutectic temperature of ZK60 alloy, there occurred an abrupt fall of strength and material failed without any plastic deformation owing to incipient melting of eutectic at the grain boundaries.

No significant effect of ER has been observed on YS of both alloys at all the tested temperature. The same holds for tensile elongation in AZ80 alloy which remains the same for both extrusion ratios and increases linearly from 200°C onward reaching a peak value of 520% at 350°C. However for ZK60 a pronounced effect of ER has been observed, with higher ER of 44 resulting in much higher elongation value. This effect is attributed to large inhomogeneous microstructure for lower extrusion ratio hindering the grain boundary sliding mechanism, generally accounted for superplastic behavior. Also the tensile elongation for ZK60 alloy does not increase linearly and rather shows maxima at 250°C after which there is a gradual fall. Dispersion of fine eutectic in the extrudate is accounted for quite higher elongation values at near eutectic testing temperature of 350°C, for which as cast alloy fractured without any plastic deformation.

Results of fatigue tests have been presented in form of SN curves in Fig. 7. In as cast condition both alloys showed similar response and failed to run for one million cycle even at very low stress amplitude of 10 MPa (Fig. 7a).

In comparison extruded (Ex) and post extrusion swaged (Ex + Sw) alloys showed remarkable fatigue life improvement with post extrusion swaged AZ80 and ZK60 alloys showing ten million cycle endurance stress of 165 and 170 MPa respectively (Fig. 7b, c). This drastic high cycle fatigue improvement is attributed to profound improvement of yield strength achieved during extrusion and swaging.

No significant effect of relatively higher SDF of 1.39 in ZK60 with ER12 has been observed on high cycle fatigue life in comparison to ZK60 with ER 44 for which it was 1.13. However after extrusion and also after post extrusion swaging the fatigue response is still almost the same for AZ80 and ZK60 alloys (Fig. 7b, c) despite ZK60 (Ex + Sw) having 50 MPa higher yield strength in tension than AZ80 (Ex + Sw). This effect could be due to be the layered structure observed in AZ80 alloy where harder mix eutectic of Al and Zn with Mg stops the crack advancement, as observed in the extruded AZ31 alloy [9].

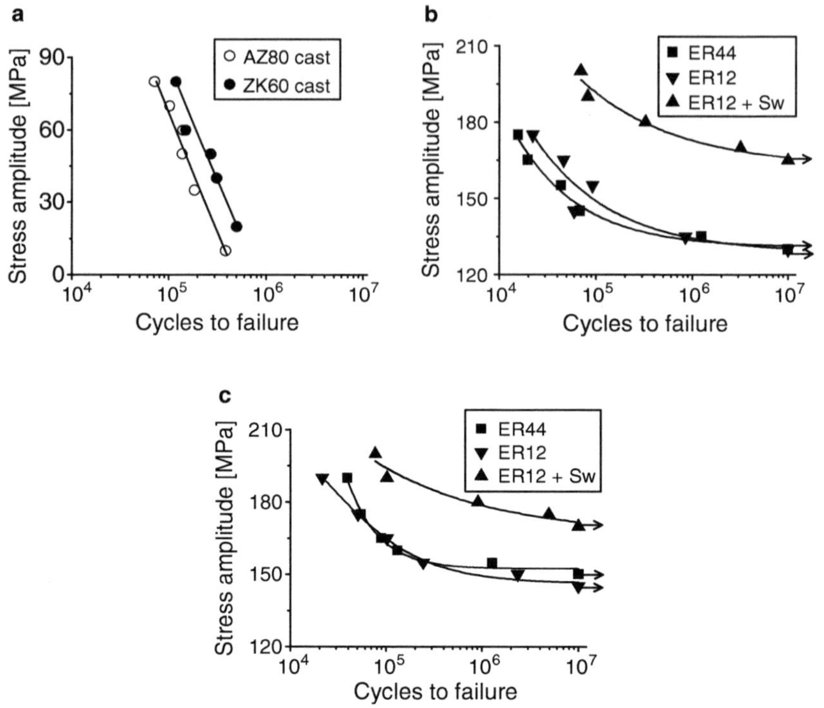

Fig. 7 SN curves of the **a** as-cast alloys, **b** Ex and Ex+Sw AZ80 alloys and **c** Ex and Ex+Sw ZK60 alloys

4 Conclusions

- ZK60 alloy shows poor recrystallization ability and large unrecrystallized grains have been observed after extrusion whereas AZ80 alloy fully recrystallized under similar conditions. Higher extrusion ratio improves the microstructure by reducing the size of such grains.
- No significant effect of extrusion ratio has been observed on microstructure and mechanical properties of AZ80 alloy. In comparison, lower extrusion ratio results in higher strength differential and lower ductility especially over 200 °C for ZK60 alloy, however no significant effect of ER has been observed on fatigue life.
- Swaging after extrusion significantly improves the fatigue life but results in higher strength differential.

References

1. Sillekens W H, Bohlen J (2004) The MAGNEXTRUSCO project: European community research on hydrostatic extrusion of magnesium In: Kainer K U (ed) Magnesium. Wiley-VCH Verlag, Weinheim

2. Leeflang M A, Zhou J, Duszczyk J (2007) Effect of billet temperature and ram speed on the behavior of AZ31 extrusion. In: K U Kainer (ed) Magnesium. Wiley-VCH Verlag, Weinheim
3. Ming Z, Jianyong C, Fulin Y, Xia W (2005) Evaluation of magnesium extrusion production. Mater Sci Forum 488–489:483–486
4. Murai T, Matsuoka S-I, Miyamoto S, Oki Y (2002) Effects of extrusion conditions on microstructure and mechanical properties of AZ31B magnesium alloy extrusions. J Mater Process Tech 141:207–212
5. Shahzad M, Eliezer D, Gan W, Yi S, Wagner L (2007) Influence of extrusion temperature on microstructure, texture and fatigue properties of AZ80 and ZK60 magnesium alloys. Mater Sci Forum 561–565:187–190
6. Kleiner S, Uggowitzer P J (2004) Mechanical anisotropy of extruded Mg-6%Al-1%Zn alloy. Mater Sci Eng A 379:258–263
7. Ocenasek V, Slamova M (2001) Resistance to recrystallization due to Sc and Zr addition to Al-Mg alloys. Mater Charact 47:157–162
8. Barnett M R, Keshavarz Z, Beer A G, Atwell D (2004) Influence of grain size on compressive deformation of wrought Mg-3Al-1Zn. Acta Mater 52:5093–5103
9. Nan Z, Ishihara S, Goshima T, Nakanishi R (2004) On sharp bend in SN curve of AZ31 extruded Mg alloy. In: Proceedings of the 15th European Conference of Fracture, Stockholm, Sweden

Session V
Applied Fracture Mechanics

Investigations on Fracture of Collector Copper Lamellas

Liviu Marsavina, Nicolae Faur, Romeo Susan-Resiga, and Radu Negru

Abstract The performance of engineering components and structures is strongly influenced by the interaction between design, manufacture and materials. This paper presents the fracture of copper lamellas from a circular collector. The investigated circular collector, having 315 copper lamellas, is one of the main parts of an electric motor. The fracture of the copper lamellas was observed in normal operation. This unexpected fracture required an investigation of the fracture origin, in order to improve the initial design. The numerical results of the Finite Element Analysis on the stress field in the copper lamellas for the operating regimes, and the stress concentration effects are shown. A study of crack initiation, based on Notch Stress Intensity Factor approach was performed in order to estimate if the conditions of crack initiation are meet. Finally a study of crack propagation will present comparatively the numerical obtained crack path against the one observed in-service.

Keywords Copper lamellas · Notch Stress Intensity Factors · Crack initiation · Crack propagation

1 Introduction

The performance of engineering components and structures is strongly influenced by the interaction between design, manufacture and materials [11]. Often the failure of the components is due to local stress concentration. Starting with the methods of the theory of elasticity and solid mechanics, many efforts were focused in the past to analytically determine the stress field around notches, holes and other stress raisers. Recently Finite Element Analysis was proved to be a powerful tool for determining the stress distribution.

L. Marsavina (✉), N. Faur, R. Susan-Resiga, and R. Negru
"POLITEHNICA" University of Timişoara, Mechanical Engineering Faculty, Blvd. M. Viteazu, Nr. 1, Timişoara 300222, Romania
e-mail: msvina@mec.upt.ro, faur@mec.upt.ro, resiga@mh.mec.upt.ro, radun@mec.upt.ro

S. Pantelakis, C. Rodopoulos (eds.), *Engineering Against Fracture: Proceedings of the 1st Conference*,

A very useful approach for applying the Linear Elastic Fracture Mechanics is to use the Notch Stress Intensity Factor (NSIF). Different methods have been proposed for determining the NSIF, starting with the closed from expression for blunt notches of [5]. Lazzarin and Tovo [14] gave a solution dealing with sharp and V-notches, the crack and the blunt crack being particular cases of the general solution valid for any combination of notch tip radius and opening angle. Filippi et al. [8] and Lazzarin and Filippi [13] proposed explicit formulas for the NSIF in the vicinity of U- and V-shaped notches. Another approach to characterize the stress field around notches is the critical distance method proposed by Taylor [18]. It has been proposed by Boukharouba et al. [4] and El Minor et al. [6] to characterize the fracture conditions for a notched specimen by using the actual stress gradient at the notch root. This stress gradient for round notches is different from the crack tip stress gradient. This method has been applied in the present study to assess the fracture criteria of copper lamellas.

The failure of an electromotor collector is investigated, and the paper presents how design mistakes could influence the integrity of the electromotor. Once the fracture criterion is fulfilled a numerical simulation of the crack propagation was carried out in order to find the crack path, and to compare it with the in-service occurred crack.

2 The Collector Assembly

The failure of the 1,600 kW electromotor Fig. 1 was caused by the cracking of copper lamellas, Fig. 2. The cylindrical collector of the electromotor has 315 copper lamellas separated by 315 cat silver isolations. The two tightening rings acting on the "dove-tail" of lamellas are fixing the sectors and keep them compact in service. The compression of the two rings is produced by 16 M20 bolts. The collector is mounted in steady position providing a compression which would not let the lamellas and the isolations between them to move on radial direction under the centrifugal forces [15].

The regular operating regime of the motor is of 1,000 rpm, but initially collector was dynamically formed at 1,200 rpm. A heating of the collector assembly will take place due to the 11.545 kW losses in the commutator (sector-brushes). Due to difference in the coefficient of thermal expansion of steel bolts and copper lamellas this heating produces strains and additional stresses, both when the motor is running and when, after a long operating period, it is stopped.

The design of the lamellas was based on the strength condition in bending: at the middle of the lamella and at the end of the dove-tail, after Alexeev [2].

The copper lamellas (Cu 99.97) properties are indicated in Table 1, after Shigley [17].

Tests carried on at Strength of Material Laboratory from Politehnica University of Timisoara on plate specimens made from in-service copper lamellas, shown lower values than prescribed ones for the yield stress (218 MPa) and ultimate tensile strength (275 MPa).

Fig. 1 Collector assembly

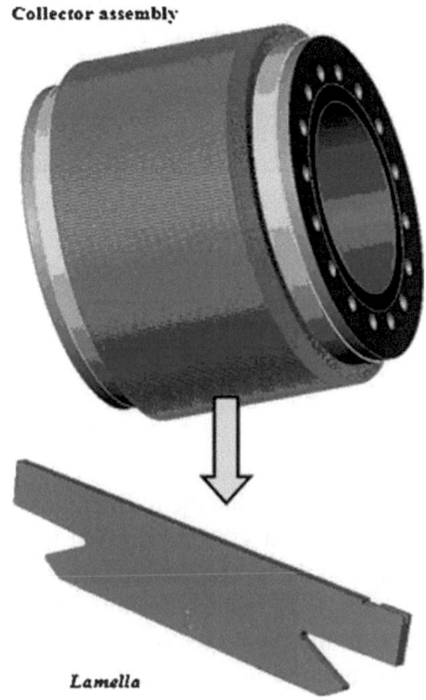

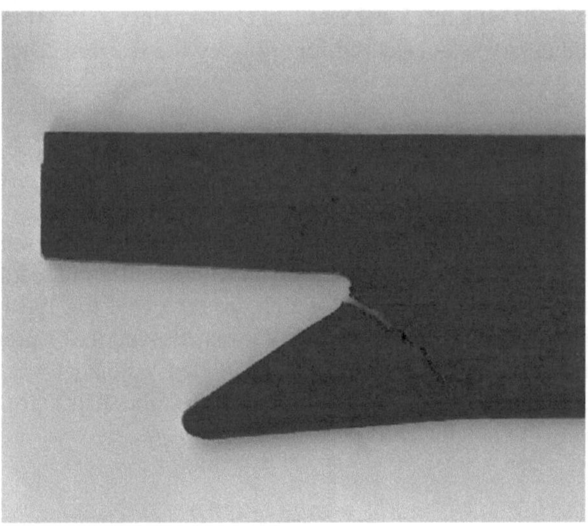

Fig. 2 Cracked copper lamella

The values shown in Table 1 and experimentally determined correspond to room temperature. A thermal analysis shows that the normal working temperature of the collector is 120°C in normal conditions and 135°C in overload [15]. In literature

Table 1 Copper properties

Properties	Symbol and measurement units	Value
Density	ρ, kg/m^3	8,940
Young's modulus	E, MPa	(1.1–1.2) 10^5
Poisson's ratio	νg –	0.34–0.36
Yield stress	σ_{ys}, MPa	250–330
Ultimate tensile strength	σ_r, MPa	280–380
Coefficient of thermal expansion	αg1/°C	(1.65–1.76) 10^{-5}

it can be found that the ultimate tensile strength decreases to 210 MPa at 100°C, and at 180 MPa at 200°C [12]. On the other hand [12] is mention that for copper the creep should be considered only at temperatures around 5,000°C.

Less data could be found for the fracture toughness of copper and copper alloys. In the databases of AFGROW code values between 20 and 75 MPa m0.5 are indicated, the higher values corresponding to copper – beryllium alloys [9]. Investigating the variation of the fracture toughness of four copper alloys from nuclear industry in [1] it was shown that the fracture toughness decreases from 80–100 MPa m0.5 at 25°C to 7–20 MPa m0.5 at 250°C. In consequence a 15 MPa m0.5 value for the fracture toughness was considered for applying the fracture criteria.

3 The Notch Stress Intensity Factor

Considerable effort was devoted to find explicit formulas to describe the elastic stress field around notches in plates or round bars under different loading conditions. Different approaches were considered which are reviewed for example in [4], and [14]. The notch stress intensity factor (KIρV) was adopted in the present study as a crack initiation criteria for a crack starting at rounded V-notch.

Creager and Paris [5] proposed an expression of the KIρ V for a blunt notch which considering the notch radius ρ:

$$K_{I\rho}^{V} = \sigma_{y,\max}\frac{1}{2}\sqrt{\pi\rho} \tag{1}$$

where σy,max represents the maximum normal stress on the notch bisector.

A derivation of Eq. (1) is suggested in [4] on the form:

$$K_{I\rho}^{V} = \sigma_y(x_C)\sqrt{\pi x_C}\frac{\left(2+\frac{\rho}{x_C}\right)^{3/2}}{2\left(1+\frac{\rho}{x_C}\right)} \tag{2}$$

where xC represents a characteristic distance related to the grain size, and σy(xC) the normal stress corresponding to the critical distance on the notch bisector. A similar expression was proposed in the same reference:

$$K_{I\rho}^{V} = \sigma_y(x_C)\sqrt{2\pi}\left(x_C + \frac{\rho}{2}\right)^{\alpha} \tag{3}$$

where α represents a parameter depending of the notch angle ψ:

$$\alpha = 0.5 - 0.089\frac{\psi}{\pi} + 0.442\left(\frac{\psi}{\pi}\right)^2 - 0.853\left(\frac{\psi}{\pi}\right)^3 \tag{4}$$

El Minor et al. [6] proposed to determine the KIρV based on a stress σy(xef) corresponding to an effective distance xef.

$$K_{I\rho}^{V} = \sigma_y(x_{ef})\sqrt{2\pi x_{ef}} \tag{5}$$

The effective distance represents the inner limit of the singularity field, and can be determined from the finite element analysis, as the limit between constant value of maximum stress and linear variation in a Log (σy) versus Log (r) plot.

The criterion for crack initiation is expressed in the form:

$$K_{I\rho}^{V} \geq K_{IC} \tag{6}$$

As a result, the crack initiates when the KIρV reaches the fracture toughness of the material KIC.

4 Numerical Analysis of the Stress Field in Copper Lamella

The design of the collector was done following Alexeev's methodology, originally published in 1954. The design criteria's were the bending stress at the middle of the lamella and the bending of the dove tail. Unfortunately, the stress concentration from V-round notch from the end of the dove tail was not taken into account. The fracture of lamellas shows that the crack initiates in the stress concentration area and propagates on a curved path due to dove tail bending, Fig. 2.

The stress and strain distribution in the copper lamella was done by finite element analysis using FRANC2DL code developed at Cornell University [10]. Based on symmetry half of lamella was modeled. The mesh consists in 7,040 quadratic 6 node triangular elements connected in 14,361 nodes. The lamella was fixed in the symmetry axis and load on the dove tail. The loadings were calculated according with tightening of the bolts, the temperature variation combined with different thermal expansion of steel bolts and copper lamellas, and centrifugal force due to rotation of the collector. Six working regimes were identified:

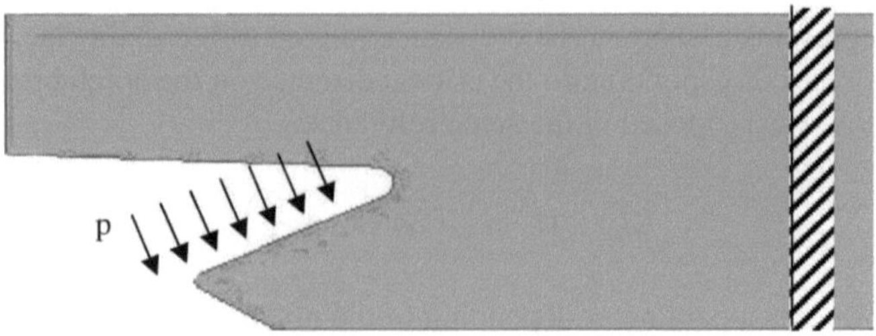

Fig. 3 Lamella with boundary conditions

I. Cold forming – stress produced only by the bolts tightening
II. Bolts tightening and rotation with 1,000 rpm
III. Bolts tightening and rotation with 1,200 rpm corresponding to dynamic forming of the collector assembly
IV. Bolts tightening, rotation with 1,000 rpm and temperature variations for normal working conditions (lamellas 120°C and bolts 60°C)
V. Bolts tightening, rotation with 1,000 rpm and abnormal temperature variations for lamellas 134.80°C (sparks)
VI. Stopped collector with loadings due to bolts tightening and temperature variations

For all these regimes the axial and radial forces were determined and applied uniform distributed on the dove tail surface ($35 \times 4.6\,\mathrm{mm} = 161\,\mathrm{mm}^2$) where the tightening ring acts, Fig. 3.

The stress distributions for the normal working regime (IV) are shown in Fig. 4.

It can be observed in Fig. 4 the stress concentration around the notch. The maximum normal stress σx was obtained on the symmetry axis and is produced by the bending of lamella. The maximum normal stress σy and equivalent stress σ ech are produced around the stress concentrator. The maximum numerical results for all six regimes are summarized in Table 2. It can be observed that for the IV (normal working conditions) and V (abnormal conditions) the equivalent stress is higher than the ultimate tensile stress of copper (380 MPa), which shows the design mistake.

5 Crack Initiation

The crack initiation assessment was performed on the basis of formulae (1), (2) and (5). The north radius was $\rho = 3.4\,\mathrm{mm}$ and the grain size was determined to be $xC = 100\,\mu\mathrm{m}$. The effective distance was determined according to El Minor et al. [6]. Stress distribution on the bisector of the round V-notch was plotted at in log-log coordinates, Fig. 5. Three distinct regions could be observed:

- Region 1 delimitated by the xef where the maximum stress σy,max is practically constant

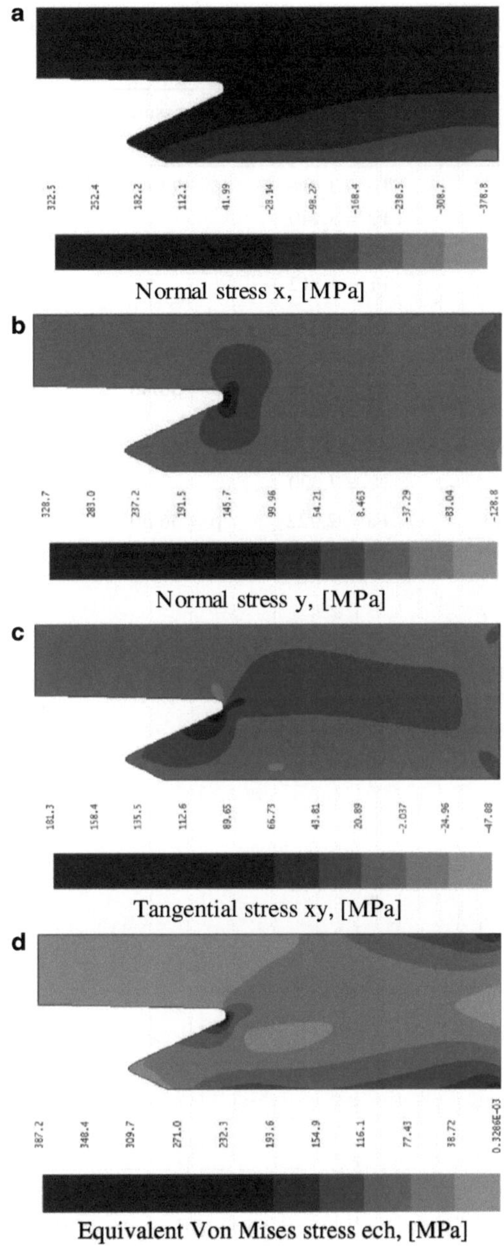

Fig. 4 Stress distributions in lamella

- Region 2 dominated by the notch stress intensity factor, where the Log(σy) is a straight line
- Region 3 far from the stress concentrator

Table 2 Maximum stresses from finite element analysis

Regime	Loads		Maximum stresses			
	Loads (N)	Pressure (Mpa)	$\sigma_{x,max}$ (Mpa)	$\sigma_{y,max}$ (Mpa)	$\tau_{xy,max}$ (Mpa)	$\sigma_{ech,max}$ (Mpa)
I	Fa = 1,986 Fr = 3,440	p = 24.67	156.5	159.5	88.0	187.8
II	Fa = 2,778.2 Fr = 4,812	p = 34.51	218.9	223.1	123.1	262.8
III	Fa = 3,146.7 Fr = 5,450	p = 39.09	247.9	252.7	139.4	297.6
IV	Fa = 4,094 Fr = 7,091	p = 50.86	322.5	328.7	181.3	387.2
V	Fa = 4,329.5 Fr = 7,500	p = 53.79	341.2	347.7	191.8	409.6
VI	Fa = 2,922 Fr = 5,061	p = 36.30	230.2	234.6	129.4	274.6

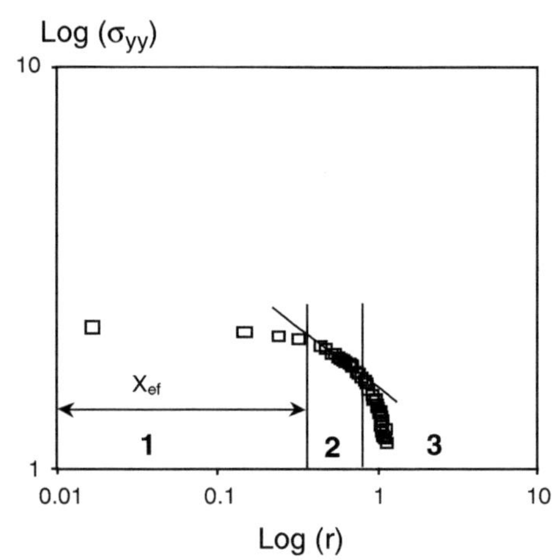

Fig. 5 Normal stress distribution on the bisector of the notch for the regime IV

Figure 4 allows the identification of the effective distance which has the value xef = 0.49 mm for regime IV, and the corresponding stress σy, ef = 325.4 MPa. The obtained results for the other regimes are presented in Table 3, together with the values of the Notch Stress Intensity Factor based on the three approaches.

It can be observed that for all three approaches the notch stress intensity factor KIρV > 15 MPa m0.5 for regimes IV and V. This indicates that the crack initiation condition is fulfilled for these regimes.

Table 3 Notch stress intensity factors values

Regime	$\sigma_{y,max}$ (Mpa)	K_{Ip}^{V}, (1) ($MPa\,m^{0.5}$)	K_{Ip}^{V}, (2) ($MPa\,m^{0.5}$)	x_{ef} (mm)	$\sigma_{y,ef}$ (Mpa)	K_{Ip}^{V}, (5) ($MPa\,m^{0.5}$)
I	159.5	8.24	8.72	0.450	157.9	8.40
II	223.1	11.53	12.20	0.465	220.9	11.94
III	252.7	13.06	13.82	0.486	250.2	13.83
IV	328.7	16.99	17.98	0.490	325.4	18.06
V	347.7	17.97	19.02	0.493	344.3	19.16
VI	234.6	12.12	12.83	0.478	228.7	12.53

6 Crack Propagation

Once the crack initiates will propagates on a curved path. The discrete crack modeling was used for crack propagation which is implemented in FRANC2DL [10]. Discrete modeling of crack propagation is an iterative process with four major steps in each simulation step [19]. A flow chart of the process is shown in Fig. 6.

The input to the process is the initial geometry of the model to be analyzed. The first step in the simulation loop is to perform a stress analysis, in order to compute the displacements, stresses, and strains to the imposed boundary conditions. Singular elements implemented in FRANC2DL were used to model the singularity at the crack tip.

The second step is the evaluation of the Stress Intensity Factors that governs the fracture process. The Stress Intensity Factors were calculated using modified crack closure integral [16].

The third step in the simulation loop is to perform an assessment to determine whether a condition of failure or arrest has occurred. Typical failure criteria are: the critical value of stress intensity factor or crack opening being reached, or the crack tip reaching a free surface of the model. The crack arrest is defined as the point at which the crack tip velocity drops to zero. If failure or arrest is not attained, the fourth step is the finite extension of the crack. This implies three sub steps. First, the fracture parameters, evaluated in the second step are used to predict the new crack direction. Second the geometry of the model is explicitly modified to reflect the new crack configuration. Third, if necessary the object is remeshed so that a new stress analysis can be performed. The simulation loop is now complete, and begins again with a new stress analysis. Crack growth can be modeled using remeshing algorithms that update the mesh only locally to the propagating crack [3].

The direction of crack propagation was based on the Maximum circumferential stress criterion [7], which gives the crack propagation angle in the form:

$$\Delta\theta_C = 2\tan^{-1}\left[\frac{1-\sqrt{1+8\left(K_{II}/K_I\right)^2}}{4\left(K_{II}/K_I\right)}\right] \tag{7}$$

where K_I and K_{II} are the stress intensity factors.

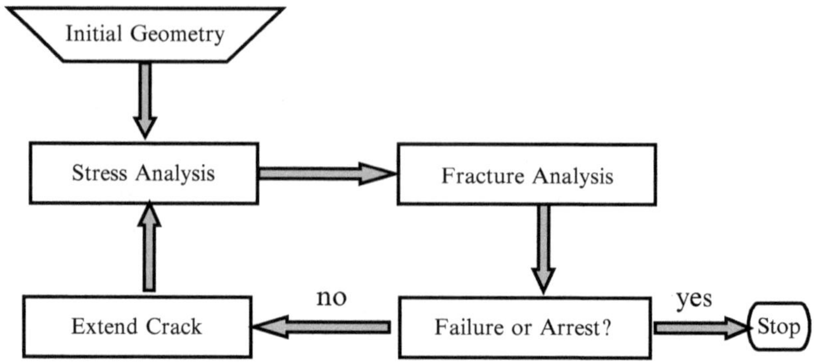

Fig. 6 A flow chart of the simulation loop used for discrete modeling of crack propagation

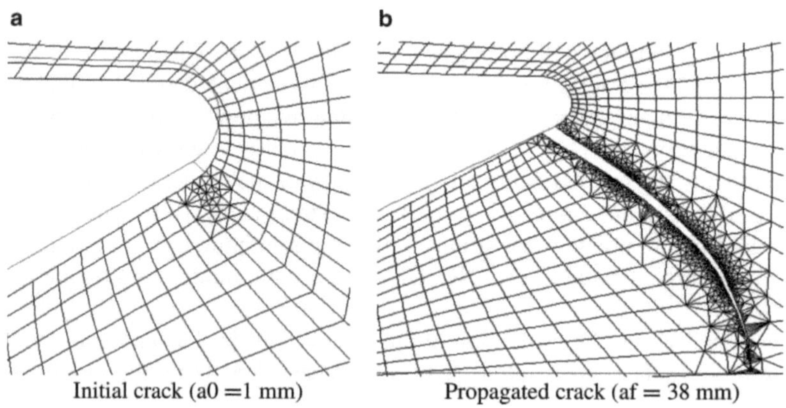

Fig. 7 Simulated crack path from initial crack to final crack

A study of the crack extension increment and was performed and the crack propagation paths obtained from simulation were compared with those obtained in service. A 0.5 mm crack increment was found to give good results.

Figure 7 presents the deformed mesh obtained for a simulation starting from the initial crack of a0 = 1 mm and after 74 propagation steps af = 38 mm.

A comparison between in-service observed crack and simulated crack path is shown in Fig. 8. It can be observed that for the chosen crack propagation parameters a good agreement was obtained.

7 Conclusions

The results of the Finite Element Analysis presented in Table 1 show that for the working regime IV and V the maximum stresses ahead of the notch are higher than the ultimate tensile strength of the copper. The experimental values determined on

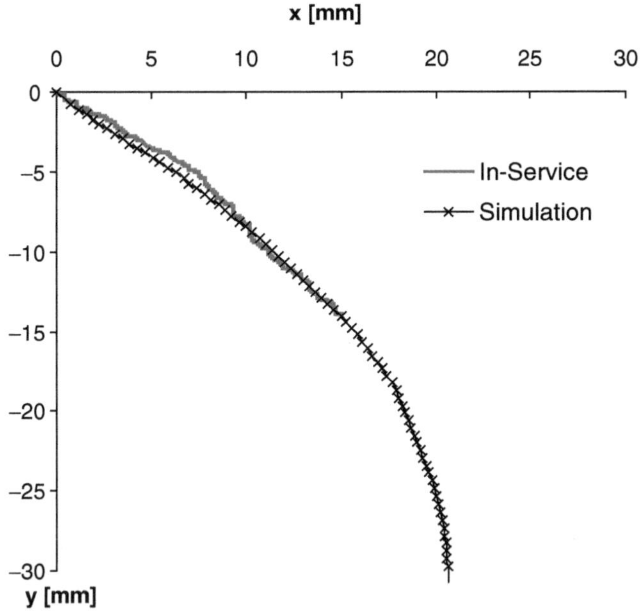

Fig. 8 Comparison between in-service observed crack and simulated crack path

specimens realized from lamellas were lower ($\sigma r = 275\,MPa$) that those prescribed in the literature. For these values the stresses corresponding to the III-rd working regime are higher than ultimate tensile stress. However, this regime occurs for a very short period of time, only for dynamic forming of the collector.

For the assessment of crack initiation the three different formulas were considered for calculating the NSIF. All three approaches give similar results for the NSIF. Rel. (1) takes into account only the maximum stress and the notch radius, being the simplest to use. The second approach, based on rel. (2), requires supplementary the determination of grain size, xC. Rel. (5) needs the values of the effective distance xef and the stress corresponding to this distance $\sigma y,ef$ on the notch bisector. These quantities were determined using a Finite Element Analysis with a relatively fine mesh around the notch.

From Table 2 it can be observed that for the regimes IV and V corresponding to normal and abnormal working regimes of the collector, the crack initiation criterion is fulfilled, and the crack initiates.

The numerical simulation of crack propagation, based on discrete crack propagation, give a crack path very close to the one observed is service. It was confirmed that the maximum circumferential stress criteria for crack propagation angle and the 0.5 mm crack increment were correctly used in the crack propagation simulation.

Finally, we can conclude that:

- The crack starts to initiate from the round – V notch.
- This stress concentration was not taken into account in the design phase, based on Alexeev [2].
- The numerically obtained crack path agrees well with the in-service cracked lamellas.
- The lamellas design methodology has to be significantly revised in order to account for the actual stress concentration.

Acknowledgments The authors express their gratitude to ANCS who support part of this work trough Research for Excellence program 202/2006.

References

1. Alexander D.J., Zinkle S.J., Rowcliffe A.F.: Fracture toughness of copper-base alloys for fusion energy applications. Journal of Nuclear Materials. 271 & 272, 429–434 (1999)
2. Alexeev A.E.: Construcţia maşinilor electrice. Ed. Energetică de stat, Bucureşti (1954) (in Romanian)
3. Bittencourt T.N., Wawrzynek P.A., Ingraffea A.R., Sousa J.L.: Quasi automatic simulation of crack propagation for 2D LEFM problems. Engineering Fracture Mechanics. 55 (2), 321–334 (1996)
4. Boukharouba T., Tamine T., Niu L., Chehimi C., Pluvinage G.: The use of Notch Stress Intensity Factor as a fatigue crack initiation parameter. Engineering Fracture Mechanics. 52 (3), 503–512 (1995)
5. Creager M., Paris P.C.: Elastic field equations for blunt cracks with reference to stress corrosion cracking. International Journal of Fracture Mechanics. 3, 247–252 (1967)
6. El Minor H., Kifani A., Louah M., Azari Z., Pluvinage G.: Fracture toughness of high strength steel using the notch stress intensity factor and volumetric approach. Structural Safety. 25, 35–45 (2003)
7. Erdogan F., Sih, G.C.: On the crack extension in plates under plane loading and transverse shear. Journal of Basic Engineering. 85, 519–527 (1963)
8. Filippi S., Lazzarin P., Tovo R.: Developments of some explicit formulas useful to describe elastic stress fields ahead of notches in plates. International Journal of Solids and Structures. 39, 4543–4565 (2002)
9. Harter J.A.: Afgrow users guide and technical manual. Wright Patterson Air Force Base, Final report (1999)
10. Iesulauro E., FRANC2D/L a Crack Propagation simulator for plane layered materials. Cornell University, Ithaca, NY (2002)
11. James M.N.: Design, manufacture and materials; their interaction and role in engineering failures. Engineering Failure Analysis. 12, 662–678 (2005)
12. Kudashov D.V., Martin U., Heilmaier M., Oettel H.: Creep behaviour of ultrafine-grained oxide dispersion strengthened copper prepared by cryomiling. Materials Science & Engineering. A 387–389, 639–642 (2004)
13. Lazzarin P., Filippi S.: A generalized stress intensity factor to be applied to rounded V-shaped notches. International Journal of Solids and Structures. 43, 2461–2478 (2006)
14. Lazzarin P., Tovo R.: A unified approach to the evaluation of linear elastic fields in the neighbourhood of cracks and notches. International Journal of Fracture. 78, 3–19 (1996)

15. Resiga Susan R., Marsavina L., et al. Analysis of the direct current-brush motor C 990/400-6-2, 1600 kW, 750 V, 1000 rpm with the thermo-mechanical stresses evaluation in the collector and the analysis of the ventilation-cooling system running. Research report, POLITEHNICA University of Timisoara, Timisoara (2005)
16. Rybicki E.R., Kanninen M.: A finite element calculation of stress intensity factors by a modified crack closure integral. Engineering Fracture Mechanics. 9, 931–938 (1977)
17. Shigley T.: Mechanical engineering design. McGraw-Hill (1980)
18. Taylor D.: A mechanistic approach to critical – distance methods in notch fatigue. Fatigue and Fracture of Engineering Materials and Structures. 24 (4), 215–224 (2001)
19. Wawrzynek P.A., Ingraffea A.R.: Discrete modeling of crack propagation: theoretical aspects and implementation issues in two and three dimensions. Cornell University, Ithaca, NY (1991)

The Regularities of Fatigue Crack Growth in Airframes Elements at Real Operation Conditions

Igors Pavelko and Vitalijs Pavelko

Abstract The results of analytical and experimental researches concerning predicting of fatigue crack growth in the operating conditions are presented. First of all the main factors causing a fatigue damage initiation and growth are analyzed and divided to two groups. Common conditions of fatigue damage precise predicting are established. The problem of fatigue crack growth at the stresses of variable amplitude was analyzed and an approach of description of this process is performed. Two examples present the efficiency of this approach. Theory of fatigue crack growth indication and the crack growth indicator (CGI) are developed. There is planned and executed a flight experiment using CGI located on two aircraft An-24 and An-26. Results of crack growth in CGI at operational load allowed to evaluate the parameters of generalized Paris-Erdogan law and statistical properties of crack increment per flight.

Keywords Fatigue crack growth · Operation conditions · Indication fractography

1 Introduction

Fatigue failure is generally considered to be a three-stage process. First stage involves the initiation of a crack from a defect and the subsequent growth of the crack along some favorably orientated direction in the microstructure. Eventually the crack will become sufficiently large that the microstructure has a reduced effect on the crack direction and the crack will propagate on average in a plane normal to the maximum principal stress direction. At the present time the accuracy of a crack initiation predicting is relatively poor. But it is sufficient for SHM planning and optimizing. There are many methods for predicting of fatigue damage initiation.

I. Pavelko and V. Pavelko (✉)
Professor, Aviation Institute of Riga Technical University, Latvia
e-mail: {Igors.Pavelko, Vitalijs.Pavelko}@rtu.lv

S. Pantelakis, C. Rodopoulos (eds.), *Engineering Against Fracture: Proceedings of the 1st Conference*,

The often used method is on the basis of the local strain calculation at the given spectra [1, 2]. The strain-based approach is especially appropriate for ductile airframe materials such as aluminum and titanium. Using simple Neuber's rule can calculate local stresses and strains in critical points of structural element as the function of nominal stresses. Using the strain–life curve and Miner's rule, the total fatigue damage can be calculated and compared with critical one.

$$\frac{dl}{dN} = C(\Delta K)^m \tag{1}$$

Where **ΔK** is the range of the stress intensity factor, **C** and **m** are material coefficients. Usually m lies between 2 and 7. This relationship can be used to predict the remaining lifetime of a structural component, if the stress amplitude remains approximately constant. There are known others, more complex models used a stress intensity factor as main motive factor of fatigue crack growing.

At the same time the precise predicting of the fatigue crack growing is difficult problem, because there are many effects causing large range of the rate of fatigue crack growth.

First of all if the stress amplitude varies, then the growth rate may depart markedly from the simple power law. The single overloads can reduce the crack growth rate drastically. There are a number of models for description of overload effect. First physical models [3–5] explain this effect on the basis of residual and active stresses interaction in plastic zone of the crack front. Recent results for modeling of crack growing at the load with variable amplitude are some versions of these models. The more perspective approaches of the crack growing at variable amplitude load are methods for predicting flight-by-flight crack growth [6–13]. These methods use the load spectra for one flight as a unit of generalized periodic load. As a result all effects of overloading are accumulated in a fatigue crack increment for one flight and generalized Paris-Erdogan law (or similar formulas) can be used for predicting of crack propagation. Obviously better accuracy of predicting can be achieved. But in all cases it is necessary to reduce real flight loads to some idealized spectra for its using in fatigue tests or analysis. It is common disadvantage of all these approaches. Furthermore for example, small changes in the concentration of corrosive agents in the environment can also produce very different results in comparison with predicted ones.

Therefore there are many researches for direct definition of fatigue resistance of materials and structural elements. Many devices were patented with this purpose.

In this paper the results of analytical and experimental researches concerning predicting of fatigue crack growth in the operating conditions are presented. First of all the main factors, causing a fatigue damage initiation and growing were analyzed and divided to two groups. Common conditions of fatigue damage precise predicting are established. The problem of fatigue crack growth at the stresses of variable amplitude was analyzed and an approach of description of this process is performed. Theory of fatigue crack growth indication and the crack growth indicator (CGI) are developed. There is planned and executed a flight experiment using CGI located on two aircraft An-24 and An-26.

2 Some Problems of Fatigue Damage Predicting

2.1 Description of Common Conditions Causing Fatigue Damage

It is known the most essential factor causing the fatigue failure is the level and distribution of variable stresses during mechanical loading. Below the concept of a local critical point of fatigue failure are using. It is some point for which there is small region δW that completely surrounded this point and the probability of occurrence of a fatigue micro crack in the this point is not below in comparison with other points of the region δW. It is obvious that the critical point can be isolated, can belong also to a line or a surface with the uniform probability of fatigue failure.

There are concomitant factors important for process of fatigue failure: technological, statistical, material properties, quality of a surface, an environment, etc.

Let there are two various critical points to which all concomitant factors are identical. Then it is possible to approve, that fatigue resistance in these two points of dynamically loaded structure M_1 and M_2 is alike, if it is possible to choose two coordinate systems with the beginning in each critical point so that the corresponding components of a tensor of stresses in regions δW_1 and δW_2 are equal, i.e.

$$\sigma_{ij}^{(1)}(x-x_1,y-y_1,z-z_1,t)=\sigma_{ij}^{(2)}(x-x_2,y-y_2,z-z_2,t) \qquad (2)$$
$$(i,j=1,2,3)$$

Let under the same conditions of identity of concomitant factors of the ratios of corresponding components of a tensor of stresses are constant, i.e.

$$\sigma_{ij}^{(1)}(x-x_1,y-y_1,z-z_1,t)=k_s\sigma_{ij}^{(2)}(x-x_2,y-y_2,z-z_2,t). \qquad (3)$$

Where k_s is a constant. In other words, there is full similarity of the stress state in any moment of dynamic loading of structural element.

Then it is possible to approve there is a common characteristic of fatigue resistance at the given spectrum of load. It can be cyclic loading, the program, random stationary process, some periodic combination of these spectra. Let the spectrum of loading P(t) is defined by one parameter of intensity σ, i.e.

$$P(t)=\sigma\cdot\overline{P}(t) \qquad (4)$$

Where $\overline{P}(t)$ is some dimensionless function defining structure of loading process on time t.

Then the common characteristic of fatigue resistance D(s) at the given spectrum of load connects some parameter D of intensity of fatigue damage with parameter σ of intensity of loading. For a cyclic load it is Veller's curve $N(\sigma)$, connecting number of cycles N with amplitude of stresses. For random stationary process it is function $T(\sigma)$ connecting the time of destruction with a standard deviation of stresses. Each of these functions concerns to some critical point, and the condition (3) can be considered as a condition for a component of stresses tensor at two various parameters of intensity of loading σ_1 and σ_0. In this case a constant $k_s = \sigma_1/\sigma_0$.

A variety of concomitant factors, uncountable set of variants of the stress state and processes of their change as a function of time do not allow to create the general theory of fatigue crack initiation, suitable for predicting of fatigue strength of real structural component. But if some restrictions for separate classes of designs will be accepted, then such theory can be created. For example the applied theory of fatigue failure of rivet joints [14] was performed. It allows obtaining the uniform local fatigue curve for the isolated fixing point, being a basic element of various types of rivet joints.

2.2 A Fatigue Crack Predicting at the Complex Spectrum of Load

The situation essentially becomes simpler for a stage of growth of fatigue cracks. First of all remarkable properties of asymptotic representation of elastic stress in front of a some mode crack provides the description of distribution of stresses as functions of coordinates by means of unique parameter – the stress intensity factor. Therefore the condition of similarity of stress (2) in two points of front of a crack is carried out automatically, if the ratio of corresponding of the stress intensity factors on time does not vary. Thus, for cracks of the first mode, for example, the condition of full similarity of the stress state is transformed to a simple view

$$k_s = \frac{K_I^{(1)}}{K_I^{(0)}} \tag{5}$$

Thus if the set function $\overline{P}(t)$ of structure of loading process is stationary in sense of its parameters identity at all possible intervals of loading, the uniform characteristic of fatigue failure for the given loading spectrum (cyclic loading, stationary random process, periodic program loading) can be presented in the form of generalized Paris-Erdogan law

$$\frac{dl}{d\tau} = C_s(k_s\overline{K})^{m_s} = C_s^*\overline{K}^{m_s} \tag{6}$$

Where Cs and ms parameters cracking resistance of a material at the set function of loading structure obtained at some base loading intensity σ_0.

$$\overline{K} = \frac{K_I^{(0)}}{\sigma_0}$$

It is relative stress intensity factor independent of a level of external load. It defines character of effect of length of a crack to stresses intensity near front of a crack. The factor k_s considers influence of intensity of the given loading in comparison with basic process. In the formula (6) τ is the parameter of an operating time which can be estimated in flight hours, number of flights, number of cycles of program periodic load or other units.

Thus, using the formula (6) it is possible to predict growth of the crack in a structural element, if conditions of all concomitant factors are satisfied and external loading has similar structure with basic process, but different intensity. Below it is shown the results of generalization of data about fatigue crack growth of at test of samples of a material 7075-T651 by thickness 12.7 MM and width 152.4 mm [15]. The loading program spectrum was similar with an operational spectrum of the airplane. The maximal stresses in a spectrum were 169, 211.8 and 254 MPa. If to accept the basic parameter of loading intensity $\sigma_0 = 169$ MPa, then $k_s = 1.0$, 1.253, 1.503 accordingly. For each level of loading intensity the direct processing of results of tests "the length of a crack – number of blocks of loading" has allowed to define value of the constant C_s^*, and under the formula $C_s^* = C_s \cdot K_s^{m_s}$ to define these constants for the second and third level of loading, using value C_s^* of the first level. It was accepted average value of a parameter $m_s = 3.2$. Results of processing are presented in Table 1.

Other example: Fatigue test of samples from a titanic alloy Ti-6Al-4 V with thickness 6.35 mm and width 152.4 mm was executed at the program loading [15]. Each block of program contained 17 levels and cyclic load and modeled aircraft load in operation during one flight. At conservation of internal structure of the loading block the test has been lead at four values of a stresses 386.6, 421.8, 456.9 and 492.1 MPa. Average value of a parameter ms = 3.6 was accepted. Results of calculations are presented in Table 2.

Thus, if the program block of loading is used as unit of an operating time predicting, a fatigue crack average growth rate estimated by the law (6) satisfied to data of test with accuracy (6–8)%. It allows to assume, as at use of the information on growth of a fatigue crack under action of real operational loadings it is possible to obtain enough reliable value of generalized law (6) constants and to use them for predicting remaining lifetime of structural elements.

Table 1 Theoretical parameters of the fatigue crack growth with experiment

Maximum stress (MPa)		169.00	211.8	254
k_s		1.0	1.253	1.503
$C_s^* \cdot 10^4$	Experiment	1.907	4.2814	7.6655
	Theory	1.907	3.928	7.025
Error, %		0	−8.25	−8.35

Table 2 Theoretical parameters of the fatigue crack growth with experiment

Maximum stress (MPa)		386.6	421.8	456.9	492.1
k_s		1	1.091	1.182	1.273
$C_s^* \cdot 10^4$	2.592	3.802	4.768	5.955	7.6655
	2.592	3.548	4.730	6.179	7.025
Error, %		0	−6.703	−0.797	3.77

3 Indication of Fatigue Crack Growth

3.1 Common Theory

At the same time the simulation of a flight loading spectrum in laboratory test does not guarantee exact definition of characteristics of cracking resistance in real operation (change of duration of flight, flight weight, weather and climatic conditions, rare random overloads, an environment), because it is difficulty to consider all factors of operation in the program of fatigue tests with the ordered, limited number of loading levels. Therefore for increase of reliability of predicting and efficiency of aircraft operation it would be desirable to obtain the experimental data on growth of fatigue cracks directly during operation. Such attempts are known. In [16, 17] results of data processing about growth of cracks in operation in a covering of a wing of the C-130 aircraft are stated. For crack reconstruction the crack growth rate was measured between two sequential inspections. Paris' law exponent was accepted 3, but constants were defined by numbers of cycles and length of cracks for each one.

However, the cracks in primary structural elements are not admissible at regular flights of civil aircraft. If the crack is found out, the structural element either should be repaired or replaced. So long observation of a growing crack that necessary for reliable definition of cracking resistance constants is not possible at operation.

Therefore an idea of inclusion in dynamically loaded structure some device for indication and measurement of a crack is the logical conclusion of this situation. There are many solution of different aspects of this problem (for example, US Patents No.3, 979,949; 5,319,982; 6,983,660; 5,614,680; 5,816,530; 6,443,018; Japan Patent 46359-78; SU 1504548-1989; 938093-1982 and others) [18–27]. All type of these devices must be subjected the same load but its destruction must not influence to the fatigue lifetime of a structure. These are common properties.

Common disadvantage of all these solutions is low reliability of fatigue damage predicting connected with that of other significant difference of initiation and development of fatigue damage in comparison with original structure. Here the version of device [20] is used for definition of characteristics of crack resistance in real operation of aircraft. This device is free from mentioned disadvantage because it is created in complete accordance with general theory stated above. The general theory allows to solve a problem of obtaining of material constants at the set function of loading structure. It is obvious the device must have a sensitive element containing a fatigue crack. Such device is named by the crack growth indicator (CGI). Constants of material for some structural element can be correctly obtained at following conditions:

1. All concomitant conditions (the mark of a sensitive element material, technology of its mechanical and thermal processing, thickness, etc.) should be same, as at an investigated structural element.
2. CGI should be fixed on a structural element so that to have the same history of loading, defined by function $\overline{P}(t)$ of structure of loading process.

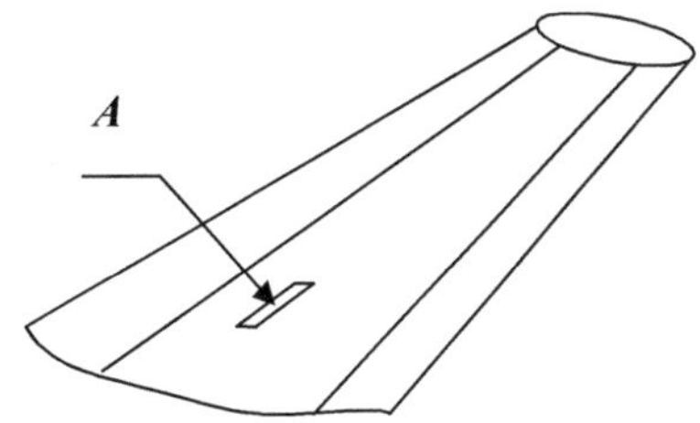

Fig. 1 The scheme for CGI mounted at the wing panel of aircraft

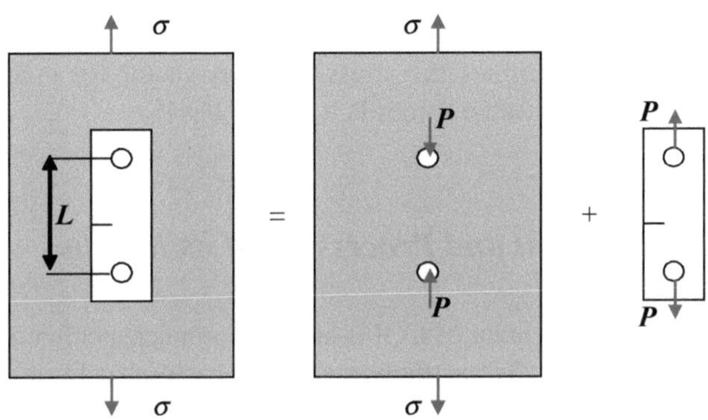

Fig. 2 Crack Growth Indicator mounted at a structural element: scheme of interacting forces

In this case growth of a crack in CGI will be caused by action real operational load and influence of all concomitant factors (temperature, humidity, a chemical compound of environment, etc.). Figure 1 shows principal scheme of CGI mounted at the wing panel of aircraft (Unit A). The scheme of structural element and CGI interaction is presented in Fig. 2.

Disregarding nonlinear effect, it is possible to accept the operational stress σ(t) and the force P(t) acting on the CGI are proportional. Therefore, the stress intensity factor K(t) on a crack front in CGI is proportional to σ(t). It means both structural element and CGI have the same function $\overline{P}(t)$ of structure of loading.

The stress intensity factor $\overline{K}$ is some function of a crack length. For this function determination the energetic method of mechanics can be used. First of all interaction force P (Fig. 2) can be defined by the formula

$$P = \frac{\sigma L}{E_0}(\lambda_0 + \Delta\lambda_i) \tag{7}$$

Where λ_0 is structural element, indicator without crack and fastening device common elastic compliance; $\Delta\lambda_i$ is elastic compliance increment, caused by the growing of a crack, L is base of CGI, E_0 is elasticity module of structural element material.

The elastic compliance λ_0 can be calculated theoretically or obtained experimentally. Elastic compliance increment $\Delta\lambda_i$ can be determined in accordance the theorem of displacements reciprocity [21]. Effect of a crack can be expressed by displacement increment Δi in direction of appropriate generalized force Q_i

$$\Delta_i = \Delta\lambda_i Q_i \tag{8}$$

As a result elastic compliance increment is defined by known formula of the linear fracture mechanics

$$\Delta\lambda_i = \frac{2}{Q_i}\int_s \left[\int_s \frac{(1-\nu^2)}{E}\left(K_I^2 + K_{II}^2\right) + \frac{1+\nu}{E}K_{III}^2\right] dS \tag{9}$$

Where E and E_0 are elasticity modules of CGI and structural element materials, δ is sensitive element thickness.

The function φ(l) determines the stress intensity factor for crack in sensitive element at action of concentrated forces P, it is K = Pφ(l).

3.2 Flight Experiment and Processing of its Results

In this case the sensitive element of CGI is around compact specimen of aluminium alloy D16T (Fig. 3a) with 5 mm thickness and a basic size D = 54 mm. Others geometric relations and basic function K = Pφ(l) can find out in [22] for such configuration of isolated sensitive element. Using this function and formula (10) it is possible to define the function "Relative stress intensity factor – Crack length". The analytical approach for determination of elastic compliance λ0 was created for a thin-walled structural element and two-points fastening of CGI. For experimental determination of initial elastic compliance λ0 the special device (Fig. 3b) was designed and made. It allows also direct experimental obtaining of function "Relative stress intensity factor – Crack length". Figure 3c presents a flat sample (Al alloy sheet) prepared for experimental evaluation of this function. CGI sensitive element can be joined with structural component by additional strain multiplier. Desirable degree of multiplication can be obtained by the increasing of CGI installation base L (distance between the CGI fastening points to structural component).

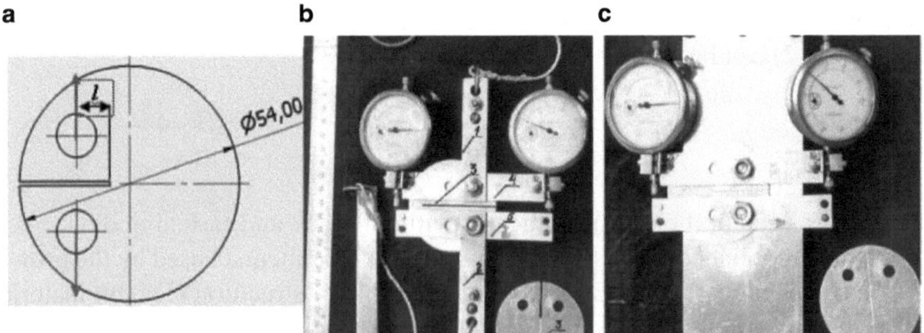

Fig. 3 Round compact shape of sensitive element of CGI **a**, CGI with a device for experimental evaluation of stress intensity factor by elastic compliance method **b** and a sample prepared for the function "stress intensity factor – crack length" experimental evaluation

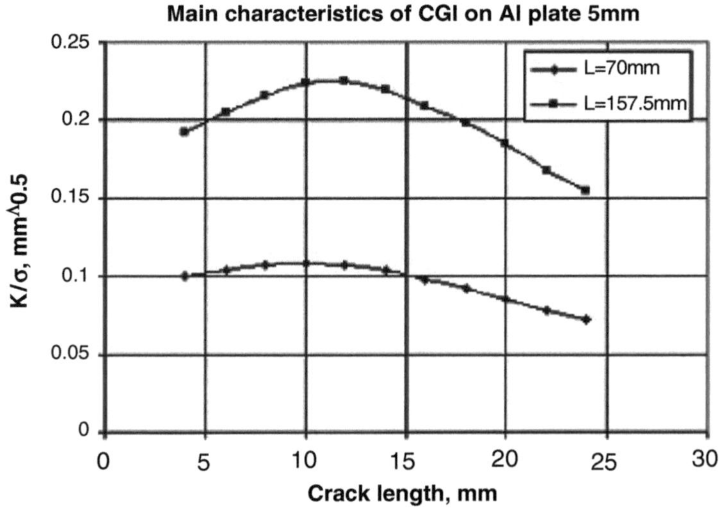

Fig. 4 CGI main characteristics

Figure 4 presents CGI typical main characteristics "Relative stress intensity factor – Crack length" for two L. They were made for special flight experiment described below. It is seen, for fixed base of CGI installation L the stress intensity factor variation is small in wide interval of crack length. Therefore a fatigue crack growth rate in this interval of crack length also will be relatively small. Otherwise if the base L will be increased, then both the stress intensity factor and a growth rate of a crack increases.

There was planned and executed a flight experiment using two aircraft An-24 and An-26 in operation. CGI was used for the evaluation of parameters of law (6) for bottom panel of central part of a wing. This part is located between fuselage and a compartment of engine. CGI was attached to the back longeron (here there are maximum nominal stresses in section of a wing). It was attached by bolts using the holes of the joint of bottom technological panel and the longeron.

Previous analysis of CGI influence to structural elements of wing was executed. Before installation to the aircraft the preliminary initiation of a fatigue crack was executed in laboratory by relatively low level of cyclic loading which could not cause the effect of crack braking.

The range of operating time of CGI sensitive element was from 5 to 300 flights. After taking down the sensitive element was prepared for fractography analysis of a crack surface. It is known the quantitative fractography was successfully used for coupons to generate crack growth curves under flight spectrum loading [10–12,23,24]. Fractography research was carried out by the electronic microscope. During the flight experiment the parameters of aircraft loading in each flight were strictly fixed: take-off weight, payload and fuel, flight range and duration. So the opportunity to estimate effect of these parameters to the crack growth rate was achieved. Typical view of a crack surface of CGI after 294 flights is presented in

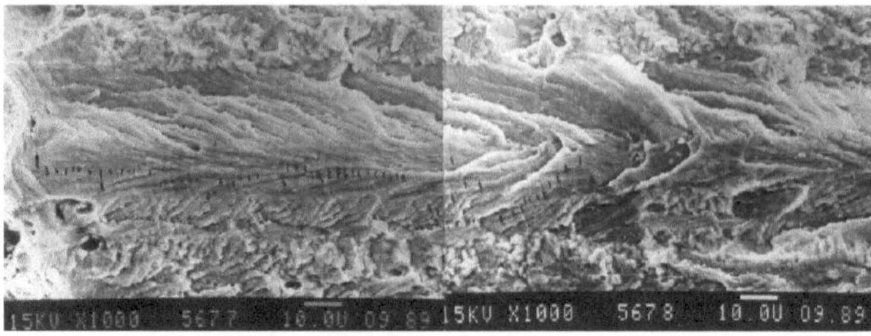

Fig. 5 View of a crack surface in CGI after 294 flight

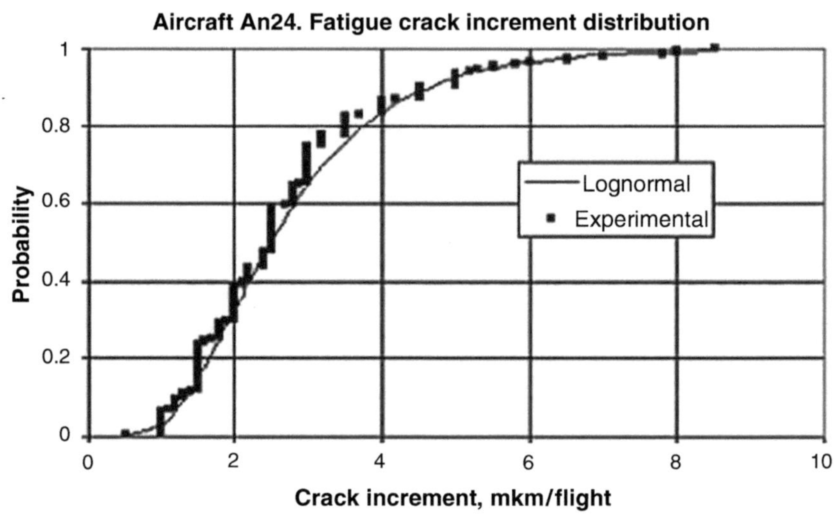

Fig. 6 Comparison of distribution of crack increment per flight

the picture (Fig. 5). The bands caused by the flights could be distinguished on the fracture surface and an accurate reconstitution of the crack growth curve could be made. The crack increments occurring in the flights were measured (it is possible to see the marks for identifying each flight).

The attempt of an evaluation of the statistical characteristics of a crack growth was undertaken. Below the results of statistical analysis of CGI in 231 flights experiment, are presented. The crack growth was fixed precisely in 202 flights.

The measurement of width bands was executed with use of a special technique with maximum error no more than 10%. The estimated average increment is equal $2.72 \cdot 10^{-3}$ mm/flight, and standard deviation $1.44 \cdot 10^{-3}$ mm/flight. The constant of generalized Paris-Erdogan law $C_s^* = 0.0102$ and m = 2.5, if relative stress intensity factor is measured in $\sqrt{m}$ and a crack growth rate in millimeter per flight. The statistical hypothesis about lognormal distribution low of a crack increment was carried out with positive result. Figure 6 presents the lognormal distribution function (continues line) and points of empirical cumulative distribution.

4 Conclusions

The fatigue damage having significant influence on remaining strength and remaining lifetime is very difficultly predicted, especially at a stage of initiation of a fatigue crack. Therefore, if the high accuracy of a prediction of remaining lifetime of some structural element is needed, then conditions of testing must correspond as much as possible with conditions of real operation. The proposed approach of this problem solution assumes such theoretical analysis and experiment at which the stress state in critical points of samples and its variation for time would be same as in an investigated structural component. This is main condition of reliable predicting of fatigue damages initiation and its development. The second condition assumes identity of concomitant factors which are not the direct cause of fatigue degradation, but can significantly accelerate it. If these conditions are exactly satisfied, then it provides practically reproduction in experiment of the investigating phenomenon. In this case the experimental mechanical properties of a material will be exact and can provide a reliable prediction of remaining lifetime of a structural component. It is obvious also, that the area of use of these characteristics is limited, especially, for initiation of a crack. However, at the careful analysis this area can be widespread to separate classes of critical places of a structure.

It is shown the fatigue crack growth allows obtaining more common solution in comparison with the initiation of a crack. The variant of effective solution offered above assumes evaluation of the fatigue crack resistance properties of material in conditions of real operational load at action of all concomitant factors influencing intensity of fatigue damage development. The approach assumes use of special device, fatigue crack growth indicator, for which all necessary conditions for determination of fatigue crack growth are satisfied practically precisely. On the other hand, the combination of CGI with fractography of a fatigue crack surface allows decreasing necessary tests, terms of their carrying out and increasing accuracy and reliability of results.

Acknowledgments This research was induced by the authors' participation in 6FP research project AISHA (Aircraft Integrated Structural Health Assessment). The authors are grateful to European Commission for financial support and all partners for scientific and technological collaboration.

Especially the authors are grateful to the Kiev Mechanical Plant (Antonov Design Bureau) for collaboration in the flight experiment organization and high quality of fractographic investigation.

References

1. J.C. Newman Jr. A crack-closure model for predicting fatigue crack growth under aircraft spectrum loading. In: J.B. Chang and C.M. Hudson, Editors, Methods and models for predicting fatigue crack growth under random loading, ASTM STP 748, American Society for Testing and Materials, Philadelphia, PA, 1981, pp. 53–84.
2. N.E. Dowling. Mechanical behavior of materials, engineering methods for deformation, fracture and fatigue, Prentice-Hall, Upper Saddle River, NJ, 1993.

3. O. Wheeler. Spectrum loading and crack growth. Journal of Basic Engineering, Volume 94, 1972, pp. 181–186.
4. J. Willenborg, R. Engle and R. Wood. Crack growth retardation model using an effective stress concept. Air Force Flight Dynamics Laboratory Report, Technical Report AFFDL-TM-71-1-FBR.
5. J. Bannentine. Fundamentals of metal fatigue, Wiley, New York, 1980.
6. R.A. Pell, P.J. Mazeika and L. Molent. The comparison of complex load sequences tested at several stress levels by fractographic examination. Engineering Failure Analysis, Volume 12, Issue 4, August 2005, pp. 586–603.
7. L. Molent, R. Jones, S. Barter and S. Pitt. Recent developments in crack growth assessment, International Journal of Fatigue, Volume 28, Issue 12, 2006, pp. 1759–1768.
8. M. Levy, A.S. Kuo, K.P. Grube. A practical method for predicting flight-by- flight crack growth in Fighter Type Aircraft for damage tolerance assessment. ICAS-80, 19.2, 1980, pp. 666–675.
9. D. Broek and S. Smith. The prediction of fatigue crack growth under flight-by-flight loading. Engineering Fracture Mechanics, Volume 11, Issue 1, 1979, pp. 123–141.
10. W. Zhuang, S. Barter and L. Molent. Flight-by-flight fatigue crack growth life assessment. International Journal of Fatigue, Volume 29, Issues 9–11, September–November 2007, pp. 1647–1657.
11. R.A. Pell, P.J. Mazeika and L. Molent. The comparison of complex load sequences tested at several stress levels by fractographic examination. Engineering Failure Analysis, Volume 12, Issue 4, August 2005, pp. 586–603.
12. O. Pártl and J. Schijve. Reconstitution of crack growth from fractographic observations after flight simulation loading. International Journal of Fatigue, Volume 12, Issue 3, May 1990, pp. 175–183.
13. J. Schijve, M. Skorupa, A. Skorupa, T. Machniewicz and P. Gruszczynski. Fatigue crack growth in the aluminium alloy D16 under constant and variable amplitude loading. International Journal of Fatigue, Volume 26, Issue 1, January 2004, pp. 1–15.
14. V. Pavelko. Stress stat in the rivet joints: the applied theory of the fatigue fracture. ICAS 2006-4.3.1, 25th Congress of the International council of the Aeronautical Sciences, Hamburg, Germany, 3–8 September, 2006.
15. D. Broek and S. Smith. The prediction of fatigue crack growth under flight-by-flight loading. Engineering Fracture Mechanics, Volume 11, Issue 1, 1979, pp. 123–141.
16. W.S. Johnson, R.A. Heller and J.N. Yang. Flight inspection data and crack initiation times. Proceeding 1977 Annual Reliability and Maintainability Symposium, 1977, pp. 148–154.
17. R.A. Heller and G.H. Stevens. Bayesian estimation of crack initiation times from service data. Journal of Aircraft, Volume 15, Issue 11, 1978, pp.794–798.
18. J.P. Gallagher and H.D. Stalnacer. Developing normalized crack growth curves for tracking damage in aircraft. Journal of Aircraft, Volume 15, Issue 2, 1978, pp. 114–120.
19. J.P. Gallagher et al. Tracking potential crack growth damage in US Air Force Aircraft. Journal of Aircraft, Volume 15, Issue 7, 1978, pp. 435–442.
20. SU Patent 1290412. A. Kondratiev and V. Pavelko. Method of definition of structure fatigue lifetime. 1985.
21. I. Pavelko and V. Pavelko. The valuation of stiffness of rods with cracks. Informator inst. tech. wojsk lot., Warszawa, 1997,– pp. 105–110.
22. J.C. Newman Jr., Stress intensity factor and crack-opening displacements for round compact specimens. International Journal of Fracture, Volume 17, Issue 6, December 1981, pp. 567–578.
23. Joonpyo Hong and J. Gurland. Direct observation of surface crack initiation and crack growth in the scanning electron microscope. Metallography, Volume 14, Issue 3, September 1981, pp. 225–236.
24. J. Siegl, J. Schijve and U.H. Padmadinata. Fractographic observations and predictions on fatigue crack growth in an aluminium alloy under miniTWIST flight-simulation loading. International Journal of Fatigue, Volume 13, Issue 2, March 1991, pp. 139–147.
25. J.P. Gallagher et al. Tracking potential crack growth damage in US Air Force Aircraft. Journal of Aircraft, Volume 15, Issue 7, 1978, pp. 435–442.

26. F. Pfeiffer. Bestimmung des Schädigungsverlaufs bei Dauerschwingbeanspruchung mittels Ermüdungsindikatoren.IfL-Mitteilungen, Volume 17, Nr 6, 1978, s.220–223.
27. I. Pavelko and V. Pavelko. The valuation of stiffness of rods with cracks. Informator inst. tech. wojsk lot., Warszawa, 1997, pp. 105–110.

Session VI
Engineering Applications

Fracture in Electronics

William J. Plumbridge

Abstract Although the World's largest industrial sector, Electronics has received relatively little attention with regard to mechanical failure of its equipment. Design strategies against fracture are advanced for applications, such as aerospace, automotive, pressure vessel and engine applications. So a fundamental question arises as to the extent to which these can be applied to Electronics. The paper describes the special characteristics of electronics equipment in service, and outlines common modes of failure, with emphasis on thermomechanical fatigue. The particular challenge of size is considered when behaviour determined from conventional bulk samples is no longer representative of actual components. Two case studies are presented that illustrate the special nature of electronics. It is suggested that as miniaturisation continues, fracture will become a major cause of failure.

Keywords Miniaturisation · Solders · Size effects · Time dependent behavior · Service conditions for electronics

1 Introduction

In many ways, the discipline of Electronics Engineering can be regarded as distinct from the more traditional Engineering sub-disciplines, such as Civil, Mechanical, Structural, Materials and Aeronautical. Admittedly, the manufacturing aspects of electronics equipment production are at the cutting edge of technology – often in advance of those involved with more conventional products. However, with regard to performance, a chasm has generally existed, and only recently has the importance of topics like Fracture, Structural Integrity, Mechanical Design and Lifetime Prediction, been recognised within the electronics community. The reasons for this are:

W.J. Plumbridge
Professor, The Open University, UK
e-mail: W.Plumbridge@open.ac.uk

S. Pantelakis, C. Rodopoulos (eds.), *Engineering Against Fracture: Proceedings of the 1st Conference*,

- The advent of Surface Mount Technology (SMT), which involves direct attachment of components to printed circuit boards (PCBs), necessitates a structural role for soldered joints (interconnections) in addition to being conductors of heat and electricity.
- Continuous miniaturisation has eliminated the luxury of large safety factors and introduced a requirement for much greater efficiency in mechanical design.

Electronics is entering an era in which topics, such as fatigue, creep, fracture, crack propagation and other mechanical failure modes are becoming as important as they are in pressure vessels, gas turbines and structural frameworks. Moreover, there are additional challenges facing electronics.

The question as to '*to what extent can design against fracture and the assurance of structural integrity, developed in the traditional sectors for comparatively' large-scale applications, be employed to ensure reliability in electronics equipment?*' remains largely unanswered. The present paper addresses this enquiry and aims to introduce the 'Challenge of Electronics' to practitioners in fracture and structural integrity in other fields. After an initial consideration of the key element of electronics equipment, the printed circuit board and its components, those factors that constitute the special characteristics of electronics are discussed and subsequently illustrated with case studies.

2 Fundamentals of Electronics Equipment

The heart of electronic equipment is the printed circuit board which is usually a reinforced polymeric material, such as laminated paper phenolic or glass fibre reinforced epoxy. It is the base for the circuitry, provided by copper tracking, and the various electronic components, such as resistors, capacitors, integrated circuits and so on, which are mounted upon it. The properties of PCBs are strongly anisotropic.

Mounting techniques may be classified as Through Hole Technology or Surface Mount Technology. In the former, the PCB contains holes, and components have wire terminations or stiff lead frames. Components are attached to the circuitry by soldering – a process involving the formation of intermetallic compounds (IMCs) but no localised melting of the items being joined. It is performed when the PCB is exposed to a wave of molten solder contained in a bath (wave soldering). Alternatively, in surface mount technology, the components are mounted on the surface of the PCB instead of in holes. Surface Mount Devices (components) are placed automatically onto the PCB to which solder paste (a mixture of solder powder, flux, and binder agent) has been applied. The PCB then traverses through a furnace and the solder paste melts and re-solidifies (reflow soldering). The temperature-time profile of this process is critical.

Recent years have seen an increase in performance in terms of chip speed and number of components per unit area. A consequence has been the requirement for larger numbers of interconnects (i.e. more pins and finer pin pitch) and various strategies have provided this [1]. These developments have been associated with more arduous demands, in terms of performance requirements and more extreme temperatures, emanating both from the external environment and internal heat generation.

3 The Special Challenges in Electronics

Attainment of structural integrity and reliability in electronics equipment is impeded by several factors, additional to, or more severe than, those experienced in traditional applications. The following paragraphs outline the most significant of these.

3.1 The Range of Materials Involved

All classes of materials (metals, ceramics, polymers and composites) may be found in electronics equipment. Table 1 indicates this wide variety and the extensive variation in values of properties which can be influential in governing performance in service. Such a broad combination is unusual in engineering structures and components, particularly when the materials operate in close juxtaposition. As well as the absolute property values, differences between materials in contact in a joint can be significant.

3.2 A New Generation of Alloys

While developments in most engineering situations are associated with established materials, such as stainless and low alloy steels, or with iterative modifications in titanium or nickel-base alloys, a step change has been imposed upon solder alloys. The abolition of lead in solders for environmental reasons has necessitated development of an entirely new generation of alloys, about which information on their properties is limited, particularly under conditions likely to be encountered in service.

Over the last decade, alloys such as Sn-0.7Cu (mass percent), Sn-3.5Ag and Sn-3.8 Ag-0.7Cu have emerged as replacements for the traditional solder alloy (Sn-37 Pb). In microstructural terms, a conventional two-phase alloy has been replaced by a solid solution alloy, strengthened by fine particles. To date, the properties of the new lead-free alloys have been found to be at least as good as those of the Sn-37Pb.

Irrespective of whether traditional or new alloys are employed, solders are generally exposed to more severe temperatures than most other engineering alloys. In terms of the homologous temperature, Th (the current or operating temperature relative to the melting temperature in Kelvin) solders operating at around ambient temperature experience conditions equivalent to nickel-base superalloys at some thousand degrees higher.

Table 1 Typical values of properties of materials in electronics

Modulus			
Material	$E/GN\,m^{-2}$	*Material*	$E/GN\,m^{-2}$
Silicon carbide, SiC	450	Gold	82
Alumina, Al_2O_3	390	Aluminium	69
Nickel	214	Silver	76
CFRP	70–200	Tin and alloys	14
Platinum	172	Fibreglass (glass-fibre/epoxy)	35–45
Copper	124	GFRP	7–45
Copper alloys	120–150	Lead and alloys	14
Polyethylene, high-density	0.7	Polycarbonates	2.6
Silicon	107	Polypropylene	0.9
Zinc and alloys	43–96	Polyethylene, low-density	0.2
Thermal expansion			
Material	*Coefficient of thermal expansion* $(\times 10^{-6} K^{-1})$	*Material*	*Coefficient of thermal expansion* $(\times 10^{-6} K^{-1})$
Silver	19.0	Nickel	13.3
Copper	16.5	Alumina	8.8
Gold	13.8	E Glass	5.5
Aluminium	23.6	Silicon	2.5
Brass	20.0	Nylon	80–90
Tin	23.5	Polyethylene	60–220
Lead	29.0	Polystyrene	50–85
Solder (60/40)	24	Epoxy resin	26
Fracture toughness			
Material	*Fracture toughness* $MNm^{-3/2}$	*Material*	*Fracture toughness* $MNm^{-3/2}$
Ductile metals (Cu, Ag, Al, solders)	80–350	Nylon	3
Aluminium alloys	20–45	Epoxy	0.3–0.5
Fibreglass	42–60	Silicon carbide	3
Polypropylene	3	Alumina	3–5
Polyethylene	1–2		

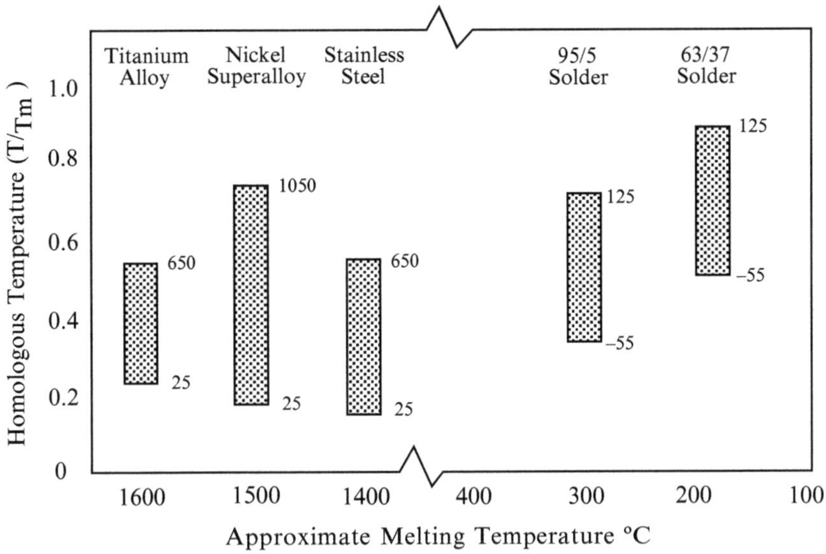

Fig. 1 Typical homologous temperatures for engineering alloys

The significant consequences for stress analysis and structural integrity include (Fig. 1) [2]:

- High strain rate sensitivity – variations in yield and ultimate strengths may be found although these changes are alloy and temperature specific [3]. Associated with this is an uncertainty concerning the value of the elastic modulus. A variation of about a factor of 5 exists in published values for Young's modulus, with those obtained from monotonic tests being particularly susceptible [4].
- Mechanical properties are also highly sensitive to temperature. For example, the strength of both lead-free and lead-containing solders falls by a factor of about 3 between 75° C and −10° C [3,5,6].

Microstructural instability – cast structures at high homologous temperatures are quite unstable.

- The properties of joints, or bulk specimens, may change both during storage and in service, when the thermally imposed cyclic strains accentuate this change further. Some laboratories employ a stabilisation anneal after casting, prior to the determination of bulk mechanical properties, although such a treatment can cause an increase in minimum creep rate of seven times over that measured in the as-cast condition [7] and is not representative of the condition of the actual soldered joint.

A further potential problem has emerged with some of the recently introduced lead-free alloys, particularly with the more dilute systems which are almost pure tin. Prolonged exposure (usually years) to temperatures below 13°C results in the formation of grey tin (tin pest), surface cracking and eventual disintegration of the

Fig. 2 Total disintegration of a Sn-0.5Cu test piece after exposure at -18°C for 5 years

sample (Fig. 2) [8]. This phenomenon has been observed in large model joints but, as yet, not in actual applications. However, the timescale for which these alloys have been in service is relatively short.

4 Modes of Failure

There are probably more ways in which electronics equipment can fail during service than in most other applications. For solders, elasticity is extremely limited, and viscoelasticity, plasticity, viscoplasticity and creep deformation may occur prior to fracture. Common deformation and fracture mechanisms include:

Thermomechanical fatigue (TMF) of soldered interconnections is the most common mode of failure. Its cause is the wide disparity in the coefficients of thermal expansion, α, that exists between the material components of a joint (e.g. α for alumina is $8.8 \times 10^{-6}\,K^{-1}$, for GFRP $17 \times 10^{-6}\,K^{-1}$, and for a eutectic tin-lead solder $22 \times 10^{-6}\,K^{-1}$). Thermal fluctuations generate strains which eventually lead to crack initiation, growth and final failure. The cracks may be contained within the solder or interact with interface between the materials being joined. Because it is the softest material, solder accommodates most of the cyclic damage, and since it is thermally driven, the entire process may be best described as strain controlled fatigue in shear. Temperature profiles in service introduce dwells at various locations in the cycle and establish low cycle fatigue–creep conditions. The classic example is illustrated in Fig. 3.

Overload failures during PCB assembly may be ductile or brittle according to the materials involved. Copper, aluminium and solder elements may be stressed above

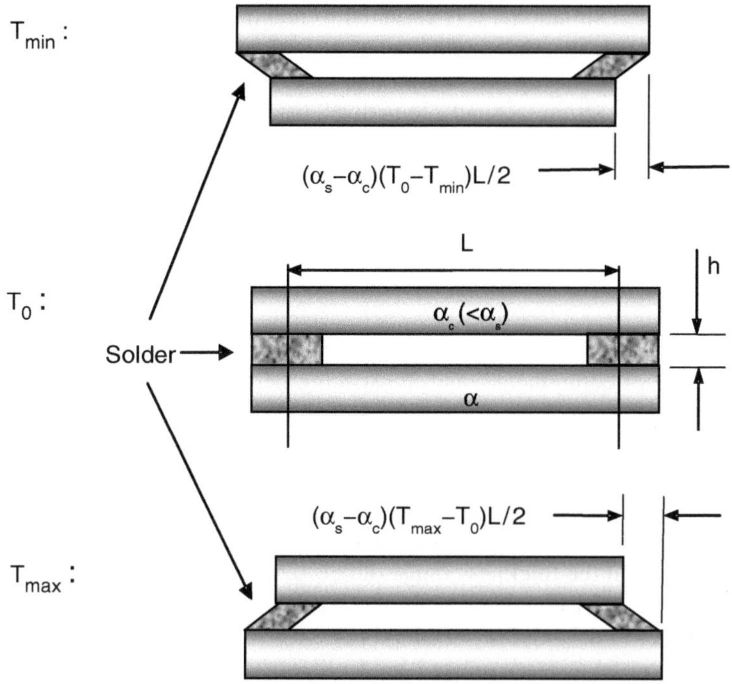

Fig. 3 Schematic model of thermo mechanical fatigue in a soldered joint

their yield point during attachment, producing potentially deleterious distortions and residual stresses. During service, similar damage may occur, often by misuse. In contrast, ceramics, glasses or polymers below their glass transition temperature may exhibit brittle failures, if their fracture toughness is exceeded. Evaluation of the dimensions and orientations of extremely small defects in already small components represents a major challenge to NDT.

An added complication is the interface, between the solder and the components being joined, that may be weaker than the adjacent materials. This vulnerability is exacerbated if the interface layer is defective, when delamination occurs. Overload problems may be accentuated by accidental impact damage during service.

Mechanical origins of *fatigue* failure range from high frequency vibrational stresses of small magnitude, as in automobile or aerospace electronics, to occasional high stress applications, as, for example, in keypad pressing. PCB flexing or twisting can transmit stress to the components and joints.

The high homologous temperatures at which solders generally operate ($Th \geq 0.6$) mean that any stress is likely to produce *creep* deformation and fracture. At the extremes, damage may be global when there is no physical constraint, or internal redistribution may result in *stress relaxation* under strain-limited conditions. In either case, microstructural damage, such as grain boundary voids and cracks, may result.

As in all design situations, it is essential to identify the dominant failure mode, and also to recognise that this might change according to the service conditions. For example, in a 'fatigue-creep interaction' the latter mechanism is favoured by a high temperature, a slow strain rate and a long dwell period in the cycle. Generally, one or the other process is dominant, and a physical interaction between the two damaging mechanisms is the exception rather than the rule.

4.1 Size Effects

The continuing miniaturisation of electronics equipment is associated with a reduction in size of components, and the spacing between them, together with their improved performance. A transition from 'micro' to 'nano' dimensions is underway. Even in the former domain, properties determined from bulk samples may not reflect actual behaviour. Some trends have emerged regarding 'size effects' but many exceptions exist and the overall situation remains unclear. An increase in strength, a fall in ductility, and often in modulus, and are usually observed with diminishing size [9–11]. Within the thickness range of 0.1–0.5 μm, the value of Young's modulus of gold films is fairly constant (53–55 GPa) although considerably smaller than that obtained for bulk samples (79 GPa) [12].

Under compression electroplated gold pillars (0.4–7.5 μm diameter) exhibit a substantial size effect for diameters below about 2 μm [13]. The 0.2% proof strength of evaporated gold film (0.8 μm thick) is almost double that measured in bulk samples [14]. Evaporated gold films possess lower ductility but greater strength than electroplated film (thickness regime 0.5–2.5 μm) [15]. In the 'nano' regime, Young's modulus is independent of the diameter of gold wires in the range 0.04–0.25 μm, while the yield stress falls by some 40% over a similar range [16]. More significantly, the yield strength of the thin samples is between 10 and 50 times greater than those measured on bulk specimens.

Grain size effects on monotonic behaviour can be marked [17, 18]. With gold films ranging from 0.8–2.0 μm thickness and a grain size over 1.5 μm, the tensile response is similar to that of large-scale test specimens, but thickness and strain rate effects are absent. Specimens with finer grain sizes (0.1–0.5 μm) exhibit a strain rate effect, with yield and tensile strength increasing with increasing strain rate. Further, no grain size effect is observed. These differences are attributed to a change in deformation mechanism when the average grain size is about 1 μm.

Fatigue performance is also affected by specimen size. For foils of 20 and 100 μm thickness, fatigued in fluctuating tension, the fatigue life, Nf, increases as the thickness diminishes [19]. Similar finding have been reported for cold rolled foil (12–35 μm thick) fatigued in bending, but electrodeposited foil (18 μm thick) exhibits fatigue properties similar to the bulk material [20]. In contrast, both electrodeposited and wrought foil (25–35 μm) exhibit similar low cycle fatigue behaviour to that observed in the bulk [21]. Cyclic crack growth in foils thinner than 150 μm often exhibits retardation or arrest, possibly due to interaction between the crack and grain boundaries [22].

In general terms, size effects have been attributed to grain or twin boundaries, the presence of a substrate or an oxide layer and, when strain gradients exist, the constraint provided by surrounding material. In all cases, the smaller the volume (thinner material), the higher the strength. The key factor is not solely the foil thickness, but the ratio of the thickness, t, to the material grain size, g [15]. In essence, the number of grains across the foil thickness governs the degree of texture (anisotropy), the number of available slip systems, and ultimately, the fracture mechanism. A g/t ratio close to, or above unity promotes a size effect. For a fixed ratio and texture, a pronounced increase in life with thickness is observed, and the Basquin expression describes the results [22]. Similarly, the fracture strain, ε_f, is dependent on the thickness via a power law relationship [9]

$$\varepsilon_f = 3t^{0.35} \tag{1}$$

(ε_f in %, t in μm). For example, the fatigue life of 250 μm foil under a stress amplitude of 70 MPa is around 80 times greater than that of a foil 125 μm thick. The differences in life diminish as the stress amplitude falls (fatigue lives increase).

Continued miniaturisation will also see a growing significance of surface oxide layers and surface slip steps, and a new generation of 'defects' will have to be accommodated in analysis. A major consequence is that unless constitutive equations utilise appropriate values for these terms and their associated constants in the context of the application being modelled, the reliability of their predictions is fragile.

The factors considered here, pertaining to electronics equipment, are not necessarily new, but they often present a higher level of challenge than for other engineering applications. For example, the inconsistency in reported modulus values exceeds that observed for steels and nickel-based alloys at their common high temperature regimes [23]. Similarly, the strain rate sensitivity is generally greater [24]. This point is reinforced when it is recognised that cracks are likely to be in the 'short' category, and that actual service cycles will contain dwells to promote even further time-dependent behaviour.

5 Case Studies

To illustrate the points raised above, two problems are presented, based upon real situations. Some details have been changed for confidentiality reasons. The first case is the common situation, in which the element experiencing fracture is the focus of attention. In the second example, the underlying cause of fracture is addressed.

5.1 Connecting Tape Fractures

Failure of electronic equipment, containing ceramic tiles and retained within aluminium packages, occurred when the gold tape connecting adjacent tiles fractured.

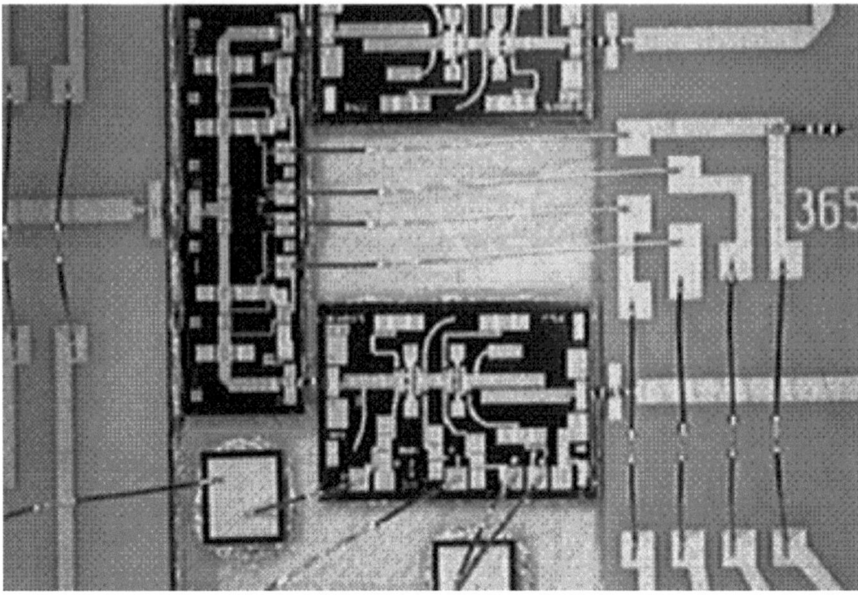

Fig. 4 Gold connecting tapes experiencing fatigue failure

Fracture had occurred by fatigue, arising from thermal cycling experienced in service and the mismatch between the coefficients of expansion in adjacent materials. The tape was in the form of a loop attached by thermo-compression bonding to the substrate near its ends, and the tiles were in both in-line and offset configurations (Fig. 4). The tape fractures occurred at either the apex of the loop or adjacent to the attachment sites. A range of tape dimensions was employed (width 20–500 μm, thickness 6–13 μm). Typical operating temperatures were in the range −40°C to 165°C.

As is often the case, the problem can be approached from the Materials and the Mechanics perspectives, although the latter is more significant in the present situation.

Gold is a face centred cubic (fcc) metal that melts at 1,063°C. Its selection for electronic applications is largely based on its thermal and electrical properties, and the functions required by the application. Apart from jewellery and dentistry, its use as a structural material is limited, as is the amount of information pertaining to its mechanical behaviour. The homologous temperatures associated with service ranged from 0.17 to 0.33, suggesting that time-dependent effects, such as creep or strain rate sensitivity, would be small in conventional bulk samples.

A situation in which a pre-deformed sheet is subjected to cyclic displacement at its ends may be described as *strain-controlled fatigue about a mean strain.* An important parameter is the mean strain ratio, Rs, which is the ratio of the minimum to maximum strain in the cycle. (This is equivalent to the stress ratio, R, under stress controlled conditions.) When elasticity prevails, expressions for fatigue life, such as the Goodman or Gerber equations can be applied.

Initially, the tape was in the cold-rolled condition. During attachment, it was bent into shape, with a radius of curvature determined by the tape length and the separation of the attachment points. The top surfaces layers of the upper loop were in tension, and the under surface in compression. During this initial bending, both stress and strain would increase linearly with distance from the neutral axis until plastic deformation occurs. If perfectly elastic-plastic behaviour is assumed, the maximum stress will be equal to the yield stress of the work hardened material. Any further work hardening would increase this value.

Unloading prior to fixing the ends (springback) will return the sheet to zero stress but with a permanent strain, equivalent to the sheet curvature. It is envisaged that this will be the starting point in service. If however, the gross curvature of the tape is imposed during attachment, and elastic springback is not permitted, the tape will experience a stress (equivalent to that needed to produce the gross deformation).

In service, the displacement caused by the temperature cycling will effectively alter the radius of curvature of the tape. As the fixture points move apart, the radius increases, and vice versa. The bending strain within the tape depends upon the radius of curvature, R, and the distance, d, from the neutral axis. For a symmetrical beam, the maximum value of d is t/2 where t is the beam thickness. Focussing upon the apex of the loop, since the attachment points will probably be associated with stress concentration and ignoring edge and corner effects, if the temperature range, δT (Tmax–Tmin) produces a lateral displacement, δ, between the attachment points,

$$\Delta = \delta T.\delta.L \tag{2}$$

Where $\delta\delta\delta$ is the difference in coefficient of thermal expansion of the materials involved, and L is a geometrical factor dependent on the dimensions of the component. For uniform temperature cycling about ambient, T0, the radii of curvature of the tape are R0, R1 and R2 respectively for the ambient, maximum and minimum temperatures. At the maximum temperature, the substrate blocks are closest together and the curvature is a maximum. The maximum tensile strain, εt, in the outer layers of the tape is given by $\varepsilon t = t/2R0$ and the compressive strain is of equal magnitude, giving a strain range, $\Delta\varepsilon$, of t/R0. If $\Delta\varepsilon$ is in the high strain regime, cycling in service, will 'cycle out' the mean stress during the early stages. When the service strain ranges are elastic, the mean stress is likely to be retained and will affect fatigue life via Goodman/Gerber expressions. Failure occurs predominantly by Mode1 crack opening, initiating from the upper (tensile) surface of the tape. More detail on the application of fracture mechanics in electronics has been presented elsewhere [25].

With the tiles in off-set configuration, lateral (or shear) displacements also occur during thermal cycling, creating a biaxial straining situation. Of course, the reality is more complex, and out-of-plane movement establishes multi-axial strains. It is likely that fatigue crack initiation occurs at the upper corners of the tape section at the apex of its loop. Subsequent propagation until fracture takes place under all possible modes of crack opening *(Mode 1 – in-plane, tensile, Mode 11 – in-plane shear, Mode 111 – out-of-plane shear)*. Analysis would be extremely challenging.

5.2 *Fracture in Glass*

Problems were encountered in a high volume process, when glass was damaged following the attachment of electrical connectors. The joining process involved soldering the connector feet to silver prints or black ceramic ink on the glass. Heating and cooling rates during this process were 1–3°C s^{-1} and 2–4°C s^{-1} respectively. Within a short period (up to a few days) crazing and cracking of the glass occurred in the vicinity of the connection region (Fig. 5). The onset of these difficulties was associated with a change in solder alloy introduced to meet legislation for lead-free solders.

Due to production and performance constraints, addressing the problem via the glass was seen as a last resort. The consequences of changing the type of glass, or thermal treatment to make it more resistant were regarded as too costly, particularly for a mass production process.

The established solder was a Pb-25Sn-10 Bi-3Ag alloy (mass per cent) that had melting range from 163°C to 236°C. Potential replacement alloys included:

Bi-42Sn-1Ag (Liquidus 140; Solidus 139°C); Sn-3.8Ag-0.7Cu (Liquidus 220; Solidus 217°C); Sn-20In-2.8Ag (Liquidus 187; Solidus 175°C).

A series of mechanical tests revealed that:

- At room temperature, the 'strength' as represented by the stress necessary to achieve 2% strain was in the order:

Fig. 5 Damage to glass in the vicinity of an electrical connector

Bi-42 Sn-1Ag (80 MPa) > Sn-3.8 Ag-0.7 Cu (60) > Sn-20In-2.8Ag (47) > Pb-25Sn-10Bi-3Ag (39).

- All alloys exhibit the expected behaviour of a reduced 'pseudo' modulus with slower strain rate and increased temperature. Values of modulus obtained by mechanical deformation are uncertain due to the high homologous temperature and the greater opportunity for time-dependent processes to occur.
  - At room temperature and a strain rate of $10^{-1}\,s^{-1}$, values for apparent modulus were:
  - Bi-42 Sn-1Ag (23 GPa) < Pb-25Sn-10Bi-3Ag (31) < Sn-3.8 Ag-0.7 Cu (45) < Sn-20In-2.8Ag (50)
- Stress relaxation is substantial in 300s, and at room temperature, when expressed as a percentage of the initial stress value, the order is:
  - Bi-42 Sn-1Ag (80) > Pb-25Sn-10Bi-3Ag (79) > Sn-20In-2.8Ag (64) > Sn-3.8 Ag-0.7 Cu (48).
  - The majority of the stress relaxation occurs within the first 30s. The effect of temperature on stress relaxation is as expected, in that the initial stress is reduced and the relaxation process is more rapid as the temperature is increased.
- Measurements of the coefficient of thermal expansion were in the order (at room temperature):
  - Bi-42 Sn-1Ag ($17.1 \times 10^{-6}{}^{\circ}C^{-1}$) < Sn-3.8 Ag-0.7 Cu (22.3) < Sn-20In-2.8Ag (23.8) < Pb-25Sn-10Bi-3Ag (27.6).
  - Assuming a typical value of $8.5 \times 10^{-6}{}^{\circ}C^{-1}$ for the CTE of glass, the mismatch strains in the Pb-25Sn-10Bi-3Ag are the highest, and over twice those associated with the Bi-42 Sn-1Ag alloy, yet it is the former that has performed satisfactorily. Converting theses strains to stress, via the modulus, indicates higher stress levels, by a factor of three, in the lead-containing alloy.

Thermal analysis indicated that the Sn-20In-2.8 Ag and the Pb-25Sn-10Bi-3Ag alloys were unusual, in that they contained low melting point phases occurring at about 115°C and 110°C respectively. With these alloys, the situation is even further complicated when the temperature exceeds the transition. For the vast majority of applications, such an event would be deemed premature failure and the data would be disregarded. The apparent success of the lead-containing alloy relies on the formation of a LMP, which accentuates the loss in strength as temperature increases.

6 Concluding Comments

Design against fracture and the assurance of structural integrity in electronics equipment constitute a large and growing challenge, as miniaturisation continues. Some of the problems are specific to electronics applications, such as the need to accommodate an entirely new generation of solder alloys, while others are generally similar to, but more acute, than those encountered in other design situations. These latter include the higher homologous temperatures of operation and the consequent level of time-dependency in mechanical behaviour and the microstructural instability. Perhaps, most demanding of all is the effect of size on the values of key mechanical properties, and the difficulties posed to any analysis employing a defect tolerant approach. Incorporation of appropriate values of temperature, strain rate, and size into constitutive equations is absolutely essential.

The debut of the 'nano' era coincides well with applications in electronics (The World's largest industrial sector). The potential for Fracture specialists is enormous but it is important not to forget that in some circumstances, solutions may be obtained by more basic means.

References

1. W J Plumbridge, R J Matela and A Westwater, Structural Integrity and Reliability in Electronics – Enhancing Performance in a Lead-free Environment, Kluwer, Dordrecht/Boston/London, 2003, Chapter 7.
2. W J Plumbridge, 'The Gas Turbine Laptop – A New Era for Solders', Materials World, 1995, 4, 422–424.
3. W J Plumbridge and C R Gagg, Journal of Electronic Materials, 1999, 10, 461.
4. W J Plumbridge, Journal of Materials Science, 1996, 31, 2501.
5. K P Jen and J N Majerus, Journal of Engineering Materials Technology, 1991, 113, 475.
6. S Vaynman and M E Fine, Solder Joint Reliability, Theory and Applications (Ed J H Lau), Van Nostrand Reinhold, New York, 1991, 333–360.
7. W J Plumbridge, Y Kariya and C R Gagg, Seventh Symposium on Microjoining and Assembly Technology in Electronics, MATE 2001 Conference, Yokohama, February 2001.
8. W J Plumbridge, Journal of Materials Science and Engineering, 2007, 18, 307.
9. B Weiss and A Hadrboletz, Fatigue 2002. Proceedings of 8th International Conference, June 2002, Stockholm (Ed A F Blom) Warrington, UK, 2002 (Emas Pub.).
10. G Khatibi, A B Betzwar-Kotas, V Groger and B Weiss, Fatigue and Fracture of Engineering Materials and Structures, 28, 723, 2005.
11. D T Read, International Journal of Fatigue, 1998, 20, 203.
12. H D Espinosa and B C Prorok, Journal of Materials Science, 2003, 38, 4125.
13. J R Greer, W C Oliver and W D Nix, Acta Materialia, 2005, 53, 1821.
14. M T Lin, K S Shiu and C J Tong, DTIP 2007, EDA Publishing, Stresa, Italy.
15. K Timpano – Private Communication.
16. B Wu, A Heidelberg and J J Boland, Nature Materials, 2005, 4, 525.
17. R D Emery and G L Povirk, Acta Materialia, 2003, 51, 2069.
18. R D Emery and G L Povirk, Acta Materialia, 2003, 51, 2079.
19. M Judelewiez, H U Kunzi, N Merk and B J Ilschner, Materials Science and Engineering, 1994, A186, 135.
20. H D Merchant, M G Minor and Y L Liu, Journal of Electronic Materials, 1999, 28, 998.
21. S Hong and R Weil, Thin Solid Films, 1996, 238, 175.
22. A Kolas, A Hadrboletz, G Khatibi and B Weiss, Fatigue 2002. Proceedings of 8th International Conference, June 2002, Stockholm (Ed A F Blom), Warrington, UK, 2002 (Emas Pub.).
23. W J Plumbridge, Soldering and Surface Mount Technology, 1999, 11, 8.
24. W J Plumbridge, Materials at High Temperatures, 17, 381, 2000.
25. W J Plumbridge and Y Kariya, Fatigue and Fracture of Engineering Materials and Structures, 2004, 27, 723.

Improving the Crashworthiness of Aluminium Rail Vehicles

Donato Zangani, Mark Robinson, and George Kotsikos

Abstract An experimental and modelling programme of work have been undertaken to predict the performance of aluminium welds in rail vehicles under highly dynamic loading conditions and provide design guidelines to reduce the likelihood of the occurrence of weld unzipping. Modelling of weld unzipping in large rail structures is a challenging task since it requires to deal with material instability, to take into account the uncertainties in material parameters and to address the problem of mesh resolution which together pose severe challenges to computability. The proposed methodology to the prediction of weld failure is based on the validation of the numerical models through correlation with laboratory scale tearing tests. The tearing tests were conducted on samples taken from real rail extrusions with the purpose of obtaining the failure parameters under dynamic loading and understanding the effect of weld material composition on joint behaviour. The validated material models were used to construct a FEA simulation of the collision of an aluminium rail car and investigate the effect of both joint geometry and welding techniques on the failure mechanism. Comparisons of the model with the failures observed in an aluminium rail vehicle that was involved in a high speed collision, have shown that it is possible to model the phenomenon of weld unzipping with good accuracy. The numerical models have also been used as a tool for the optimisation of joint design to improve crashworthiness.

Keywords Crashworthiness · Weld unzipping · Aluminium · Modelling

D. Zangani
D'Appolonia SpA, Via S. Nazaro 19, 16145 Genova, Italy
e-mail: donato.zangani@dappolonia.it

M. Robinson and G. Kotsikos (✉)
NewRail – Centre for Railway Research, Newcastle University, Stephenson Building, Newcastle upon Tyne, NE1 7RU, UK
e-mail: {m.a.robinson, george.kotsikos}@ncl.ac.uk

S. Pantelakis, C. Rodopoulos (eds.), *Engineering Against Fracture: Proceedings of the 1st Conference*,

1 Introduction

The use of aluminium alloys in rail vehicle manufacture has introduced a number of advantages, namely good corrosion resistance, lightweight and superior surface finish. Furthermore, the use of double skinned aluminium extrusions that can be welded together to form the vehicle body has enhanced the efficiency of the manufacturing and assembly methods. Double skinned closed cell extruded sections have an inherently excellent resistance to impact loading that contributes to the crashworthiness of modern rail vehicles. In fact, it has been found that a double skinned train carriage made of aluminium longitudinal hollow extrusions behaves like a "rigid body" during a collision. For this reason, impact energy absorbing zones (crumple zones) are introduced at either end of the carriage to absorb the impact energy that would otherwise be transferred to the passengers, crew and equipment.

In recent accidents involving rail vehicles of aluminium construction, the highly dynamic loads sustained by the structure during a collision introduce a failure mechanism of unstable crack growth along the heat affected zone (HAZ) to weld metal (WM) interface, also known as "weld unzipping" [1,2]. Understanding this fracture mechanism and designing aluminium joints to eliminate it are therefore of gresa importance to improving the crashworthiness performance of aluminium rail vehicles.

An inherent problem with fusion welding of heat treatable aluminium alloys, such as the 5,000 and 6,000 series used in train construction, is that the heat input during the welding process subjects the material to a localised solution heat treatment which alters the microstructure at the weld region and results in a reduction of the mechanical properties there, compared to the parent material (undermatched weld). This weld undermatch can become a critical problem. Recent 'real life' collisions have revealed a critical failure mode in rail vehicles where the weldments appear to fail by fast fracture along the heat affected zone The term 'weld unzipping' is commonly used to describe this type of failure.

This paper investigates the fracture process in aluminium MIG welded sheets, obtained from real rail vehicle extrusions, to assess the parameters contributing to the observed fast fracture failure (weld unzipping). In addition, the parameters necessary for improving the modeling of failure during collisions of rail vehicles are determined and the effect of improvements in welded joint design to avoid the weld unzipping investigated.

2 Experimental

The material considered in this work was the 6005 T6 aluminium alloy, commonly used for train manufacture in the UK. Aluminium closed cell extruded sections, representative of a typical profile used in train manufacture were supplied by Alcan. The sections were welded together by an automated metal inert gas (MIG) process. Two types of weld were produced:

(a) MIG welds made with an aluminium – silicon filler wire (4043)
(b) MIG welds made with an aluminium – Magnesium filler wire (5356)

The thickness of the welded plating was 3.5 mm. Mechanical property characterization of the welded sections (i.e. parent plate, HAZ and WM) included hardness measurements, tensile, impact and fracture mechanics tests. The results of the hardness scans across the weld profiles are given in Fig. 1. The results show that the hardness of the weld metal is almost half that of the parent plate. There is no difference in the hardness values between the Al-Mg and Al-Si welds.

Due to the small thickness of the test section it was not possible to obtain standard Charpy impact test specimens. For this reason a small scale Zwick impact tester was used. Impact tests were carried out on both parent plate and the welds. The starter notch was placed as close as possible to the weld metal–HAZ interface. For the parent plate samples the notch was machined in the extrusion direction. The results are given in Table 1.

It is evident that the impact strength of the welds is substantially lower than that of the parent plate and the ones produced with the Al-Si filler wire exhibiting the lowest impact strength, less than half that of the parent plate.

In order to investigate the dynamic tearing stability of the welds a series of fracture mechanics tests were carried out based on the 'Energy dissipation rate' approach [3–5]. A major advantage of the energy dissipation approach is that it

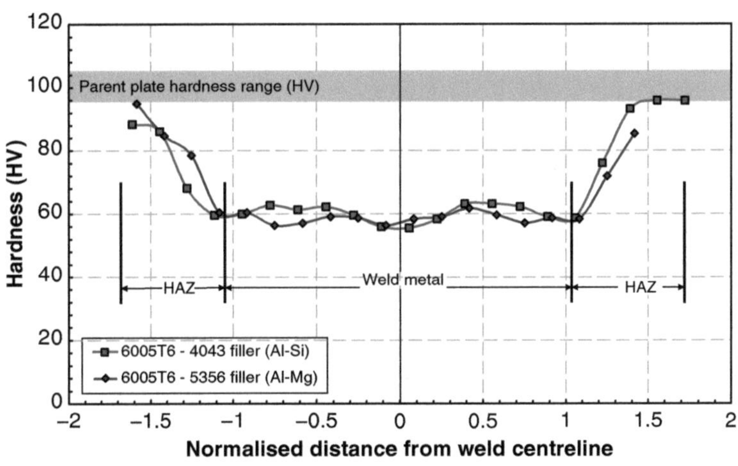

Fig. 1 Hardness measurements across the aluminium weld profiles

Table 1 Impact test results

Material	Energy absorbed (J)
6005 T6 – Parent plate	0.72
6005 T6 – Weld – Al-Si (4043) filler	0.32
6005 T6 – Weld – Al-Mg (5356) filler	0.55

does not need an increase in toughness with crack growth (the R curve concept) to explain stable tearing. Increase in tearing resistance with crack growth sometimes occurs due to shear lip development, but in thin specimens where fully slant fracture is developed after only a few millimetres of crack growth, this explanation is not available. The procedure is similar to the J integral test approach and utilises relatively small specimens to predict tearing instability. For these tests centre cracked tension (CCT) specimens were used. The specimen width was 100 mm and the length approximately 250 mm. A notch, 40 mm long, was machined in the centre of the sample incorporating a widened section to accommodate a clip gauge for the measurement of the crack opening displacement upon tearing. Typical traces from the fracture mechanics tests are shown in Fig. 2. Plot (a) is for a weld prepared with Al-Mg filler (5,356) and plot (b) from a weld prepared with an Al-Si filler (4,043).

a

b

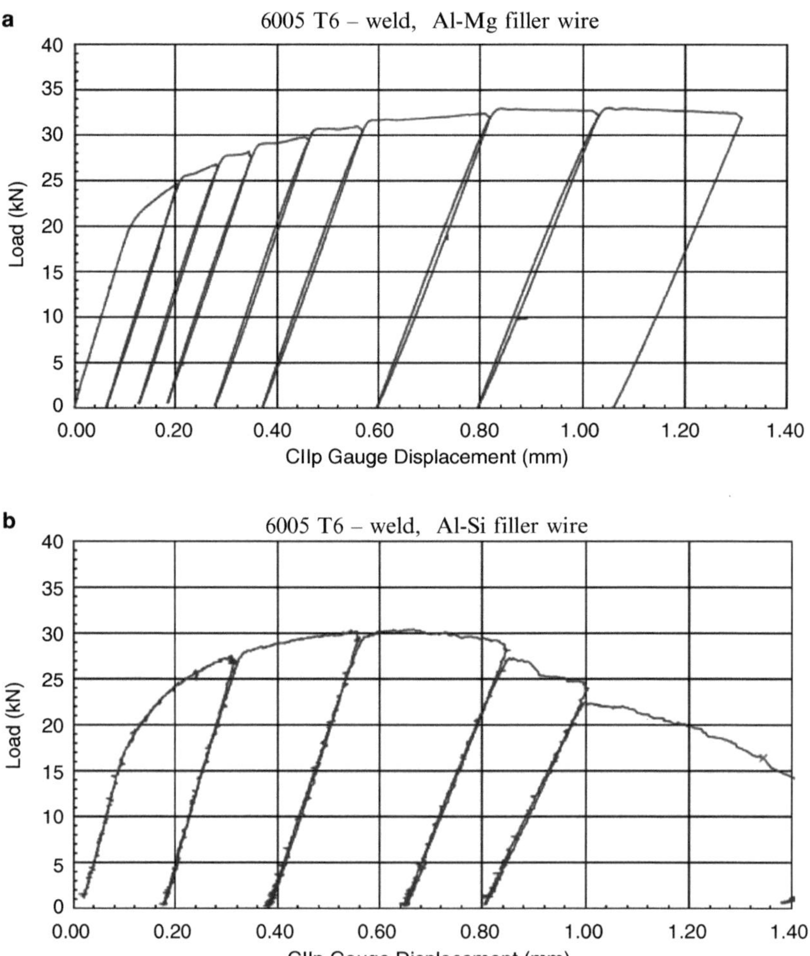

Fig. 2 Typical traces of fracture mechanics tests on CCT specimens. **a** Weld produced with Al-Mg filler wire, **b** weld produced with Al-Si filler wire

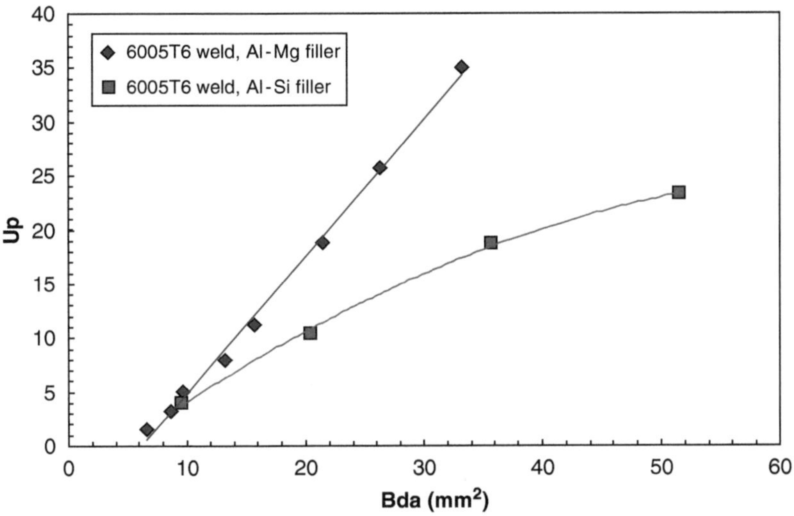

Fig. 3 Plot of dissipated energy for crack advance in 6005T6 welds

In fracture mechanics the energy dissipation rate is defined as:

$$D = \frac{dUp}{Bda} \tag{1}$$

where, dU_p is the increase in dissipated energy for crack growth da and B is the specimen thickness. A plot of the dissipated energy for various amounts of crack growth as obtained from the tests is given in Fig. 3.

The results show that the welds produced with an Al-Mg filler wire exhibit a much higher energy dissipation rate that those produced with an Al-Si filler wire. The calculated values of the energy dissipation rate, D, are:

$$6005\text{T6} - \text{Al-Mg weld D} = 1{,}270\,\text{kJ/m}^2$$
$$6005\text{T6} - \text{Al-Si weld D} = 350\,\text{kJ/m}^2$$

In addition to D, the tearing resistance index, T, of the two types of welds was calculated. This was found to be:

$$6005\text{T6} - \text{Al-Mg weld T} = 0.030$$
$$6005\text{T6} - \text{Al-Si weld T} = 0.011$$

The results indicate that the welds produced with an Al-Mg filler wire possess a value of D almost four times that of the welds produced with an Al-Si filler wire and a tearing resistance index three times higher. It has to be mentioned though that although the advantage of using an Al-Mg filler wire is immediately apparent, the result of an energy dissipation rate of $350\,\text{kJ/m}^2$ is still a reasonable value. It can be argued that this value is high enough to prevent elastic brittle fractures.

In a collision though there is a short period of extremely high input energy rate for which the difference between the D values in the two types of welds tested could be significant.

To aid with the development of the FE modeling work, Crack Tip Opening Angle (CTOA) Tearing Tests were also carried out on extrusions taken from real railway vehicle double skinned extrusions. A modified double cantilever beam (DCB) specimen was designed to conduct the fracture tests so that a crack initiator could be centred in one of the relevant zones: parent metal, HAZ and weld. Samples had dimensions equal to 200 mm length, 130 mm high, with a gauge thickness of 2 mm. Figure 4 shows the test set-up.

Figure 5 shows the typical behaviour of the crack growth in samples with weld produced with Al-Mg filler wire and crack initiator positioned at the centerline of

Fig. 4 CTOA tearing test set-up

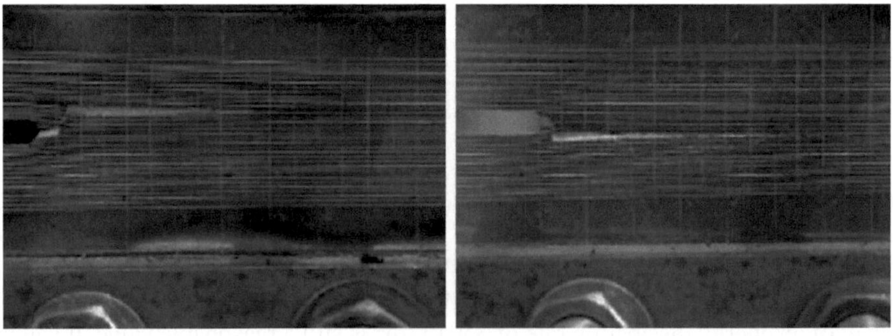

Fig. 5 Crack growth in samples with weld produced with Al-Mg filler wire and crack initiator centred in the weld

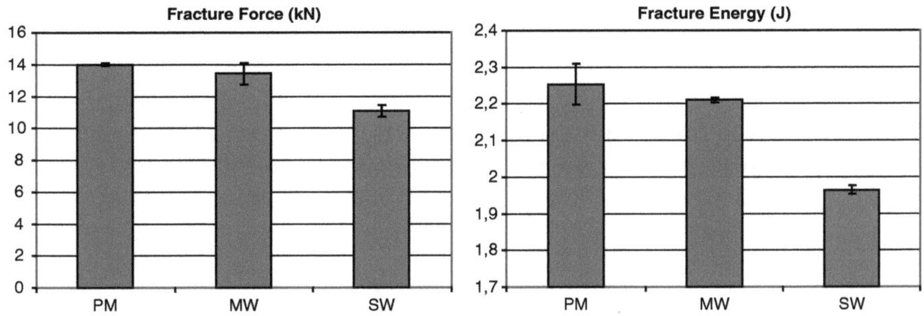

Fig. 6 Comparison of performance of Al-Mg and Al-Si welds with parent material

the weld. It was seen that the crack quickly deviated to the interface between the weld material and the HAZ. A comparison of the results in terms of the peak value of the fracture force and the fracture energy required to propagate the crack in the parent metal (PM), centerline of the weld (MW) and the interface between the weld metal and HAZ (SW) is provided in Fig. 6.

It is evident that the weakest part is the weld metal HAZ interface which is also the region where weld unzipping takes place in real life failures.

3 Modeling

Modeling activities have been developed with the aim of defining reliable material models able to represent with good accuracy the progressive failure behaviour shown by the structures under study subjected to highly dynamic loading conditions.

The weld obtained with aluminium-magnesium filler wire has been considered in the simulations, mainly because it is the preferred filler wire used in manufacture today and also because of the superior performance of these welds in comparison to one prepared with aluminium-silicon filler wire.

Two material models have been considered to model the progressive fracture in the material: a linear plasticity model with maximum strain failure criteria and the Gurson material model. In the first model considered, failure condition is based on the accumulated plastic strain in the structure, according to the following relationship:

$$\varepsilon^{p}_{eff} = \int_{0}^{t} \left(\frac{2}{3} \dot{\varepsilon}^{p}_{ij} \dot{\varepsilon}^{p}_{ij} \right)^{1/2} dt \leq \varepsilon^{p}_{f} \tag{2}$$

being ε^{f}_{p} the effective plastic strain and εfp the plastic strain to failure in an uniaxial tensile test. The second material model considered is the Gurson model, a more sophisticated approach which uses a state variable plasticity law, in which the void volume fraction is an explicit state variable and the yield stress is taken to be a function of the volume fraction of voids in the material. The Gurson flow function is defined as:

$$\Phi = \left(\frac{\sigma_e}{\sigma_y}\right)^2 + 2f^* q_1 \cosh\left(\frac{3q_2\sigma_H}{2\sigma_e}\right) - 1 - (q_1 f^*)^2 = 0 \tag{3}$$

Where σ_e is the von Mises equivalent stress, σ_y is the yield stress, σH is the hydrostatic stress component, f* the effective void volume fraction, defined as:

$$f^* = \begin{cases} f & \text{if } f \le f_c \\ f_c + (1/q_1 - f_c)(f - f_c)/(f_E - f_c) & \text{if } f > f_c \end{cases} \tag{4}$$

Being q_1 and q_2 constant material parameters, f_c the critical void volume fraction, f_E the damage parameter corresponding to the void volume fraction at failure. The additional parameters required to describe the nucleation of new voids and the growth of the existing voids are the initial void volume fraction, f_0, and the void volume fraction of nucleating particles, f_N. Final failure of the porous ductile solid occurs by void coalescence at a certain volume fraction of voids. The values of the Gurson parameters used in the simulations are reported in the following Table 2. Figure 7 shows the comparison of the numerical prediction with the experimental results.

Table 2 Values of the material parameters for the Gurson model

Material zone	q_1	q_2	f_c	f_0	f_N	f_{f0}
6005A-T6	1.5	1	0.008	0.012	0.005	0.25
Al-Mg filler weld	1.5	1	0.015	0.020	0.01	0.25
HAZ	1.5	1	0.100	0.200	0.03	0.25

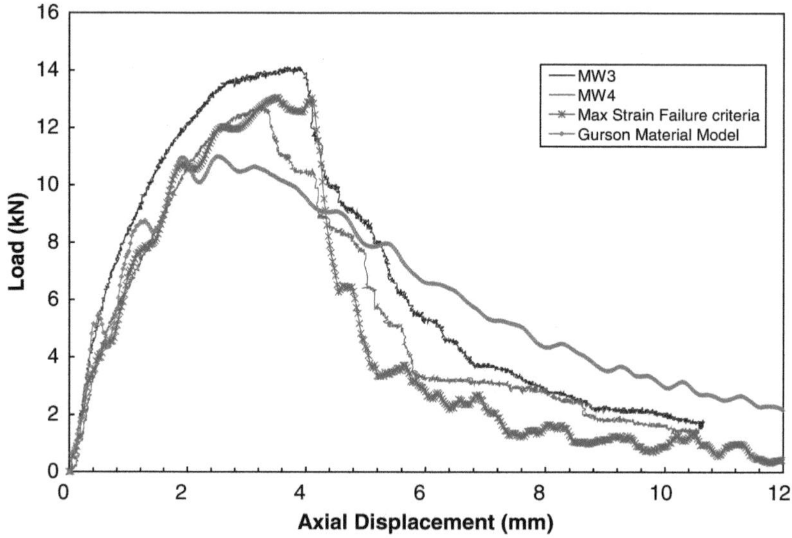

Fig. 7 Load required for the crack propagation in tearing test of aluminium-magnesium weld: experimental results and numerical predictions

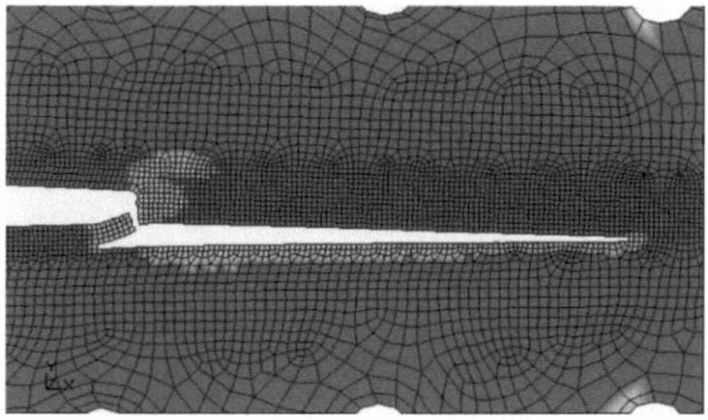

Fig. 8 Numerical simulation of weld unzipping

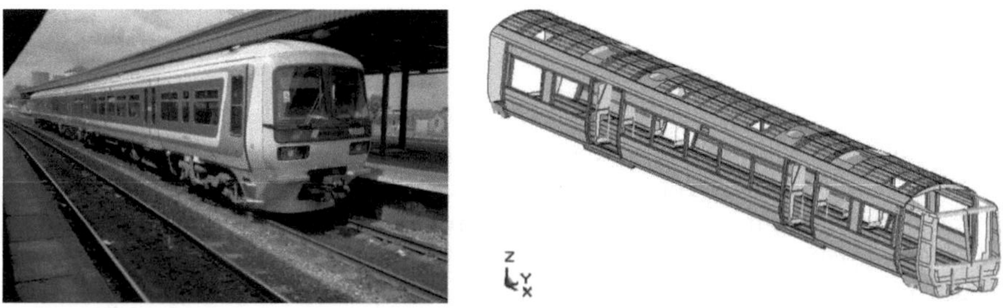

Fig. 9 Bombardier vehicle class 165 and its FEM model

The approach based on the use of the maximum strain failure criteria provides a good estimate of the maximum fracture force and the rapid decrease of the load due to the weld unzipping. The Gurson Model underestimates the maximum fracture load and overestimate the load required to propagate the crack in the material. Both the models are able to describe the propagation of the crack and the weld unzipping, but the maximum strain model is also able to predict the deviation from the weld centre line to the weld metal/HAZ interface, as shown in Fig. 8.

The above-defined material failure models have been used in simulations of a full front coach of the Bombardier Class 165 rail vehicle. The vehicle and the computer model are shown in Fig. 9. Various impact scenarios have been investigated with the focus on the longitudinal welded joints in the car body. Figure 10 shows a typical result of the simulations for the vehicle with standard welded joints under an impact speed at 20 kph against a flat obstacle (left) and the result of the simulation for a new joint design that make allowance for the strength loss in the weld region (right). The weld lines are highlighted in red (dark grey). In the figure on the left it is clearly visible that the weld elements immediately behind the impact area have been removed from the FE model as a result of failure of the longitudinal welded joints immediately behind the area of impact. With the improved joint design (on the right of Fig. 10) the welded joints immediately behind the impact area have not fractured.

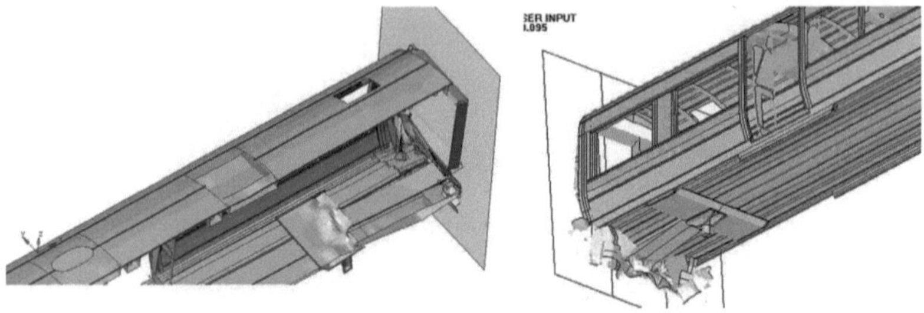

Fig. 10 Simulation of full-vehicle impacts

4 Welded Joint Design

The simulations carried out show that weld unzipping would still take place during a collision, even with weld produced with aluminium-magnesium filler wires.

Another important aspect contributing to the failure is the inherent strength undermatching (weld strength lower than the parent plate strength) found in aluminium welds, as a result of the heat input during the welding process.

What matters when a collision takes place is whether the structure spreads the impact energy or concentrates it in a specific region (the weld in this case). The input (impact) energy in a structure with a strength undermatch, may channel all this energy to the weld region.

This will have to be taken up by the energy dissipation rate which implies extensive crack growth. The above results support the argument that the "weld unzipping" that has been observed in recent train collisions [1] can be a result of low tearing resistance index of the weld material and the strength undermatching at the welded joints.

A way of improving the impact behaviour of the joints would be to modify the local geometry, for example thicken the plating near the weld region to compensate for the weld undermatching.

The modelling was extended to modify the profile of the weld joints to compensate for the strength undermatching by locally thickening the plate at the weld region as well as achieving a full penetration weld.

The joints were tested for their impact behaviour through a simulated high velocity impact test on closed cell welded extrusions. This comprised of an air cannon firing a weight to one end of the joint whilst the other end was clamped on a solid base. The impact load, displacement and strain on the joint near the weld region were recorded and high speed video recorded the failure process. The set up is shown schematically in Fig. 11. In the figure a schematic of the idea behind the joint geometry modification is shown. The exact detail of the geometry are subject to commercial confidentiality and will not be given here.

The effect of the joint modification was to spread the impact energy throughout the section, rather than confine it to the weld region. The experimental results

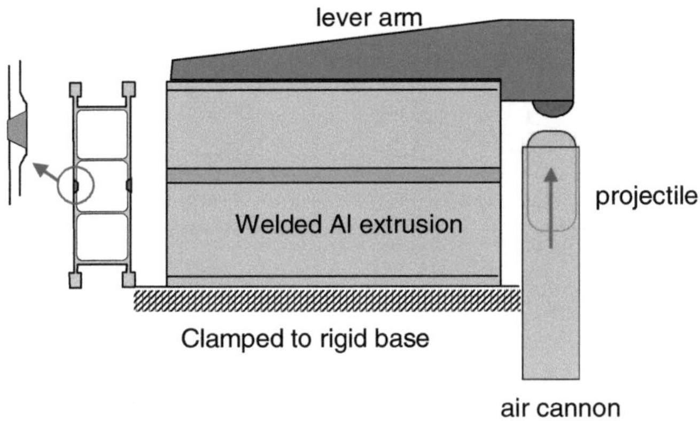

Fig. 11 Test set-up to assess effects of joint geometry on weld unzipping

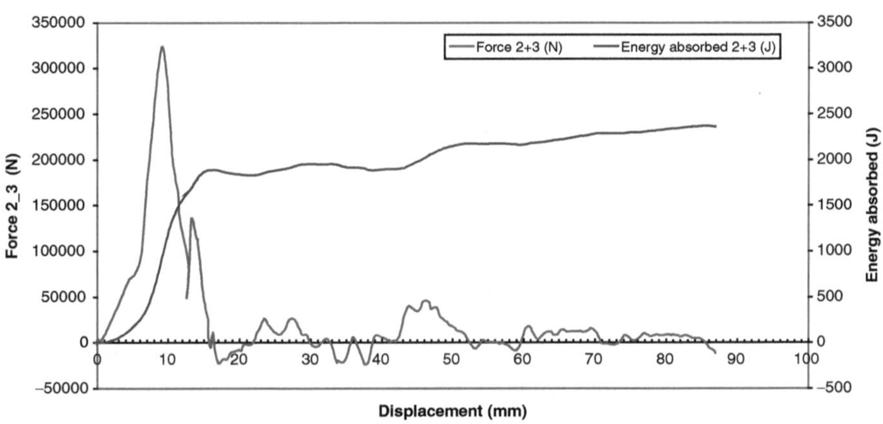

Fig. 12 Typical trace from impact test

available, showed that thickening of the section alone is not sufficient to suppress weld unzipping as excessive localised thickening can reduce the absorbed impact energy.

The mechanical property and fracture mechanics test results, have been used for the tuning of the material models used in the simulation of the collision of a rail vehicle. A FE model of the front coach of Bombardier Class 165 rail vehicle has been constructed and subjected to head on collision scenarios. The simulation results have demonstrated a notable improvement in the failure mode with the new joint design that makes allowance for the strength reduction in the weld region. The failure mode changes form weld unzipping to buckling of the structure on impact (Fig. 13).

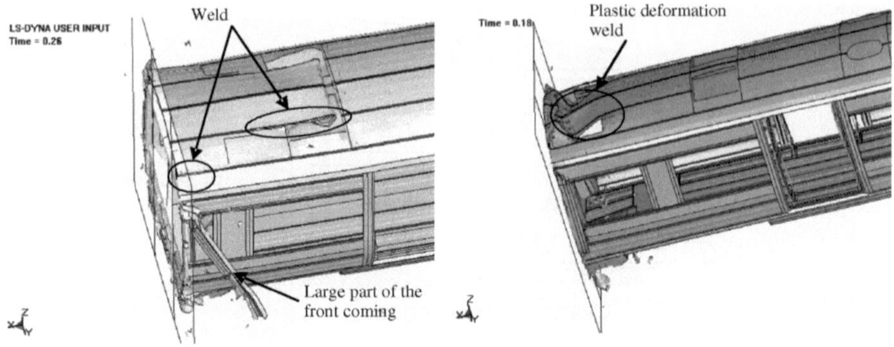

Fig. 13 Collision simulation of class 165 rail vehicle with standard weld geometry (*left*) and modified weld geometry (*right*)

5 Conclusions

Fracture mechanics tests and tearing tests on real scale extrusions have demonstrated that the use of Al-Mg filler wires in MIG welding results in superior performance welds in terms of strength, ductility and fracture toughness, compared to Al-Si filler wires. Another important aspect of the failure is the strength undermatching (weld strength lower than the parent plate strength) found in aluminium welds. The input (impact) energy in a structure with a strength undermatch, may channel all this energy to the weld region. This will have to be taken up by the energy dissipation rate which implies extensive crack growth. The above results support the argument that the "weld unzipping" that has been observed in recent train collisions [1] can be a result of low tearing resistance index of the weld material and the strength undermatching at the welded joints. A way of improving the impact behaviour of the joints would be to thicken the plating near the weld region to compensate for the weld undermatching. The effect of this modification to the weld joint would be to spread the impact energy throughout the section, rather than confine it to the weld region. The experimental results available, in particular the results of the tearing tests, have been used for the tuning of the material models used in the simulation of the collision of a rail vehicle. A FE model of the front coach of Bombardier Class 165 rail vehicle has been constructed and subjected to head on collision scenarios. The simulation results have demonstrated a notable improvement in the failure mode with the new joint design that makes allowance for the strength reduction in the weld region.

Acknowledgements The present work was carried out within the European collaborative project ALJOIN. Financial support from the European Commission and technical contributions from the ALJOIN project partners TWI, Bombardier Transportation, Alcan, and DanStir are gratefully acknowledged.

References

1. Gay R., M. Robinson, J. Yates, D. Zangani, 2003, ALJOIN – crashworthiness of joints in aluminium rail vehicles, European Railway Review, 9(3), 63–67.
2. Rt Hon Lord Cullen, 2001, The Ladbroke Grove Rail Inquiry – Part 1 Inquiry. HSE Books, Norwich.
3. Sumpter J. D. G., 2004, The energy dissipation rate approach to tearing instability, Engineering Fracture Mechanics, 71, 17–37.
4. BS 7448-4, 1997, Method for determination of fracture resistance curves and initiation values for stable crack extension in metallic materials.
5. ASTM E561-98, 1998 Standard practice for R-curve determination.

Information Fusion in Ad hoc Wireless Sensor Networks for Aircraft Health Monitoring

Nikos Fragoulis, Vassilis Tsagaris, and Vassilis Anastassopoulos

Abstract In this paper the use of an ad hoc wireless sensor network for implementing a structural health monitoring system is discussed. The network is consisted of sensors deployed throughout the aircraft. These sensors being in the form of a microelectronic chip and consisted of sensing, data processing and communicating components could be easily embedded in any mechanical aircraft component. The established sensor network, due to its ad hoc nature is easily scalable, allowing adding or removing any number of sensors. The position of the sensor nodes need not necessarily to be engineered or predetermined, giving this way the ability to be deployed in inaccessible points. Information collected from various sensors of different modalities throughout the aircraft is then fused in order to provide a more comprehensive image of the aircraft structural health. Sensor level fusion along with decision quality information is used, in order to enhance detection performance.

Keywords Structural health monitoring · Wireless sensor networks · Information fusion

1 Introduction

Structural health monitoring, or SHM, techniques are gaining the support of airframe manufacturers, airliners, and regulators. SHM incorporates into the aircraft structure itself nondestructive inspection (NDI) technologies currently used in manual inspections. SHM sensors currently being developed or evaluated [1] can find fatigue damage, hidden cracks in hard-to-reach locations, disbonded joints, erosion, impact damage, and corrosion, among other defects commonly encountered in aging

N. Fragoulis, V. Tsagaris, and V. Anastassopoulos (✉)
Electronics Laboratory, Department of Physics, University of Patras, Patras, Greece
e-mail: vassilis@upatras.gr

S. Pantelakis, C. Rodopoulos (eds.), *Engineering Against Fracture: Proceedings of the 1st Conference*,

aircraft. Ultimately an integrated network of sensors could monitor not only structural elements, but also the health of electronics, hydraulics, avionics, and other systems.

Using a network of sensors, the measured physical parameters can be transmitted to a central processing unit using wiring and multiple wiring harnesses for further prognostic analysis, or engine health prediction. These wiring and wiring harnesses can increase overall system weight and cost, and can reduce overall system reliability. Therefore, the need for a system for providing various aircraft engine parameters without the need of wiring is desirable. To this direction Ad hoc wireless sensor networks (WSNs) seems an attractive solution, since they eliminate the need of the extensive wiring and at the same time exhibit some other interesting and very useful features, such as infrastructure independency, node positioning capability, and power awareness.

On the absence of wiring, employing techniques to power up sensor nodes using energy harvesting methods is a critical task. Using energy harvesting methods, sensors systems can be powered by converting structural stresses (strain energy harvesting) into electrical power via piezoelectric and other transducers. This also includes the use of radio frequency (RF) tagging technology to remotely read data from the sensors.

The amount of information transmitted by each sensor node is very limited, and concerns only a very specific part of the aircraft mechanical structure. In addition sensors of different modalities provide different kind of information. Therefore, the overall information provided by the totality of sensors is needed to be combined using information fusion techniques, in order to form a more comprehensive image of the aircraft structural health.

The fusion and monitoring process can be carried out either on-board or to a remote site on which can be transmitted on-demand via telemetry to enable continuous, wireless transmission of data. Additionally a web site can be programmed to interrogate the data and use preset thresholds to provide continuous green light/red light information regarding structural health. In this mode of operation, it may be desirable to incorporate electronics to analyze data in-situ and extract decisions. In this way the need to transmit large raw data files is eliminated.

Finally, a sensor level decision fusion process may be employed in order to extract decisions based on the totality of the sensors. The overall decision performance can be further enhanced by programming each sensor node to transmitting decision quality bits. In such a functioning scheme, serialized sensor level fusion approach should be used, since it fits better to the multi-hop packet transmission employed in ad hoc networks.

2 Wireless Sensor Nodes

In addition to the sensor itself, a sensor node in a wireless sensor network, should contain some additional components which they solely depend on the nature of the application in which the sensor node is indented to be used. However, the general-

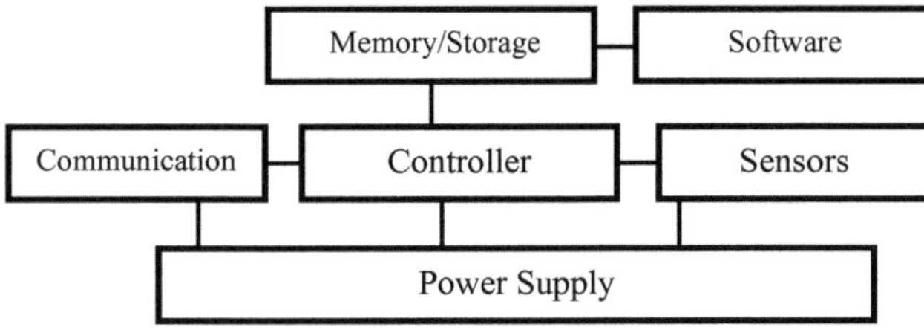

Fig. 1 General Architectural Diagram of a smart Sensor

ized components of a wireless sensor node are a controller, a memory device, one or more sensors, and a communication device as shown in Fig. 1 [2].

A qualitative description of these components is as follows:

- **Controller:** A controller is a central part of a wireless sensor node. It processes all the data it receives from memory, sensor, or communication device. It is also capable of executing arbitrary code from the memory. In other words, a controller is a Central Processing Unit (CPU) of a node.
- **Memory:** This is the main resource for storing programs and intermediate data coming from the sensors or the communication device. The size of the memory depends on the application of the sensor.
- **Sensors:** Sensors that interface to the physical world as they monitor the critical events in the aircraft and report the data to the controller. For aircraft structural health monitoring, sensors that has been described in the previous section must be used.
- **Communication:** The communication device is used to send data to other nodes to form a sensor network. We are using wireless communication in our case. The communication device is equipped with a transceiver which can take data from a controller to send it over the air and vice versa.
- **Power Supply:** A sensor node gets the energy to perform the computation and communication from a power supply. In our case the power is to be harvested from the environment via piezoelectric or other transducers.

A sensor node like the one just described is referenced as smart sensor and if Signal Conditioning and Digital Signal Processing modules are present, the corresponding sensor is referenced also as a virtual sensor. Finally the sensor that is able to make decisions (detect a phenomenon or not) are categorized as a detector.

3 Sensors for Structural Health Monitoring

The development of inexpensive, reliable sensors capable to provide information regarding aircraft structural health is certainly a challenge. This type of sensors should be typically mounted on aircraft structures where flaws are expected to form.

Sensor technologies that must be considered include flexible eddy-current arrays, capacitive micromachined ultrasonic transducers, piezoelectric transducers that can interrogate materials over long distances, acoustic emission sensors, embedded fiber optics, nickel strip magnetostrictive sensors, and conducting paint whose resistance changes when cracks form underneath.

One promising SHM sensor is the Comparative Vacuum Monitoring (CVM) sensor [3]. It is consisted of a thin, self-adhesive rubber patch, ranging from dime- to credit-card-sized, that detects cracks in the underlying material. The rubber is laser-etched with rows of tiny, interconnected channels or galleries, to which air pressure is applied. Any propagating crack under the sensor breaches the galleries and the resulting change in pressure is monitored.

4 Wireless Sensor Network

The rationale behind employing a Wireless Sensor Network (WSN) for the interconnection of the sensor nodes, is the need to eliminate wiring and wiring harnesses. In addition an ad hoc wireless network approach, eliminates also the need of an pre-existed or spatially pre-determined infrastructure and in this way new sensors, can easily participate to the established network without any need for extra programming.

Another interesting feature of ad hoc wireless networks is the fact that any wireless node takes part in data routing. In this way the data packets is needed to be transmitted only to a neighbor (usually the nearest) and finally reach their destination via multiple hops.

The transmission power required by each sensor node increases as the square of the distance between the node and the destination. Therefore, multiple short message transmission hops require less power than one long hop. In fact, if the distance between source and destination is R, the power required for single-hop transmission is proportional to R^2. If nodes between source and destination are taken advantage of to transmit n short hops instead, the power required by each node is proportional to R^2/n^2 [4]. This is a strong argument in favor of ad hoc networks, since due to the fact that transmission happens only to neighbor nodes, RF transmission power is kept at low levels limiting interference problems and lowering overall energy consumption.

In addition software power management techniques can greatly decrease the power consumed by RF sensor nodes. TDMA (Time Domain Multiple Access) protocol is especially useful for power conservation, since a node can power down or 'sleep' between its assigned time slots, waking up in time to receive and transmit messages.

There are many sensor networks on the market today. In order to provide compatibility, in 1993 the IEEE and the National Institute of Standards and Technology (NIST) began work on a standard for Smart Sensor Networks. IEEE 1451, the Standard for Smart Sensor Networks was the result. The objective of this standard is to make it easier for different manufacturers to develop smart sensors and to interface those devices to networks.

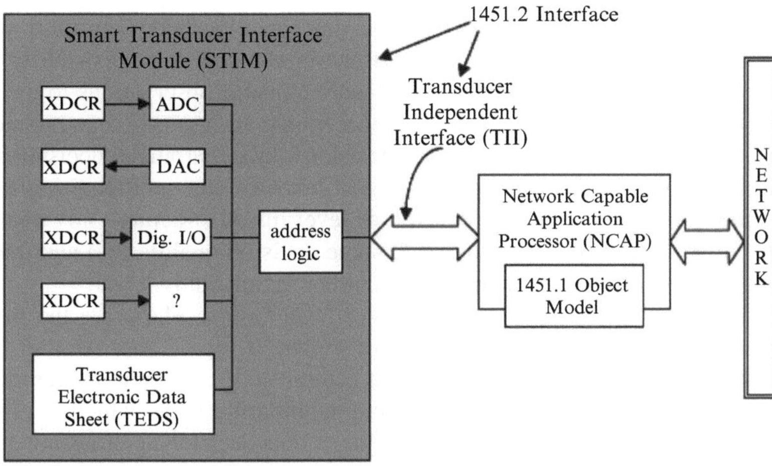

Fig. 2 The IEEE 1451 Standard for smart Sensor Networks

In Fig. 2 the basic architecture of IEEE 1451 is indicated [5]. Major components include STIM, TEDS, TII, and NCAP as detailed in this figure. A major outcome of IEEE 1451 studies is the formalized concept of a Smart Sensor. A smart sensor is a sensor that provides extra functions beyond those necessary for generating a correct representation of the sensed quantity. Included might be signal conditioning, signal processing, and decision-making/alarm functions. A general model of a smart sensor is shown in this figure. Objectives for smart sensors include moving the intelligence closer to the point of measurement; making it cost effective to integrate and maintain distributed sensor systems; creating a confluence of transducers, control, computation, and communications towards a common goal; and seamlessly interfacing numerous sensors of different types.

5 Information Fusion

Data fusion, is generally defined as the use of information processing techniques that combine data from multiple sources and gather that information in order to achieve inferences, which will be more efficient than if they were achieved by means of a single source [6–9].

The choice of the sensor network and the associated information processing methods that provide intelligence in the proposed approach, strongly depends on the characteristics of the employed sensors. In the case of a wireless sensor network employed for aircraft structure health monitoring it is expected that multiple sensors contribute with observations, measurements or decisions to a central fusion node in order to perform situation assessment. This approach is the extension of a single sensor system, where the detection-estimation process is an internal part of the sensor and we are now based in multisensor systems that can allocate detection-estimation between the sensors and the fusion node. However someone can choose between a centralized decision structure, where all sensors transmit observations to a central fusion node, and a sensor level fusion structure where each sensor is autonomously processing the observations and transmit decisions. In a centralized decision structure data are transmitted in raw form, being of large volume and requiring a wide bandwidth transmission channel. In a sensor level fusion structure, only decisions are transmitted (possibly augmented with some decision quality bits) and thus the required transmission channel bandwidth could be significantly lower.

Two major topologies are used for sensor level fusion. These are the parallel topology and the serial topology, which are shown in Fig. 3. In a parallel topology, decisions are transmitted directly to a central node and then are fused. In a serial topology decisions are transmitted to a neighboring node, which in turn has the ability either to combine this decision with its own decision and transmit a combined decision, or just to forward this decision together with its own decision. Serial

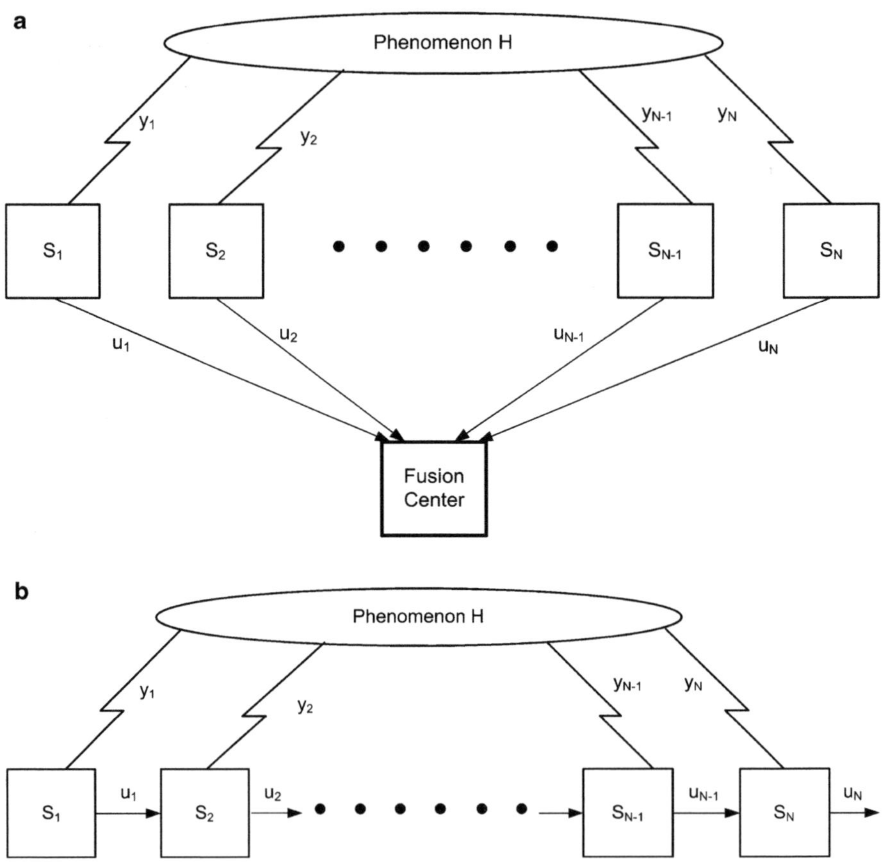

Fig. 3 The two major fusion topologies: parallel topology **a** and serial topology **b**

topology has an advantage in our case, since fits better with the multi-hop packet routing which is common to ad hoc networking.

Alternative sensor implementations include hard-decision or soft-decision sensors. Hard-decision sensors choose a single-hypothesis decision at the sensor and that decision, alone, is reported to the fusion center. On the other hand, Soft-decision sensors, partition the observation space into multiple detection regions, each representing a different decision threshold. Then, the detection decisions are quantized and the detection report contains a measure of confidence of the decision. This is slightly increases the data bits transmitted together with each decision, but also increases significantly the overall decision confidence.

Another important property of sensor network and the integrated intelligence it has, is the sensor management methods that incorporates. In multiple sensor systems the choice of complementary sensors which individually achieve robust detection and discrimination should be enhanced with proper sensor management in order for these sensors to optimize their collective performance relative to the application criteria.

Sensor management requirements have a large variety due to the unique characteristics of sensors and the application, however the sensor management includes:

- Spatial management and spatial sensor distribution especially for situation assessment purposes.
- Mode management, such as apertures, signal waveforms, power levels or processing techniques.
- Temporal management since in most cases sensors must be synchronized with other sensors or with events in the surveillance environment.

A step further towards a reliable fusion process could be to include data association and state estimation. Data association is the process of relating individual sensor measurements (data) with other measurements in order to derive if they have a common source and possible correlation. Association is implemented as an m-ary decision (correlation) process that quantifies the spatial (and/or attributes) relationships between the source data. State estimation is the process of mathematically determining an estimate of the state using discrete sensor measurements often referred as observation.

6 Summary

In this paper the general concept of a system for aircraft structural health monitoring is proposed. This system can automatically process real-time data, assess structural condition, and signal the need for human intervention. Such system would be able to detect incipient damage before catastrophic failures occur.

Such a system is based on wireless sensor networks. The wireless network consists of appropriate sensors distributed throughout the aircraft. Each sensor consisted of sensing, data processing and communicating components and could be easily embedded in any mechanical aircraft component. The established sensor network, due to its ad hoc nature is easily scalable, and power aware, while the position of the sensor nodes need not necessarily to be predetermined.

The functionality of such a system strongly depends on the existence of miniaturized mountable and embeddable sensors able to monitoring structural health.

Information collected from various sensors of different modalities throughout the aircraft is finally combined through process known as information fusion in order to achieve inferences, which will be more efficient than if they were achieved by means of a single source. In this way a more comprehensive image of the aircraft structural health is obtained.

References

1. Structural health monitoring systems accepted by Boeing, validated by airlines, Sandia National Laboratories, http://www.sandia.gov/news/resources/releases/2007/aircraft.html, 18 July 2007.
2. A. D. Afrashteh, S. Bland, F. M. Gondal, R. K. Kapania, A. Mishra, R. D. Moffitt and R. R. Wilson, Embedded Wireless Sensors for Aircraft/Automobile Tire Structural Health Monitoring, IEEE SECON, Reston, VA, 2006.
3. Use of Composite Materials, Health Monitoring and Self-Healing concepts, to Refurbish our Civil and Military infrastructure. Sandia Report, SAND207–5547, September 2007.
4. F. L. Lewis, Smart Sensor Networks, Smart Environments: Technologies, Protocols, and Applications, (ed.) D. J. Cook and S. K. Das, Wiley, New York, 2004.
5. R. Frank, Understanding Smart Sensors, 2nd ed., Artech House, Norwood, MA, 2000.
6. P. K. Varshney, Multisensor Data Fusion, Electronics and Communication Engineering Journal, 9, pp. 245–253, 1997.
7. B. V. Dasarathy, Sensor Fusion Potential Exploitation-Innovative Architectures and Illustrative Applications, Proceedings of the IEEE, 85, pp. 24–38, 1997.
8. D. L. Hall and J. Llinas, An Introduction to Multisensor Data Fusion, Proceedings of the IEEE, 85, pp. 24–38, 1997.
9. G. A. Lampropoulos, V. Anastassopoulos and J. F. Boulter, Constant False Alarm Rate Detection of Point Targets Using Distributed Sensors, Optical Engineering, The Journal of SPIE, 37(2), February 1998.

Roll Forming of AHSS: Numerical Simulation and Investigation of Effects of Main Process Parameters on Quality

Konstantinos Salonitis, John Paralikas, and George Chryssolouris

Abstract The roll forming process is one of the main processes of producing straight profiles in many industrial sectors. The introduction of Advanced High Strength Steels (AHSS), such as the DP and TRIP-series, into the production of roll-formed profiles has emerged new challenges. The combination of a higher yield strength with a lower total elongation of AHSS, brings new challenges to the roll forming process. In the current study, the numerical simulation of a V-section profile has been implemented. The effect of the main process parameters, such as the roll forming line velocity, rolls inter-distance, roll gap and rolls diameter on quality characteristics is investigated.

Keywords Cold roll forming · Modeling · AHSS · Longitudinal strains · Residual strains · Dimensional accuracy

1 Introduction

Cold roll forming (CRF) is a continuous bending operation, during which a long strip of metal is passed through consecutive sets of rolls, or stands, each performing only an incremental part of the bend, until the desired cross-section profile is obtained [15]. The introduction of Advance High Strength Steels (AHSS) into the production of roll-formed profiles has accentuated the existing processing problems and faces new challenges. These problems pertain to defects in production, as well as to limitation in surface quality and dimensional accuracy. The principle difference of AHSS from the conventional steel, HSLA and HSS lies into their microstructure [7]. They combine higher yield strength with lower total elongation.

K. Salonitis, J. Paralikas, and G. Chryssolouris (✉)
Laboratory for Manufacturing Systems and Automation, Department of Mechanical Engineering & Aeronautics, University of Patras, Patras, Greece
e-mail: {kosal, jparali, xrisol}@lms.mech.upatras.gr

S. Pantelakis, C. Rodopoulos (eds.), *Engineering Against Fracture: Proceedings of the 1st Conference*,

This imposes the challenge for generated higher loads on the rolls during roll forming, as well as a greater impact on defects and springback issues. Defects caused during the roll forming process could involve bowing, cambering, twisting, edge waving, corner buckling and edge cracking and splitting [6]. These kinds of defects and quality characteristics are affected mainly by various types of deformations. The main types of deformations are transversal elongation and bending, longitudinal elongation and bending as well as shear in strip's plane and thickness direction [6]. The only desirable deformation for the bending of the profile is the transversal bending, while the others are characterized as redundant deformations and are responsible for the existence of main defects during a roll forming process.

The scope of the current study is to provide an adequate model for roll forming of AHSS materials using numerical simulation. The model will be adequate for the investigation of the effect of the main process parameters on major redundant deformation and consequently on defects. Major redundant deformations that have been investigated on the V-section are longitudinal strains, at strip edge, and the distribution of longitudinal and transversal elongations, at the final profile, after the final roll station.

2 Background and Previous Work

During roll forming, the material in the flange travels a greater distance through successive rollers. This causes longitudinal strain in the flange, which should not enter into the plastic region in order for large longitudinal residual stresses in the finished profile to be avoided [9]. These residual stresses are responsible for major defects, such as cambering, bowing and twisting. The bigger the residual strain is the higher the defect rate of the finished product becomes [9]. Specific tolerances and quality standards could be set for the production of the V-section. Tighter tolerances require more roll stations [6]. Consequently tooling cost increase and affects product's cost. During the previous years, the CAE systems for tooling design used "rules of thumb" for the roll forming process analysis. FE-analysis has been used for validating the design from CAE tools. The disadvantage of FE-analysis is the large demanding CPU time, as special profiles need up to 40 roll forming stations [10]. Several attempts have been made for predicting defects and redundant deformations through simulations by using numerical methods and? the roll forming process.

Jeong [8] prepared a model for analyzing and predicting the longitudinal strains and the numerical values of camber and bow in the final product of two different flower patterns. Bui [2] simulated the cold roll forming process, and the results for longitudinal strains and displacement trajectories were compared with the experimental ones available in literature. Chiang [3] investigated how the flange length and the web width of a U and V shaped profile influence the average peak longitudinal strain at flange. He concluded that the average peak of the longitudinal strain is increased almost linearly with the increase of the flange length, though, without a significant change during increasing the web width. Fei-chin Jan [4] modeled the cold roll forming process with explicit dynamic method into LS-Dyna, in which the rollers were stationary and assumed as rigid. He examined how the friction forces affected the waviness surface of a Z-profile. Zhu [14] investigated the influence of the flange length, material thickness, and bed angle increment on the longitudinal membrane strain at flange edge. Lindgren [11] created a model to predict the longitudinal membrane strain in the flange and to analyze the contact between the tools and the strip, using the MARC/MENTAL FE tool. Lindgren [12] investigated the change in the longitudinal peak membrane strain at the flange edge and the deformation length when the yield strength increased. This is important since it can be used for determining the number of forming steps and the distance among them when designing the cold roll forming machine. The result from the simulations shows that the longitudinal peak membrane strain decreases and the deformation length increases when the yield strength is increased. Panton [13] states that the longitudinal peak strain occurs when the strip is in first contact with the rolls. The peak strain should not enter into the plastic region, as the longitudinal plastic strain will give a residual stress that causes defects on the profile such as wave edges and longitudinal curvature.

3 Modeling of Roll Forming Process

A V-section profile was selected for the implementation of the current study, with a total bending angle of 550 (Fig. 1). Five roll stations were initially designed for the roll forming of the V-section profile. For modeling the whole roll forming line and the strip, two different element types were used. SHELL163 elements were used for the rigid rolls, while SOLID164 elements for the deformable strip. SOLID164 is an

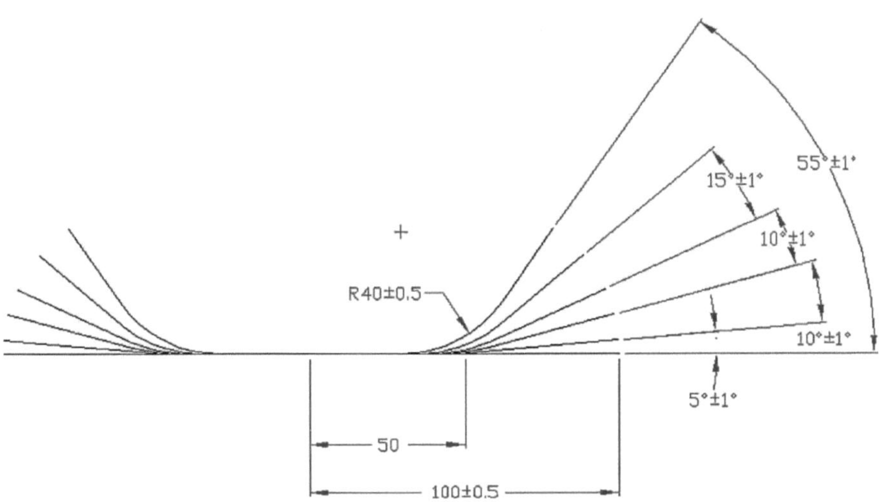

Fig. 1 Flower of the V-section profile (middle layer)

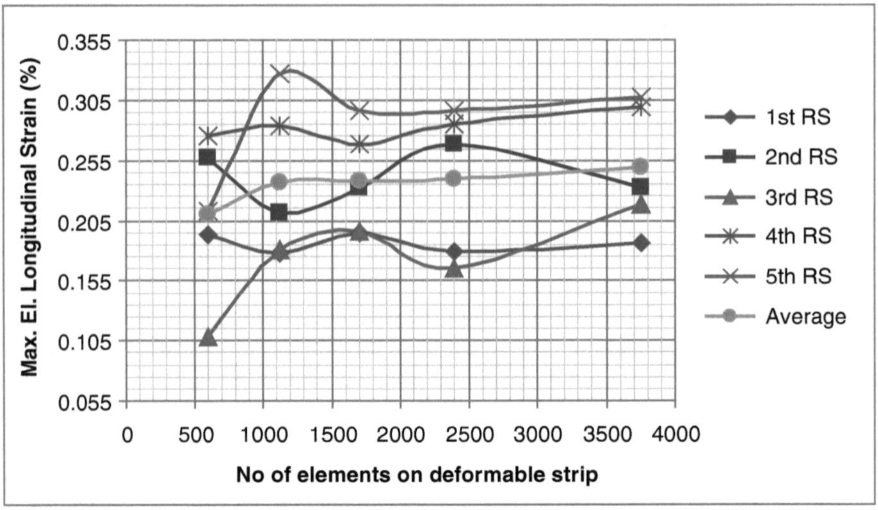

Fig. 2 Maximum Longitudinal strain% at each roll station vs. number of strip elements

eight-node brick element, and it uses reduced integration plus a viscous hourglass control for faster element formulation [1].

For the material model the isotropic hardening material model was chosen. In this material model, the center of the yield surface is fixed but the radius is a function of the plastic strain [5]. For the rolls, the rigid material model of the Ls-Dyna was chosen, with all rotations and all translations to be constrained. Thus, the rolls were assumed as stationary (not rotating cylinders) and rigid. Approximation of the effect of the rotating rolls on modeling could be the input of the rolling friction coefficient between the rolls and the strip. The material used in the current study was the AHSS DP600-HDG. DP600 steel has a yield limit of 0.2% at 417 MPa, which corresponds to a true strain of 0.005087. The hardening parameter of DP600-HDG is 0.136 and the tangent modulus corresponds to 17,500 MPa. For the optimization of the meshing size of the elements on the deformable strip, the meshing convergence procedure was followed. For almost all the roll stations, and the average of the maximum longitudinal values, the model seemed to be converged for 1,700 elements on strip and above. It was concluded that for 1,700 elements on strip and above, the results for the maximum attainable longitudinal elastic strain, at each roll station, were not changed significantly (Fig. 2).

4 Results and Discussion

A parametric study was implemented for the major roll forming process parameters. These parameters cover roll forming line speed, rolls stations inter-distances, rolls gap and rolls diameter.

4.1 Roll Forming Line Velocity

The roll forming line speed is a major process parameter, as by changing the speed, the productivity of the line is mainly affected. The issue for all roll-formers is to keep the line speed of their machines to maximum. The current study examines two different line speeds (1 and 0.75 m/s) on V-section profile in order to investigate how this change in speed affects redundant deformations and consequently, the product's quality. Total longitudinal strains (%) along the roll forming direction reveal that the residual longitudinal strain at edge reduces with the reduced velocity of the line (Fig. 3).

The line speed variation mainly affects the residual longitudinal strains at edge as well as the distribution of strains along the profile. Maximum longitudinal strain occurs at strip's edge, and maximum transversal strain at corner portion (Fig. 4).

4.2 Rolls Stations Inter-distance

Geometrical considerations play a major role in roll forming line set-up and the prevention from defects and dimensional inaccuracies. Increasing the distance between the roll stations leads to more length for deformation along the sequential bending angles, and more progressive deformation to occur. A disadvantage of increasing the rolls' inter-distances could be the increase in the total length of the roll forming line. Elastic longitudinal strains at edge, should not overcome the yield limit of the material because this would introduce plastic strain (Fig. 5). If strains surpass the yield limit of the material, an extensive plastic strain will occur, in longitudinal direction, meaning residual longitudinal strains (Fig. 6). By increasing the inter-distance, the

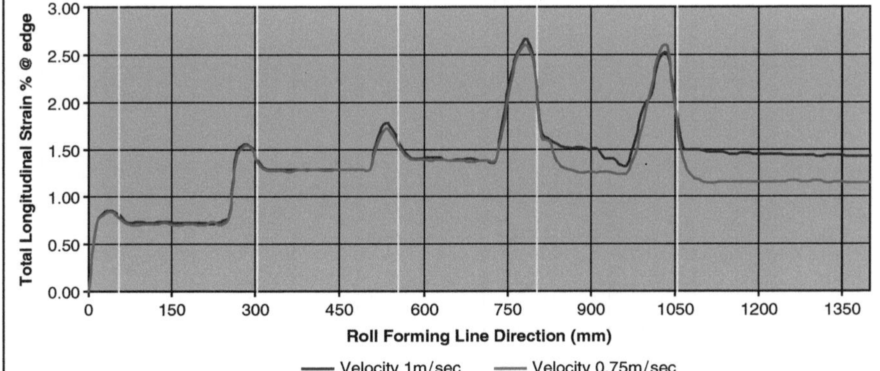

Fig. 3 Total longitudinal strain% at strip edge for different roll forming line speeds

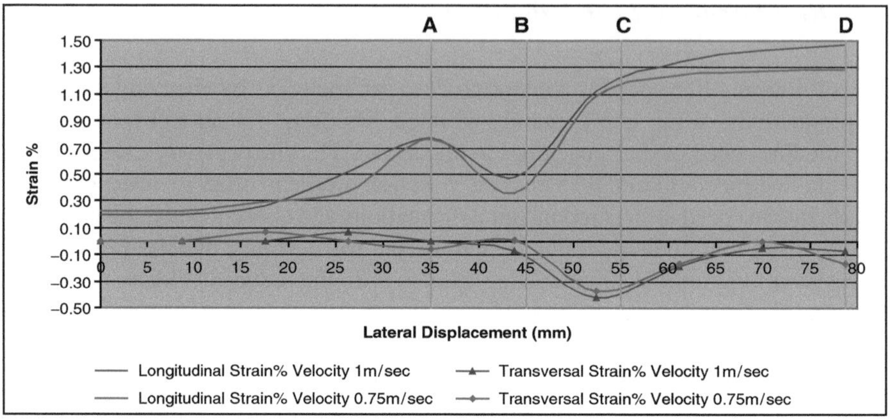

Fig. 4 Longitudinal and transversal strains% distribution on profile for different velocities after the exit from the final roll station

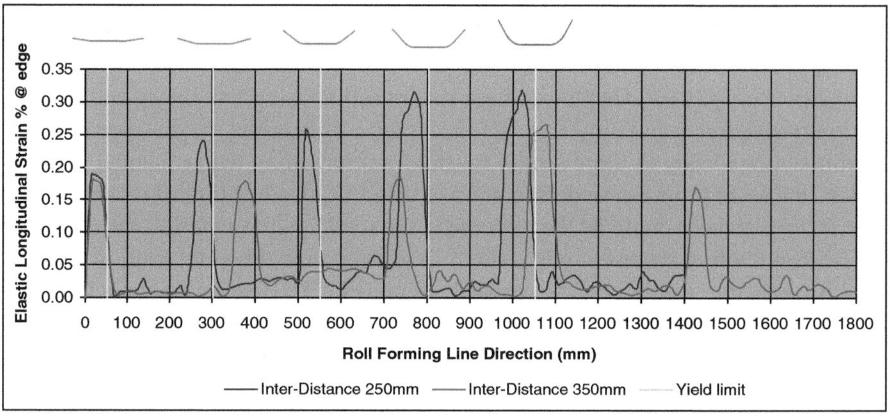

Fig. 5 Elastic longitudinal stain% at flange edge for different roll stations inter-distances

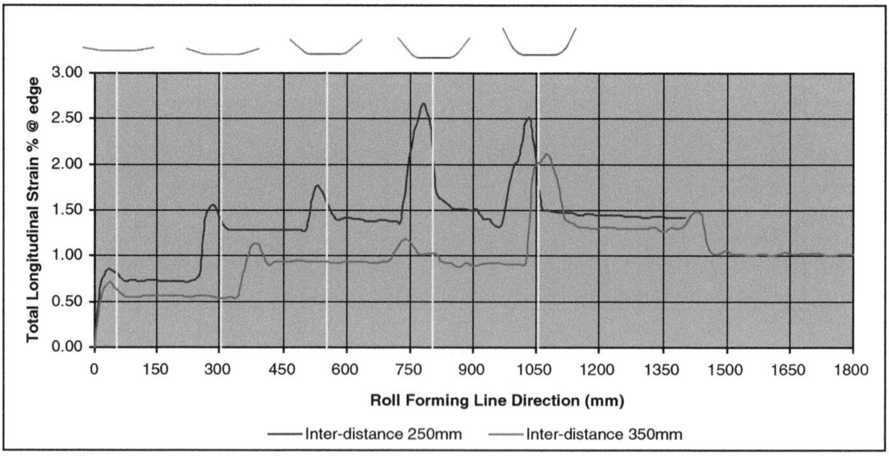

Fig. 6 Total longitudinal strain% at strip edge for different roll stations inter-distances

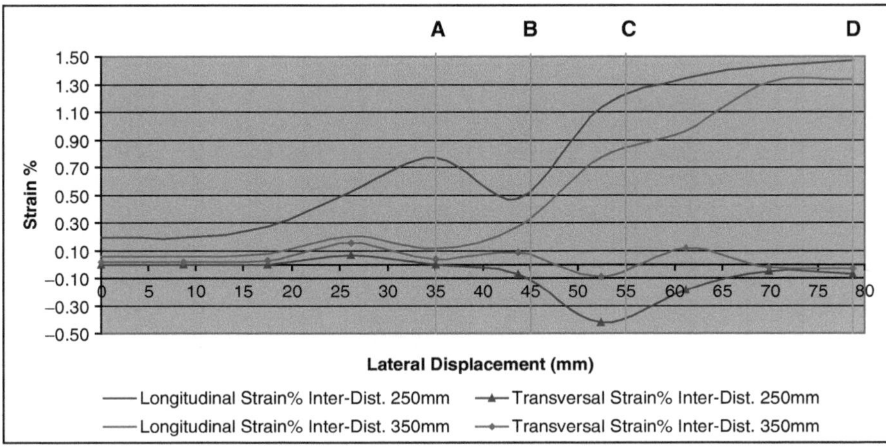

Fig. 7 Longitudinal and transversal strains% distribution on profile for different rolls stations inter-distances after the exit from the final roll station

residual longitudinal strains are reduced. The lower magnitude of peaks and a more uniform distribution can be achieved by increasing the rolls inter-distance (Fig. 7).

4.3 Roll Gap

Roll gap is the clearance between the rolls and the strip, as the latter passes through the former. It is measured in a vertical displacement of the clearance on the web. In the current study, the "tight" (0.0 mm) and the "loose" (0.6 mm) roll gaps are investigated. There seems to be a difference in the strain peaks, at greater bending angles, as wells as in the residual longitudinal strains, at the end of the line (Fig. 8). The different roll gaps also affect the distribution of the longitudinal and transversal strain along the final profile (Fig. 9). Significant is the increase in the transversal strain at the start of the bend (point A).

4.4 Rolls Diameter

The selection of the rolls' diameter is affected by numerous factors, such as roll forming machine dimensions, the material to be formed, the section depth, the rolls' surface speed, etc. The roll diameter plays a major role in the accumulation of the longitudinal plastic strain on the strip edge along the successive roll stations (Fig. 10). The speed differential between passes creates a tension in the section material and eliminates the possibility of overfeeding between passes. In this study,

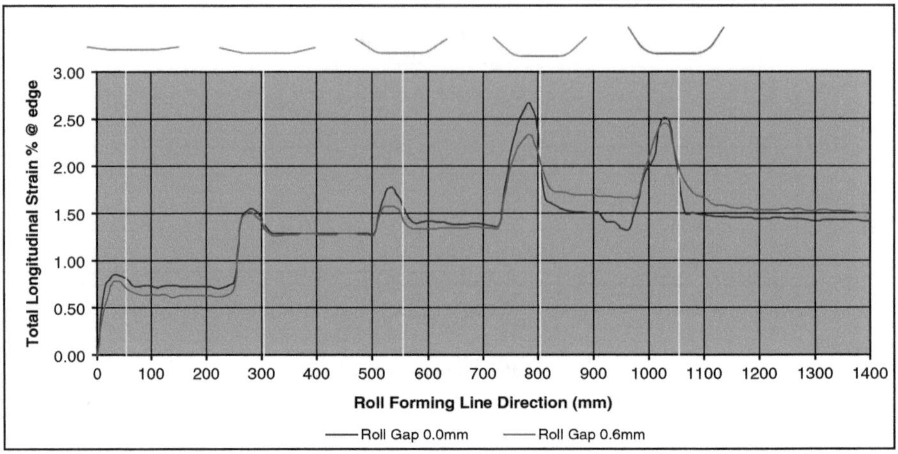

Fig. 8 Total longitudinal strain% at strip edge for different roll gaps

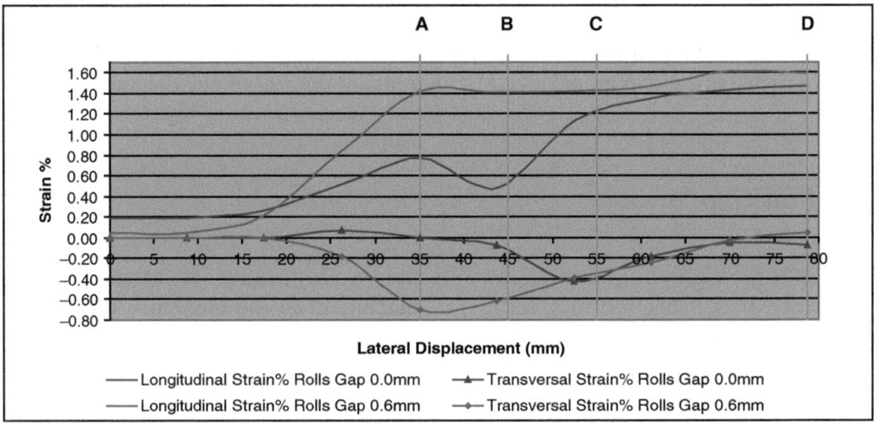

Fig. 9 Longitudinal and transversal strains% distribution on profile for different roll gaps after the exit from the final roll station

two different diameters have been investigated into simulating the roll forming line of a V-section. This residual strain is stored onto the strip and exists as residual strain after the roll forming process. With increasing the rolls' diameter the residual longitudinal strain is reduced. Moreover, the distribution of the longitudinal and transversal strains in the final produced profile, is affected by alternating the roll diameter (Fig. 11).

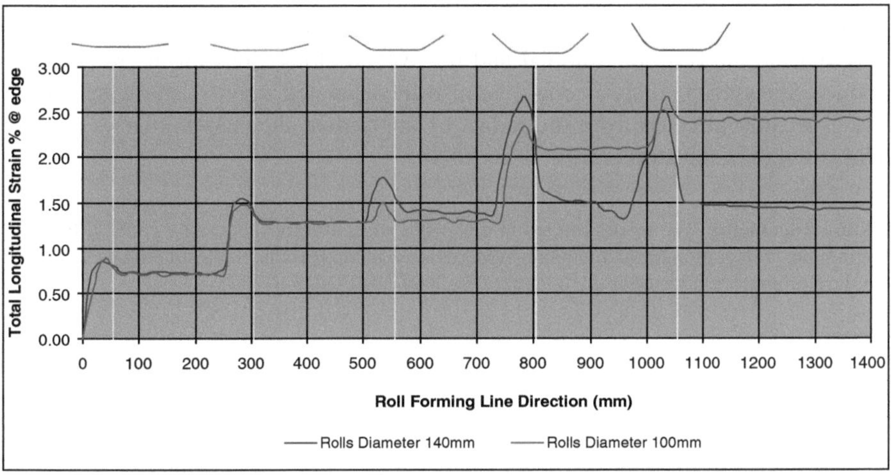

Fig. 10 Total longitudinal strain% at strip edge for different rolls diameter

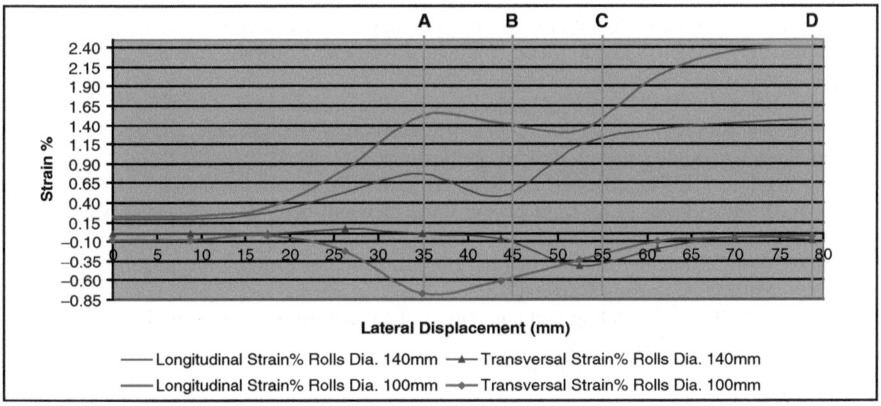

Fig. 11 Longitudinal and transversal strains% distribution on profile for different rolls diameter after the exit from the final roll station

5 Conclusions and Future Work

The modeling of the roll forming of a V-section profile has been presented in the current study. The scope of this modeling was to predict longitudinal and transversal elongations. The maximum magnitude and distribution of longitudinal strain along the profile produced, plays a major role in the presence of a defect and dimensional inaccuracies of the profile. Further study could be done for the influence of the parameters affecting the product design, such as material type, width of flange and web, material type, maximum profile height, etc., on redundant deformations and the distribution of them on the profile. Moreover, an optimization of process parameters could be implemented for reducing these redundant deformations and the uniform longitudinal and transversal elongations on the roll-formed product. Springback analysis could be also incorporated into the study, as springback is mainly affected from the radius of the corner, the strip thickness and the yield strength of the strip [7].

Acknowledgements The work reported in this paper was partially supported by CEC/FP6 NMP Programme, "Integration Multi-functional materials and related production technologies integrated into the Automotive industry of the future" (FP6-2004-NMP-NI-4-026621).

References

1. ANSYS Release 10.0 (2005) ANSYS LS-Dyna User Guide, Southpointe, 275 Technology Drive, Canonsburg, PA 15317, www.ansys.com
2. Bui QV, Ponthot JP (2007) Numerical simulation of cold roll-forming processes, J. Mater. Process. Tech., doi:10.1016/j.jmatprotec.2007.08.073
3. Chiang KF (1984) Cold roll forming. ME thesis, University of Auckland
4. Fei-chin Jan (2000) Simulation of cold roll forming process. Thesis, University of Pittsburgh
5. Hallquist JO (2006) LS-Dyna Theory Manual, Livermore Software Technology, Livermore, CA
6. Halmos GT (2006) Roll Forming Handbook, CRC Press, New York
7. International Iron & Steel Institute (2006) Advanced High Strength Steel (AHSS) Application Guidelines, Committee on Automotive Applications
8. Jeong SH, Lee SH, Kim GB, Seo HJ, Kim TH (2008) Computer simulation of U-channel for under-rail roll forming using rigid plastic finite element methods, J. Mater. Process. Tech., doi:10.1016/j.jmatprotec.2007.11.130
9. Kim YY (2002) Buckling analysis and buckling limit of strain on roll forming process. Thesis of Degree of M.Sc., Sogang University
10. Lindgren M (2005a) Modeling and simulation of the roll forming process. Licentiate thesis, Lulea University of Technology
11. Lindgren M (2005b) Finite element model of roll forming of a U-channel profile, Dalarna University, Sweden
12. Lindgren M (2007) Cold roll forming of a U-channel made of high strength steel, J. Mater. Process. Tech., doi:10.1016/j.jmatprotec.2006.12.017
13. Panton SM, Zhu SD, Duncan JL (1992) Geometric constraints on the forming path in roll forming channel sections, Proceedings of the Institution of Mechanical Engineers. Part B. Journal of engineering manufacture, vol. 206, 2:113–118
14. Zhu SD, Panton SM, Duncan JL (1996) The effects of geometric variables in roll forming a channel section, Proc. Inst. Mech. Eng. 210:127–134
15. Wick C, Benedict J, Veilleux R (1984) Tool and manufacturing engineers handbook, Volume 2 Forming, SME, Dearborn, MI

Session VII
Scale Effects and Modeling

Dependency of Micro-mechanical Properties of Gold Thin Films on Grain Size

Yong-Hak Huh, Dong-Iel Kim, Dong-Jin Kim, Hae-Moo Lee, and Jun-Hyub Park

Abstract Micro-tensile properties of Au thin films with different grain sizes were measured using micro-tensile tests and the effect of grain sizes of the films on the properties was investigated. The micro-tensile specimens 2 mm long and 100 μm wide were fabricated by electromachining process using the films, deposited on the silicon by sputtering technique, with six different thicknesses ranging from 0.18 to 1.0 μm. TEM observations showed the films with the respective different thickness had different grain sizes. As grain size increased, the yield strength decreased and followed the Hall-Petch relation. Furthermore, the yield strength also decreased with increasing the film thickness. Elastic modulus for the gold films was not affected with increase of the film thickness, but the value of the modulus for the film 0.18 μm was higher than the values for the thicker film. Tensile strength also showed the dependency of the film thickness.

Keywords Micro-tensile test · Gold · Thin film · Grain size · Thickness · ESPI (Electronic Speckle Pattern Interferometric)

1 Introduction

Thin films are widely employed in microelectromechanical system (MEMS) and integrated microelectronic components. Information on their properties may be essential in design and manufacture of these systems and components as well as in evaluation of reliability of the systems, since they are subjected to various mechanical loading conditions.

Y.-H. Huh (✉), D.-I. Kim, D.-J. Kim, and H.-M. Lee
Department of Metrology for Quality Life, Korea Research Institute of Science and Standards, Science Town, Taejon, 305-600, Korea
e-mail: yhhuh@kriss.re.kr, yuilmuil@kumho.co.kr, djkim@kriss.re.kr, hmlee@kriss.re.kr

J.-H. Park
Department of Mechatronics Engineering, Tongmyong University, Republic of Korea
e-mail: jhyubpark@korea.com

S. Pantelakis, C. Rodopoulos (eds.), *Engineering Against Fracture: Proceedings of the 1st Conference*,

Over the past decade, several testing techniques have been suggested for determining micro-mechanical properties of thin films. Typically, micro-tensile, beam bending and bulge tests were developed for free-standing film specimens and nano-indentation test was also introduced for the films on the substrates. Strain/deformation generated during these tests could be measured by employing the noble non-contacting techniques, such as ISDG (Interferometric Strain Displacement Gage), DIC (Digital Image Correlation), and laser interferometry, etc. [1–5]. With these precise and reliable testing techniques, the properties for the film materials reduced to the micron and sub-micron scale can be determined. Espinosa and Prorok [4] showed the size effect on the mechanical properties of gold thin films with thickness of 0.3, 0.5 and 1.0 μm using membrane deflection experiments, where the thickness of the films affected the stress–strain behavior significantly. Similarly, for the electroplated copper film, increase of the film thickness caused the decrease of the elastic modulus and yield strength [6].

In this study, the micro-tensile properties of the gold thin film with various thicknesses were investigated. The properties of the films were measured using the micro-tensile testing system and ESPI (Electronic Speckle Pattern Interferometry) technique for measuring the in-plane tensile strain during tensile loading. The thickness of the film was related to the grain size of the film and the effect of grain size of the film was discussed.

2 Experimental Procedure

2.1 Specimen and Testing System

Gold thin films, used in this study, of different thicknesses ranging from 0.18 to 1 μm were deposited on the silicon wafers using sputtering technique. The micro-sized tensile specimens were fabricated by electromachining process. Figure 1 shows a typical shape of the specimens. The specimens were prepared in length of 2 mm and width of 100 μm.

The microstructures of the deposited films were observed by TEM and the average rain size of the films were determined from the TEM pictures according to the linear intercept procedure [7]. Table 1 shows the average grain sizes of the films.

Micro-tensile tests were carried out using the micro-tensile testing system developed in this study. This system consisted of a micro-tensile loading system and an ESPI system for measuring the in-plane tensile strain. The micro-tensile loads was applied to the specimen with a high-resolution linear actuator which was operated in closed loop with a stroke resolution of 4.5 nm and had a load cell with a load capacity of 500 mN. The in-plane tensile strains during loading were continuously measured by analyzing the interferometric fringe patterns acquired by ESPI system. The loading system with ESPI system is presented in Fig. 2. Using the cameras installed in micro-tensile loading system, the specimen was aligned and installed on the grips with UV adhesives. Tensile test was carried out in the displacement control of 25 μm/min.

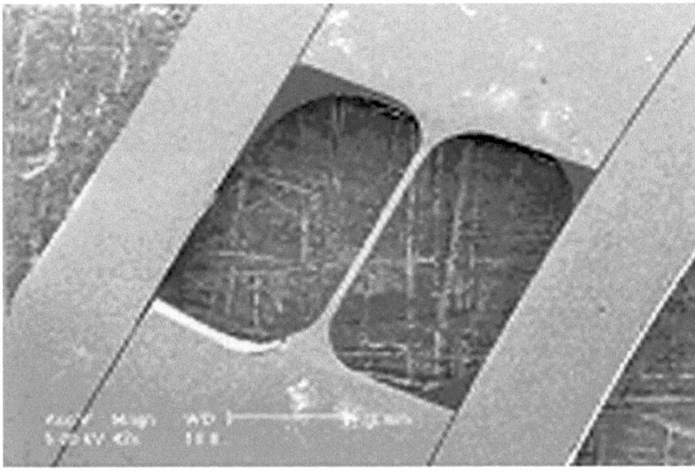

Fig. 1 Typical shape of micro-tensile testing specimen

Table 1 Thickness and average grain size of the gold films

	Thickness of film (μm)	Average grain size (nm)
Batch 1[a]	0.18	21
	0.5	43.7
	1.0	166.6
Batch 2[a]	0.31	57.8
	0.68	90.3
	0.95	106.9

[a]Gold thin films were deposited in two different sputtering processes.

2.2 *Measurement of Tensile Strain*

For ESPI technique, tensile strains of the film can be determined by quantitatively analyzing the interferometric fringe patterns of the film surface.

The speckle patterns of the strained film surface were generated by exposing the dual laser beams, which were divided from a He-Ne laser and were symmetric with respect to the observation direction, with the same intensity on the film surface. Image of the speckle pattern was focused on the screen of the CCD camera and then the images were sequentially taken in real time through a frame grabber and digitally stored. According to the subsequent strain measurement algorithm [5], the captured patterns could generate the interferometric fringe patterns and the spacing between the fringes of the pattern were then calculated. With the measured spacing and optical arrangement of ESPI system, in-plane tensile strain corresponding to the deformation of the specimen can be measured by using the relationship as the following:

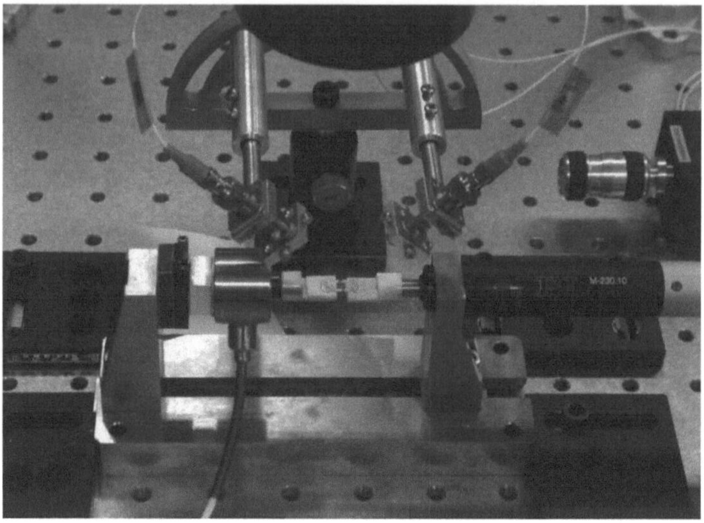

Fig. 2 Photograph of micro-tensile testing system

$$\varepsilon = \frac{\lambda}{2D_p \cos\theta}, \qquad (1)$$

where λ and θ represent the wave length of laser light source (632.8 nm for He-Ne laser) and incident angle of dual illumination lights, respectively, and D_p means spacing between two fringes.

3 Results and Discussion

3.1 *Micro-tensile Stress–Strain Curves*

Micro-tensile stress–strain curves for the gold thin films with different thickness were obtained using the micro-tensile specimens as shown in Fig. 1. Figure 3 represents an example of the micro-tensile stress–strain curves experimentally obtained for the gold thin film and the interferometric fringe patterns corresponding to the respective strains of the film. From the fringe patterns as shown in Fig. 3, tensile strains with tensile loads could be determined using Eq. 1 and the uniformity of the tensile stress over the testing section could be checked. As can be seen in Fig. 3, the fringe patterns appeared relatively clearly to be normal to the loading axis and the parallel fringes were generated with applying tensile stress. This means that the tensile stress is uniformly distributed across the cross-section normal to the loading axis. These parallel fringe patterns as shown in Fig. 3 were observed for all thin films tested in this study.

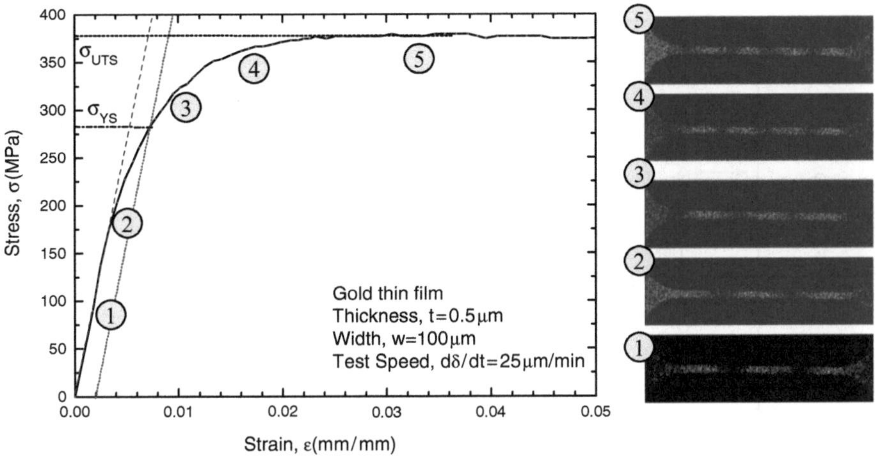

Fig. 3 Typical stress–strain curves and ESPI fringe patterns with various stress levels for Au thin film

As shown in Fig. 3, the stress–strain curve for the film shows relatively significant plastic flow. Therefore, from the curve, the yielding strength (0.2%), tensile strength and elastic modulus could be defined as defined in conventional bulk materials. However, plastic regime of the stress–strain curve for the film with relatively thick thickness is found to be larger than that for the film with thin thickness. Size effect on the plastic deformation was also shown in membrane deflection test [4].

3.2 *Micro-mechanical Behavior*

Figure 4 shows the elastic modulus, yield strength and tensile strength determined from the stress–strain curves for the gold thin films with various thicknesses. In this study, to investigate variation of mechanical properties of the films with thickness, the tensile testing specimens with six different thicknesses ranging from 0.18 to 1.0 μm were used. The gold films were deposited on the silicon wafer in two different sputtering machines. So, the respective films were hereinafter referred as batch 1 and batch 2 as expressed in Table 1. As can be seen in Fig. 4, elastic modulus for batch 1 was nearly identical with that for batch 2, but the modulus for the film 0.18 μm thick was a little higher than for the thicker films. Theoretically, the elastic modulus should not be affected by the film thickness or grain size. For the electroplated copper films, this behavior was partially explained by density difference [6]. Values of the elastic modulus measured for the gold films were ranged from 46 to 69 GPa. These values are nearly similar to those reported for gold thin films [4,8,9], but is significantly lower than the value of 78 GPa for bulk Au.

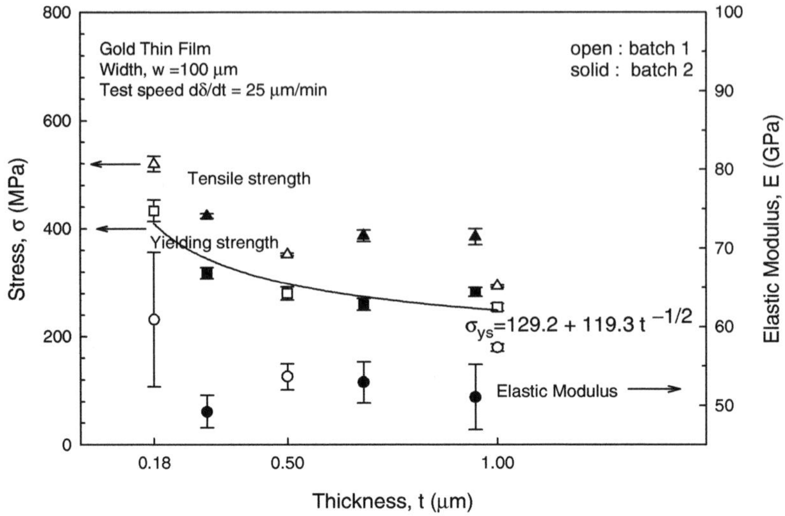

Fig. 4 Variation of micro-mechanical properties with the film thickness for the gold films

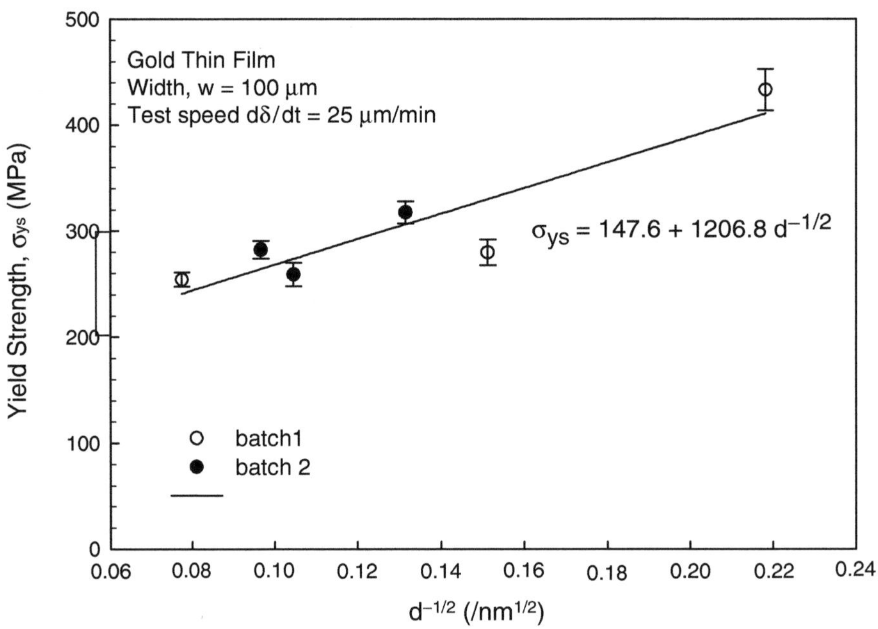

Fig. 5 Dependency of the yield strength on the grain size

Figure 4 shows a remarkable dependency of the yield strength for the gold films on the film thickness. The yield strength decreased with an increase in film thickness. The grain size of the gold film was measured as described in Table 1. Figure 5 represents variation of the yield strength with grain size of the film. As shown in Fig. 5, the yield strength can be linearly related with $d^{-1/2}$, where d is the grain size. For the gold thin films used in this study, the yield strength can be said to follow the classical Hall-Petch relationship, so the dependency of the strength on the grain size may be presented as follows:

$$\sigma_{YS} = 147.6 + 1206.8d^{-1/2} \qquad (2)$$

where d is the thin film grain size in nanometer and σ_{YS} is the yield strength in MegaPascal. According to the Wei and Hutchinson approach [10], the yield strength can be related with the film thickness instead of the grain size as the scaling parameter if the grain size can be compared to the film thickness. As shown in Fig. 4, the relationship between the strength and the film thickness for the gold thin films used can be fitted as follows:

$$\sigma_{YS} = 129.2 + 119.3t^{-1/2}, \qquad (3)$$

where t and σ_{YS} are the thin film thickness in μm and the yield strength in MPa, respectively.

Similar to the dependency of the yield strength on the grain size and film thickness, the tensile strength was also dependent on the film thickness. As can be seen in Fig. 4, the tensile strength for the gold films decreased with an increase in the film thickness.

4 Summary and Conclusions

Micro-mechanical properties for Gold films of varying thicknesses were evaluated using micro-tensile testing. The tensile stress–strain curves for the films were determined measuring tensile strains during tensile loading with in-plane ESPI techniques. Grain sizes of the respective films were provided from the assessment of the microstructures of the films taken by TEM. As grain size increased, the yield strength decreased and followed the Hall-Petch relation. Furthermore, the yield strength also decreased with increasing the film thickness. Elastic modulus for the gold films was not affected with increase of the film thickness, but the value of the modulus for the film 0.18 μm was higher than the values for the thicker film. Tensile strength also showed the dependency of the film thickness.

References

1. Sharper W.N., Jr., Yuan B. and Edwards R.L. (1997) A New Technique for Measuring the Mechanical Properties of Thin Films, J. Microelectromech. Syst. 6: 193–199
2. Sutton M.A., Wolters W.J., Peters W.H., Ranson W.F. and McNeil S.R. (1983) Determination of Displacements Using an Improved Digital Image Correlation Method, Image Vision Comput. 1: 133–139

3. Read D.T. (1998) Young's Modulus of Thin Films by Speckle Interferometry, Meas. Sci. Technol. 9: 676–685
4. Espinosa H.D. and Prorok B.C. (2003) Size Effects on the Mechanical Behavior of Gold Thin Films, J. Mater. Sci. 38: 4125–4128
5. Huh Y.-H., Kim D.I., Kim D.J., Park P., Kee C.D. and Park J.H. (2004) Application of Micro-ESPI Technique for Measurement of Micro-tensile Properties, Key Eng. Mater. 270: 744–749
6. Volinsky A.A., Vella J., Adhihetty I.S., Sarihan V., Mercado L., Yeung B.H. and Gerberich W.W. (2001) Microstructure and Mechanical Properties of Electroplated Cu Thin Films. Mater. Res. Soc. Symp. 649: Q5.3.1–Q5.3.6
7. ASTM E112-96 Standard Test Methods for Determining Average Grain Size Linear Intercept Procedure
8. Nix W.D. (1989) Mechanical Properties of Thin Films, Metall. Trans. A 20A: 2217–2245
9. Okuda S., Kobiyama M. and Inami T. (1999) Mechanical Properties and Thermal Stability of Nanocrystalline Gold Prepared by Gas Deposition Method, Mater. Trans. JIM 40(5): 412–415
10. Wei Y. and Hutchinson J.W. (1997) Nonlinear delamination mechanics for thin films, J. Mech. Phys. Solids 45: 1137–1159

Fracture and Failure in Micro- and Nano-Scale

Costas A. Charitidis

Abstract Indentation and scratch in micro- and nano-scale are the most commonly used techniques for quantifying thin film and systems properties. Among them are different failure modes such as deformation, friction, fracture toughness, fatigue. Failure modes can be activated either by a cycle of indentation or by scratching of the samples to provide an estimation of the fracture toughness and interfacial fracture energies. In the present study, we report on the failure and fracture modes in two cases of engineering materials; that is transparent SiOx thin films onto poly(ethylene terephthalate) (PET) membranes and glass–ceramic materials. The SiOx/PET system meets the demands regarding scratch-resistance, wettability, biocompatibility, gas transmission, or friction, while maintaining the bulk characteristics of PET (such as easy processing, good mechanical properties, reasonably low permeability to oxygen and carbon dioxide gases (barrier properties), and good chemical coupling with antibacterial coatings). Glass–ceramic materials, since their first accidental production in the mid fifties by S.D. Stookey, have been used in a vast area of applications, from household cooktops and stoves, to missile nose cones and mirror mounts of orbital telescopes and from decorative wall coverings to medical applications. The fracture modes, namely transgranular and intergranular modes in glass–ceramic materials have paid less attention in literature comparing with ceramic materials. In the former case the crack paves its way irrespectively of the direction of the grain boundaries, i.e., the interfaces between the different phases. In the latter case the crack preferentially follows them, i.e., debonds the interfaces.

Keywords Fracture · Failure · Nanoindentation · Nanoscratch · SiOx films · Glass ceramic materials

---

C.A. Charitidis
National Technical University of Athens, School of Chemical Engineering,
9 Heroon Polytechniou Str., Zographos, Athens, Greece GR-157 80
e-mail: charitidis@chemeng.ntua.gr

S. Pantelakis, C. Rodopoulos (eds.), *Engineering Against Fracture: Proceedings of the 1st Conference*,

1 Introduction

Materials (i.e. hard thin films) are usually used to enhance reliability and performance of components in mechanical applications. However, there is a high risk of crack formation through the film or failure at the interface between the film and the substrate. To study failure mechanisms and to characterize mechanical properties of coated systems, nanomechanical techniques such as nanoindentation and nanoscratching have widely been accepted as effective experimental methods. The measurements are usually controlled by the coating material itself, by the substrate upon which the coating is deposited and by the interfaces which bond the system together. The experimental procedures and analytical models to extract the film properties such as hardness, elastic modulus as well as interfacial fracture and fracture toughness have been described in several papers [1–3]. Indentation of coated systems showed that soft substrate yields the formation of annular primary cracks, while hard substrate favors the formation of radial corner cracks [4].

The role of nano and microcracks on mechanical properties are not yet well established. The failure mechanism of coatings during scratch tests is very complex and depends on many factors, such as the hardnesses of the coating and the substrate, the coating thickness, and the scratch indenter radius, the deposition techniques and, the substrate material and the scratch method [5].

The failure of coatings is also determined by the deformation of both the substrate and the coating. When the coating is not thick enough, a high compressive stress field caused by the indenter during a scratch test may be transferred to the substrate, and hence the scratch response is controlled by plastic deformation of the substrate. Coatings with the same hardness may exhibit different resistances to plastic deformation, and the deformation of a hard coating is strongly dependent upon the combination of its hardness (H) and elastic modulus (E).

The SiOx/PET system meets the demands regarding scratch-resistance, wettability, biocompatibility, gas transmission, or friction, while maintaining the bulk characteristics of PET (such as easy processing, good mechanical properties, reasonably low permeability to oxygen and carbon dioxide gases (barrier properties), and good chemical coupling with antibacterial coatings). The SiOx onto PET improves gas barrier properties and in combination with the flexibility and the relatively low price of the plasma techniques used to deposit SiOx onto PET, makes it very interesting as a tool for the packaging industry for food and medical device packaging.

To study failure modes in hard thin films, SiOx thin films onto PET substrates were subjected to indentation stressing and nanoscratching. This case is of high interest due to the mismatch of the hardness and elastic modulus of the SiOx film and PET substrate. The objective of the present work is to provide an improved understanding of the mechanical response of the SiOx/PET system.

Fracture, being the local separation of a piece of material, under the action of stress, comprises one of the mechanisms of energy absorbtion, for instance during indentation, impact or wear. Cracking can be undesirable, in the cases where it causes catastrophic failure, while desirable for the proper function of specific applications, like armour plates. In the present study, we report on the fracture modes in micro scale, namely transgranular and intergranular modes. In the former case the crack paves its way irrespectively of the direction of the grain boundaries, i.e., the interfaces between the different phases. In the latter case the crack preferentially follows them, i.e., debonds the interfaces [6]. These fracture modes have been extensively studied in the case of ceramic materials [7–9], while they have not received the same attention in the case of glass–ceramic materials. The intergranular fracture is connected to high toughness, while transgranular to high strength [10].

2 Experimental

2.1 *SiOx/PET System*

The examined PET membranes (thickness ~12 μm) were industrially supplied, treated using mechanical stretching that is usually applied in industrial scale. An industrial roll-to-roll vacuum deposition coater was used to produce SiOx films 50 nm thick.

Hardness and reduced elastic modulus of the coatings were assessed by means of a nanoindentation system (Hysitron TriboLab system). A pyramidal diamond Berkovich tip with a total included angle of 142.3 and a radius of curvature of approximately 100 nm has been used for both indentation and scratch measurements. A trapezoidal loading–unloading ramp was used. The maximum load is reached after 5 s, kept constant for 5 s and decreased to zero during the same time duration. The H and the E were determined from the experimental load–displacement curve using the Oliver and Pharr model [11] and the values from more than ten indents were averaged for each sample. Scratch measurements have been conducted using a two-dimensional force–displacement transducer. This transducer allows the measuring of both lateral force and normal displacement. The total scratch length is fixed at 10 μm. Scratch measurements have been realized at a normal force of 10,000 μN and a constant normal loading rate of $33.33\,\mu Ns^{-1}$.

Tests were performed in a clean-air environment with a relative humidity of approximately 25%, while the temperature was constantly kept at 20°C (ambient temperature). The samples to be measured were placed into the cabinet for several hours prior to testing, in order to reach equilibrium with the thermal mass inside the cabinet.

2.2 *Glass–Ceramic Materials*

Glass–ceramic materials have been produced from the thermal treatment two vitreous products. IS1 product contains 50 wt% of a rich in PbO and Fe2O3 incinerated sludge, 35 wt% SiO2 and 15 wt% Na2O. IS2 product contains 60 wt% of the same incinerated sludge, 25 wt% SiO2 and 15 wt% Na2O.The resulting glass–ceramic materials have different final morphologies. IS1 glass–ceramic is made up of a vitreous matrix with homogeneously dispersed rectangular shaped Pb8Fe2O11 crystallites. IS2 glass–ceramic is made up of a vitreous matrix with inhomogeneously dispersed oblong Fe2O3 crystallites of varying widths [12, 13].

The modes of fracture were determined by observing the paths of crack propagation with respect to the crystallite/amorphous matrix interfaces by optical microscopy. All indentation induced cracks were produced on flat and polished surfaces of the thermally treated products with a Vickers diamond indenter. This was necessary since an indentation on a rough surface cannot be observed clearly by high magnification optical microscopy, due to the particularly narrow depth of focus. Polishing was achieved by wet mechanical grinding with SiC grinding papers and alumina (Al2O3) pastes. An Anton-Paar MHT-10 microhardness tester was utilized [14] attached on a Zeiss Axiolab-A optical microscope. The indentation parameters were set at: load $=$ 1–3 N, duration $=$ 10 s and loading rate $= 0.2\,\mathrm{N/s}$. The duration of 10 s is a typical time, that is, long enough for the onset of plasticity, while the load range was selected in order to produce indentations with clearly defined radial cracks.

3 Results

3.1 SiOx/PET System

Figure 1 shows two representative indentation curves for a maximum normal load approximately equal to 90 and 450 μN. The load hysteresis and residual penetration depths indicate that plastic deformation occurred even with loads as low as 90 μN. Unloading produces a highly nonlinear load response characteristic of a viscoelastic material, for the normal load of 450 μN.

Both H and E are increased in the surface/near surface region (2.2 and 12 GPa, for SiOx film 50 nm thick). For higher contact depths, the H (E) values tend to decrease approaching the bulk PET H and E values (0.25 and 3.2 GPa, respectively). The bulk H and E values of PET are in good agreement with measurements in the literature [15].

Friction is a complex phenomenon, which involves asperity interactions involving adhesion and deformation (plowing). Figure 2 shows the coefficient of friction as a function of normal load obtained from nanoscratch tests performed with a maximum normal load of 10,000 μN for SiOx 50 nm thick onto PET membranes and uncoated PET. The coefficient of friction increased with increasing normal load to a maximum value ($\sim$0.2) corresponding to the transition load ($\sim$3,000 μN). The behavior of coefficient of friction of the uncoated PET, in the same load range, was found to decrease (from a maximum value $\sim$0.6) to a value of $\sim$0.3. The cohesive strength of the SiO3,000x/PET system is enhanced comparing to that of uncoated

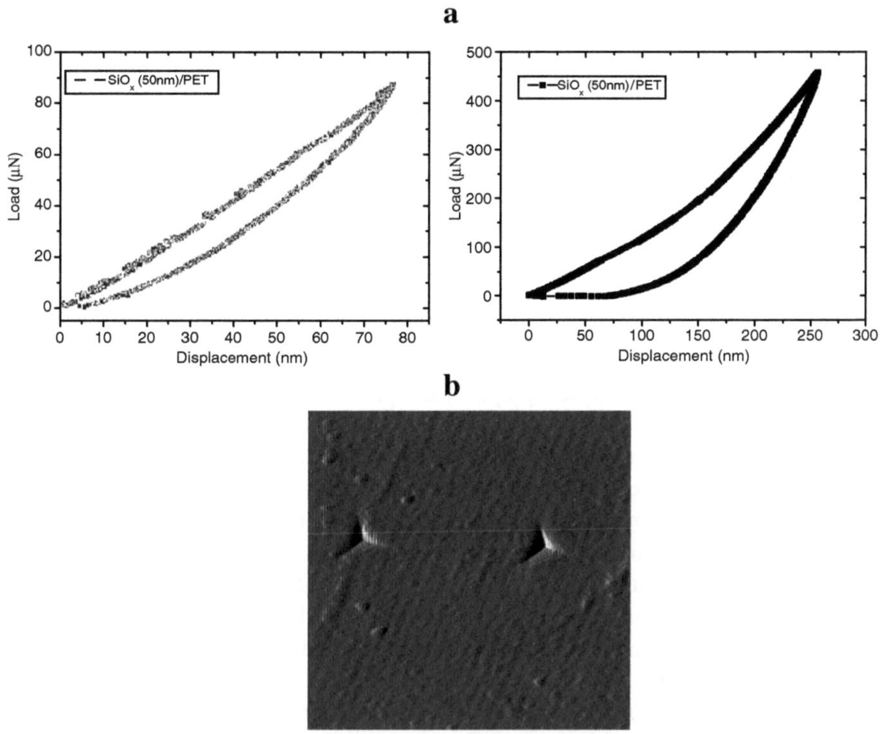

Fig. 1 **a** Evolution of the applied load vs. Displacement upon nanoindentation cycle, for loads 90 and 450 μN, **b** Scanning Probe Microscopy (SPM) imaging of the nanoindentations for load 450 μN

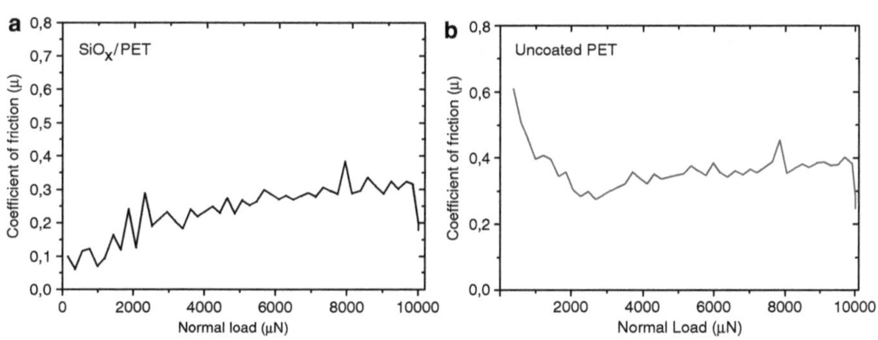

Fig. 2 Coefficient of friction profile, as a function of the normal load applied for **a** SiOx/PET and **b** uncoated PET

PET, and as a result, the friction force needed to overcome the adhesion of the system will increase, leading to the increase of coefficient of friction. The high-load range for SiOx/PET reveals higher coefficient of friction (μ ∼0.2–0.3). Taking into account that the deformation of SiOx/PET was essentially elastic for maximum normal loads 50–1,500 μN (Fig. 1) the low coefficient of friction in this regime is considered to be a manifestation of surface adhesion forces. The scratches produced on SiOx/PET at loads greater than 3,000 μN indicated that plastic deformation was dominant and that the higher coefficient of friction in this regime was mainly due to the plowing process occurring at the nanometer and micrometer scales.

3.2 Glass–Ceramic Materials

Figure 3 depicts an optical micrograph of the path of an indentation-induced radial crack. The indentation was made on the IS1 glass–ceramic product. This micrograph depicts a mixed crack propagation pattern. In places indicated by the arrows numbered by 1 the crack is deflected by the crystallites. More specifically, the crack propagates following the boundaries of the separated crystallites causing interfacial debonding, i.e., by separating the amorphous from the ceramic phase. In places indicated by the arrows numbered by 2 the crack propagates in straight a line as it cuts through the microcrystallites, without being deflected by the crystallite/amorphous matrix interfaces. Consequently, the fracture mode is both transgranular and intergranular for IS1 product.

In IS2 product, the morphology is different, with randomly oriented elongated crystallites of various widths dispersed in the remaining glass matrix. Whether these crystallites will act as a barrier to crack propagation depends on their width and relative orientation with respect to the direction of the crack propagation. In Fig. 4 all radial cracks, except the lower one, follow mixed mode propagation. The crack follow for a short distance the crystalline/glass matrix interface (intergranular mode) and then they cut through it (transgranular mode). This interchange between the two

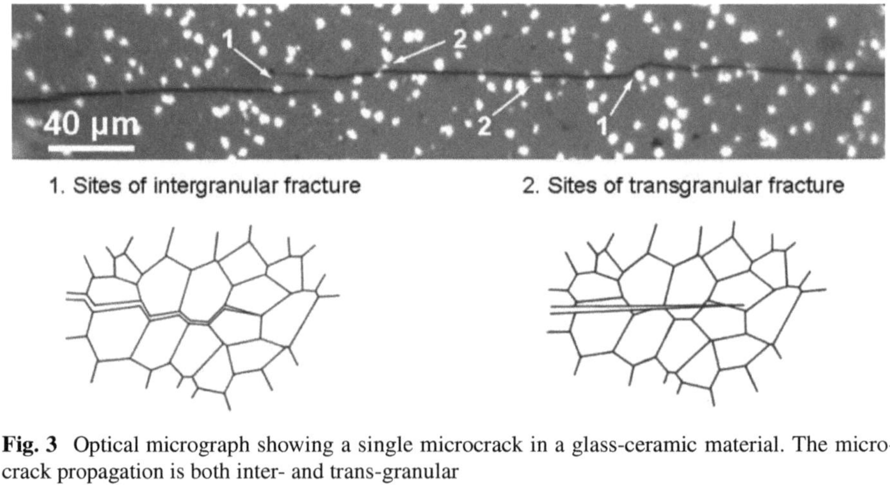

Fig. 3 Optical micrograph showing a single microcrack in a glass-ceramic material. The microcrack propagation is both inter- and trans-granular

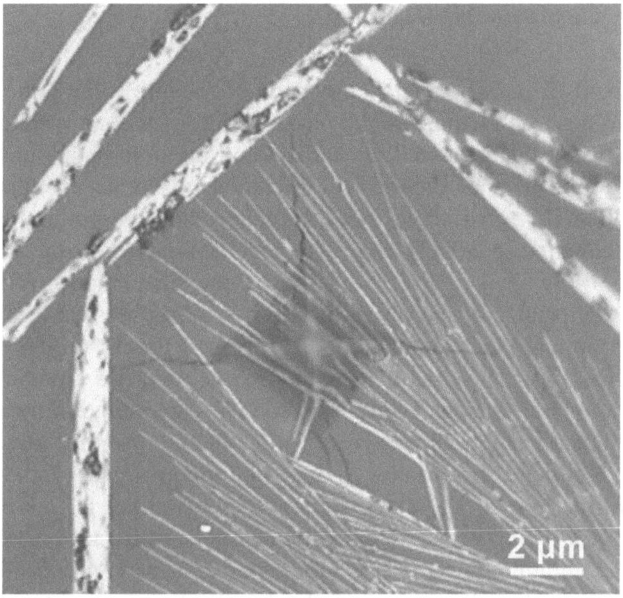

Fig. 4 When the microcrack propagates perpendicularly with respect to the needle-like crystallites the cracking occurs in a transgranular manner. When the microcrack propagates at 45° with respect to the needle-like crystallites the cracking occurs in a mixed manner (both trans- and intergranularly). The crack itself has the shape of a broken line

propagation modes is most clearly observed in the upper crack, where the crack path has a step-like appearance. The lower crack finds a thicker elongated crystallite and is been annihilated.

4 Discussion–Conclusions

4.1 SiOx/PET System

Results from the nanomechanical and nanotribological characterization of the SiOx/PET have revealed significant differences in the nanomechanical properties of the surface/near surface region. SiOx/PET system exhibited higher hardness, elastic modulus and elastic recovery than uncoated PET membranes under the same normal load, implying that SiOx films sustain higher scratch induced stresses and are effective as protective coating material. SiOx/PET system revealed a transition from elastic to plastic deformation accompanied by a change of the dominant friction mechanism from adhesion to plowing was observed with increasing normal load. The results of the present work illustrate that significantly different friction properties can occur at different ranges of loads.

4.2 Glass–Ceramic Materials

For both IS1 and IS2 glass ceramics the mode of fracture is the same although the shape and composition of the ceramic phase is different. As a result, the different coherency between the separated crystallites and the amorphous matrix does not seem to play dominant role. The mode of fracture is mixed, i.e. is both transgranular and intergranular. The two above cases, although they present similar behaviors, they should be treated separately in order to reach an interpretation.

In the case of IS1 glass ceramic, the reason of the mixed mode of fracture is possible to stem on the size of the crystallites. It has been observed in the same glass ceramic a predominant transgranular mode of fracture for crystallites of sizes lower than one micron [16]. For sizes higher than several microns a predominant intergranular mode of fracture has been observed. The interpretation for this behavior has been attributed to the relative sizes of a distorted crystalline zone that acts as the boundary between the crystallite and the amorphous matrix and the size of the crystallites. In the case of IS1 glass ceramic the crystallites have sizes between 2 and 5 μm, i.e. in a region where no predominant mode of fracture is expected.

In IS2 glass ceramic, the mixed crack propagation pattern depends on the orientation of the crack with respect to the longer dimension of the crystallites. When the crack propagates parallel to the longer dimension, it follows the crystallite/amorphous matrix interface, i.e. the propagation is intergranular. When the crack propagates at right angles it ruptures the crystallites, i.e. the propagation is transgranular. This is because of their relatively small width, namely below 1 μm [17].

In Fig. 2, the orientation of the crack is approximately at 45° angle, therefore it lies at the centre between the above two extreme cases. As a result, a mixed crack propagation mode is observed in this case. Intergranular mode has been proved to enhance the toughness of a material [18] since easier interfacial decohesion provides an additional energy dissipative mechanism [19]. Although the intergranular cracks become longer, they are easily deflected by the interfaces and as a result they are spatially located with respect to a straight transgranular crack. On the other hand, transgranular mode is favourable when the resulting products are designed to have higher toughness. Due to the mixed fracture mode both IS1 and IS2 glass ceramics are expected to show mechanical performance that is between a performance of a predominantly strong and predominantly tough material.

Acknowledgements The author thanks Dr. Klaus Noller, Head of Department Materials Development, Fraunhofer Institute Process Engineering and Packaging, Freising, for providing the SiOx/PET samples.

References

1. Tsui TY, Vlassak J, Nix WD (1999) Indentation plastic displacement field: Part I. The case of soft films on hard substrates. J. Mater. Res. 14(6): 2196–2203
2. Kriese MD, Moody NR, Gerberich WW (1998) Effects of annealing and interlayers on the adhesion energy of copper thin films to SiO2/Si substrates. Acta Metall. Mater. 46: 6623–6630

3. Xiaodong L, Dongfeng D, Bhushan B (1997) Fracture mechanisms of thin amorphous carbon films in nanoindentation. Acta Mater. 45: 4453–4461
4. Bethmont D, Karimi A (2001) Proceedings of the Second International Conference on European Society for Precision Engineering and Nanotechnology, A. Balsamo et al. (Eds.) Augusta Edisioni Mortarino, vol. 1, Turin, 2001, p. 144
5. Luo F, Gao K, Pang X, Yang H, Qiao L, Wang Y (2008) Characterization of the mechanical properties and failure modes of hard coatings deposited by RF magnetron sputtering. Surf. Coat. Technol. 202(14): 3354–3359
6. Strnad Z (1986) Glass-Ceramic Materials, Elsevier Science, New York
7. Flinders M, Ray D, Anderson A, Cutler RA (2005) High-toughness silicon carbide as Armor. J. Am. Ceram. Soc. 88: 2217–2226
8. Zhou Y, Brito ME, Yang JF, Ohji T (2003) Fracture-mode change in alumina-silicon carbide composites doped with rare-earth impurities. J. Am. Ceram. Soc. 86: 1789–1792
9. Sun X, Li JG, Guo S, Xiu Z, Duan K, Hu XZ (2005) Intragranular particle residual stress strengthening of Al2O3–SiC nanocomposites. J. Am. Ceram. Soc. 88: 1536–1543
10. Cahn RW, Haasen P, Kramer EJ (1996) Processing of Ceramics – Part I, Materials Science and Technology Series vol. 17A, VCH editions
11. Oliver WC, Pharr GM (1992) An improved technique for determining hardness and elastic modulus using load and displacement sensing indentation experiments. J. Mater. Res. 7: 1564–1583
12. Kehagias Th, Komninou Ph, Kavouras P, Chrissafis K, Nouet G, Karakostas Th (2006) Crystal phase separation and microstructure of a thermally treated vitrified solid waste. J. Eur. Ceram. Soc. 26: 1141–1148
13. Kavouras P, Komninou Ph, Chrissafis K, Kaimakamis G, Kokkou S, Paraskevopoulos K, Karakostas Th (2003) Microstructural changes of processed vitrified solid waste products. J. Eur. Ceram. Soc. 23: 1305–1311
14. Kavouras P, Komninou Ph, Karakostas Th (2004) Effect of composition and annealing temperature on the mechanical properties of a vitrified waste. J. Eur. Ceram. Soc. 24: 2095–2102; Kavouras P, Charitidis C, Karakostas Th (2006) On the parameters affecting the fracture modes in glass-ceramic materials. J Non-Cryst. Sol. 352(52–54): 5515–5521
15. Beake BD, Leggett GJ (2002) Nanoindentation and nanoscratch testing of uniaxially and biaxially drawn poly(ethylene terephthalate) film. Polymer 43: 319–327; Charitidis C, Gioti M, Logothetidis S (2004) Nanomechanical and Optical Studies on Polymeric Membranes, 'Macromolecular Symposia by Wiley-VCH', 205: 239–249
16. Charitidis C, Karakasides TE, Kavouras P, Karakostas Th (2007) The size effect of crystalline inclusions on the fracture modes in glass–ceramic materials. J. Phys-Condens. Mat. 19: 266209 1–12
17. Kavouras P, Kehagias Th, Komninou Ph, Chrissafis K, Charitidis C, Karakostas Th (2008) Interface controlled active fracture modes in glass-ceramics. J. Mater. Sci. DOI 10.1007/s10853-007-2221-6
18. Zhai H, Huang Y, Wang C, Wu X (2000) Toughening by multiple mechanisms in ceramic-matrix composites with discontinuous elongated reinforcements. J. Am. Ceram. Soc. 83: 2006–2016
19. Hull D, Clyne TW, Clarke DR (1966), An Introduction to Composite Materials, Cambridge University Press, Cambridge

Local, Semilocal and Nominal Approaches to Estimate the Fatigue Strength of Welded Joints

Bruno Atzori, Paolo Lazzarin, and Giovanni Meneghetti

Abstract This paper is concerned with the analysis and the comparison of several so-called "local approaches" to estimate the fatigue strength of welded joints. The considered approaches are based on numerical analyses or strain gauges evaluations of the stress or strain state in the vicinity of the weld toe. The notch stress intensity factor (NSIF) approach applied to fillet-welded joints, as far as the opening angle between the weld and the main plate surface is constant (e.g. 135°, typical for many fillet welds), is able to rationalise the fatigue strength data both for different joint geometries and absolute dimensions. The NSIF approach has been previously developed as an extension of the Linear Elastic Fracture Mechanics (LEFM) to open V notches and is based on the exponential local stress field around the V-notch tip. Several proposed "local approaches", although simpler to use than the NSIF approach, are based on the stress (or strain) field beyond the exponential local one. To distinguish such approaches from the NSIF-based one, we define the former ones as semilocal approaches while the latter is a local approach. The paper underlines that the ability to unify fatigue strength data in a single scatter band is not the same for the semilocal approaches if compared to the local approach.

Keywords Local approaches · Structural stress approach · Nominal stress approach · Notch-Stress Intensity Factor

1 Introduction

The weakest points for as-welded joints under fatigue loading are the toe and the root of the welds. The local approaches are usually suggested for failures starting from the toe and only this case will be considered in this paper. Moreover, the local

B. Atzori (✉) and G. Meneghetti
Department of Mechanical Engineering, University of Padova, via Venezia, 1-35131 Padova, Italy
e-mail: bruno.atzori@unipd.it

P. Lazzarin
Department of Management and Engineering, University of Padova, Stradella San Nicola, 3-36100 Vicenza, Italy

S. Pantelakis, C. Rodopoulos (eds.), *Engineering Against Fracture: Proceedings of the 1st Conference*,

approaches are usually adopted to estimate the crack initiation and short crack propagation life, and to this criterion the present paper will refer. Nevertheless as far as fatigue tests on laboratory specimens are concerned, this can coincide with the total life of the specimen (up to complete failure), as it will be seen later.

The use of a so-called hot spot strain range rather than the nominal applied stress range started in the 1960s, see e.g. the pioneering work of Haibach [1] and was developed especially for the analysis of fatigue strength of tubular connections. The items concerned with design by analysis in Division 1 of Section III as well as Division 2 of Section VIII of the 1973 edition of the ASME Boiler and Pressure Vessel Code [2] may be cited as the first design criteria which introduced the hot spot strain concept. In the Code, the hot spot strain is termed the peak stress, which is defined as "the highest stress in the region under consideration". Differently from these approaches Haibach adopted, as a parameter unifying the fatigue strength, the strain amplitude measured with a 3 mm long strain gauge positioned close to the weld toe. As clearly pointed out by Iida [3], there were at the beginning different approaches suggested for the application of the same concept, first for the possibility of using stress or strain and second for the choice of the point where to evaluate theoretically or experimentally the stress or the strain. Today, when speaking of high cycle fatigue of welded joints, an elastic behaviour is usually assumed and stress and strain can be used indifferently, since they are linked by the modulus of elasticity E for plane stress (or $E' = \frac{E}{1-\nu^2}$ for plain strain) conditions. Moreover, the hot spot stress is usually defined as the extrapolation to the weld toe of the values measured at given distances (function of the thickness of the main plate) from the weld toe, as shown in Fig. 1 [4].

The prediction of the fatigue strength by the local strain approach introduced by Haibach has been analysed by several authors, which adopted strain gauges having 0.2–3.0 mm grid length, located at distances of 0–5 mm from the weld toe. In each research the dimensions of the strain gauge and its distance from the weld toe were kept constant, differently from the hot spot approach where the distances change with the main plate thickness.

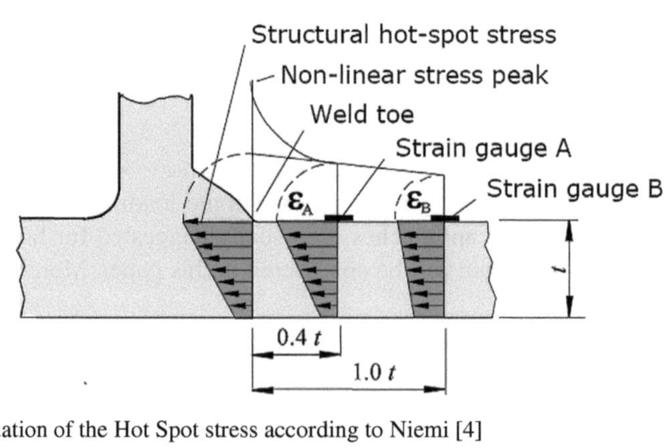

Fig. 1 Evaluation of the Hot Spot stress according to Niemi [4]

Later, a different concept was suggested by several researchers independently (Usami [5], Atzori [6–8], Verreman [9]), who proposed to consider the geometric transition from the main plate to the weld as an open V notch with apex radius $\rho = 0$. This concept was theoretically developed later by Verreman [10] and by Lazzarin and Tovo [11], who gave the stress field expressions around the weld toe in terms of the mode I and the mode II different contributions. In particular the singular term is characterised by a degree of singularity dependent on the opening angle, according to a previous solution provided by Williams for the general case of open V notches [12]. The weld geometry more frequently assumed in the applications has an opening angle of 135° and a degree of singularity of 0.326.

Alternative local approaches were proposed by Radaj [13], Taylor [14] and Lazzarin [15, 16]. For steel joints Radaj assumed an apex radius $\rho = 1$ mm for every geometry and size of the weld, in such a way to substitute the NSIF with an equivalent peak value of the stress, more familiar to many engineers. Taylor based the prediction of the fatigue strength on the stress evaluated at a critical distance, which is characteristic of the material, that he found to be equal to 0.215 mm for welded joints in low carbon steel. He also proposed to use as an alternative the mean stress along a line or an area around the weld toe, that, for the same case, he found to be equal to a length of 0.86 mm and to an area with a radius of 0.43 mm respectively. Lazzarin considered as the driving parameter for the prediction of the fatigue strength the elastic Strain Energy Density evaluated on a control volume around the notch tip, that he evaluated as the one corresponding to a radius $R_0 = 0.28$ mm for steel weldments.

However, it should be underlined that the problem of the notch effect on a welded joint is not simple, since several effects are superimposed and their combination causes the intensity of the local stress field. As an example, in the case of transversal welds the notch effects due to the welded joint itself act: in the plane of the main plate (as in the case of longitudinal stiffeners, spot welded joints etc.); in the longitudinal middle plane normal to the main plate, in all cases.

It has been already evidenced that for discontinuous welds the global stress field can be viewed approximately as the result of the superposition of the in-plane and the out-of-plane notches and that the in-plane notch effect is fully active at a distance from the transversal weld toe on the order of the main plate thickness [17, 18]. As a consequence this effect can be evaluated not only by the NSIF approach but also by the "local" strain approach and by the hot spot approach.

Recently it has been shown that also the out-of-plane notch effect can be viewed approximately as the superposition of a local sharp notch (due to the opening angle and to the weld leg length h) on a structural (or geometric) notch (due to the overall geometry of the joint), as shown in Fig. 2 [19, 20].

The stress field along the surface of the main plate of a typical welded joint can be plotted as a function of the distance from the weld toe normalised not only by the thickness of the plate [11], but also by the depth a of the V notch, as shown in Fig. 3, or by the leg length h, which is directly connected to the depth a.

From this figure the so-called local stress field clearly appears to have different features as the normalised distance from the weld toe increases, due to the nominal

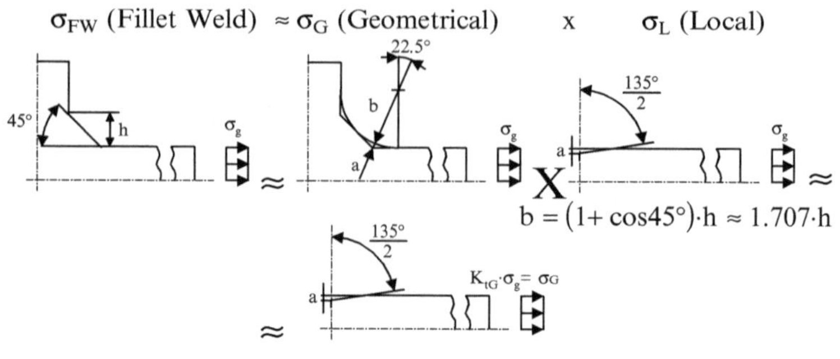

Fig. 2 Superposition of a sharp notch on a blunt notch

stress value or to the superposition of mode I and mode II contributions [11]. It should be noted that the extension of the exponential field due to the mode I only is much smaller (about one order of magnitude) along the main plate surface (about $\frac{x}{a} \leq 0.1$) than along the notch bisector (about: $\frac{r}{a} \leq 1.0$) as a consequence of the fact that along the notch bisector the normal stress field is influenced only by the mode I contribution and the shear stress field only by the mode II contribution. Since the range of applicability of the NSIF approach is limited to these small zones, we propose to apply only to them the denomination of "local stress field". Along the main plate surface and "far" from the weld toe, the stress will reach a zone not influenced by the weld joint itself but only by the overall geometry of the structure, that is a "structural stress field". The stress field bridging the mode I stress field and the structural stress field can be distinguished because it is characterised by the superposition of either mode I and mode II stress contributions. Our estimates, according to the FE results along the main plate surface, are the following for the case reported in Fig. 3:

$$\text{LOCAL STRESS (or STRAIN) FIELD}: \frac{x}{a} < 0.1$$
$$\left(\frac{x}{t} < 0.007\right);$$

$$\text{SEMI-LOCAL STRESS (or STRAIN) FIELD}: \frac{x}{a} = 0.1 \text{ to}$$
$$10.0 \left(\frac{\mathrm{x}}{\mathrm{t}} \approx 0.007 \text{ to } 0.7\right);$$

$$\text{STRUCTURAL STRESS (or STRAIN) FIELD}: \frac{x}{a} > 10$$
$$\left(\frac{\mathrm{x}}{\mathrm{t}} > 0.7\right)$$

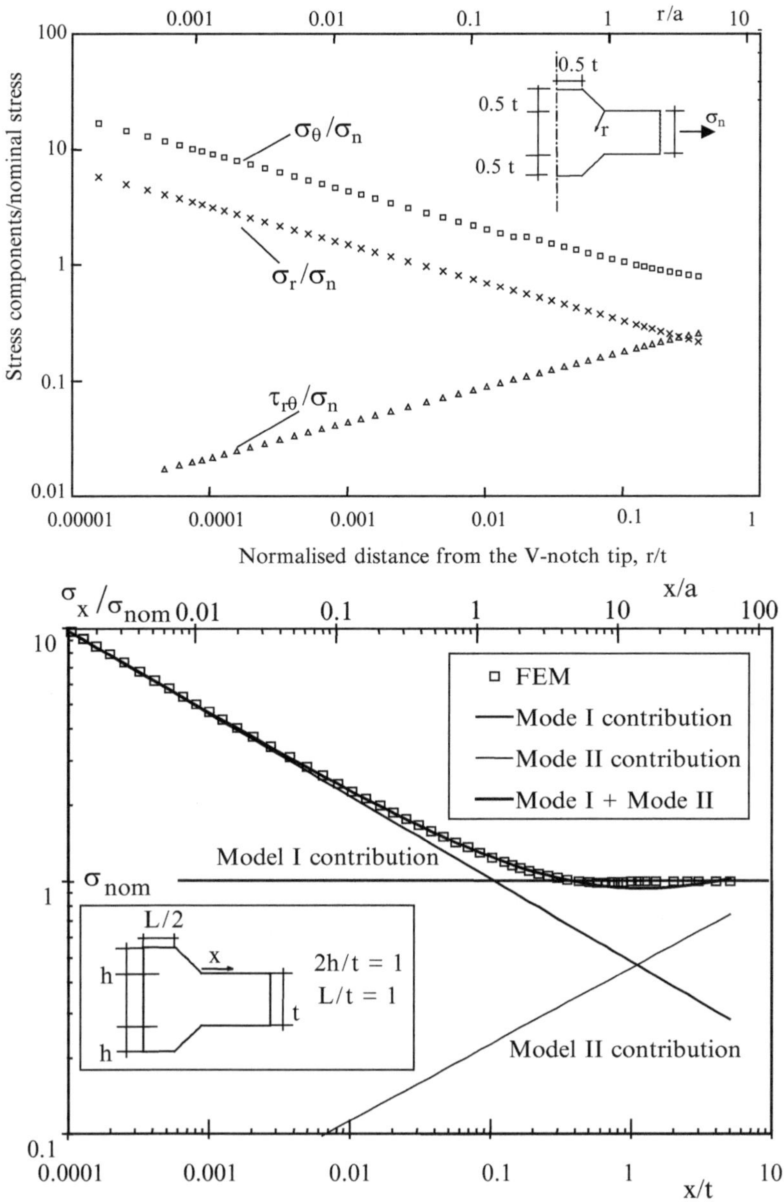

Fig. 3 Mode I stress components (σ_θ and σ_r) and mode II stress component ($\tau_{r\theta}$) along the V-notch bisector **a**; mode I + mode II stress component σ_x along the main plate surface **b** [11]

It appears clearly that the structural stress is only a better definition of the nominal stress and that they could coincide or diverge. For example for continuous transversal welds the two stresses are the same, while as an example for plates with longitudinal stiffeners they are different (since the structural stress takes into account the in-plane stress concentration effect).

On the basis of the distances used for their application we propose also to distinguish in the same way the previously recalled approaches for the fatigue strength evaluation, as:

LOCAL APPROACHES: the approaches for fillet-welded joints based on stresses or strains determined inside the zone governed by the mode NSIF;

SEMI-LOCAL APPROACHES: the practical applications of different approaches based on stresses or strains outside the zone strictly controlled by the mode I NSIF, as it happens in the "local" strain gauge approache;

STRUCTURAL APPROACHES: the Hot Spot approach and its practical applications

These definitions could be applied also to FE evaluations along the V-notch bisector, considering that in this case the local stress field will have an extension of about one order of magnitude larger than along the main plate surface.

In the following analysis of these approaches we will refer to steel welded joints. For the nominal strength we will refer to the values $\Delta\sigma_C$ of the stress range of the classes ($N = 2 \cdot 10^6$ cycles, PS = 95%) given by the Eurocode 3 [21]. For the local fatigue strength at a given distance from the weld toe we will adopt here the unified scatter-band proposed by Atzori and Meneghetti [22], as shown in Fig. 4 in non-dimensional scales.

Figure 5 shows the application of this scatter band to a set of 116 experimental fatigue test results, after evaluation by FEM of the stresses at a distance x = 0.01 mm from the weld toe along the main plate surface according to the NSIF concept [22].

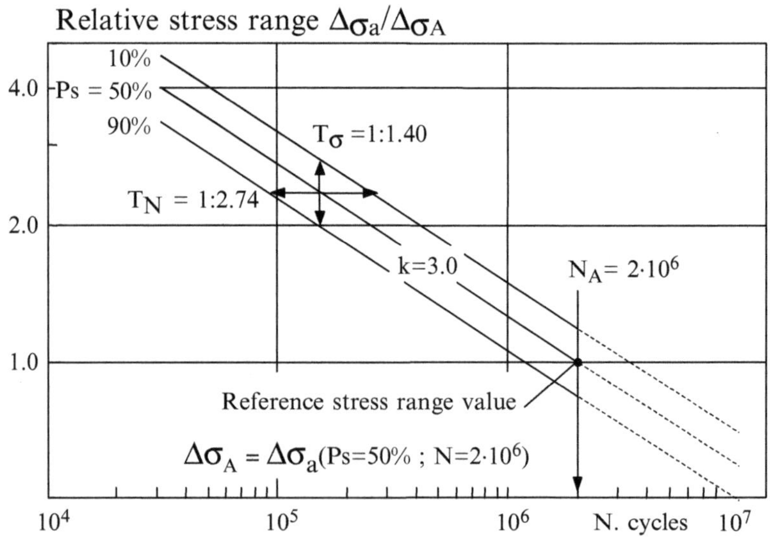

Fig. 4 Proposed unified scatter band for fillet welded joints in structural steel

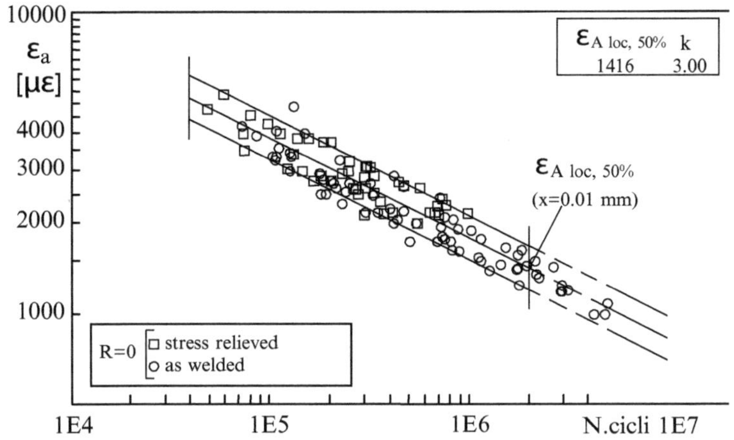

Fig. 5 Application of the proposed unified scatter band to the experimental results reported in [22]

As it is evident from the figure, for simple joint configurations the NSIF approach is able to unify the total fatigue life, since for the specimens analysed in Fig. 5 most of the fatigue life is spent in initiation and short crack propagation within the stress field of the V-notch governed by local stresses, as already clarified by Lazzarin [23]. This is not true in general for real structures or welded joints having a complex geometry, where much longer cracks may occur and a significant part of the total life can be spent to propagate the cracks outside the zone governed by the NSIF. For this reason the values given in Fig. 5 should be used in these more complex cases to estimate only the fatigue life for the initiation of a technical crack.

2 Parameters Affecting the Applicability of the Approach

In the case of as welded steel structures, the parameters which affect the applicability of the considered approach to the fatigue strength evaluation of a welded joint are:

- The value of the applied range of nominal stress $\Delta\sigma$
- The dimension of the welded joint
- The geometry of the welded joint

The influence of the applied stress range on a welded joint of a given geometry and dimension (e.g. as the one already considered in Fig. 3) is shown qualitatively in Fig. 6, considering the stress distribution along the main plate surface for several applied nominal stresses, as it would be calculated for example by means of linear elastic finite element analyses. Since the stress field simply translates in the vertical direction, the shape and the slope of the SN line will not be influenced by the

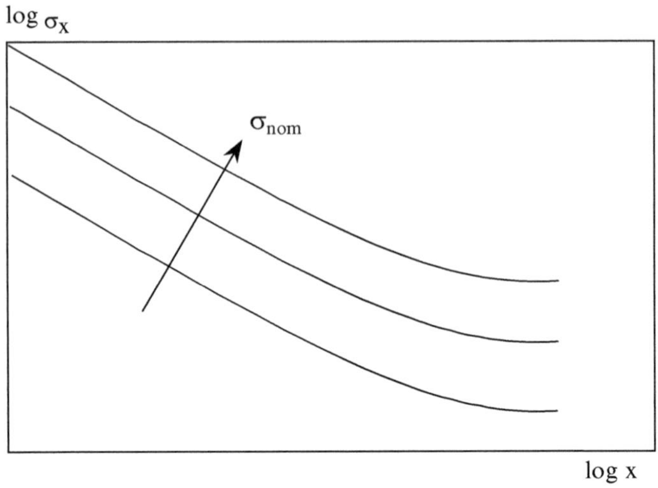

Fig. 6 Influence of applied nominal stress on a welded joint of given geometry and dimensions

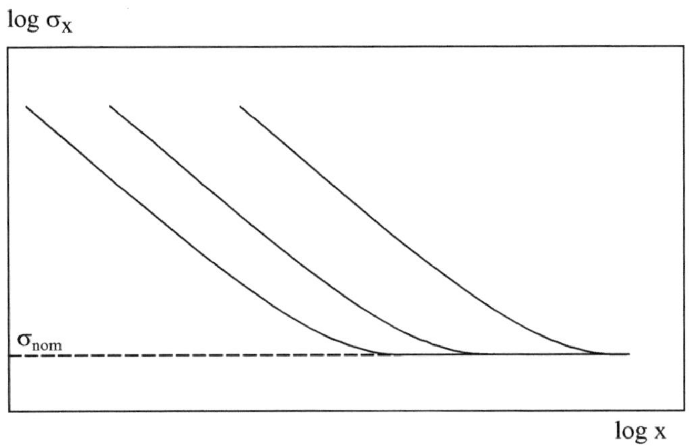

Fig. 7 Influence of absolute dimensions on the stress field of a welded joint for the same applied nominal stress

considered approach (that is by the distance x at which the stress is evaluated). Also the SN line will simply translate in the vertical direction when a different approach is applied.

The influence of a variation of the dimension of a welded joint of given geometry (size effect) is shown qualitatively in Fig. 7, where the stress distribution along the main plate surface is reported for a constant nominal stress applied to joints having the same geometry but different sizes. In this case the consequence can be seen as a translation of the stress field in the horizontal direction. The ratio of the stress values evaluated at different distances x will be the same only for the small values of the distance x where the stress field is fully governed by the mode I for all the considered sizes. Considering now the Eurocode 3, where the scale effect is considered only for thicknesses greater than 25 mm, the maximum distance from the weld toe x, at which local stresses for the application of the NSIF approach should be evaluated, will be of about 0.2 mm for the geometry reported in Fig. 3, which is much smaller that the one compatible with a strain-gauge measurement. The experimental strain approach is then clearly based on a mixed local stress field at most and cannot be expected to unify the fatigue strength data for welded joints of different sizes, but could only be applied in an approximate way for classes of size, each of them characterised by different S-N curve, as already suggested in the past on a phenomenological basis [24, 25].

Different conclusions should be taken when speaking of computational approaches, since they could be based on stress values evaluated along the notch bisector, where the local stress field is about one order of magnitude higher (in the case of $t = 25$ mm, about 2 mm). Anyway care should be taken when choosing the finite element size (since the extension of local stress field around the weld toe, i.e. the only one useful for a unifying local approach, depends both on the size and on the shape of the joints and then defines the maximum allowable size of the elements of the mesh used for the analysis). It is also evident from Fig. 7 that the size effect is not included in a Hot Spot or in a nominal stress approach, and that in this case an appropriate coefficient should be introduced for this purpose. In deriving a size effect coefficient from the diagram of Fig. 7, it should be emphasized that this diagram refers to completely similar joints, which is a condition that is hardly assured in real structures. The ratio h/t, which has been chosen equal to 0.5 in Fig. 3 is usually higher for low values of t and lower for high values of t. The comparison should then take into account that joints of the same type but of different sizes are not similar in real applications [11, 26]. Since the local stress field can be thought of as caused by the local notch of depth a, as already shown in Fig. 2, the scale effect for fatigue strength might be expressed in terms of leg length h as $\Delta\sigma_{c2} = \alpha \cdot \Delta\sigma_{c1} \left(\frac{h_1}{h_2}\right)^{0.326}$, with $\alpha = 1.0$ valid only as a first approximation.

The influence of the geometry of the welded joint is shown in Fig. 8, considering the stress distribution along the longitudinal axis on the main plate surface, for a transversal and for a longitudinal attachment, assuming the same main plate thickness, the same leg length and the same attachment length for the two joints [27]. The comparison is made by imposing the same nominal stress and the figure highlights the differences in the structural and in the local stress field. It appears that the two curves looks different since the longitudinal stiffeners generates an in-plane stress concentration effect which rises the stress field beyond the nominal value and which defines the structural stress field different from and higher than the nominal one. Conversely in the transverse attachment joint the structural stress field coincide with the nominal one.

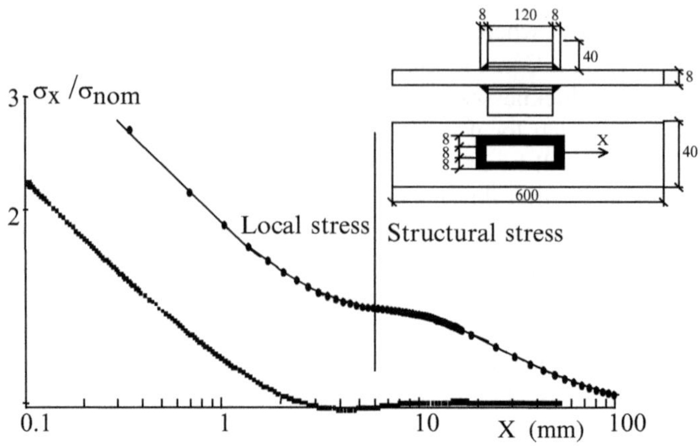

Fig. 8 Influence of joint geometry on the stress field for the same applied nominal stress [27]

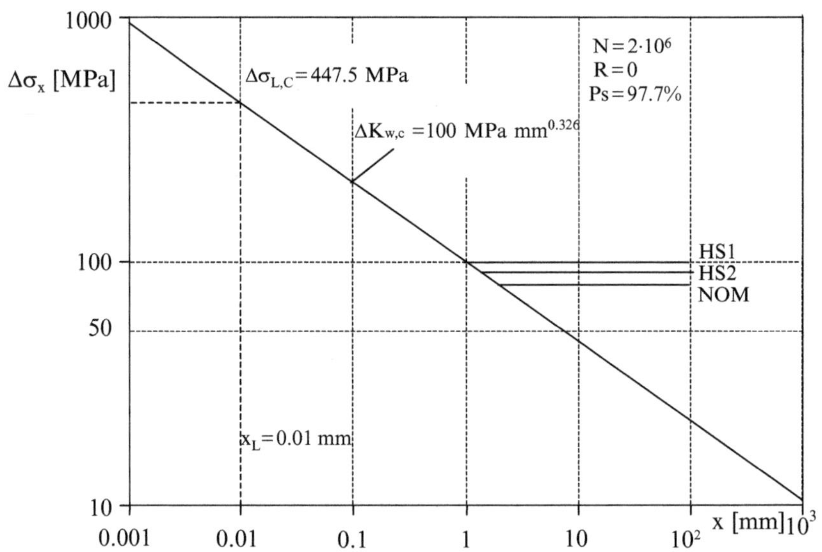

Fig. 9 Comparison of the nominal strength given by Eurocode 3 [22] for some joint geometries with the suggested values for the Hot Spot approach [4] and the corresponding critical local stress field on the surface of the main plate according to NSIF approach [28]. HS1: Hot Spot (full penetration butt and cruciform and load-carrying welds) $\Delta\sigma_C = 100\,\text{MPa}$; HS2: Hot Spot (fillet welds butt and cruciform and load-carrying welds) $\Delta\sigma_C = 90\,\text{MPa}$; NOM1: Eurocode 3 (butt and cruciform welds, longitudinal attachments not longer than 50 mm) $\Delta\sigma_C = 80\,\text{MPa}$

As a conclusion Fig. 9 shows the stress fields that can be expected for a life of $2 \cdot 10^6$ cycles on the surface of the main plate of joints of different geometries (with a thickness of the main plate of 25 mm for the nominal and hot spot values) according to Eurocode 3 and to NSIF and Hot Spot available data.

3 Conclusions

In this paper local, semi-local and nominal approaches have been distinguished in the context of fatigue strength evaluation of welded joints. According to the local approach based on the mode I NSIF, a unique fatigue design scatter band valid for as-welded fillet welded joints in construction steels having main plate thickness ranging from 6 to 100 mm can be proposed. Differently, hot spot approaches based on experimental strain measurement in proximity of the weld toe are not able to include the scale effect: this result in different design curves for different ranges of main plate thickness.

References

1. Haibach E (1968) Die Schwingfestigkeit von Schweissverbindungen aus der Sicht einer örtlichen Beanspruchungsmessung. LBF Report No. FB-77, Darmstadt.
2. Rules for Construction of Nuclear Plant Components (1973) Division 1 of Section III (Pressure Vessels) of the ASME Boiler and Pressure Vessel Code.
3. Iida K (1983) Application of hot-spot strain concept to fatigue life prediction. IIW Doc. XIII-1103-83.
4. Niemi E (2000) Structural hot-spot stress approach to fatigue analysis of welded components. IIW Doc XIII-1819-00.
5. Usami S, Kimoto H, Kusumoto S (1978) Cyclic strain and fatigue strength at the toes of heavy welded joints. Transaction of the Japan Welding Society 2: 118–127.
6. Atzori B, Dattoma V (1977) Parametric stress analysis of a cruciform joint under axial loading. Rivista Italiana della Saldatura 4: 187–196 (in Italian).
7. Atzori B, Giordano G, Pappalettere C (1983) Fatigue strength prediction of mechanical components with sharp notch effects. XI AIAS National Conference, Genova, Italy.
8. Atzori B, Pappalettere C (1986) A method for the evaluation of the strength of mechanical structures is applied to the design of a single lap joint using numerical and experimental techniques. In: Proceedings of the International Conference on Fatigue of Engineering Materials and Structures, Sheffield.
9. Verreman Y, Dickson JI, Bailou JP (1989) Generalization of the Kitagawa diagram to V-notched members. Advances in Fatigue Science and Technology/edited by C. Moura Branco and L. Guerra Rosa. Dordrecht : Kluwer, c1989, NATO ASI Series E, 159: 785–798.
10. Verreman Y, Nie B (1996) Early development of fatigue cracking at manual fillet welds. Fatigue Fract Eng Mater Struct 19: 669–681.
11. Lazzarin P, Tovo R (1998) A notch intensity factor approach to the stress analysis of welds. Fatigue Fract Eng Mater Struct 21: 1089–1103.
12. Williams ML (1952) Stress singularities resulting from various boundary conditions in angular corners of plates in extension. J Appl Mech 19: 526–528.
13. Radaj D (1990) Design and Analysis of Fatigue Resistant Welded Structures. Abington Publishing, Cambridge.
14. Taylor D, Barrett N, Lucano G (2002) Some new methods for predicting fatigue in welded joints. Int J Fatigue 24: 509–518.
15. Lazzarin P, Zambardi R (2001) A finite-volume-energy based approach to predict the static and fatigue behaviour of components with sharp V-shaped notches. Int J Fracture 112: 275–298.
16. Livieri P, Lazzarin P (2005) Fatigue strength of steel and aluminium welded joints based on generalised stress intensity factors and local strain energy values. Int J Fracture 133: 247–276.
17. Lazzarin P, Santanera A (1986) Parametric analysis of the notch effect in longitudinal stiffners. AIAS National Conference, Catania, Italy.

18. Blarasin A, Atzori B, Filippa P, Gastaldi G (1990) Design fatigue data for joining steel sheets in the automotive industry. SAE paper No 900742.
19. Atzori B, Meneghetti G (2002) Evaluation of the fatigue strength of welded joints in structural steels by means of an equivalent notch stress analysis. In: Proceedings of the International Conference on New Trends in Fatigue and Fracture, Metz, France. On CD ROM.
20. Atzori B, Meneghetti G, Susmel L (2002) Estimation of the fatigue strength of light alloy welds by an equivalent notch stress analysis. Int J Fatigue 24: 591–599.
21. EN 1993-1-9 Eurocode 3: Design of steel structures, Part 1–9: Fatigue, August 2005.
22. Atzori B, Meneghetti G (2001) Fatigue strength of fillet welded structural steels: finite elements, strain gauges and reality. Int J Fatigue 23 (8): 713–721.
23. Lazzarin P, Livieri P (2001) Notch stress intensity factors and fatigue strength of aluminium and steel welded joints. Int J Fatigue 23: 225–232.
24. Casavola C, Nobile R, Pappalettere C (2005) A local strain method for the evaluation of welded joints fatigue resistance: the case of thin main plates thickness. Fatigue Fract Eng Mater Struct 28: 759–767.
25. Casavola C, Pappalettere C (2005) Application of WEL.FA.RE. method on aluminium alloys welded joints. In: Proceedings of the 2005 SEM Annual Conference and Exposition on Experimental and Applied Mechanics.
26. Atzori B, Meneghetti G (1998) A physical interpretation of the scale effect in fatigue of fillet welded aluminium alloys. In: Proceedings of the 7th International Conference on Joints in Aluminium INALCO, Cambridge, UK.
27. Meneghetti G, Tovo R (1998) Stress gradient effect of fatigue strength of light alloy weldments. In: Proceedings of the 7th International Conference on Joints in Aluminium INALCO, Cambridge, UK.
28. Atzori B, Lazzarin P, Tovo R (1999) Stress field parameters to predict the fatigue behaviour of notched components. J Strain Anal Eng 34: 437–453.

Superficial Strength Properties Modification of 2024 Aluminum Specimens Subjected to Cyclic Loading, Detected by Nanoindentations

Konstantinos-Dionysios Bouzakis, Ioannis Mirisidis, Spiros G. Pantelakis, and Paraskevas V. Petroyiannis

Abstract Structural elements and components of planes are mainly manufactured by aluminum alloys due to weight restrictions. According to experimental results, it has been observed that specimen regions of aluminum alloys which have been stressed at various loads, potentially leading to fatigue damages, have higher hardness compared to regions in which no fatigue mechanisms have been activated. With the aid of nanoindentations, the material hardness increase due to fatigue was captured, especially when a fatigue fracture approaches and the remaining safe operational time of structural components can be estimated. Moreover, stress–strain curves associated with the superficial mechanical properties of the examined materials are determined, employing a developed FEM-supported evaluation procedure of nanoindentation experimental results. The procedure was applied to obtain the superficial strength properties of the alloy during fatigue loading. By means of these techniques, valuable conclusions on the fatigue mechanism development of various plane structural elements and components can be drawn.

Keywords Nanoindentations · Fatigue · Stress–strain curves

1 Introduction

Aircraft structural components are subjected to fatigue loading. Failure by fatigue continues to remain the most serious concern for structural failure of aircraft components despite the exhaustive amount of past research. It is difficult to predict an

K.-D. Bouzakis (✉), I. Mirisidis, S.G. Pantelakis, and P.V. Petroyiannis
Laboratory for Machine Tools and Manufacturing Engineering,
Mechanical Engineering Department, Aristotle University of Thessaloniki, Greece
e-mail: bouzakis@eng.auth.gr

Laboratory of Technology and Strength of Materials,
Department of Mechanical Engineering & Aeronautics, University of Patras, Greece

S. Pantelakis, C. Rodopoulos (eds.), *Engineering Against Fracture: Proceedings of the 1st Conference*,

accurate fatigue failure because the different physical processes, which reign to the gradual fatigue damage accumulation during the fatigue life of a metallic component, are complex and interrelated. They develop with increasing number of fatigue cycles from atomic to macro scale damage mechanisms and also entail a host of material, geometric and loading parameters. Additionally, the increasing trend in aeronautical industry to replace the existing differential by integral structures increases the need to early detect, whether there is a transfer of fatigue loads from structural elements, which are designed to sustain certain fatigue loads (e.g. stringer) on structures, which are not supposed to sustain these fatigue loads (e.g. skin).

Plasticity at the material layers, which are near the surface, at the atomic-, nano- and mesoscale is the prevailing damage mechanism at the early stages of high cycle fatigue damage accumulation. The convention adopted in the investigations for the scales is $1 * 10^{-4}$ to $1 * 10^{-3}\,\mu m$ for the atomic-scale, $1 * 10^{-3}$ to $1 * 10^{-1}\,\mu m$ for the nano-scale and $1 * 10^{-1}$ to $10\,\mu m$ for the meso-scale. It is sensible to expect that accompanying these local structural changes, there are changes in the local stress–strain properties and respective changes in hardness, which should depend on the magnitude of the fatigue stresses as well as the number of applied cycles. Micro-hardness measurements, conducted on 2024 T3 alloy specimens fatigued at reversed bending, have confirmed the increase of hardness at the micro-scale following fatigue loading [1,2]. Yet, as the changes are limited within a narrow surface material layer classical hardness measurements lack the necessary sensitivity to identify and even less to quantify this phenomenon. For instance, the depth of penetration for the square pyramid indenter used for the Vickers micro-hardness test is more than 0.5 mm. The lack of sufficient sensitivity of the performed hardness measurements has been reflected into appreciable scatter of the experimental data presented in [1], thus making it difficult to exploit the technique for practical purposes. The ability of measuring material hardness changes at the nano- and meso-scale will be exploited to develop a concept for the early detection of fatigue damage accumulation in metallic structures.

The present paper investigations are aiming to exhibit the feasibility of developing relations between the remaining fatigue life of aluminum alloy specimens subjected to constant amplitude fatigue tests and the surface hardness changes which occur during fatigue, on the basis of nanoindentations. With the intention of accomplishing the above objective, material's surface hardness investigations at the nano- and meso-scale during fatigue loading were conducted. The dependency of hardness changes on both, the number of applied fatigue cycles and value of the applied fatigue stress, were derived using a FEM supported evaluation procedure of nanoindentation experimental results. Stress–Strain curves associated to the superficial properties of the alloy were resultant and the gradual evolution of the alloy's superficial strength properties during fatigue loading could be determined.

2 Specimens' Superficial Hardness Increase at Repeatedly Loads Through Nanoindentation Measurements

In order to monitor the hardness increase after the cyclic stressing of aluminum sheet specimens at various loads, nine fatigue specimens, as shown in Fig. 1, were prepared according to the ASTM specification E466-82 from 3 mm thick bare 2024 T3 Aluminum alloy sheets. The specimens were machined in longitudinal (L) orientation relative to the rolling direction. The fatigue tests were performed according to the ASTM E466-82 specification. The tests were performed for constant stress ratio $R = 0.1$ for different applied fatigue stresses. The frequency used was 25 Hz. An overview of the performed fatigue tests is given in the table included in Fig. 2. In order to investigate the hardness increase during fatigue loading the fatigue tests were interrupted at certain numbers of fatigue cycles, which refer to different percentages of the fatigue life, and hardness measurements were carried out at different specimen locations [3]. Each fatigue test has been interrupted six times to conduct nanohardness measurements.

The determination of the specimen hardness and of the associated superficial strength properties modifications can be carried out accurately by means of nanoindentations [4–6]. The nanoindentation is a precise method to monitor continuously the course of the applied on a diamond indenter force versus the occurring penetration depth. This measurement consists of two steps; the so-called loading stage and the unloading one (see Fig. 3). During the loading stage, a load forces the diamond indenter to penetrate into the specimen. The load is gradually applied and at the same time the indentation depth is measured, as presented in the upper right part of the figure. By means of a developed FEM continuous simulation of the indentation procedure, coating stress–strain curves can be determined. The related FEM model is qualitatively illustrated in the bottom left part of the same figure. The actual indenter tip form differs from the ideal sharp one. A sufficient approximation of the cross section contour of this area is by a spline curve, with two tangents, with the magnitudes b and th as shown in figure. The parameter b is the width of the indenter tip area with form deviations and th the corresponding height.

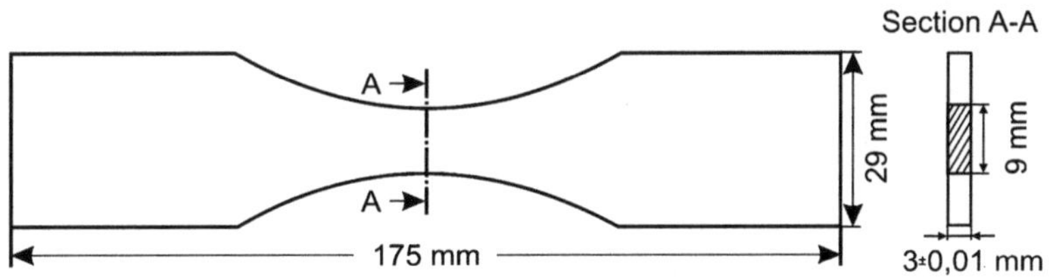

Fig. 1 Geometrical data of the examined aluminum sheet specimens

Load case	Spec. Nr.	f (Hz)	F_{max} (N)	F_{min} (N)	F_a (N)	F_m (N)	σ_{max} (N)	Overall Oper. cycles	Comment
1	1a	25	4,860	486	2,187	2,673	180	5,000,000	Interrupted
	1b	25	4,860	486	2,187	2,673	180	1,577,460	Broken
	1c	25	4,860	486	2,187	2,673	180	5,000,000	Interrupted
2	2a	25	5,400	540	2,430	2,970	200	428,600	Broken
	2b	25	5,400	540	2,430	2,970	200	919,865	Broken
	2c	25	5,400	540	2,430	2,970	200	1,349,290	Broken
3	3a	25	6,750	675	3,037	3,712	250	208,692	Broken
	3b	25	6,750	675	3,037	3,712	250	184,582	Broken
	3c	25	6,750	675	3,037	3,712	250	182,365	Broken

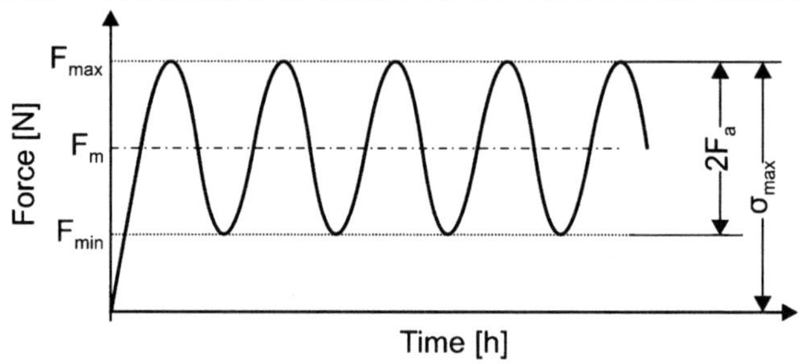

Fig. 2 Test data of the examined aluminum sheet specimens

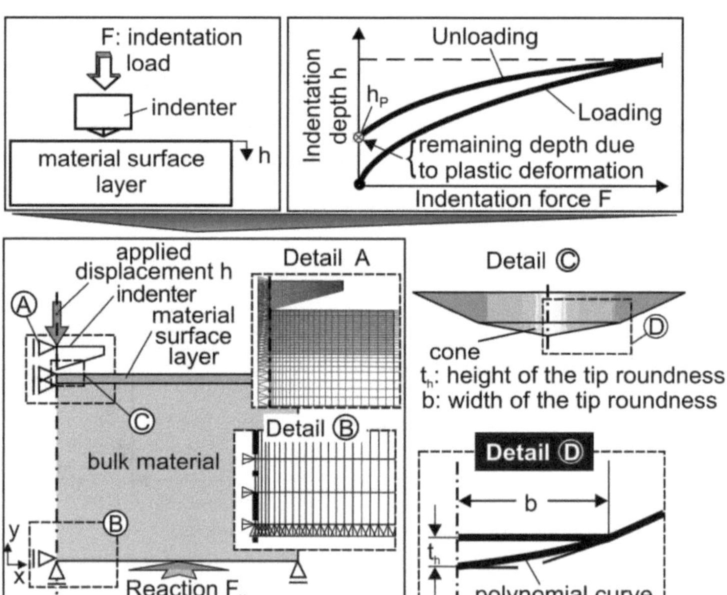

Fig. 3 FEM continuous simulation of nanohardness measurement to extract coatings stress–strain curves and description of the actual indenter tip geometry

3 Indentation Depth Alteration Versus the Operational Cycles and Local Material Stressing at Various Stress Gradients

Figure 4 shows the indentation depth versus the operational cycles in the examined specimen areas, after various operational cycles. As the operational cycles increase, the maximum indentation depth decrease and thus the superficial indentation hardness grows. The superficial indentation hardness in area 4 is at most

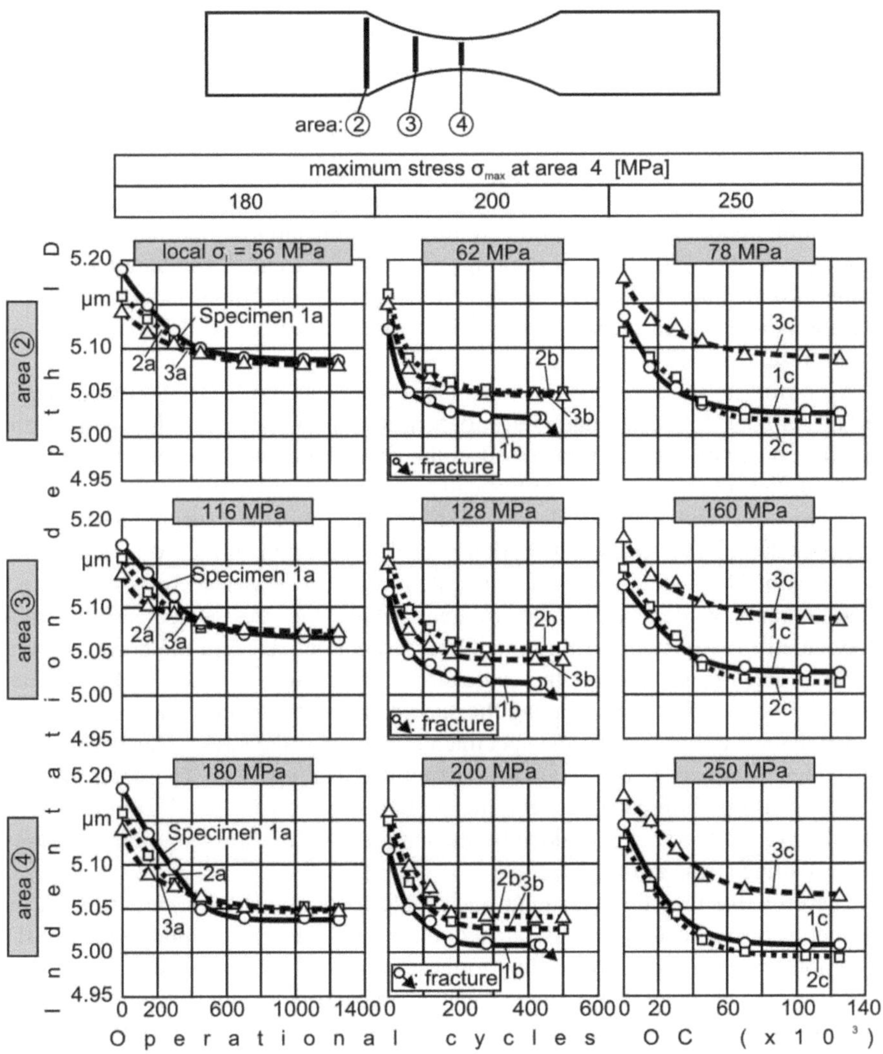

Fig. 4 Nanoindentation depth results at various operational cycles in all examined areas for all specimens

increased, in comparison to all other areas, due to the smallest cross section of that area and the specimen fracture takes place there. In this region the indentation depth decrease, i.e. the indentation hardness growth is larger, as compared to the rest-measured areas and amounts approximately to 150 nm in some specimen cases. For example in area 3, the alteration of the indentation depth before loading and after 140,000 operational cycles at a maximum stress of 250 MPa in region 4, lies between 100–120 nm, depending on the specimen case. The corresponding scatter in area 2 amounts to 80–100 nm. After a certain number of operational cycles (NSID) depending on the maximum stress, the superficial hardness is stabilized and seems to remain constant at further loadings. Hence, as the applied maximum stress increases the maximum indentation depth decreases, i.e. the indentation hardness grows. It is worth noting that, at a low local stress of 78 MPa in area 2, corresponding to a maximum stress of 250 MPa in region 4, a high indentation hardness alteration occurs, comparable to the one of region 4, after a relatively small number of 90,000 applied fatigue cycles. On the other hand, in area 3, a higher local stress of 116 MPa, corresponding to a maximum stress of 180 MPa in region 4, leads to a lower overall indentation depth alteration, stabilized after a significantly higher number of operation cycles of ca. 800,000. At the opposite areas of 2 and 3, the nanoindentation results revealed approximately the same tendencies; concerning indentation depth alterations and the corresponding deviations were always less than 10 nm, as already presented in [2].

According to these results the indentation depth alteration is stabilized after a certain number of operational cycles, depending on the occurring local stress and simultaneously on the maximum load in the critical area 4. This remarkable result needs further investigation to be physically interpreted and understood.

In order to elucidate this tendency the stress gradient in the x specimen's direction was considered. The stress distribution over the specimen surface was determined by means of FEM calculations, as illustrated in Fig. 5. The maximum stress is applied in area 4, while as draw away to areas 3 and 2 the stresses decrease. Because of symmetry reasons concerning the x and y specimen axes, in the corresponding mirrored areas to these axes, the same stress values appear. The FEM calculated equivalent stress distribution, along the specimen length direction, at various loads is also demonstrated. Due to the specimen's geometry, the stress distribution is almost linear between areas 2 and 4, in all examined loading cases. The stress gradient is defined as the difference between the applied stress $\sigma 4$ in area 4 from the corresponding one $\sigma 2$ in area 2, divided by the length l24. The stress gradient increases, as the loading stress grows.

The effect of the stress gradient and of the local applied fatigue stress on the number of operational cycles, associated to the stabilized indentation depth (NSID) is demonstrated in Fig. 6. According to these results, at a constant local stress level σmaxl, as the stress gradient increases the stabilized indentation depth SID grows as well, whereas the operational cycles up to IDA stabilization NSID decrease. On the other hand, at the same NSID, the local stress growth leads to a SID increase.

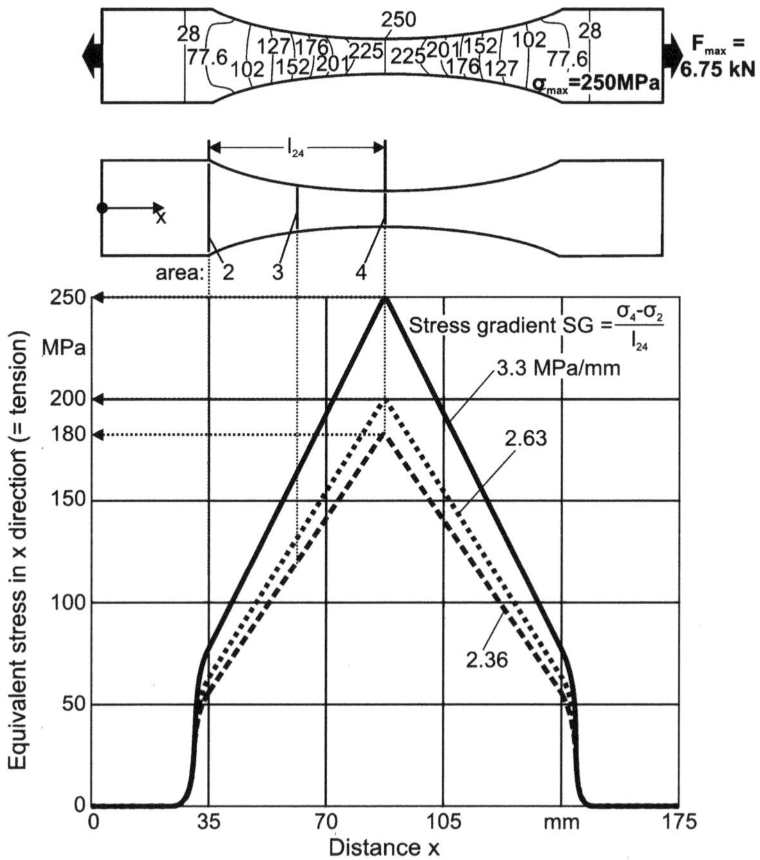

Fig. 5 FEM calculated stress distribution over x specimen direction for all loading cases

4 Superficial Strength Properties Alteration Versus the Operational Cycles and Local Material Stressing at Various Stress Gradients

With the aid of a FEM supported continuous simulation of the nanoindentation [4–6], material constitutive laws can be determined. Figure 7 illustrates determined stress–strain curves, associated to superficial material mechanical strength data, in the indicated specimen areas, at a maximum stress of 250 MPa, after 90,000 operational cycles. As approaching the critical area 4, a yield and maximum strength increase can be observed. Hereupon the corresponding yield and maximum stress growth depend on the measured indentation depth alterations. Corresponding stress–strain curves were extracted in specimen areas after various operational cycles and loads. These results show that the superficial strength properties increase and

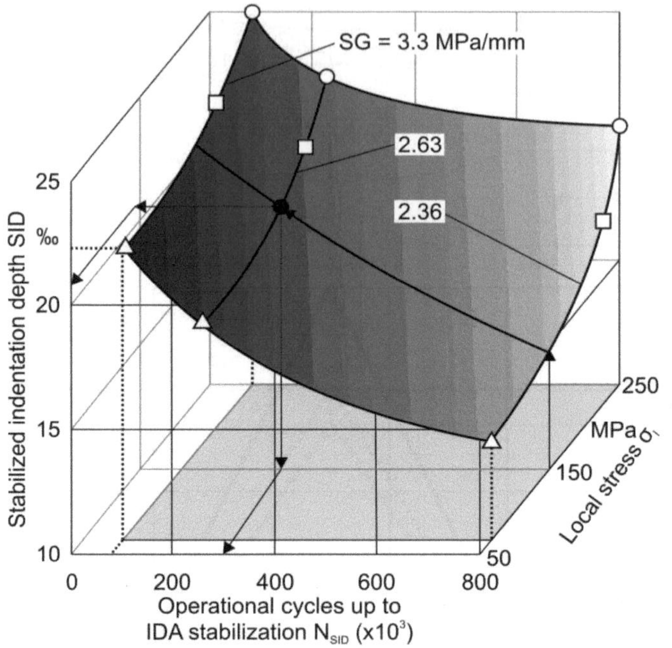

Fig. 6 Effect of local stress σlocal, NSID and stress gradient on stabilized indentation depth SID

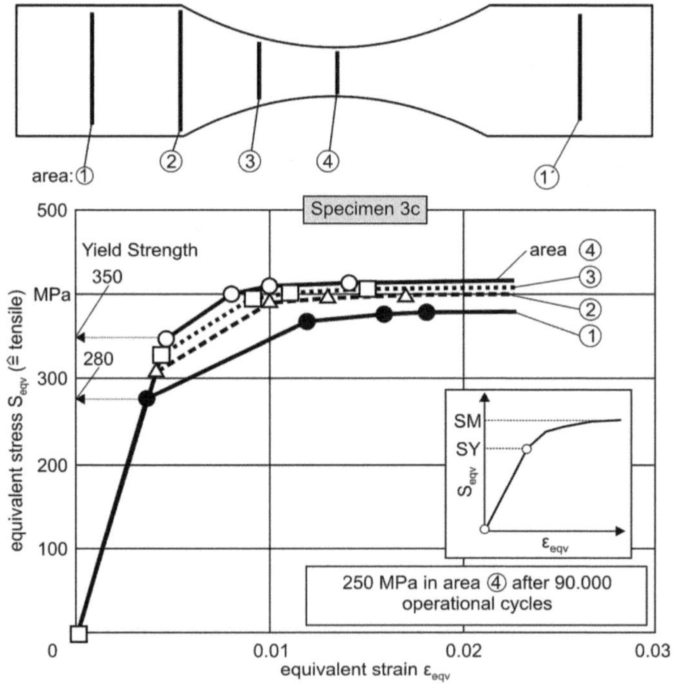

Fig. 7 Stress–strain curves in all examined specimen areas at maximum applied load

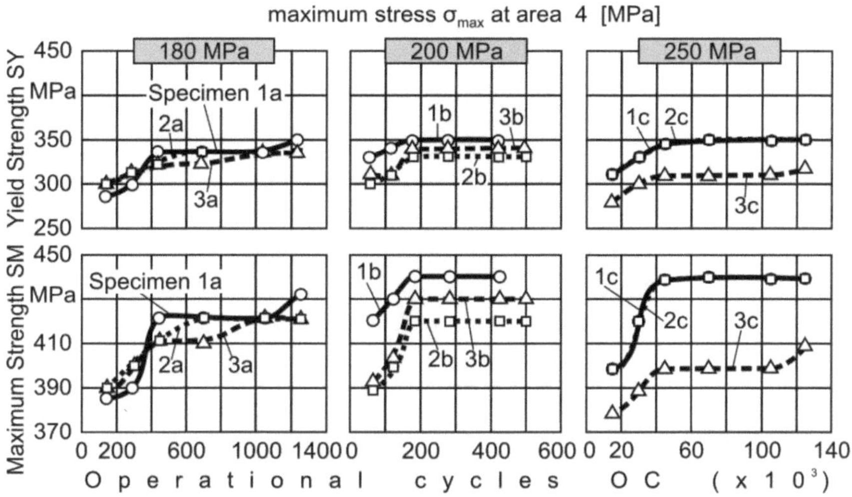

Fig. 8 Yield and maximum stress in all examined specimens for maximum applied load at area 4

furthermore are stabilized, according to the corresponding behavior of the indentation depth (see Fig. 8).

Figure 9 demonstrates the effect of the operational cycles up to IDA stabilization NSID and local stress σmaxl on the stabilized superficial yield strength SSY. According to these results, at a constant local stress level σmaxl, as the stress gradient increases the stabilized superficial yield strength SSY grows as well, whereas the operational cycles up to IDA stabilization NSID decrease. On the other hand, at the same NSID, the local stress growth leads to a SSY increase. The described effect of NSID and σmaxl on SSY is similar to the effect of the same parameters on SID (see Fig. 6). Moreover the operational cycles increase up to IDA stabilization NSID and the stress gradient reduction, decreases significantly the ratio NSID/Nfract.

5 Conclusions

In the present paper, surface hardness alteration of 2024 T3 aluminum sheet specimens subjected to fatigue loading were monitored with the aid of nanoindentations. These measurements were carried out in various specimens regions at various fatigue loads. The surface hardness of the alloy 2024 increases at different rates depending on the applied fatigue stress and number of applied fatigue cycles. The dependency of indentation hardness increase on the number of fatigue cycles is non-linear and seems to tend to a hardness saturation value. The surface hardness increase process is accelerated at higher fatigue stress amplitudes. Nanoindentation results demonstrated that the stress gradient and the local stress σmaxl have to be taken into account in predicting the number of operational cycles NSID up to

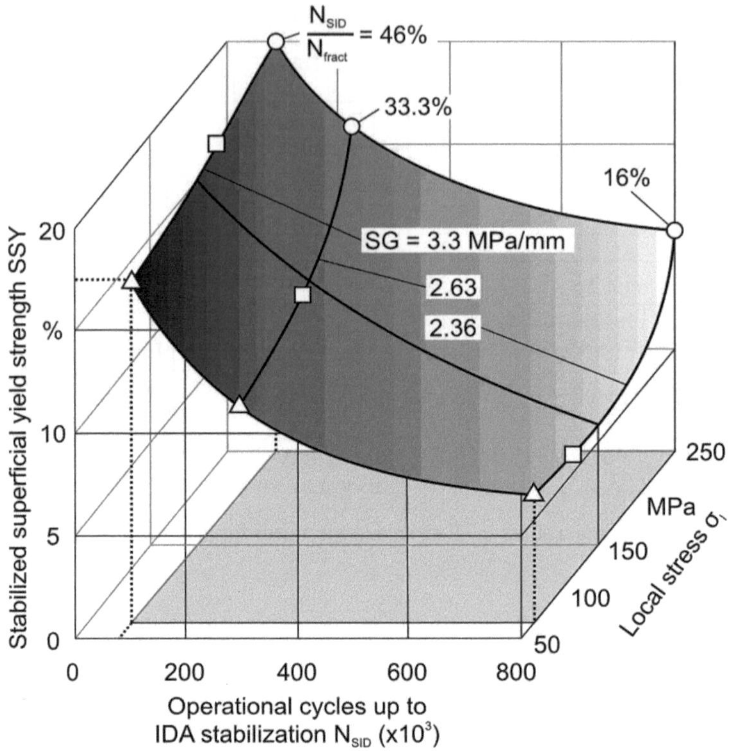

Fig. 9 Effect of local stress σmax, NSID and stress gradient on stabilized superficial yield strength SSY

IDA stabilization. Moreover the operational cycles increase up to IDA stabilization NSID and a stress gradient reduction, decrease significantly the ratio NSID/Nfract. The use of nano-hardness measurements can provide a useful methodology to early detect and quantify fatigue damage accumulation on structural elements.

References

1. Th.B. Kermanidis, S.P. Pantelakis, D.G. Pavlou (1990) Increase in hardening of 2024–T42 aluminum with fatigue stress amplitude, Theoretical and Applied Fracture Mechanics 14:43–47.
2. S.G. Pantelakis, P.V. Petroyiannis, K.D. Bouzakis, I. Mirisidis (2007) Surface hardness increase of 2024 aluminum alloy subjected to cyclic loading, Theoretical and Applied Fracture Mechanics, 48:68–81.
3. S. D. Henry, F. Reidenbach (1995) "Fatigue Data Book: Light Structural Alloys", ASM Handbook, ASM International, Materials Park, OH.
4. K.-D. Bouzakis, N. Michailidis, G. Erkens (2001) Thin hard coatings stress – strain curve determination through a FEM supported evaluation of nanoindentation test results, International Journal Surface and Coatings Technology, 142–144:102–109.

5. K.-D. Bouzakis, N. Michailidis, S. Hadjiyiannis, G. Skordaris, G. Erkens (2002) A continuous FEM simulation of the nanoindentation to determine actual indenter tip geometries, material elastoplastic deformation laws and universal hardness, Zeitschrift fuer Metallkunde, 93:862–869.
6. K.-D. Bouzakis, N. Michailidis, S. Hadjiyiannis, G. Skordaris, G. Erkens (2003) The effect of specimen roughness and indenter tip geometry on the determination accuracy of thin hard coatings stress–strain laws by nanoindentation, Journal of Materials Characterisation, 49:149–156.

Session VIII
Surface Treatments and Engineering

Characterisation of Residual Stresses Generated by Laser Shock Peening by Neutron and Synchrotron Diffraction

Alexander Dominic Evans, Andrew King, Thilo Pirling, Patrice Peyre, and Phillip John Withers

Abstract The fatigue behaviour of engineering alloys can be significantly improved through the application of mechanical surface treatments. These processes generate significant compressive residual stresses near surface by inhomogeneous plastic deformation. In the case of mechanical surface treatments such as laser shock peening, certain burnishing and rolling techniques and ultrasonic impact treatment (UIT), the compressive residual stress layer can extend to a depth of the order of millimeters, with balancing tensile stresses located deeper. Techniques to characterise the residual stresses generated by such mechanical surface treatments non-destructively are mainly limited to diffraction methods using penetrating neutron and synchrotron X-ray radiations. The application of these radiation sources is illustrated here by the characterisation of residual strain distributions in a two types of specimens treated with laser shock peening (LSP). Analyses of diffraction peak broadening provide qualitative information concerning the depth to which the plastic deformation of the treatments extends. Two case studies of laser shock peening of titanium and aluminium alloys is presented to demonstrate the capabilities of neutron and synchrotron diffraction techniques in the field of residual stress characterisation of surface engineered material non-destructively.

Keywords Laser shock peening · Neutron · Synchrotron · Diffraction · Residual stress

---

A.D. Evans (✉)
Paul Scherrer Institut (PSI), 5232 Villigen-PSI, Swizterland
e-mail: alexander.evans@psi.ch

A. King and P.J. Withers
School of Materials, Grosvenor Str., University of Manchester, M1 7HS, Manchester, UK

T. Pirling
ILL, 6 rue Jules Horowitz, Grenoble, 38042, France

P. Peyre
Laboratoire pour l'Application des Lasers de Puissance (LALP),
16 bis Avenue Prieur de la Côte dÓr, Acceuil, 94114, France

S. Pantelakis, C. Rodopoulos (eds.), *Engineering Against Fracture: Proceedings of the 1st Conference*,

1 Introduction

Laser Shock Peening (LSP) is one of several mechanical surface treatments, which have been shown to introduce compressive residual stresses to depths an order of magnitude larger than those characteristic of conventional shot peening [1]. The LSP process uses a high-energy laser pulse to vaporize locally a sacrificial ablative coating to form a plasma. An overlaying curtain of water confines the rapid expansion of the plasma. This drives a shock wave into the material, causing plastic deformation across the impact area [2], shown schematically in Fig. 1a. The deformation generates an indent the size of the impact footprint as sketched in Fig. 1b. The process is repeated across the desired area of the material, whereby impacts are arranged and overlaid until the desired coverage is obtained. The permanent plastic strain generates a misfit between the plastically deformed zone and the underlying elastically deformed region [3]. This shape misfit results in the formation of compressive residual stresses in the parallel to the plane of the surface. Balancing tensile residual stresses established to maintain stress equilibrium are located adjacent to the compressive region, often deeper within the bulk.

The improvements gained through the application of laser shock peening of fatigue and stress corrosion cracking behaviour have been widely reported for several material systems [4,5]. However the characterisation of the residual stresses and estimates of the depth of plastic work have often only been established by destructive

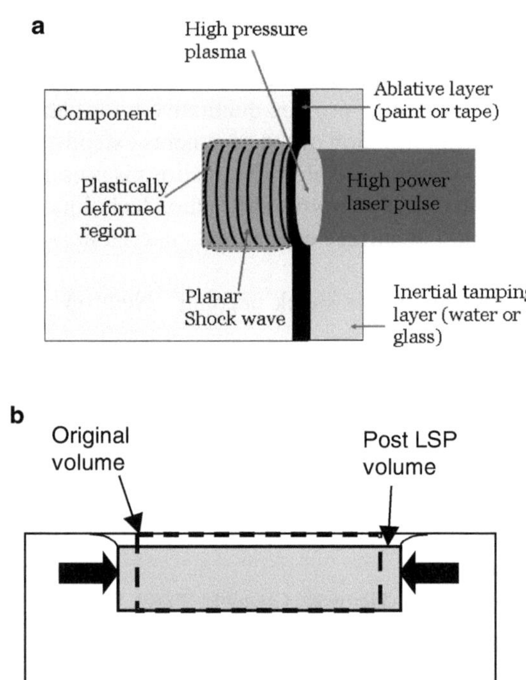

Fig. 1 Schematic of the LSP process **a** and **b** the generation of in-plane compressive residual stresses in response to a shape change caused by the plastic deformation in LSP

techniques such as laboratory X-ray diffraction combined with electrolytic layer removal or hole-drilling, slitting, and the contour method. To complement these techniques, measurements of residual strain and determined residual stresses of laser shock peened materials obtained by penetrating neutron and synchrotron radiation are presented. The elastic strain can be determined non-destructively through the analysis of the recorded diffraction patterns, enabling the macro residual stress to be calculated. In addition, the depth of plastic work can be estimated from associated diffraction peak broadening. Such measurements can even be performed on complex geometries such as airfoils, blades and disks.

2 Experimental

2.1 *Specimens and Material*

Two samples were characterised within this investigation. The first was a flat plate (Plate 1) of Ti-6Al-4 V (dimensions 45 × 45 × 7.7 mm), which had been laser shock peened on one side by Metal Improvements Company, Earby, UK. By rastering the laser, individual $3 \times 3\,\text{mm}^2$ square impacts were arranged side by side, overlapping by 10%. The same pattern was then overlaid with an offset of 50% in both directions of the edges of the impact as seen in Fig. 2. Therefore the majority of the peened area has essentially been double peened. The energy of the laser pulse was 180 J. and the pulse length 18 ns. Prior to treatment, the material had been hot rolled and creep flattened below the β-transus and exposed to a slow cooling rate, giving an equiaxed microstructure that was not observed in light microscopy to de significantly altered, however modification on the sub-grain level has been reported [6] where increased dislocation density has been observed.

The second sample (Disk 1) was a 30 mm diameter disk and a thickness of 12.8 mm as shown in Fig. 3. The disk was produced in the aluminium alloy AA

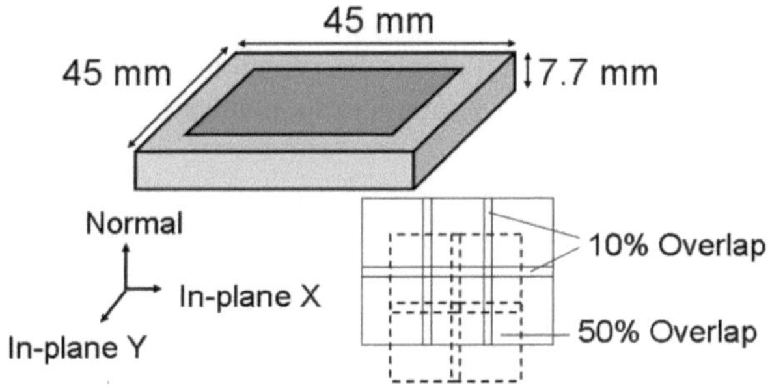

Fig. 2 Schematic of the laser shock peened Ti-6Al-4 V plate (Plate 1) and the pattern of overlapping $3 \times 3\,\text{mm}^2$ square impacts to produce an array which covered a large proportion of one face

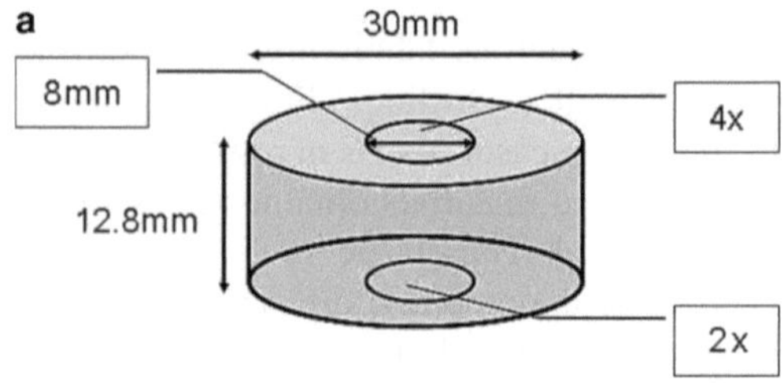

Fig. 3 Schematic of the laser shock peened AA 7075 T-7351 disk (Disk 1) which had been treated at the central position four times repetitively on one face and two times on the opposite face

7075-T7351. The laser shock peening was performed by Dr. Patrice Peyre at the French CNRS laboratory for laser science Laboratoire pour l'Application des Lasers de Puissance (LALP), Acceuil, France. The sample was laser shock peened at the centre on both the circular faces of the disc. One side was laser shock peened repetitively twice in the same position while the other face was peened four times. The laser impact footprint was 8 mm diameter. The laser was an Nd-glass laser operating at $\lambda = 1.06\,\mu m$ and consisted of an oscillator and four amplifier stages. Laser peening was performed using a sacrificial opaque coating (aluminium adhesive tape) and a transparent tamping layer (2–5 mm of flowing water). The power of the laser impacts were 3–4 GW/cm^2 (8 mm spot, 20 J energy, 15 ns pulse duration).

2.2 Residual Stress Determination by Neutron Diffraction

The residual stress states of the LSP Ti-6Al-4 V plate (Plate 1) were determined by neutron diffraction. Residual strains were measured in three orthogonal directions using monochromatic neutron diffraction. In the case of the time of flight strain measurements, triaxial stresses were calculated in Plate 1 from the two measured directions of strain. This was done by assuming an isotropic in-plane strain state. The strains were measured as a series of depth profiles taking the surface as the origin. The residual strains between respective measurement points were interpolated to allow the residual stress state at a particular point with that profile to be calculated from the triaxial strain measurements.

Residual strains were measured along the principal directions given by the geometry of the plate. The orthogonal stress components are then given by:

$$\sigma_i = \frac{E_{hkl}}{(1+\nu_{hkl})(1-2\nu_{hkl})}\left[(1-\nu_{hkl})\,\varepsilon_i + \nu_{hkl}(\varepsilon_j+\varepsilon_k)\right] \tag{1}$$

Where E is the elastic modulus, ν the Poisson's ratio and i, j and k are the axes of the stress/strain tensors. E and ν must be representative of the polycrystalline plane on which the strain is being measured i.e. the diffraction elastic constants. Stresses calculated from strains measured by time of flight neutron diffraction were based on the bulk elastic constants $E = 115\,GPa$ and $\nu = 0.33$ [7]. Stresses calculated from measurements using monochromatic neutron radiation were made using the $(10\bar{1}1)$

plane which has diffraction elastic constants very close to the bulk values [8]. Near to the surface, the plane stress condition can be assumed (zero stress normal to the sample surface):

$$\sigma_i = \frac{E_{hkl}}{\left(1 - \nu_{hkl}^2\right)} \left[\varepsilon_i + \nu_{hkl}\varepsilon_j\right] \tag{2}$$

This stress condition was used for both samples. In order to calculate the absolute values of strain in the three directions, the unstressed lattice parameter d_0 must be determined; this was determined by both far-field measurements and corrected by stress balance of the in-plane stress components. The far-field measurements were made laterally well beyond the peened region, scanning through the whole thickness and then taking an average. This measurement of d_0 differed by only 40×10^{-6} from that obtained by varying it until stress and moment balance criteria are obeyed for the in-plane directions.

The monochromatic neutron diffraction measurements were made on the instrument D1A (now superseded by SALSA diffractometer) at the ILL, Grenoble, France. The ILL is a reactor neutron source, where a continuous white beam of thermal neutrons is produced. From the reactor, a polychromatic beam of thermal neutrons travel along a Ni guide to the D1A monochromator. D1A is a powder diffractometer adapted for strain measurements. The diffracted neutrons are detected using a position sensitive detector (PSD). The residual strain is calculated from the detection of the 2θ angular shift of the diffraction peak position. The wavelength used was 2.993 Å, as selected from the polychromatic beam by a Ge(111) monochromator, giving a diffraction angle (2θ) of 84° for the $(10\bar{1}1)\alpha$-Ti reflection. A schematic of the experimental set-up is shown in Fig. 4. The gauge volume height was determined by a cadmium masks fixed on the plate surface. The height was 15 mm for X and Z direction and 1 mm for Y. The X and Y-directions describe the two in-plane directions while the Z direction stands for the normal direction.

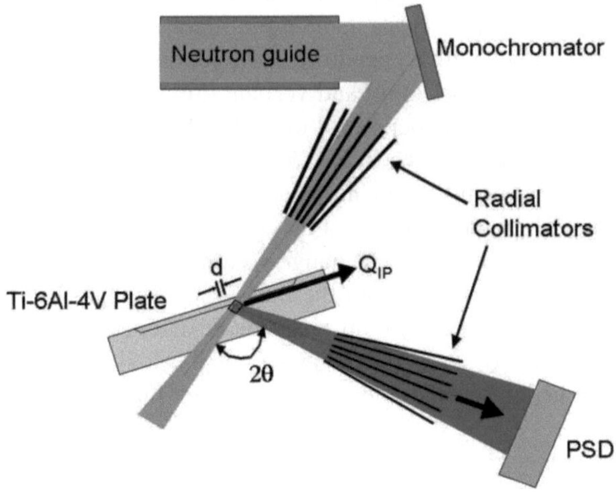

Fig. 4 Schematic of the experimental arrangement on D1A measuring the in-plane strain component. The plate is scanned along a path perpendicular to the surface plane

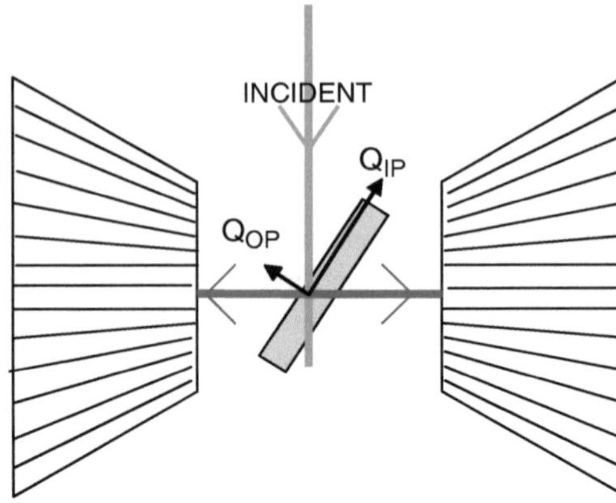

Fig. 5 Schematic of the experimental set-up on ENGIN-X. The two scattering vectors shown correspond to the two directions of the measured strain; IP (in-plane) and OP (out-of-plane/normal)

The peak position and full width at half maximum (FWHM) was determined by fitting the diffracted intensity using a Gaussian peak function, as shown in Fig. 5.

The time-of-flight (TOF) neutron diffraction measurements were made on the ENGIN-X instrument at ISIS, UK. ISIS is a pulse spallation neutron source. The pulse contains neutrons having a range of wavelengths which travel at their own characteristic speeds. From the time taken for the neutron to travel from the source to the detector, the wavelength, λ can be inferred [9].

$$\lambda = \frac{ht}{m_n L} \tag{3}$$

Where m_n is the mass of the neutron, t is the time of flight to travel the source to detector L and h is Plank's constant. From the wavelength, the d-spacing of the diffraction planes is derived using Bragg's law. Since there is a continuous band of wavelengths, the Bragg condition is satisfied for a range of lattice planes at a set angle. Therefore the whole diffraction pattern can be analysed, in our case using GSAS software [10]. The GSAS TOF peak shape used, function 3, is a convolution of two back-to-back exponentials with a pseudo-Voigt function [10]. The experimental set-up used slits to define the horizontal and vertical dimensions of the incident beam and radial collimators with a focus of 1 mm to define the diffracted beams. The instrument is schematically shown in Fig. 5. Radial collimator banks are placed at $2\theta \pm 90^\circ$ to the incident beam, allowing detection of diffracted beams in two directions. The sample was mounted such that the diffracted beams gave the strain in one in-plane and the normal direction. The gauge volume used was $1 \times 1 \times 15\,\text{mm}^3$. The residual stresses were calculated from the measured residual elastic strains. Since Ti-6Al-4 V alloy is predominantly comprises the hexagonal α phase, it has two different lattice parameters (a and c). Both are used to calculate the residual strain in reference to unstrained lattice parameters a_0 and c_0, giving the residual strains of the two lattice parameters ε_a and ε_c. The average strain in the α-titanium has been calculated from Eq. (4), suggested by Daymond et al. [11] weighted by multiplicities of the reflections

$$\varepsilon_{\text{average}} = (2\varepsilon_a + \varepsilon_c)/3 \tag{4}$$

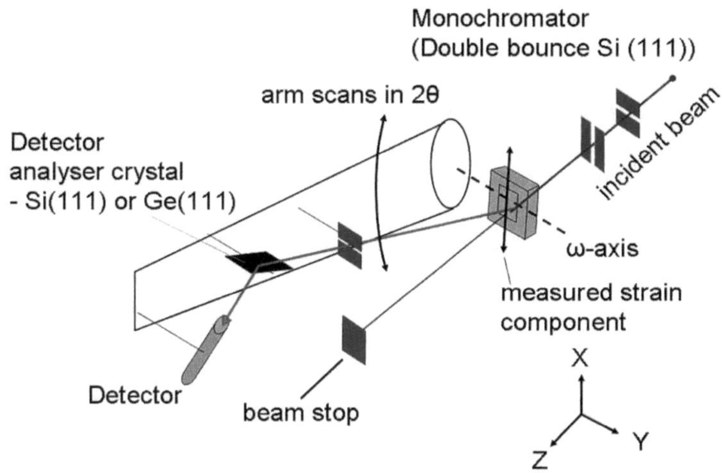

Fig. 6 Schematic of the instrumental arrangement on ID31 at the ESRF when measuring residual strain depth profiles in the laser shock peened Ti-6Al-4 V plate and the direction of strain measured

The residual stress was then calculated from the averaged residual strains.

The effective position of the measurement gauge volume was corrected for partial filling as it penetrates, which is necessary to in order to determine the near surface residual stress. The effective position was found a geometric model which calculates the centre of gravity of a partially immersed square cross section of the sampling gauge volume, neglecting absorption. The spurious strain [9] associated with the partial immersion of the sampling gauge volume present in the absence of analyser crystal optics was corrected using the "folding method". This involves measuring the full depth profile as the samples is passed through the beam and repeating the measurement after rotating the sample by 180 (so that the entrance surface becomes the exit surface and vice versa). By averaging the two sets of measurements the surface effects cancel out leaving only the true strain profile as schematically shown in Fig. 6. The determined correction (the true strain profile minus the measured strain profile) was obtained and subsequently applied to the other measured profiles.

2.3 *Residual Strain Measured by Monochromatic X-Rays*

Monochromatic synchrotron x-ray diffraction measurements were made on the laser shock peened plate in transmission on the high resolution powder diffractometer ID31 at the ESRF, France. The energy of the monochromatic synchrotron beam was 60 keV ($\sim$0.2 Å). The $(10\bar{1}1)$ reflection was measured at a diffraction angle of (2θ) of $\sim 5.3^\circ$. The experimental arrangement is shown in Fig. 6. The samples were rotated by 2.7° from the perpendicular direction of the incident beam using the ω rotation, ensuring the the scattering vector lay along the X axis of the sample coordinate system. The diffraction peak is recorded by scanning the point detector through 2θ. The beamline has a very low instrumental contribution to the width of the diffraction pattern. The high angular resolution is obtained by the combination of a double bounce Si(111) monochromator which selects a very narrow wavelength range from the incident polychromatic beam and the use of single crystal (Si or Ge available) analysers before the detector. The analyser crystals further diffracted the diffracted intensity into the detector. They have a dual role ensuring that only the diffracted intensity is detected (since other scattering is not in phase), reducing background and giving a narrow peak width and they eliminate the position sensitivity of the centre of the diffracted intensity [12]. This is important when measuring the near surface residual stresses in plates in transmission, where the gauge volume defined by the slit optics is only partially immersed within the sample. In this case, the centre of the diffracted intensity is offset from the centre of the total gauge volume, which is centred on the central axis of the instrument. Without analyser crystal optics, this shift in the position of the centre of gravity of the diffracted intensity would be observed as a peak shift and hence determined as a spurious strain. The analyser crystal ensures that diffracted intensity from displaced positions within the gauge volume is still measured at the same diffraction angle.

Synchrotron measurements of the radial residual strain were also performed on ID31. The energy of the monochromatic synchrotron beam was 60 KeV (0.20677 Å wavelength). The (311) diffraction plane was measured at a diffraction angle (2θ) of ~9.7°. The incident beam size was 0.2 × 0.2 mm, defined by automated slits. The secondary slits were also set to 0.2 mm, giving a gauge length of ~2.4 mm. The stress free lattice spacing d_0 was taken from an average of far field measurements 1 mm inwards from the edge of the disk at mid height. The residual stress state was calculated from the depth profile of the radial residual strain. It was done so by assuming the in-plane residual strain state to be isotropic for both geometries within the peened region, i.e. $\varepsilon_x = \varepsilon_y$ or $\varepsilon_{radial} = \varepsilon_{hoop}$ and the through thickness stress state is assumed to be governed by the plane stress criterion, σ_{rmz}, $\sigma_{axial} = 0$. Hence the residual stress was calculated assuming equi-biaxial plane stress. The elastic constants used for the calculation of residual stress were the elastic modulus and Poisson ratio of the (311) reflection; E = 64.9 MPa and ν = 0.35 [13].

3 Results and Discussion

3.1 Residual Stress Determination of LSP Ti-6Al-4 V Plate

The triaxial residual strain depth profiles for the laser shock peened Ti-6Al-4 V plate measured using monochromatic neutron diffraction is shown in Fig. 7. The in-plane strain is essentially transversely isotropic at all depths (Z). The characteristic peening profile of decreasing levels of compressive residual strain from a maximum near surface decreasing to zero occurs to a depth of 2–3 mm or so with elastic bending at

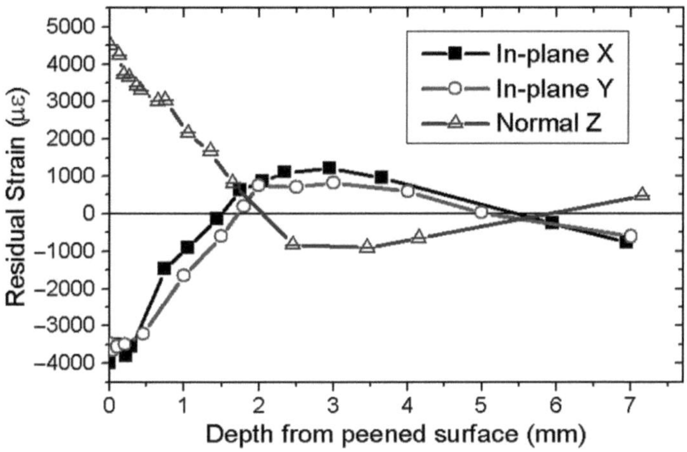

Fig. 7 Residual strain depth profiles for the three orthogonal residual strain components measured using monochromatic neutron diffraction

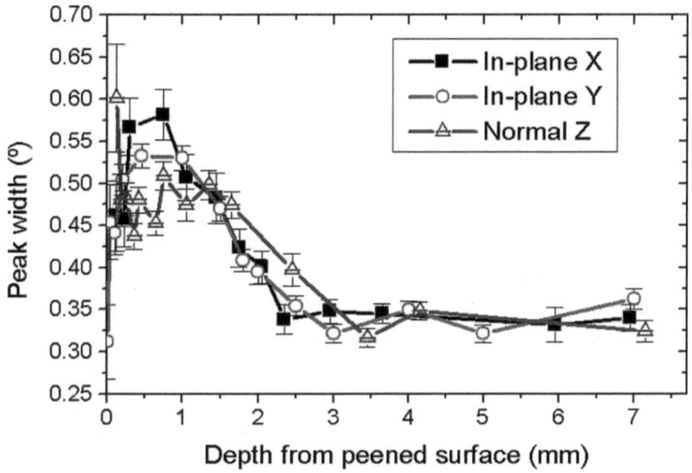

Fig. 8 Depth profile of the full width at half maximum (FWHM) corresponding to the three orthogonal directions of residual strain measured

greater depths. There is some evidence that the extent of bending is slightly greater in the X direction compared to the Y direction. The strain normal to the surface is mostly the result of the Poisson contraction of the in-plane residual strains.

The diffraction peak width profiles of the three measured orthogonal directions in the plate are shown in Fig. 8. The in-plane profiles show very similar depths of peak broadening. The normal direction exhibits slightly greater depth of peak broadening and slightly lower magnitudes (except at the surface where the error is high). In both in-plane directions, the peak width broadens over a depth of approximately

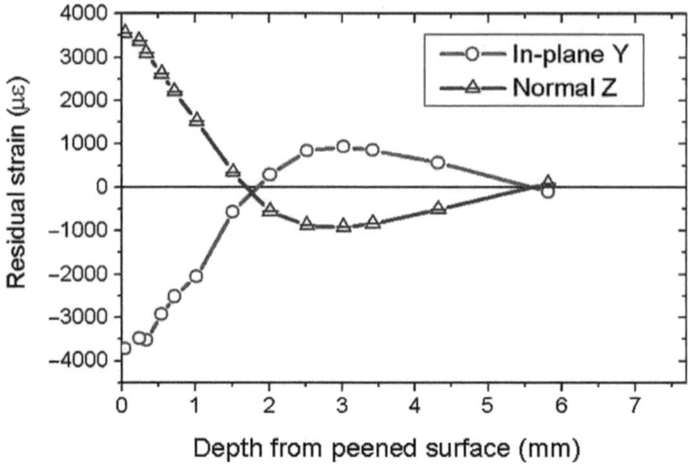

Fig. 9 Residual strain depth profiles of the in-plane strain component Y and the normal component Z measured by TOF neutron diffraction

2.5 mm. This depth corresponds to the position of maximum tensile residual stress in the depth profiles and corresponds to the onset of the linear solely elastic bending portion of the profiles. The profiles for both in-plane directions do not exhibit much variation well within the uncertainty of the error of the Gaussian fit of the peak (not shown for clarity).

The residual strain profile of the in-plane Y direction and the normal Z direction measured by time of flight neutron diffraction in the same plate are shown in Fig. 9. The profiles exhibit a similar trend to that observed with monochromatic neutron diffraction. The compressive residual strain field extends to 2 mm in depth with the tensile maximum at 3.5 mm in depth. The remaining depth is placed into linear elastic bending. The normal strain is the near mirror image of the in-plane strain profile. The biaxial residual stress profile was calculated assuming isotropic in-plane strain in addition to plane stress. This assumption is reasonable considering the strains measured using monochromatic neutron diffraction, Fig. 8. The triaxial stress state is shown is Fig. 10, is also calculated assuming isotropic in-plane strain. The normal stress exhibits a compressive dip in the near surface region, roughly corresponding to the depth of maximum tensile residual stress. The normal stress decreases towards zero at the surface as expected by the plane stress condition which must be exactly obeyed at the very surface. This result indicates that the residual stress state generated by laser shock peening of flat plate geometries is biaxial plane stress. In this case where the treated region is position centrally ensuring a uniform border of constraining material, the stress state is equi-biaxial plane stress. The magnitude of the maximum compressive stress generated using these laser shock peening parameters in 8 mm thick plate is approximately 70% of the yield strength of the material (880 MPa) [8].

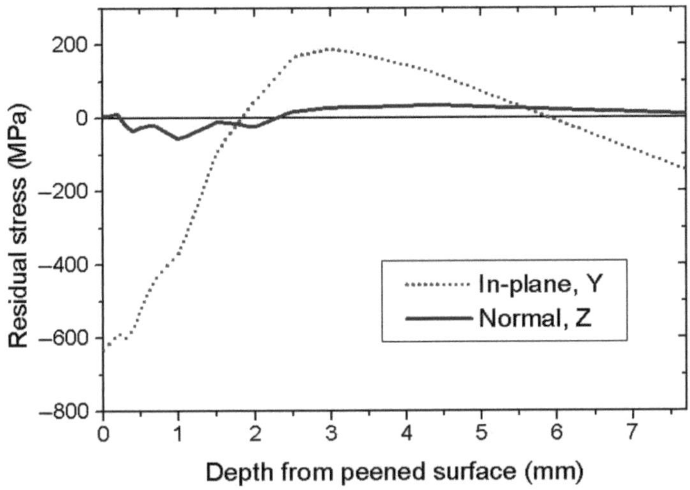

Fig. 10 Residual stress depth profiles of the in-plane strain component Y and the normal component Z calculated for the triaxial stress condition

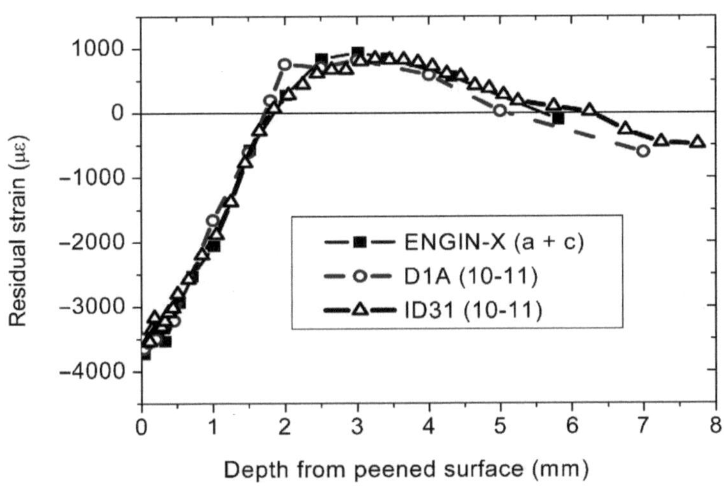

Fig. 11 Comparison of the in-plane residual strain depth profiles measured by TOF neutron diffraction (ENGIN-X), monochromatic neutron diffraction (D1A) and monochromatic synchrotron diffraction (ID31)

3.2 Synchrotron Measurement of In-Plane Residual Strain

Monochromatic synchrotron diffraction measurement of the in-plane (Y component) of the residual strain is compared to both neutron diffraction measurements in Fig. 11. The magnitude of the maximum compressive residual strain shows good agreement, as does the position of the inversion from compression to tension and the inflection point marking the maximum balancing tension. Since the diagonal of the cross-section of the sampling gauge volume as defined by the optics is similar for the three cases, the proportions of the strain profiles were expected to show only minor variations. However the vertical height of the gauge volume, which is parallel to the treated surface is markedly different for the both neutron measurements compared to the synchrotron. The vertical height for both neutron measurements was 15 mm, giving a matchstick shaped sampling gauge volume. In order to define a diagonal of the gauge cross section of 1.4 mm, similar to that obtained in the neutron measurements, the vertical dimension of the gauge volume was only 50 μm. The third direction corresponding to the long edge of the neutron gauge was 2 mm in this case. Hence the sampling volume in the case of the synchrotron measurements was significantly reduced compared to that of the neutron measurements. The fact that the profiles exhibit good agreement indicates that the in-plane stress state is quite uniform over the length scale of several laser shock peening impacts. In addition the good agreement of the profiles near surface suggests that the spurious strains associated with partial immersion of the sampling gauge volume that afflicts the neutron measurements due to the absence of an analyzer crystal are accounted for correctly.

The comparison of the diffraction peak width depth profiles of monochromatic neutron (D1A) and synchrotron (ID31) is shown is Fig. 12. The profiles exhibit clear correlation with respect to the general shape of the profile, showing strong diffraction peak broadening over the initial 2.5 mm depth beneath the treated surface. The instrumental contribution to the diffraction peak width of the high resolution powder diffractometer ID31 is very low, nominally instrumental contribution to the FWHM of around 0.003°2θ. Hence the increased broadening over the baseline value is indicated to arise from size and strain effects. These are a convolution of the effect that plastic deformation is believed to have on the effective size of a coherently scattering domain (on the grain or sub-grain level) and the strain associated with the stress fields arrange long range crystal defects, i.e. dislocations [14].

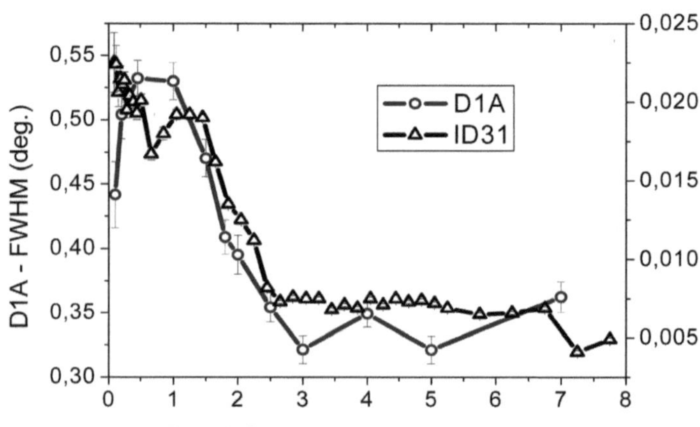

Fig. 12 Comparison of the diffraction peak broadening (FWHM) depth profiles of monochromatic neutron (D1A) and monochromatic synchrotron diffraction (ID31)

3.3 Characterisation of Residual Strains in AA7075 LSP Disk

The residual strain profile for the LSP AA 7075-T7351 disk shown in Fig. 13 reveals that the side peened repetitively four times exhibits a layer of compressive residual strain to a depth of 2 mm. The peak balancing tensile residual strain is located around 2.5 mm in depth, decreasing to near zero strain 5 mm beneath the surface. In contrast the opposite face which had been peened two times has the transition from compression to tension at around 1 mm beneath the surface. The balancing tensile residual strain exhibit a peak shape, but has a rather flat profile at ~150 με. The greater scatter in the shape of the profile may suggest that the average grain (sub-grain) size increases in this region. This is confirmed by observations of the variations of the integrated diffracted intensity.

The residual stress profiles calculated assuming isotropic (equi-biaxial) in-plane residual strain and the plane stress condition through thickness are shown in Fig. 14. The profiles are plotted as a function of depth with respect to their own treated surface for clarity. The depth of compression of the four times peened surface is clearly shown to be almost double that observed for the surface peened twice. However the magnitude of the maximum compressive residual stress located at the surface, is almost identical for the two cases. The magnitude of the maximum compressive residual stress is less than half of the yield strength of the alloy which is 435 MPa [8].

The two-dimensional map of residual strain of the half the diameter of the circular impact of the surface peened four times and the surrounding untreated material is shown in Fig. 15. The compressive residual strain of the peening impacts is clearly seen extending from the centre along the radial direction to approximately 7 mm, beyond the edge of the impact at ~4 mm. In depth along axial direction of the disk, the stress state becomes nominal around 2 mm in depth and becomes slightly tensile

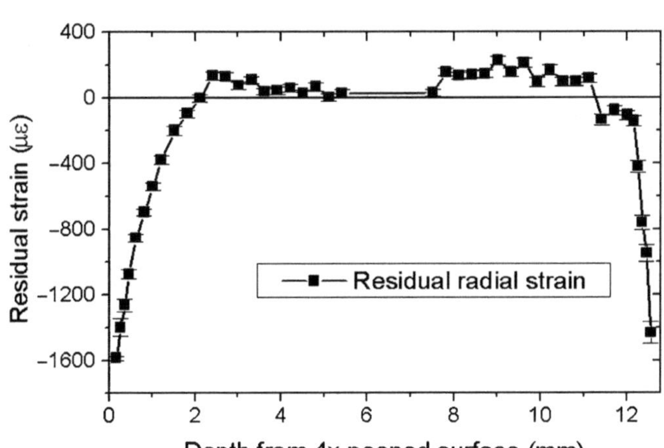

Fig. 13 Through thickness residual strain profile of the radial strain component of the Al LSP disk from the side peened four times (*left*) to the side peened twice (*right*) measured with monochromatic synchrotron diffraction

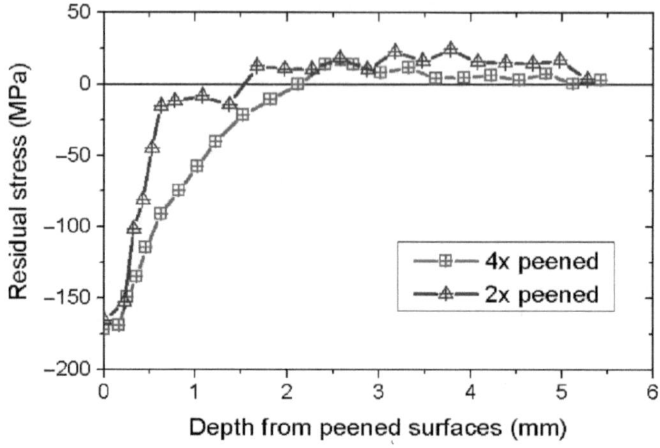

Fig. 14 Residual stress profiles for the two opposite faces of the disc calculated assuming plane stress condition and isotropic in-plane strains

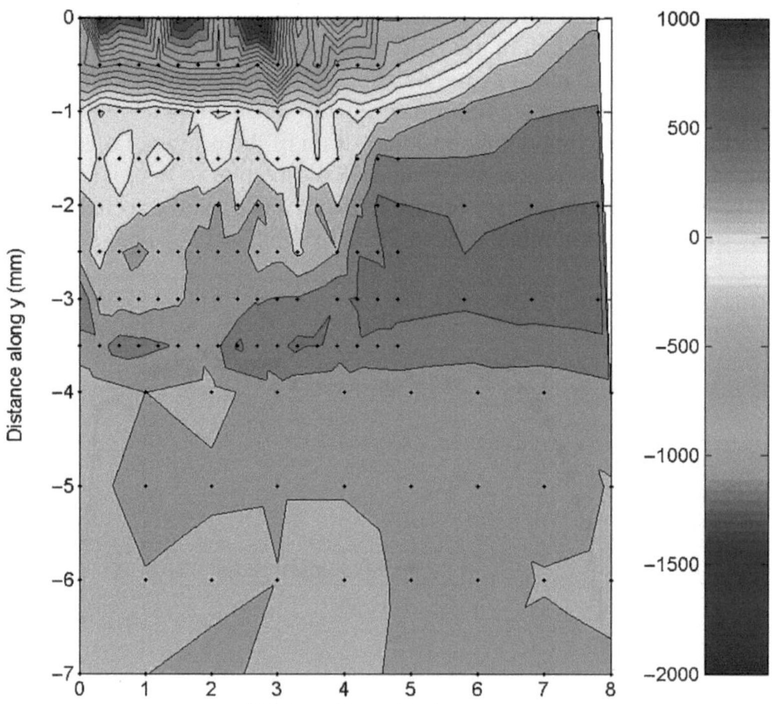

Fig. 15 Two-dimensional residual strain distribution of the radial strain component for the Al Disk over the radius (4 mm) of the four time laser shock peened face as measured on ID31 using synchrotron radiation. The scale is residual strain (x10^{-6})

at a depth of approximately 3 mm. The particularly striking aspect of the map is that the bulk of balancing tensile residual strain located not beneath the compressive region, but laterally from it, towards the edge of the disc. The largest tensile strains are found in the range 2–3 mm beneath the surface, although the tensile region extends almost to the surface.

4 Conclusions

The residual stain distributions generated by laser shock peening in the two aerospace alloys Ti-6Al-4 V and AA 7075-T7351 has been successfully characterized non-destructively using penetrating neutron and high energy synchrotron X-ray radiation. Residual stresses in Ti-6Al-4 V have been determined from the measurement of residual strain along the orthogonal directions of the strain tensor using neutron diffraction. The calculated residual stress state has revealed an isotropic plane stress condition generated in the flat plate geometry. These measurements have revealed the magnitude and depth of compression generated by laser shock peening in these alloys. The reported measurements have shown that the process parameters used for these studies generated compressive residual stresses which are approximately 70% of the yield strength for Ti-Al-4 V alloys but significantly lower for the AA 7075-T7351 sample. While the difference in sample geometries has a strong influence on the resulting measured residual stresses as a proportion of their respective yield strengths, the discrepancy may arise from the need to further optimise the process parameters to generate greater levels of compression. In addition the position and magnitude of the balancing tensile residual stresses have been determined. Such measurements may assist with the understanding of the fatigue failure of mechanically surface treated materials, where subsurface crack initiation may occur. Complementary high energy synchrotron diffraction has been exploited to validate neutron diffraction measurements and their associated corrections. Secondly synchrotron X-ray provides the higher spatial resolution required to map out the residual strain distribution generated by impacts at a single location. These measurements have shown that balancing tensile residual strains can be located laterally of the compressive region. This has implications for the treatment of thin samples such as airfoils where the balancing tension is likely to be located laterally of treated region. Overall the combination of the presented techniques allows the complete residual stress distributions to be characterised non-destructively and offers future opportunity to study the associated plastic strain through the analysis of the diffraction peak shape.

Acknowledgements The authors would like to greatly acknowledge the institutes for the provision of neutron (ILL and ISIS) and synchrotron (ESRF) experimental beam time. The instrument scientists E. Oliver (ENGIN-X, ISIS) and M. Brunelli (ESRF) are thanked for assisting in the experiments.

References

1. Hammersley, G., Hackel, L. A., Harris, F. (2000). Surface prestressing to improve fatigue strength of components by laser shot peening. Optics and Lasers in Engineering 34: 327–337
2. Peyre, P., Fabbro, R., Merrien, P., Lieurade, H. P. (1996). Laser shock processing of aluminium alloys. Application to high cycle fatigue behaviour. Materials Science and Engineering A 210(1–2): 102–113
3. Korsunsky, A. M. (2005). On the modelling of residual stresses due to surface peening using eigenstrain distributions. Journal of Strain Analysis 40(8): 1–8
4. Wagner, L. (1997). Mechanical surface treatments on titanium alloys: Fundamental mechanisms. Surface Performance of Titanium. Eds. Gregory, J.K., Rack, H.J. and Eylon, D. The Minerals, Metals and Mining Society TMS, Cincinnati, pp. 199–215. ISBN 0-87339-402-X
5. Rodopoulos, C. A., Romero, J. S., Curtis, S. A., de los Rios, E. R., Peyre, P. (2003). Effect of controlled shot peening and laser shock peening on the fatigue performance of 2024-T7351 Aluminium alloy. Journal of Materials Engineering and Performance 12(4): 414–419
6. Ruschau, J. J., John, R., Thompson, S. R., Nicholas, T. (1999). Fatigue crack nucleation and growth rate behavior of laser shock peened titanium. International Journal of Fatigue 21: 199–209
7. Butler, B. D., Murray, B. C, Reichel, D. G., Krawitz, A. (1989). Elastic constants of alloys measured with neutron diffraction. Advanced in X-ray Analysis 32: 389–395
8. Brandes, E. A., Brook, G. B. (Eds.) (1992). Smithells Metals Reference Book. Oxford, Butterworth-Heinemann
9. Hutchings, M. T., Withers, P. J., Holden, T. M., Lorentzen, T. (2005). Introduction to the characterisation of residual stress by neutron diffraction. Boca Raton, FL/London, Taylor & Francis
10. Larson, A. C., Von Dreele, R. B. (2004). General Structure Analysis Software (GSAS), Report No. LAUR 86–748, Los Alamos National Laboratory, USA
11. Daymond, M. R., Bourke, M. A. M., Von Dreele, R. B., Clausen, B., Lorentzen, T. (1997). Use of Rietveld refinement for elastic macrostrain determination and for evaluation of plastic strain history from diffraction spectra. Journal of Applied Physics 82: 1554–1562
12. Withers, P. J. (2003). Use of synchrotron X-ray radiation for stress measurement. Eds. M. E. Fitzpatrick and A. Lodini. London, Taylor & Francis
13. Eigenmann, B., Macherauch, E. (1996). X-ray investigations of stress states in materials. Materialwissenschaft und Werkstofftechnik 27: 426–438
14. Ungár, T., Gubicza, J., Ribárik, G., Borbély, A. (2001). Crystallite size distribution and dislocation structure determined by diffraction profile analysis: principles and practical application to cubic and hexagonal crystals. Journal of Applied Crystallography 34: 298–310

Dry Ice Blasting – Energy-Efficiency and New Fields of Application

Eckhart Uhlmann, Robert Hollan, and Adil El Mernissi

Abstract Cleaning technology changes from a mere necessity and cost factor to an integral part of the industrial value creation chain, either as a process step in manufacturing, e.g. surface pre-treatment for coating or joining, or as a cleaning tool for service and maintenance of machine components and equipment. Conventional technologies are based either on mechanical, watery or chemical methods and/or substances.

Dry ice blasting and CO_2-snow blasting as highly flexible and environmentally friendly alternatives for established cleaning technologies are already introduced. With the described developments this technology is enabled for energy efficient cleaning and pre-treatment processes.

Keywords Cleaning · Pre-treatment · Dry ice · Blasting

1 Introduction

Cleaning technologies has become an integral part of the industrial value creation chain, either as a process step in manufacturing, e.g. surface pre-treatment for coating or joining, or as a cleaning tool for service and maintenance of machine components and equipment. Conventional technologies are mostly based on mechanical or chemical methods with water or chemical substances. These will be reduced to a minimum to meet current or envisaged legal frameworks as well as technological and economical requirements. Dry ice blasting and CO_2-snow blasting are highly flexible and environmentally friendly alternatives. It is a by-product of the chemical industry. Therefore, the blasting process does not contribute additionally to the greenhouse effect.

E. Uhlmann, R. Hollan (✉), and A. El Mernissi
Institute for Machine Tools and Factory Management, Technical University Berlin, Berlin, Germany
e-mail: Hollan@iwf.tu-berlin.de

S. Pantelakis, C. Rodopoulos (eds.), *Engineering Against Fracture: Proceedings of the 1st Conference*,

The acceleration by compressed air has different disadvantages, e.g. low energy efficiency. Up to now without a pre-acceleration the mechanical acceleration by centrifugal wheel blasting wasn't suitable for sensitive blasting media. Adhesive and coating technology have been established in several fields. While the cohesive strength is relatively high, the adhesion, the weakest link, has to be improved. Consequently, the pre-treatment by dry ice blasting was investigated.

2 Solid Carbon dioxide

Carbon dioxide (CO_2) is either stored in fluid form in high or in low pressure tanks at 20°C or at −20°C [1]. At ambient conditions (1 bar) carbon dioxide is either gaseous or solid depending on the temperature. It is non-toxic, non-corrosive and non-abrasive. Furthermore carbon dioxide is non-conducting and chemically inert.

Due to a sudden expansion of the liquid stored carbon dioxide it is cooled down to −78.5°C because of the Joule-Thomson-effect and solid carbon dioxide is generated [2,3]. This snow is pressed through a mould and finally forms the cylindrical dry ice pellets used for blasting. The parameters of the pellets (density, hardness, shape) are influenced by the conditions during their production (e.g. degree of compaction), transport and storage as well as their age. Another field of application is the food industry where dry ice blocks are used for cooling.

3 Dry Ice Blasting

State-of-the-art, dry ice blasting is based on compressed air. A compressor, a dehumidifier and various filter devices are needed to dry and clean the air used to accelerate the dry ice pellets. The sublimation without any residues is an advantage of dry ice that avoids remaining media in the structure of the workpiece [4]. It allows a flexible cleaning even of sensitive or structured surfaces while highly adhering or hard contaminants are difficult to remove. Despite these advantages the maximum allowed work place concentration for gaseous carbon dioxide must not be exceeded.

Dry ice blasting is based on a combination of different effects. A thermal mechanism occurs as a local cooling down effect at the impact point, a mechanical effect caused by the impact and a partial sublimation of the pellet take also place [5]. Owing to this, elasticity is lost and the coating embrittles and shrinks, different thermal expansion coefficients, the kinetic energy of the particles and the air stream contribute to the removal. When the adhesive energy is exceeded by this combined thermo-mechanical mechanism the coating chips off.

4 Centrifugal Wheel Blasting with Sensitive Blasting Media

4.1 Mechanical Acceleration

For conventional centrifugal wheel blasting durable blasting media like sand, glass or steel is used because of the loads of the mechanical acceleration. A number of 3,000 rpm at a wheel diameter of ca. 35 cm resulting in blasting velocities of ca. 70 m/s is sufficient for deburring of metal workpieces, removing of casting residues or surface hardening.

The carbon dioxide pellets have a comparably low hardness and sublime at any point of impact. To reach blasting velocities of ca. 300 m/s, comparable to compressed air blasting, a higher number of revolutions per minute is needed. Therefore the accelerating components have been adapted to sensitive blasting media.

4.2 Centrifugal Wheel Blasting Prototype

At the Institute for Machine Tools and Factory Management IWF together with the Fraunhofer IPK a prototype has been developed [6]. Compared with conventional wheel blasting devices the main improvement is a pre-acceleration chamber. This device, shown in Fig. 1, will be patented in due course. The core components are manufactured by selective laser sintering (SLS), which is usually used for rapid

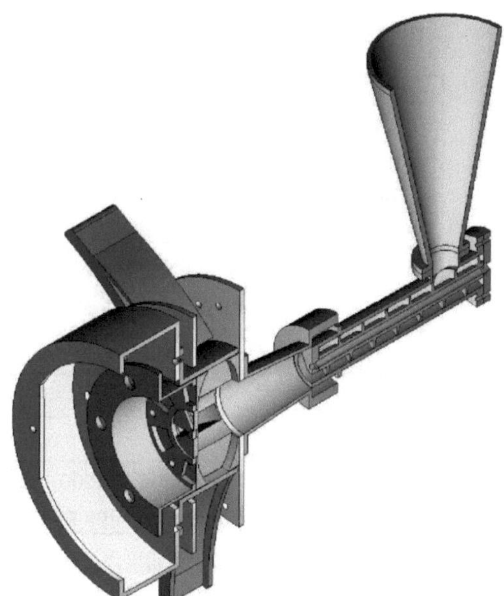

Fig. 1 Patented prototype for centrifugal wheel blasting with sensitive blasting media

prototyping. This manufacturing process enables complex three dimensional geometries to be generated which are necessary to reduce the mechanical stresses, e.g. impacts and friction, on the blasting medium to a minimum.

The device consists of a storage and supply unit, an acceleration unit and a clamping device for the actuation shaft. The screw conveyer is state-of-the-art regarding dry ice blasting devices. The acceleration bushing and the blades are of conventional wheel blasting devices. The acceleration bushing is used to speed up the blasting media to the speed of the inner blade edge to reduce mechanical loads feeding the blades. To reduce the mechanical stress during the centrifugal acceleration of the pellets after feeding, a centrifugal pre-acceleration bushing was constructed. Furthermore, the inner contour of the acceleration bushing is adapted and chamfered.

4.3 High Speed Camera Investigations

Owing to this pre-acceleration bushing almost unbroken pellets were observed at 2,400 rpm using a high speed camera. Tests with three different blasting wheel diameters have been made. The results of the high speed camera investigations of different numbers of revolutions per minute are promising and shown in Fig. 2.

In general the results correlate to the calculated velocity of the diameter and the number of revolutions per minute. The highest blasting velocity of nearly 63 m/s was measured at 2,400 rpm for a diameter of 396 mm. Actually test at a higher number of revolutions per minute are planned. Therefore a second prototype with up to 6,000 rpm an a possible blasting wheel diameter of more than 500 mm is constructed.

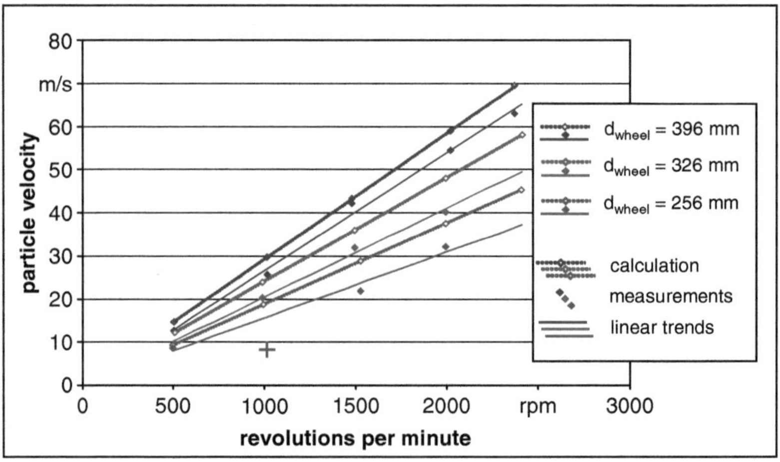

Fig. 2 High speed camera investigation results of wheel blasting with different diameters – theoretical calculation and measurements

5 Surface Pre-treatment

Surface pre-treatment is required for improving the adhesive strength to the effect of an optimum adhesion of adhesive bonds and coatings. Surface pre-treatment, however, usually requires the use of solvents and reactive chemicals. The use of such substances becomes more and more restricted by law. Therefore, it is necessary to search for new, environment-friendly processes. Dry ice blasting is deemed to be an environment-friendly and eco-efficient process. To investigate the applicability of this process for the pre-treatment a technological comparison was carried out between dry ice blasting and alternative pre-treatment processes regarding tensile lap-shear strength. Two cold hardening two-component epoxy resin adhesives EP1 and EP2 were used as test adhesives. For the determination of tensile lap-shear strength, rigid-to-rigid bonded assemblies according to DIN EN 1465 were used. Pre-treatment was realized by dry ice blasting, two types of CO_2-snow blasting (ACP and Mycon), laser processing, ultrasonic defatting with ethanol, corundum blasting as well as chemically according to DIN EN 13887.

For closer investigation of influencing factors of pre-treatment by dry ice blasting on the relevant adhesive surface characteristics, roughness and surface topography as influencing factors of the mechanical adhesion, specific surface, mechanical activation and chemical composition as influencing factors of the specific adhesion of AlMg3 after dry ice blasting were determined. First, the surface energy as precondition of the becoming of adhesion was analyzed.

5.1 Comparison of Dry Ice Blasting with Alternative Pre-treatment Processes

Figure 3 shows the tensile lap-shear strength of AlMg3 samples for two adhesives EP1 and EP2 in the unprocessed state and after pre-treatment with the tested processes. The tensile lap-shear strength and thus the adhesive strength for the adhesive EP2 did not show any significant changes after defatting, chemical pre-treatment, laser blasting, and both types of CO_2-snow blasting in contrast to those of the unprocessed state. It increases, however, from $16.1\,N/mm^2$ to $18.5\,N/mm^2$ after dry ice blasting and to $20.3\,N/mm^2$ after corundum blasting. In the case of the adhesive EP1, tensile lap-shear strength increases from $12.3\,N/mm^2$ to $19.6\,N/mm^2$ after chemical pre-treatment, to $19.3\,N/mm^2$ after CO_2-snow blasting according to ACP, to $19.2\,N/mm^2$ after CO_2-snow blasting according to Mycon, to $20.8\,N/mm^2$ after dry ice blasting as well as corundum blasting.

Apart from a few exceptions, the results of tensile lap-shear strength show that pre-treatment especially with dry ice blasting and CO_2-snow blasting are as good as, or even better than the chemical treatment or corundum blasting.

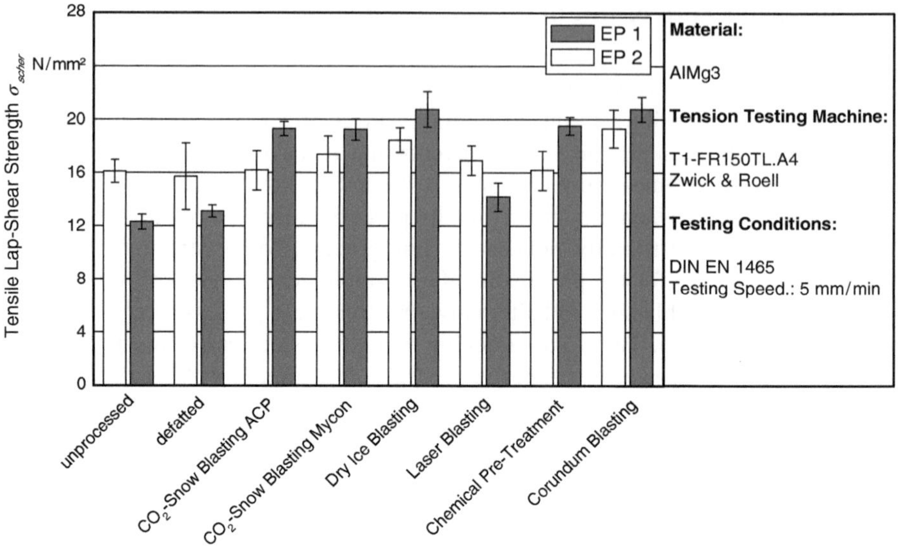

Fig. 3 Comparison of tensile lap-shear strength after pre-treatment with the tested processes for the adhesives EP1 und EP2

5.2 *Adhesive Surface Characteristics*

There is a number of theories to clarify the effects of pre-treatment by dry ice blasting on the adhesive surface characteristics [7]:

- Roughing and creation of undercuts facilitating a mechanical clamping of the adhesive
- Changes in the surface topography and zone
- Enlargement of specific, e.g. active surfaces
- Changes in the near-surface areas (mechanical activation, changing of the chemical composition) and
- Enlargement of the surface-energy and thus the improvement of the adhesive's wettability

5.2.1 Surface Energy

As precondition for the becoming of the adhesion is a good wettability of the substrate through the adhesive. In contrast to atoms and molecules, a surface or a phase interface is internally characterized by an additional specific energy, the interface energy. In order to estimate the influence of dry ice blasting on the wettability, the energy of the substrate surface is determined by contact angle measurement after the respective pre-treatment. Wettability is improved through reduced contact

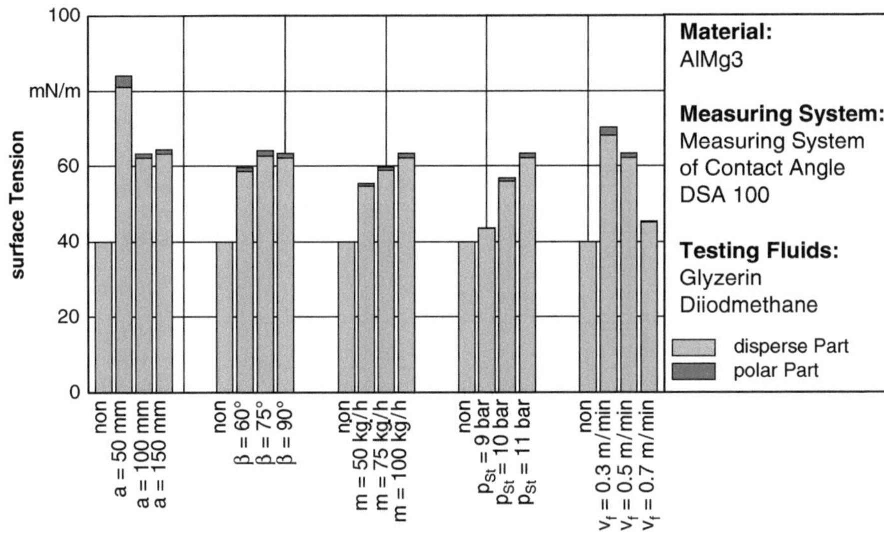

Fig. 4 Surface energy depending on relevant setting parameters

angles. The latter is a necessary but not sufficient requirement for good adhesive strength. The method of the lying drop is used for contact angle measurement.

In sum it can be stated that the total surface energy has increased after dry ice blasting independently of the setting parameters, see Fig. 4. Hereby, the polar percentage of the surface energy has also increased in each case. The polar percentage of the surface energy is decisive for the downstream bonding or coating process. The increase of the surface energy depends on the parameters.

5.2.2 Mechanical Adhesion

The mechanical adhesion is the theory of mechanical interlocking. It is based on the fact that, at the microscopic level, all surfaces are rough consisting of crevices, cracks and pores. The adhesive penetrates these features and hardens in such a way that it keys into the surfaces and forms a strong surface bond.

To estimate the influence of dry ice blasting on the mechanical adhesion, the Roughness and the surface topography are investigated. Roughness and topography investigations with static nozzle demonstrate that an increase of blasting time leads to increased roughing. The measured roughness variables arithmetical mean deviation of the assessed profile Ra, maximum height of the profile depth Rz, and total height of the profile Rt could be increased by up to 40 times, see Fig. 5. At the same time there was a strong change in topography. After 5 s of blasting time, the surfaces had a mixed structure of troughs and rolling traces. After 40 s of blasting time, a mixed structure of impact craters and troughs could be seen, see Fig. 6.

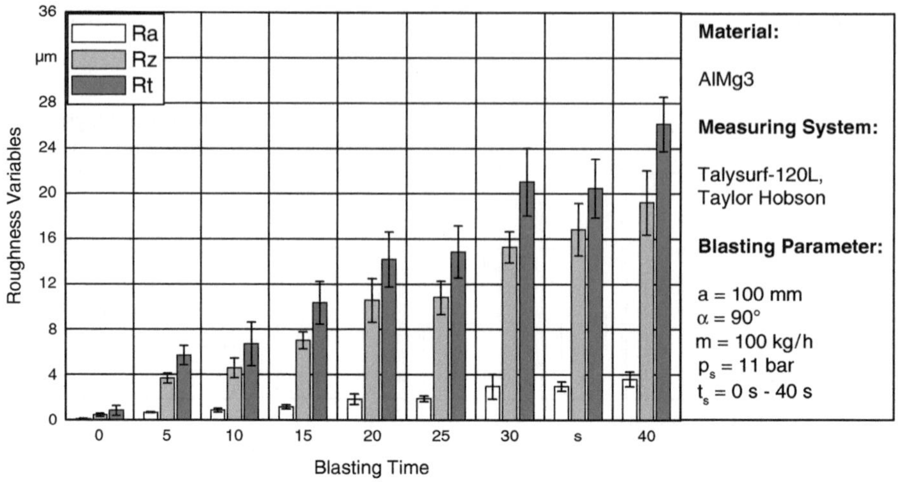

Fig. 5 Roughness variables versus blasting time

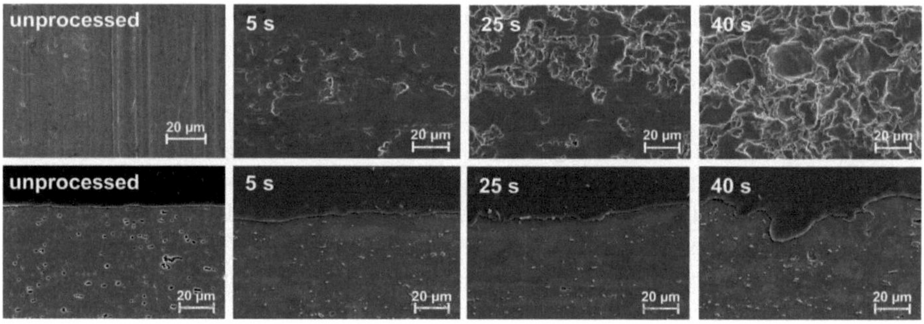

Fig. 6 Surface topography and zone versus blasting time

5.2.3 Specific Adhesion

The specific adhesion is based on chemical and physical processes. It is divided in the literature into:

- Adsorption theory
- Chemisorption theory
- Electrostatic theory and
- Diffusion theory

To estimate the influence of dry ice blasting on the specific adhesion the specific surface, the mechanical activation and the chemical composition of near-surface areas were analysed. The specific surface was investigated at the Bundesanstalt für Materialforschung und -prüfung (BAM – Federal Institute for Materials Research and Testing). A measuring instrument type ASAP 2010 by Micrometrics was used for the investigations. The instrument allows for the characterization of surfaces based on adsorption/desorption experiments with different measuring gases (typically N_2, Kr, Ar). The first measurements were carried out with Nitrogen N_2 as adsorptive agent. Hereby, no change could be detected in the specific surface. In order to verify if the absolute surface is too small for the sensitivity of the measuring instrument using N_2, further measurements were carried out with Krypton (Kr) as adsorptive agent. Here too, no change in the specific surface could be measured, see Table 1. Thus it can be deduced that dry ice blasting does not lead to an enlargement of the specific surface.

A number of scientific papers [8] have demonstrated that mechanical processing of solid bodies leads to a modification of reactivity. Through surface blasting, their reactivity is significantly increased. Mechanical processing of crystalline surfaces in the sub-microscopic range leads to a micro-plastic flow of the material. Due to the occurring high energy densities, the chemical bonding forces are overcome and fracture-free scratches develop. The following passage presents test results providing findings of the influence of dry ice blasting on plastic deformations (mechanical activation of surface-near areas) of the blasted samples. In order to investigate this influence, the residual stress before and after dry ice blasting were compared.

Plastic deformations in metals lead to grid expansions and thus to a change in macro residual stresses in the deformed area [9]. The influence of dry ice blasting on residual stresses in the material is to be investigated for AlMg3. Figure 7 shows the measured residual stresses of AlMg3 samples in dependence on the feed speed.

Dry ice blasting leads to an increase of residual compressive stresses in the near-surface area. The blasting agent impinging on the material to be blasted cause a deformation of the grid leading to an increase of residual compressive stresses. If, however, a critical blasting intensity is exceeded, the residual pressure stresses are reduced, see Fig. 7. This process, in literature called "overblasting" can be explained by increased roughness and cracks on the blasted material caused by high blasting intensity, thus stresses generated by grid deformation are reduced again.

Table 1 BET-measurements with krypton

BET-measurements with Krypton (Kr), processed $AlMg_3$-samples				
Relative Pressure	Pressure (mmHg)	Vol.adsorbed (cm^3/g STP)	Measuring time (h:min)	Saturation pressure (mmHg)
0.014888311	0.03855	−0.0003	01:08	2.58928
0.22749764	0.58906	−0.0010	01:41	2.58928
0.40665719	1.05295	−0.0014	02:23	2.58928
BET-measurements with Krypton (Kr), processed $AlMg_3$-samples				
Relative pressure	Pressure (mmHg)	Vol.adsorbed (cm^3/g STP)	Measuring time (h:min)	Saturation pressure (mmHg)
0.015723628	0.04040	−0.0004	00:59	2.56938
0.228362349	0.58675	−0.0015	01:30	2.56938
0.407510589	1.04705	−0.0022	01:52	2.56938

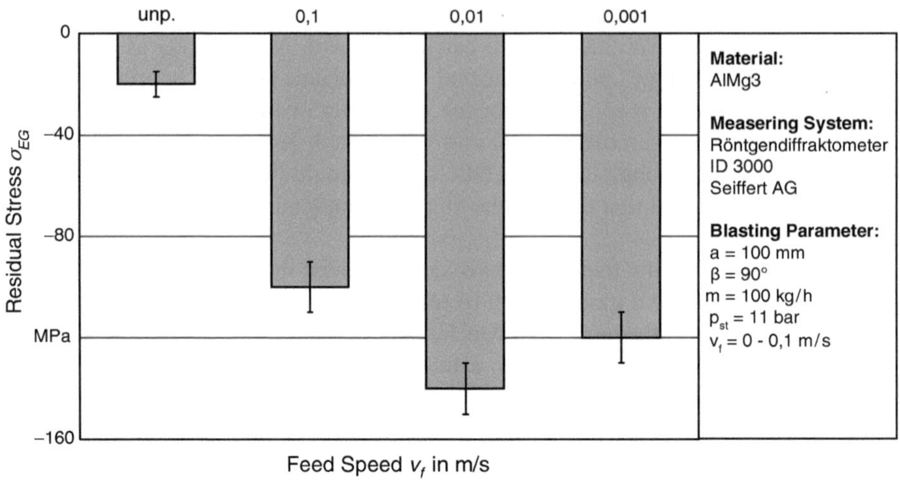

Fig. 7 Residual stress versus feed speed

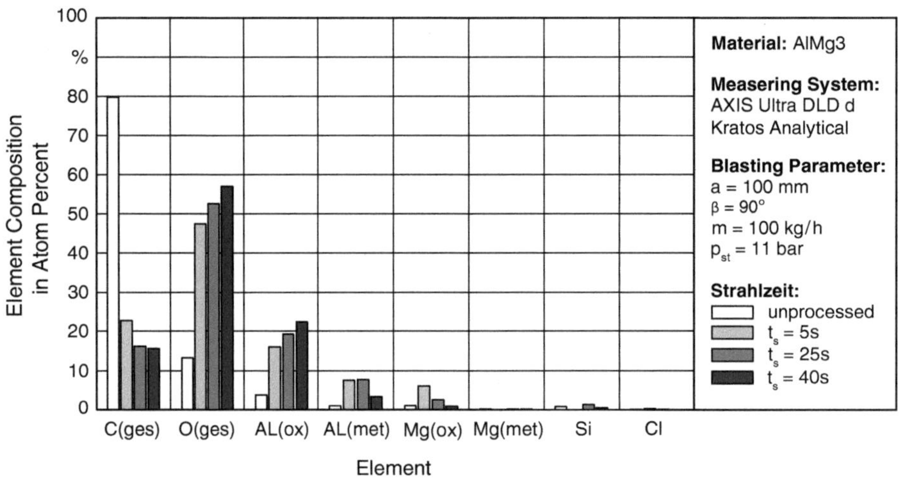

Fig. 8 Chemical composition versus blasting time

With the help of electron spectroscopy for chemical analysis it is possible to gain information on the chemical composition of the surface of a solid body. This also provides information on the bond status of elements on the surface. Thus, surfaces relevant and crucial in terms of their chemical actual state before bonding can be analysed, see Fig. 8.

The results of the analysis of the chemical composition of near-surface areas lead to the conclusion that dry ice blasting has a cleaning and re-oxidation effect. The cleaning effect was characterized by a strong reduction of the carbon fraction. Furthermore, all other contaminations such as silicon and chloride have been reduced to nearly 0 atom%. The re-oxidation effect, also conditioned by the compressed air used as acceleration medium, was characterized on the basis of a strong increase in the oxygen, aluminum oxide and magnesium oxide fraction.

6 Conclusions

As an alternative to compressed air based dry ice blasting a prototype for centrifugal wheel blasting with sensitive blasting media has been developed. The blasting velocities of almost unbroken pellets are comparable to the velocities of conventional blasting wheel devices. Due to this, centrifugal wheel blasting will become an alternative to conventional dry ice blasting with higher energy efficiency as well as a higher maximum mass flow.

The comparison of different conventional pre-treatment technologies show, that blasting with solid carbon dioxide will become a reasonable alternative. It depends on the type of application if either the flexibility of dry ice or CO_2-snowblasting based on compressed air is needed or the centrifugal wheel blasting suitable for large scale applications. In both cases the pre-treatment mechanisms of solid carbon dioxide will be part of further research.

References

1. Air Liquide, 2002, Safety Data Sheet Carbon Dioxide, Version 1.01.
2. Spur, G., Uhlmann, E., Elbing, F.: Process Study and Optimisation for Cleaning of Metal and Plastic Parts by Dry Ice Blasting, WFK 39th International Conference, Luxembourg, 1999, pp. 297–300.
3. Elbing, F., Möller, D., Ulbricht, M.: Fog Dissipation by Dry Ice Blasting: Technology and Applications, Proceedings of 2nd International Conference on Fog Collection, St. John's, 2001.
4. Uhlmann, E., El Mernissi, A., Dittberner, J.: Blasting Techniques for Disassembly and Remanufacturing, Proceedings of Global Conference on Sustainable Product Development and Life Cycle Engineering, Berlin, 2004, pp. 217–223.
5. Uhlmann, E.2, Hollan, R., Veit, R., El Mernissi, A.: A Laser Assisted Dry Ice Blasting Approach for Surface Cleaning, Proceedings of 13th CIRP International Conference on Life Cycle Engineering, Leuven/Belgium, 2006, pp. 471–475.
6. Uhlmann, E.3, El Mernissi, A., Krieg, M., Gottheil, I.: Schleuderradstrahlen mit Trockeneis, JOT - Journal für Oberflächentechnik, No. 8, 2006, pp. 60–62.
7. Brockmann, W., Geiß, P.L., Klingen, J., Schröder, B.: Klebtechnik: Klebstoffe, Anwendungen und Verfahren. 1st Edition, WILEY-VCH Verlag GmbH & Co. KGaA, Weinheim, 2005.
8. Thiessen, P.-A., Meyer, K., Heinicke, G.: Grundlagen der Tribochemie. 1st Edition, Akademie Verlag, Berlin, 1967.
9. Lepand, H.: Änderung des Dauerschwingverhaltens von Federstahl 50CrV4 durch Oberflächenverfestigen mit Strahlmittel verschiedener Härte bei unter-schiedlicher Flächendeckung. Dissertation, Technische Hochschule Clausthal, 1965.

Fatigue Life Improvement for Cruciform Welded Joint by Mechanical Surface Treatment using Hammer Peening and UNSM

Seung-Ho Han, Jeong-Woo Han, Yong-Yun Nam, and In-Ho Cho

Abstract For the improvement of fatigue strength of welded structures, mechanical post treatments have been applied in various industrial fields and have in most cases been founded to give substantial increases in their fatigue lives. These methods, generally, consist of the modification of weld toe geometry and the introduction of compressive residual stresses. In mechanical surface treatments, e.g. PHP (pneumatic hammer peening) and UNSM (ultrasonic nano-crystal surface modification), the weld profile is modified due to remove or reduce minute crack-like flaws, and compressive residual stresses are also induced. In this study, a pneumatic hammer peening procedure and a UNSM device were introduced, and a quantitative measure of fatigue strength improvement was performed. The fatigue strength at 2×106 cycles of hammer-peened and UNSM treated on a non-load carrying cruciform welded joint shows 220 and 260 MPa, respectively, which are more than two times higher than that of as-welded specimen. Especially, the surface layer in the vicinity weld toe treated by the UNSM provides nano-crystal structure created by an ultrasonic cold forging and introduces very high welding residual stress in compression.

Keywords Fatigue life improvement · Mechanical surface treatment · Pneumatic Hammer Peening · Ultrasonic Nano-crystal Surface Modification

1 Introduction

Mechanical post treatments for welded structures have been applied in various industrial fields and have in most cases been founded to give substantial increases in their fatigue strength [1–5]. These methods, generally, consist of the modification

---

S.-H. Han (✉), J.-W. Han, and Y.-Y. Nam
System Engineering Research Division, Korea Institute of Machinery & Materials, Daejeon, Korea
e-mail: seungho@kimm.re.kr, jwhan@kimm.re.kr, yynam@kimm.re.kr

I.-H. Cho
Design Mecha Co., Asan, Chungnam, Korea
e-mail: clkjh@designmecha.co.kr

S. Pantelakis, C. Rodopoulos (eds.), *Engineering Against Fracture: Proceedings of the 1st Conference*,

of weld toe geometry and the introduction of compressive residual stresses. Shot peening, as a representative technique of the compressive residual stress introduction method, has been widely used to enhance the fatigue performance in welded components by impacting it with a tool or small metal balls. Shot peening can build up compressive residual stresses of about 70–80% of the yield stress [5] though the components subjected to an axial tensile load has a less effect on fatigue strength compared to bending or twist load. Other peening techniques are hammer peening and ultrasonic impact treatment. Hammer peeing is carried out using pneumatic equipment driving several thousands impacting ratio per minute by repeatedly hammering with blunt-nosed chisel. Ultrasonic impact treatment, a recently developed technique, is also used as an effective post treatment through the impacts at ultrasonic frequency between 24–48 kHz [4]. Since both of them make it possible to give the weld, not only a favorable shape which reduces the local stress concentration, but also a beneficial compressive residual stress layer close to near surface, it has been known that the fatigue strength can be remarkably increased after the treatment.

To practical use of the peening technology in the welded components, a quantitative evaluation of fatigue strength including the improvement in the weld shape and residual stress induced by surface treatment is required. In this study, a pneumatic hammer peening procedure and a UNSM (Ultrasonic Nano-crystal Surface Modification) device were introduced, and a quantitative measure of fatigue strength improvement was performed.

2 Specimens and Procedures of Post Treatment

Several specimens of cruciform welded joint as shown in Fig. 1 were prepared. The material used in the welded joints was SM490B steel with a yield stress of 352 MPa and tensile strength of 514 MPa. The joints were manufactured using semi-automatic GMAW with CO2-gas, which is described in detail in [6].

To develop a pneumatic hammer peening (PHP) procedure, the IIW recommendation [2] was referred. PHP was carried out using a commercial pneumatic

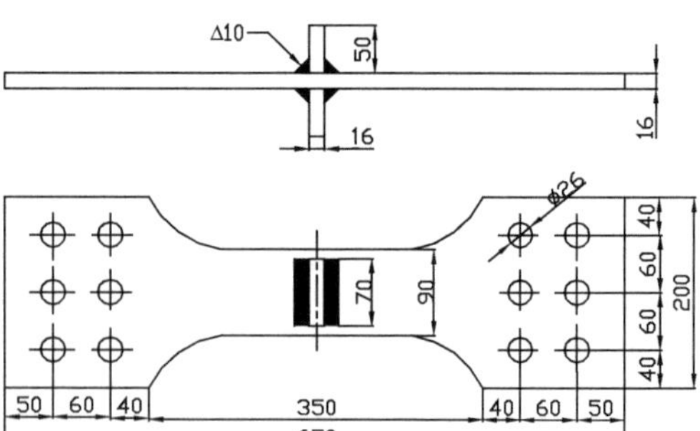

Fig. 1 Specimen configuration

chipping hammer with the rate of chipping blow of 3,600 bps, in which a variable pressure gage was installed. This pressure gage plays a role in setting up a proper pneumatic force to prevent an irregular work from the tool moving in an unsteady manner due to the striking hammer and its vibration during surface treatment. The length of indenter which strikes the work-pieces repeatedly is 200 mm. Since the quality of the treatment depends generally upon the magnitude of diameter of the tip used, a spherical shape of tip with a diameter of 6 mm in front of the indenter head was selected. Total four passes of peening over the weld toe region was performed at a tool speed similar to the welding speed. Figure 2 shows the pneumatic chipping hammer with the attached indenter and the weld toe configuration after PHP treatment, respectively.

The UNSM (Ultrasonic Nano-crystal Surface Modification) device consists of piezo-transducer (vibrator), booster (amplifier and transmitter of vibration) and horn, which are described in [7] in detail. In the end of horn, there is a ball which transmits mechanical vibration to the material. The device unit is built on a turret slide in a CNC lathe. The ball strikes 20,000 times per s, which bring severe plastic deformation to surface layers and thus induces nano crystal structure on the surface. Figure 3 shows the UNSM device working on the specimen and the weld toe configuration of the specimen after treatment. Proper static load, amplitude and number of strikes in unit area are major process variables to achieve nano crystal structure and additional effects such as improved configuration in the vicinity of

Fig. 2 Tool for PHP treatment and weld toe configuration by PHP

Fig. 3 UNSM and treated weld toe configuration after treatment

weld toe and compressive residual stress. Static load, amplitude and table speed of the CNC lathe for this experiments were selected as 10 kgf, 30 μm and 1,500 mm/min, respectively.

3 Geometry of Weld Bead and Welding Residual Stress

To quantitatively estimate a geometrical change in the weld toe by PHP and UNSM treatment, the notch radius ρ and frank angle θ at the weld toe were measured using a plaster cast with a special resin. Figure 4 shows the statistical characteristics of the ρ and θ by plotting the measured data in a cumulative probability paper. The ρ obtained from PHP has a relatively narrower scatter band and a high mean value compared to as-welded specimens with the range of ρ is 3.1–3.8 mm for PHP, 2.0∼5.68 mm for UNSM and 0.8–4.5 mm for as-welded (AW) specimens. The θ for PHP and UNSM shows an average value of 28°∼30° as an effective improvement in the weld shape. These modifications by PHP and UNSM treatments can give a favorable shape which reduces the weld toe stress concentration to enhance the fatigue performance in welded joints. Figure 5 shows microscopic images of surface layer in a longitudinal section adjacent the PHP and UNSM treated weld bead by an optical microscope. A uniform plastic deformation due to the treatments at the boundary of the heat affected zone removed the weld defects, in which the grains with the sizes of several micrometers for PHP and several hundred nanometers for UNSM in direction to thickness of the specimen are much denser at the surface.

The welding residual stress in vicinity of the weld toe was measured by means of a hole-drilling method. Figure 6 shows the measured welding residual stresses

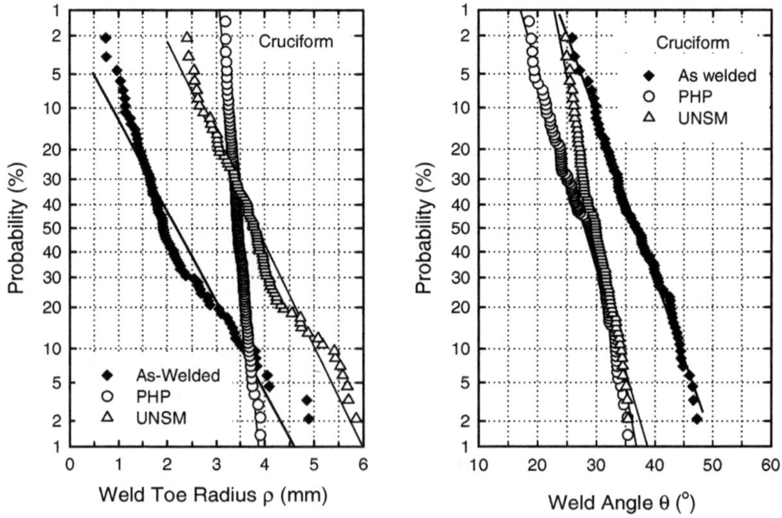

Fig. 4 Probability of ρ and θ

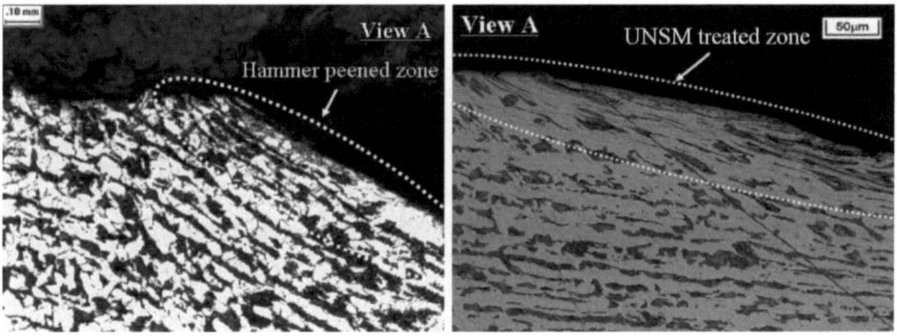

Fig. 5 Optical inspection for weld toe by PHP and UNSM

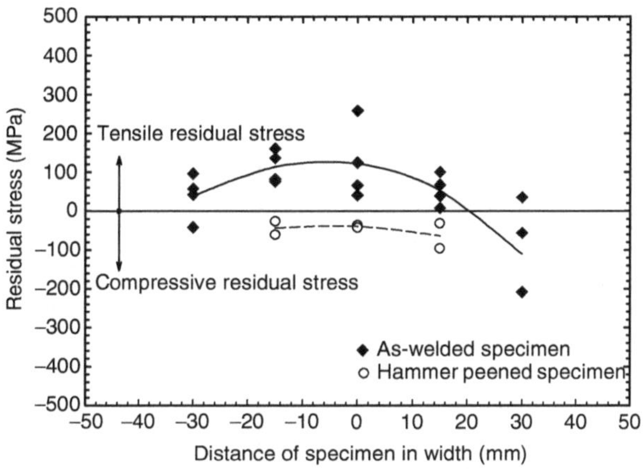

Fig. 6 Distribution of welding residual stress measured by hole-drilling method

in perpendicular direction of weld bead by PHP compared to that of AW specimen referred by the authors [8]. It was observed that the residual stress at the mid of weld bead is changed from 120 MPa for AW specimens to −40 MPa for PHP specimens.

Compressive residual stress is of most importance factor in fatigue durability. A measurement using X-ray diffraction method with a layer removal technique was carried out at the front of weld toe treated by UNSM. Figure 7 shows the distribution of residual stress along the depth. Although the residual stress at top surface shows −50 MPa, the stress is decreased significantly under −240 MPa at the depth of 0.1 mm. Then the compressive residual stress disappears at the depth of 1.5 mm.

4 Calculation of Fatigue Life

To calculate the fatigue life for the PHP treated welded joint, a notch strain approach and linear fracture mechanics was applied. In the case of UNSM, the experiments are being carried out, so that the analytical approach to calculate fatigue life is

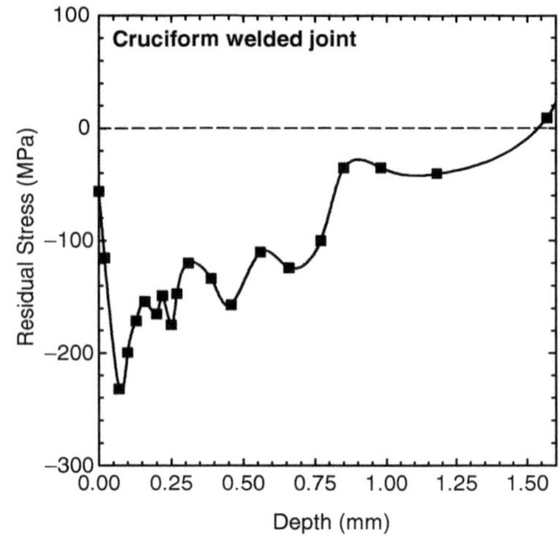

Fig. 7 Distribution of welding residual stress for UNSM measured by X-ray diffraction

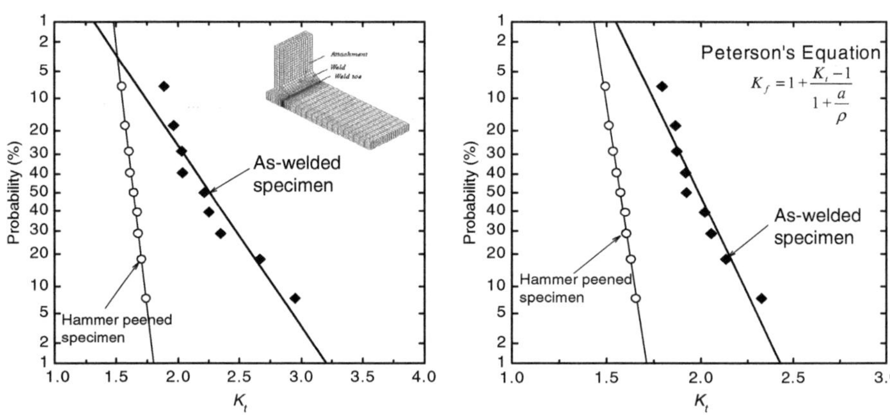

Fig. 8 Probability of K_t and K_f

planned in near future. To apply the notch strain approach [9], under maximum nominal stress S_{max} with the residual stress σ_r, the relation between the maximum notch stress σmax and the maximum notch strain ε_{max} can be calculated to be as $\sigma max \varepsilon max = \{KfS max/(1-\sigma_r/\sigma_{max})\}2/E$, where E is the elastic modulus and K_f, as the fatigue notch factor, is evaluated via the stress concentration factor K_t calculated from a finite element analysis considering the geometrical parameters shown in Fig. 4 and Peterson's formula [10]. The calculated K_f of PHP and AW specimens for the cruciform welded joints are plotted using a Gaussian cumulative paper as shown in Fig. 8. The calculated K_t and K_f for PHP specimen have much less values and a narrower scatter band than those of AW specimen. The fatigue crack initiation life NI, considering σr as the mean stress effect, can be subsequently obtained by an equation as proposed by Smith et al. [11].

On the other hand, the fatigue crack propagation life NP was calculated using fatigue crack propagation evaluation program developed by the authors [8]. This program permits the calculation of the stress intensity factor in front of the crack under the mechanism of mutual interaction, coalescence and growth in multiple surface cracks via the Mk-factor, a weld toe magnification factor, and an estimation of the fatigue crack propagation life per each crack increment via Paris's equation. The solver of the program provides a statistical analysis of the fatigue crack propagation life via a Mote-Carlo simulation. Total fatigue life NT corresponding to failure probabilities of 2.3%, 50% and 97.7%, therefore, can be calculated by the summation of those of NI and NP, respectively.

5 Results and Discussion

Fatigue tests for the cruciform welded joint in Fig. 3 was conducted under a constant amplitude loading conditions with $R = 0.1$ and the loading frequency of 10 Hz in a servo-controlled hydraulic test machine with a capacity of 500 kN. The tests were carried out until complete failure of the specimens or until a maximum life of 2×106 cycles.

Figure 9 shows the results obtained from fatigue tests for PHP and AW specimens. In the case of UNSM, three specimens were tested, which run out at 2×106 cycles. The fatigue life by PHP treatment at $\Delta S = 220$ MPa was found to be lengthened tenfold and the improvement in fatigue strength at 2×106 cycles was approximately 200% as compare to that of the AW specimens. The run-out specimen treated by UNSM shows also higher endurance limit of $\Delta S = 260$ MPa, which

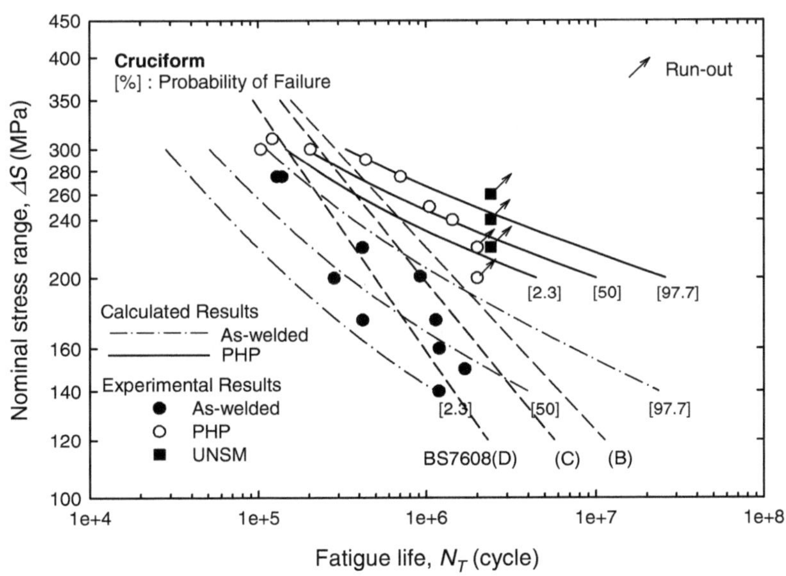

Fig. 9 S-N curves for cruciform joints

can be caused by the compressive residual stress induced by UNSM treatment. In comparison with welding detail categories according to BS7608, it was confirmed that the fatigue strength level of the PHP specimen in the region of high cycle fatigue is higher than that of class B and C.

The calculated $\Delta_S - N_T$ curves corresponding to the failure probabilities of 2.3%, 50% and 97.7% were also plotted in Fig. 9. It can be seen that the fatigue life calculated is similar to that of experimental results, which gives a possibility to predict analytically a fatigue life for PHP specimens with the measured data for the bead shape and residual stress. Finally, the test results for PHP demonstrate that the enhancement in fatigue strength has mainly a beneficial effect at a lower stress range, whereas has little at a high stress range. This phenomenon indicates that the effect of a beneficial compressive residual stress on fatigue resistance are vanished by a relaxation of residuals stress as results of repetition of cycle load, as well as the crack propagation life occupies the entire fatigue life in the high stress range.

6 Summary

The improved weld shapes and the residual stress induced were measured after treating the weld toe region of cruciform joint via PHP and UNSM by the application according to the developed procedures, and the calculated fatigue life for the case of PHP was verified through the fatigue tests. The verification showed a possibility to predict analytically fatigue life of the welded joints by treated by PHP. The fatigue life of the specimen treated by PHP at $\Delta S = 220\,\text{MPa}$ was increased ten times, and the improvement in fatigue strength at 2×106 cycles was approximately 200% as compare to the as-welded joints. In the case of UNSM, fatigue endurance limit shows higher value of $\Delta S = 260\,\text{MPa}$.

Acknowledgement The authors are grateful for the financial support provided by the Ministry of Science and Technology (MOST) of Korea.

References

1. Haggensen, P. J.: Fatigue Improvement Techniques. Welding in the World, Vol. 47, pp. 43–63 (2003)
2. Haggensen, P. J. and Maddox, S. J.: IIW Recommendations on Post Weld Improvement of Steel and Aluminium Structures. IIW Commission XIII, XIII-1815-00 (2001)
3. Roy, S., Fisher, J. W. and Ben, T. Y.: Fatigue Resistance of Weld Details enhanced by Ultrasonic Impact Treatment(UIT). International Journal of Fatigue, Vol. 25, pp. 1239–1247 (2003)
4. Statnikov, E. S.: Application of Operational Ultrasonic Impact Treatment Technologies in Production of Welded Joints. IIW Doc. XIII-1667-97 (1997)
5. Kirkhope, K., Bell, R., Caron, L., Basu, R. and Ma, K.: Weld Detail Fatigue Life Improvement Techniques. Part 1: Review. Marine Structures Vol. 12, pp. 447–474 (1999)
6. Han, J. W., Han, S. H., Shin, B. C. and Kim, J. H.: Fatigue Crack Initiation and Propagation Life of Welded Joints. Key Engineering Materials Vol. 297–300, pp. 781–787 (2005)

7. Cho, I. H., Song, G. H., Kim, C. S., Nobuhide A., Combs, A., Park, J., Suh, C. M., Park, J. H. and Pyoun, Y. S.: Nano Structured Surface Modification of Tool Steel and Its Beneficial Effects in Mechanical Properties. Journal of Mechanical Science and Technology Vol. 19, No. 11, pp. 2151–2156 (2005)
8. Han, S.H. Han, J. W. and Kim, J. S.: Estimating the fatigue life of welded joints considering statistical characteristics of weld toe shapes and multiple collinear surface cracks. International Fatigue Congress 2006, Atlanta, U.S.A., (2006)
9. Lawrence F.m Burk, J. and Yung, Y.: Influence of Residual Stress on the Predicted Fatigue Life of Weldments. ASTM STP 776, pp. 33–43 (1982)
10. Peterson, R.: Notch Sensitivity. McGraw-Hill, New York, pp. 293–306 (1959)
11. Smith K., Watson, P. and Topper, T.: A Stress-Strain Function for the Fatigue Damage: Journal of Materials, Vol. 5, No. 4, pp. 767–778 (1970)

The use of Ultrasonic Impact Treatment to Extend the Fatigue Life of Integral Aerospace Structures

Chris A. Rodopoulos and James Bridges

Abstract The work presents extensive experimental data proclaiming the potential enhancement on the fatigue life of integral structures using the Esonix Ultrasonic Impact Treatment.

Keywords Esonix Ultrasonic Impact Treatment · Residual stresses · Laser beam welding

1 Introduction

Besides attractive mechanical properties, especially in fatigue and strength, integral structures are claimed to offer cost and weight savings [1]. Joining technologies like friction stir welding and laser beam welding have recently been identified by leading aircraft manufacturers as "key technologies" for fuselage and wing manufacturing [2, 3]. Although many industries are looking at the processes, problems associated with the damage tolerance performance of integral structures remain. In particular, research works dealing with the fatigue crack propagation in welding direction or perpendicular to it, found more or less pronounced effects of the residual stresses especially in the case of near threshold loading [4, 5]. As a result damage tolerance methodologies based on material data can lead to unprecedented errors. In many works the problem was attributed to the presence of residual stresses [6–8]. Tensile residual stresses have been found to be responsible for the premature initiation of fatigue cracks, the mitigation of the crack initiation sites, the increase in the number of catastrophic cracks and hence the probability for crack coalescence, etc.

C.A. Rodopoulos (✉)
Laboratory of Technology and Strength of Materials, Department of Mechanical Engineering and Aeronautics, University of Patras, Patras, GR26500, Greece; Materials and Engineering Research Institute, Sheffield Hallam University, Howard Str., S1 1WB, UK
e-mail: Rodopoul@mech.upatras.gr

J. Bridges
Applied Ultrasonics Inc., 5871 Old Leeds Road, Suite 201, Birmingham, AL 35210, USA

S. Pantelakis, C. Rodopoulos (eds.), *Engineering Against Fracture: Proceedings of the 1st Conference*,

In order to minimize the harmful effects of the tensile residual stresses, surface engineering treatments have been employed. Controlled shot peening and laser shock peening have been found to provide some enhancement in fatigue resistance [7, 8]. In principle it was found that surface treatments can redistribute the residual stresses to the level of obtaining reversal of their direction. Controlled shot peening is considered as the most broadly available industrial process. Yet the induced residual stresses and therefore the potential of redistribution is limited to a shallow process zone, around 0.3–0.7 mm. As a result, in a through-thickens crack, only partial effect on the crack front aspect ratio is expected leading to small differences in the crack growth rate [9]. Laser shock peening on the other hand, delivers a process zone in excess of 1.5 mm and has been proven to provide significant crack growth resistance [10]. Yet the process requires significant capital investment and is characterised by high operational cost.

The Ultrasonic Impact Treatment (UIT) [11] is a technique that directly deforms the surface of materials using ultrasonic impacts. This technique fundamentally differs from contact methods of ultrasonic deformation treatment, the development of which dates back to 1950s [12]. The UIT process is employing continuous ultrasonic vibrations at the ultrasonic transducer output end strengthened with hard materials (carbide-containing alloys, artificial diamonds etc.) and being in direct and generally continuous contact with the treated surface. The process is mainly controlled by the output of the ultrasonic transducer (frequency), the selected pressure, the feed rate and the number of passes (coverage). The process can induce on request different amounts of cold work and residual stress profiles. The depth of the latter can range from 0.8 to 4 mm in aluminium alloys [13]. The process is relatively cheap compared to controlled shot peening and can achieve process rates in excess of 300 mm/min. The work that follows portrays experimental results conducted within the framework of the EC programme Daton.

2 Experimental Investigation

Several test panels, shown in Fig. 1 have been investigated. The panels were made of 6056-T651 and 2024-T3 aluminium alloys. Joining of the stringers have been made using laser beam welding. The composition and the mechanical properties of the two alloys are listed in Tables 1 and 2.

Six points have been selected for measuring the residual stresses using X-Ray diffraction. Residual stress measurements on all cases were conducted on Philips Xpert Pro XRD using Cr-Ka-radiation. Depth profiles were taken using sequential electrochemical polishing. The residual stresses were calculated using the $\sin^2 y$ method. Their location in relation to the centre line can be found in Table 3. The results are depicted in Fig. 2a, b.

Residual stress measurements on all three cases were conducted on Philips Xpert Pro XRD using Cr-Ka-radiation. Depth profiles were taken using sequential electrochemical polishing. The residual stresses were calculated using the $\sin^2 y$ method.

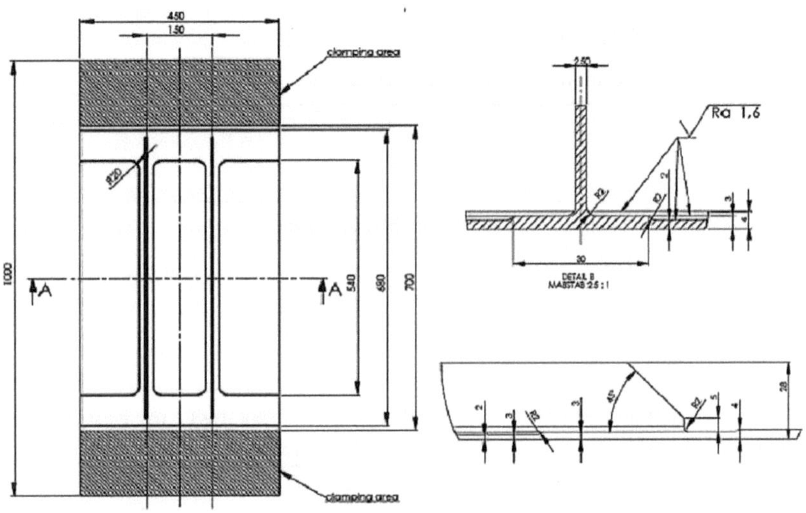

Fig. 1 Schematic representation of the two-stringer testing panel

Table 1a Chemical composition of 2024-T3 in wt%

Alloy	Si	Fe	Cu	Mn	Mg	Cr	Zn	Ti	Zr	Al
2024-T3 Min	–	–	3.8	0.30	1.2	–	–	–	–	Balance
Max	0.50	0.50	4.9	0.90	1.8	0.10	0.25	0.15	–	

Table 1b Basic mechanical properties of 2024-T3 according to ASTM E8 m-94a

Mechanical properties	Mean	99% Conf.	Std. Dev	Std. Err
0.2% Yield strength (MPa)	347.444	5.129	4.586	1.529
Tensile strength (MPa)	484.667	2.623	2.345	0.782
Elongation (%)	15.078	0.242	0.217	0.072
Fracture toughness – plane stress ($MPam^{1/2}$) for thickness 1.6 mm	135.57	8.74	3.00	1.50
Strain energy density (MJ/m^3)	70.517	0.635	0.568	0.189

Table 2a Chemical composition of 6056-T651 in wt%

Alloy	Si	Fe	Cu	Mn	Mg	Cr	Zn	Zr+Ti	Al
6056-T651 Min	0.7	–	0.5	0.4	0.6	–	0.1	–	Balance
Max	1.3	0.50	1.1	1.0	1.2	0.25	0.7	0.2	

Table 2b Basic mechanical properties of 6056-T651 according to ASTM E8 m-94a

Mechanical properties	Mean	99% Conf.	Std. Dev	Std. Err
0.2% Yield strength (MPa)	357	4.1	2.3	1.1
Tensile strength (MPa)	381	1.8	2.3	0.9
Elongation (%)	9	0.2	0.2	0.03

Table 3 Position of XRD points

Position	Distance from centre line (mm)	No.
1	10	1
2	50	2
3	70	3
	Distance from Stiffener base	
1	5	4
2	10	5
3	Top	6

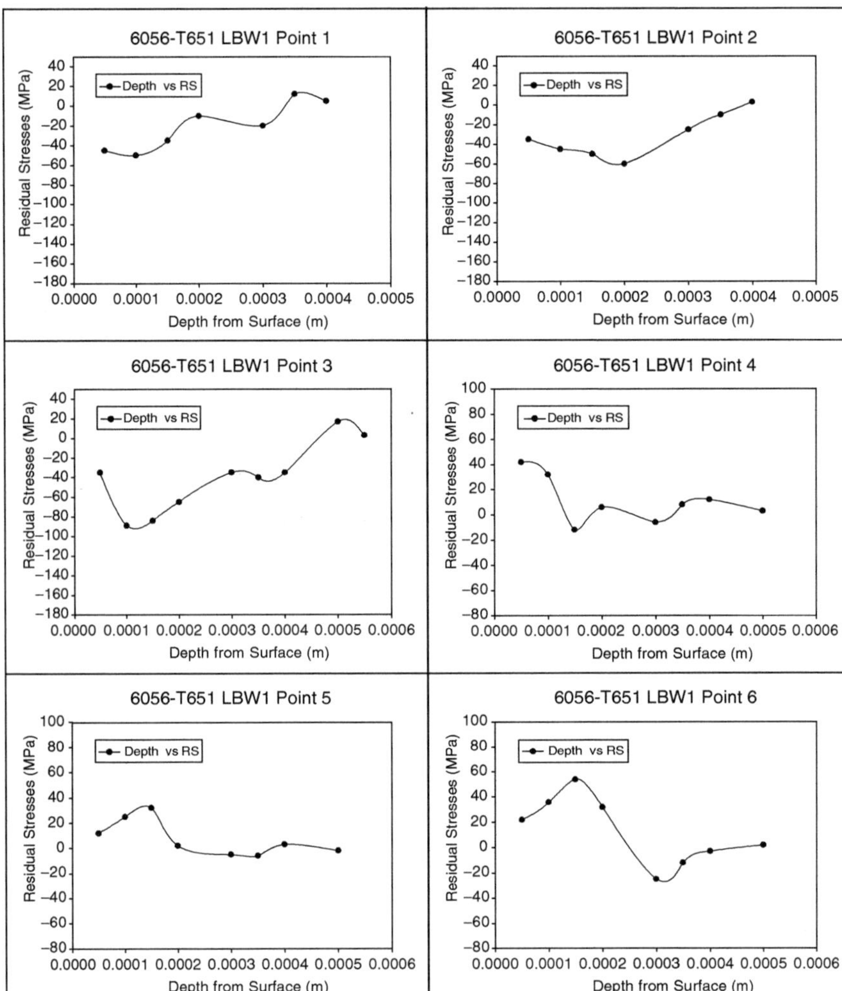

Fig. 2a Residual stress distribution profiles as a function of depth for 6056-T651 LBW panels

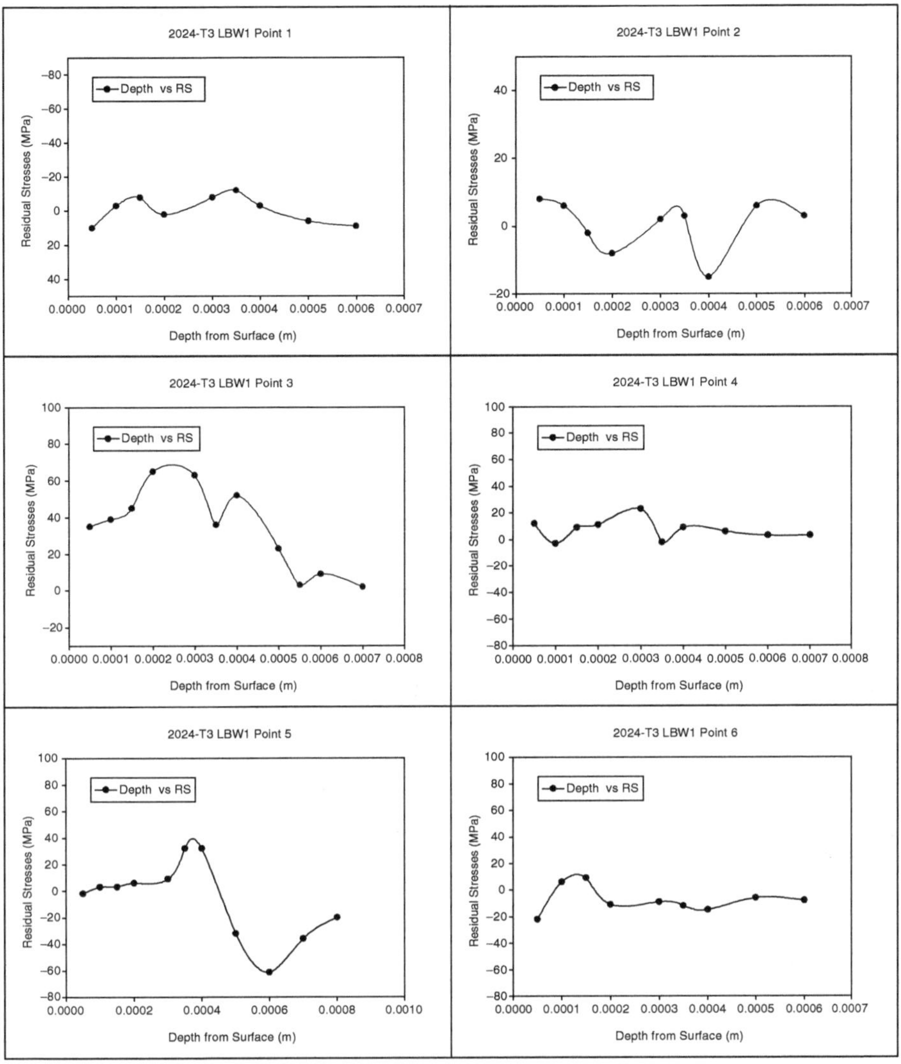

Fig. 2b Residual stress distribution profiles as a function of depth for 2024-T3 panels

The results indicate that despite the fact that laser beam welding was performed under identical conditions, the response of the material may significantly differ. Of particular interest are the points close to the stringer. In the case of 6013 point 3 exhibits significant compression while in the case of the 2024 tension has been found. Herein it is important to note that local discrepancies in the residual stress profile will most certainly affect other points. The problem is quite complex and therefore quality control can only be partially sought. To overcome such problem and in order to induce residual stresses: (a) uniform and (b) having a profile able to enhance the fatigue resistance of the panels Esonix Ultrasonic Impact Treatment was selected.

UIT treatment was performed on all panels tested according to the parameters shown in Table 4. The panels prior to treatment were cleaned in nitride-free detergent and degreaser suitable for aerospace components. Cleaning is imperative to prevent any residues from machining and handling entering the surface. Uniform clamping of the panels was performed using a vacuum table. The exerted clamping pressure was 50 MPa.

The parameters have been selected in such way as to smooth the signal from the transducer and hence protecting the uniform distribution of the residual stresses. In order to treat the fillet section between skin and stringer, the tool was set at a 60° angle. Selection of the treatment path was made using a sacrificial panel on which 12 LVDTs have been placed at locations where maximum deflection was expected. The treatment path can be found in Fig. 3.

Table 4 UIT *Esonix* process parameters

UIT *Esonix* process parameter	
Carrier frequency (KHz)	36
Pin dimension (mm)	Ø 6.3 17, R25, Titanium
Normalized impact	64 impulses
Amplitude under load (μm)	12
Pressure (kg)	15
Impact frequency (Hz)	36
Tool overlapping rate (%)	50
Coverage (%)	200
Feed rate (mm/min)	150

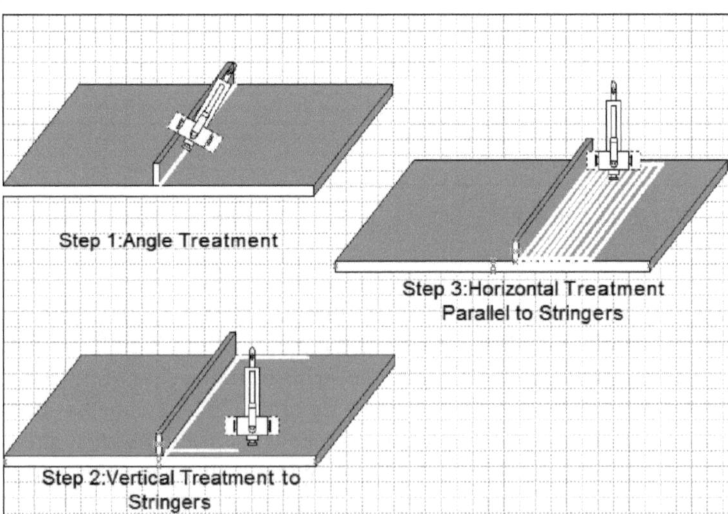

Fig. 3 Treatment paths for two-stringer LBW panels

The UIT unit was attached on a FANUC ArcMate120iBe robotic arm. The processing time for each panel was about 7 min. The time can be significantly reduced when two or more arms are used. The parameters selected for treatment have been also chosen in order to nanocrystallise the near surface area of the material and hence to provide additional protection to corrosion damage [14, 15]. The size of the treated area is shown in Fig. 4. XRD measurements of the changes in the residual stresses after UIT are shown in Fig. 5. Three different locations have been selected

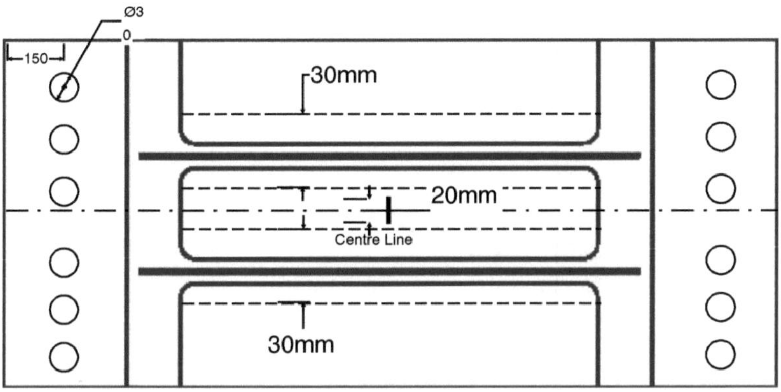

Fig. 4 UIT treated and fatigue crack growth testing panel. Not to scale

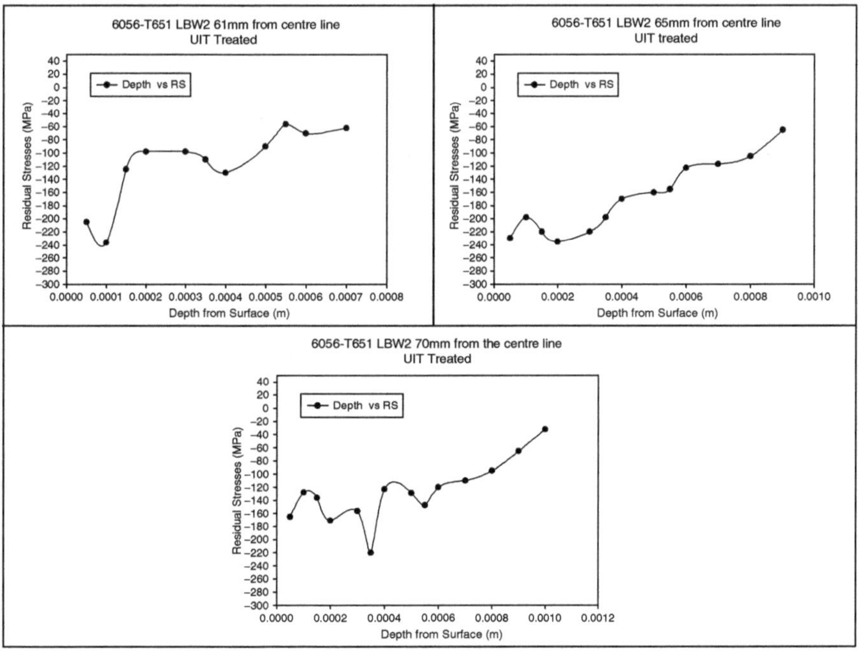

Fig. 5a XRD measured residual stresses taken from different locations of the UIT treated zone for 6056-T651 panels

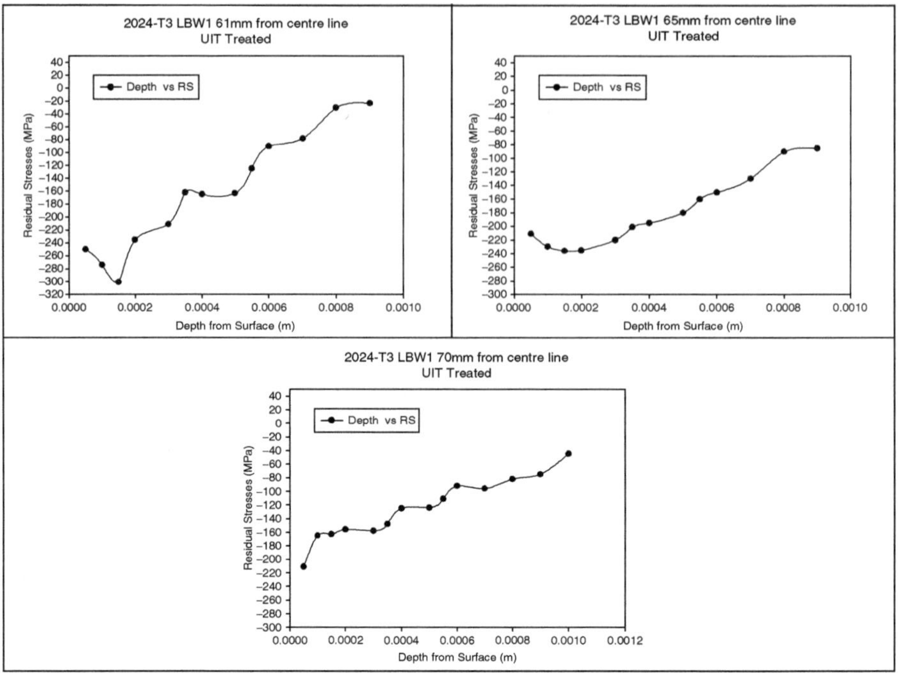

Fig. 5b XRD measured residual stresses taken from different locations of the UIT treated zone for 2024-T3 panels

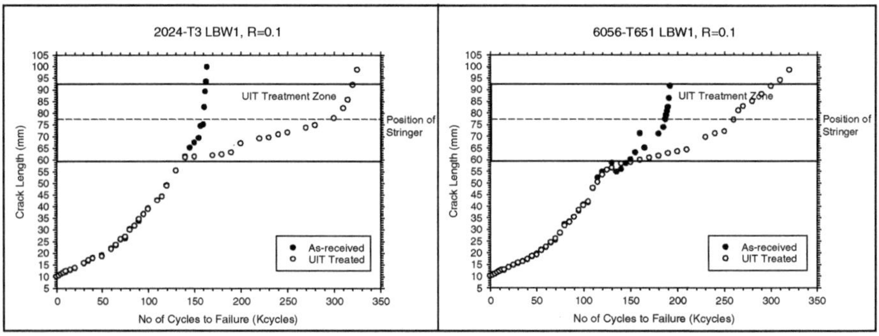

Fig. 6 Comparison of crack growth rates of 6056-T651 and 2024-T3 LBW panels with and without UIT

for analysis. Their distances from the centre line are: 61, 65 and 70 mm. The results indicate that within the UIT treated zone the residual stresses have significantly changed into compression. Their maximum depth is such as to allow enough material for equilibrium without causing tensile peaks.

Fatigue testing was performed on a 500KN MTS rig. The panels received an initial 20 mm mid section through-thickness slit using EDM with 0.2 mm wire. The maximum far-field stress was 80 MPa within the bay section and the stress ratio $R = 0.1$. During testing 10 Epsilon laser strain extensometers were used to measure strain accumulation along the crack path. The results can be used to determine the crack opening displacement. Crack measurement was performed using an in-house fully automatic apparatus comprising an Optronics CamRecord 600 high speed camera with a Vivitar macroscopic lens, an image analysis code based on IMAq by National Instruments. The apparatus delivers an accuracy of 0.1 mm. The results from the crack growth tests are shown in Fig. 6. Comparison reveals that the UIT can enhanced the fatigue resistance of the panels by a minimum of 80%.

3 Conclusions

Despite their obvious benefits, joining technologies are primarily subjected to complex damage tolerance behaviour. The presence of residual stresses and microstructural changes create a complex web of problems requiring significant attention and investment. Surface engineering treatments can potentially offer solution by mitigating quality problems associated with the joining process itself. Herein, Ultrasonic Impact Treatment has been selected as a process embracing the following benefits: (a) is industrially available; (b) has a process rate similar to that of the joining process; (c) is cheaper that laser shock peening and (d) can offer control over the treatment products.

In this work, testing was performed on panels approaching in dimensions industrial scale and hence problems associated with having to scale down residual stresses are minimised. The residual stresses from two stringer laser beam welded panels made of two different aluminium alloys were found to exhibit significant discrepancies despite the fact that the joining process was identical. The above reflects the role of the material in the development of the residual stresses.

To overcome such problems, UIT was chosen in order to treat the area around the stringers. The treatment was performed under industrial requirements in terms of setup, timing and machinery. Of primary concern was the use of readily available machinery found at any aerospace workshop. Treatment path analysis was necessary in order to prevent local deformation of the panels. The treatment delivered significant changes in the as-received residual stresses both in terms of profile and direction. Compression was achieved throughout the UIT zone. Crack growth analysis revealed that the UIT can significantly enhance the fatigue resistance of the panels and hence it holds a valid candidacy for damage tolerance enhancement.

Acknowledgement Part of this work has been financed through the European Commission under contract number AST3-CT-2004-516053. The authors are indebted to Applied Ultrasonics Incorporated for financial support and for the commissioning of an UIT unit.

References

1. R. G. Pettit, J. J. Wang and C. Toh: Validated Feasibility Study of Integrally Stiffened Metallic Fuselage Panels for Reducing Manufacturing Costs, NASA/CR-2000-209342, 2000.
2. R. Talwar, D. Bolser, R. Lederich and J. Baumann: Friction Stir Welding of Airframe Structures, 2nd International Symposium on Friction Stir Welding, TWI, UK, 2000.
3. E. Schubert, M. Klassen, I. Zerner, C. Walz and G. Sepold (2001) Light-weight structures produced by laser beam joining for future applications in automobile and aerospace industry, Journal of Materials Processing Technology, 115(1), 2–8.
4. G. Bussu and P. E. Irving: Damage Tolerance of Welded Aluminum Aircraft Structures, Proceedings of the 21st Symposium of the International Committee of Aeronautical Fatigue ICAF 2001, Rouchon, J. (Ed.), Cépaduès, 2001, 331–350.
5. R. John, K. V. Jata, K. and Sadananda (2003) Residual stress effects on near-threshold fatigue crack growth in friction stir welds in aerospace alloys, International Journal of Fatigue, 25(9–11), 939–948.
6. M. Eibl, C. M. Sonsino, H. Kaufmann and G. Zhang (2003) Fatigue assessment of laser welded thin sheet aluminium, International Journal of Fatigue, 25(8), 719–731.
7. A. Ali, X. An, C. A. Rodopoulos, M. W. Brown, P. O'Hara, A. Levers and S. Gardiner (2007) The effect of controlled shot peening on the fatigue behaviour of 2024-T3 aluminium friction stir welds, International Journal of Fatigue, 29(8), 1531–1545.
8. O. Hatamleh, J. Lyons and R. Forman (2007) Laser and shot peening effects on fatigue crack growth in friction stir welded 7075-T7351 aluminium alloy joints, International Journal of Fatigue, 29(3), 421–434.
9. T. Honda, M. Ramulu and A. S. Kobayashi (2006) Fatigue of shot peened 7075-T7351 SENB specimen – A 3-D analysis, Fatigue & Fracture of Engineering Materials & Structures, 29, 416–424.
10. C. Rubio-Gonz'aleza, J. L. Ocana, G. Gomez-Rosas, C. Molpeceres, M. Paredes, A. Banderas, J. Porro and M. Morales (2004) Effect of laser shock processing on fatigue crack growth and fracture toughness of 6061-T6 aluminium alloy, Materials Science and Engineering A 386, 291–295.
11. E. Sh. Statnikov, V. N. Vityazev and O. V. Korolkov, Study of Comparative Characteristics of Ultrasonic Impact and Optimization of Deformation Treatment Processes, 5th World Congress on Ultrasonics, Paris, France, 2003.
12. I. I. Mukhanov and Yu.M. Golubev, Strengthening Steel Components by Ultrasonically Vibrating Ball, Vestn. Mashin., No 11, 1966 (in Russian).
13. C. A. Rodopoulos, A. Th. Kermanidis, E. Statnikov, V. Vityazev and O. Korolkov (2007) The effect of surface engineering treatments on the fatigue behaviour of 2024-T351 Aluminium alloy, Journal of Materials Engineering and Performance, 16,1 30–34.
14. X. An, C. A. Rodopoulos, E. Statnikov, E. Vitaxev and O. Korolkov (2006) Study of the surface nanocrystallization induced by the esonix ultrasonic impact treatment on the near-surface of 2024-T351 aluminium alloy, Journal of Materials Engineering and Performance, 15(3), 355–362.
15. S. Pilli and C. A. Rodopoulos (2007) Use of Ultrasonic Impact Treatment as an Effective Surface Engineering Treatment to Improve the Corrosion-Fatigue Properties of Al-2024-T351, EUROMAT 2007, 10–13 Sept, 2007, DGM.

Session IX
Theoretical Fracture Mechanics and Modeling I

Analysis of Crack Patterns Under Three-Dimensional Residual Stress Field

George Labeas, Ioannis Diamantakos, and Theodoros Kermanidis

Abstract Welding joining processes such as, laser-beam, friction-stir and arc welding lead to the development of residual stresses. Cracks developed at the influenced regions grow under the effect of these residual stresses and additional attention is required to be paid to the assessment of the structural integrity. In the present work three-dimensional Finite Element Analysis is utilized for the calculation of Stress Intensity Factors at cracks in welded components. The developed residual stress fields are calculated from the numerical simulation of the thermo-mechanical process. The fracture analysis methodology is initially validated by comparison of numerical results with existing analytical solutions for the case of a flat plate under typical welding residual stresses. Finally, the numerical procedure is applied to the study of the effect of residual stress fields on SIF values at cracks in welded stiffened panels.

Keywords Residual stresses · Welded panels · Stress intensity factors

1 Introduction

Residual Stresses (RS) are the usual consequence of all welding processes (i.e. arc welding, laser-beam welding – LBW and friction-stir welding – FSW) due the thermo-mechanical treatment of the material [1]. Development of RS leads to undesired geometrical distortions [2, 3] and possibly reduction of structural integrity [4, 5]. Furthermore, the behaviour of cracks located inside or near the RS affected zone is seriously influenced; crack initiation period, crack propagation rate or angle and residual strength may be significantly affected by the presence of RS in the vicinity of the cracked area.

G. Labeas (✉), I. Diamantakos, and T. Kermanidis
Laboratory of Technology and Strength of materials, University of Patras,
Panepistimioupolis Rion, 26500 Patras, Greece
e-mail: {labeas, diamond, kermanid}@mech.upatras.gr

S. Pantelakis, C. Rodopoulos (eds.), *Engineering Against Fracture: Proceedings of the 1st Conference*,

For assessing the fatigue crack behaviour applying well accepted laws, e.g. Paris [6] and Forman [7], Stress Intensity Factors (SIF) concept is commonly used. For the calculation of SIFs under RS fields, limited works have been published. A customary method based on the superposition principle and Muskhelishvili's stress functions has been presented by Tada et al. [8] and Terada [9, 10] for the calculation of SIFs at cracks situated perpendicular to the welding bead. In the above mentioned works, only mode-I cracks are considered, while in real applications cracks at various angles with respect to the weld line usually appear.

In the present work the effect of residual stresses due to welding process on SIFs at cracks situated in welded panels is studied. A numerical methodology has been developed for the calculation of SIFs at cracks in welded components. Realistic RS fields resulting from Laser Beam Welding (LBW) have been obtained by detailed numerical simulation of the thermo-mechanical process.

The three-dimensional FE model used for the calculation of SIFs at cracked structures under the presence of RS is initially validated by the comparison of numerical results to existing analytical solutions for the case of a flat cracked plate including RS defined by an analytical relationship. Then, the validated numerical methodology is applied in the case of a stiffened panel produced by the High Speed Machining (HSM) process that introduces no significant RS in the structure and for the case of a stiffened panel produced by LBW for the assessment of the effect of RS on the calculated SIFs. Calculated SIF results are qualitatively compared to experimental measurements and useful conclusions are drawn concerning the SIF values governing the propagation procedure.

2 Cracked Plate under RS Field

2.1 Geometry and RS Field Considered

A 300 mm square plate with 1 mm thickness is considered, resulting from the welding of two identical segments. A centrally located crack of half length a is considered, as shown in Fig. 1. The crack is perpendicular to the vertical (weld-line) axis. The stress σ_y, which represents the RS field, is considered to be constant in y (parallel to the weld line) and z (through-the-thickness) directions, while its variation along x-coordinate is schematically presented in Fig. 1.

The typical RS distribution may be described by the following equation:

$$\sigma_y(x) = \sigma_{0y} \frac{1-\left(\frac{x}{c_0}\right)^2}{1+\left(\frac{x}{c_0}\right)^4} \tag{1}$$

where, σ_{0y} is a parameter defining the maximum value of the tensile residual stress and c_0 is the distance from y-axis at which the residual stress value changes from positive to negative, i.e. from tension to compression.

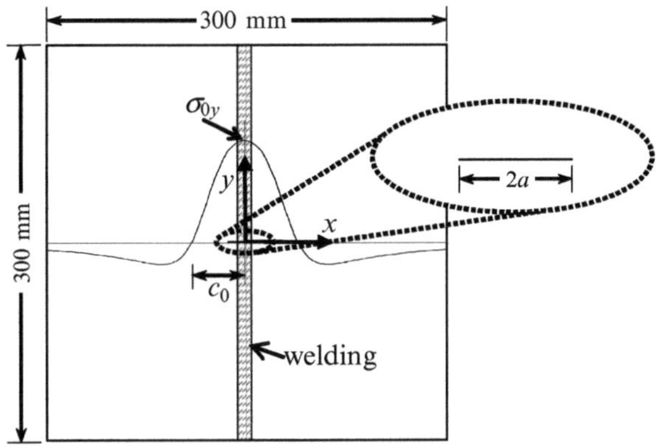

Fig. 1 Geometry and residual stress distribution considered in the investigated cracked plate

2.2 Numerical Analysis Methodology and Validation

3-D FE modelling has been used for the stress analysis and the calculation of SIFs of the cracked structure. In order to be able to simulate the stress singularity observed at the crack front (the edge of the crack in a 3-D model) when linear elastic fracture mechanics is considered, the elements around the crack front should be quadratic, with the mid-side nodes placed at the quarter points (singular elements).

SOLID95 type of element of ANSYS FE code has been used for the model development of the cracked plate. It is a higher order 3-D 20-node solid element having three degrees of freedom per node: translations in the nodal x, y, and z directions. The FE model developed for the case of a cracked flat welded plate is presented at Fig. 2a. The discretization of the area around the crack tip can be seen in the FE mesh detail of Fig. 2b.

Consequently, the RS values are calculated at element centroids according to Eq. (1) and are introduced in the FE model as initial stresses. Typical values of RS parameters $\sigma_{0y} = 200$ MPa and $c_0 = 2.5$ mm observed at welded structures have been considered in this study. Linear elastic finite element analysis and SIF are calculated using the nodal displacements in the vicinity of the crack tip. As displacements are not constant through-the-thickness of the plate, due to plane-stress/plane-strain effect, calculated SIFs also exhibit a respective variation, with SIF being maximum at the middle thickness and minimum at the surface.

In order to validate the numerical methodology, the analytical relationships of [11], providing SIFs for mode-I cracks located in a one-dimensional RS field have been used. The average value of the through-the-thickness numerically computed SIF distribution (K_{aver}) has been considered as a characteristic SIF value. In Fig. 3, computed values of K_{aver} for different crack lengths are compared to those analytically derived, indicating a very good agreement, especially for the maximum and average values of the through-the-thickness SIF.

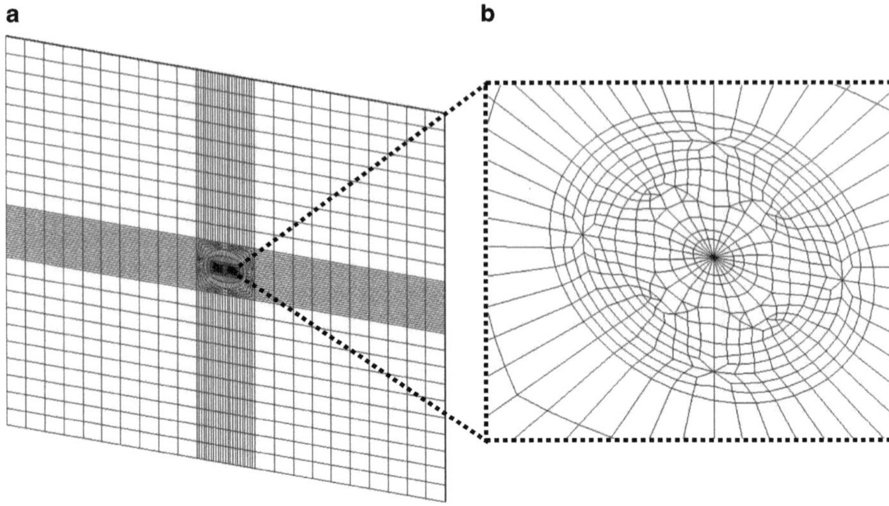

Fig. 2 **a** Typical FE model of the plate, **b** FE mesh detail around the crack tip

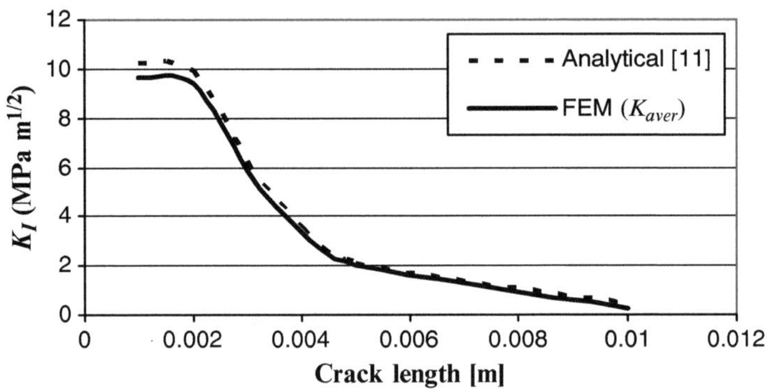

Fig. 3 Comparison of analytical and numerical SIF results for mode-I cracks located in a one-dimensional RS field

3 Analysis of Stiffened Panels

3.1 Stiffened Panel Geometry and RS Field

The basic geometrical features of the stiffened panel are presented in Fig. 4. The panel is 450 mm wide and 1,000 mm long. The skin thickness is 2 mm, except in the stiffened area where 3 mm thick sockets exist. The thickness of the stiffeners is 2.5 mm and their pitch is 150 mm. After the panel manufacturing an artificial through crack of 20 mm length is created in the panel centre vertical to the stiffeners direction. The panel is made of 6,056 T3 Al alloy.

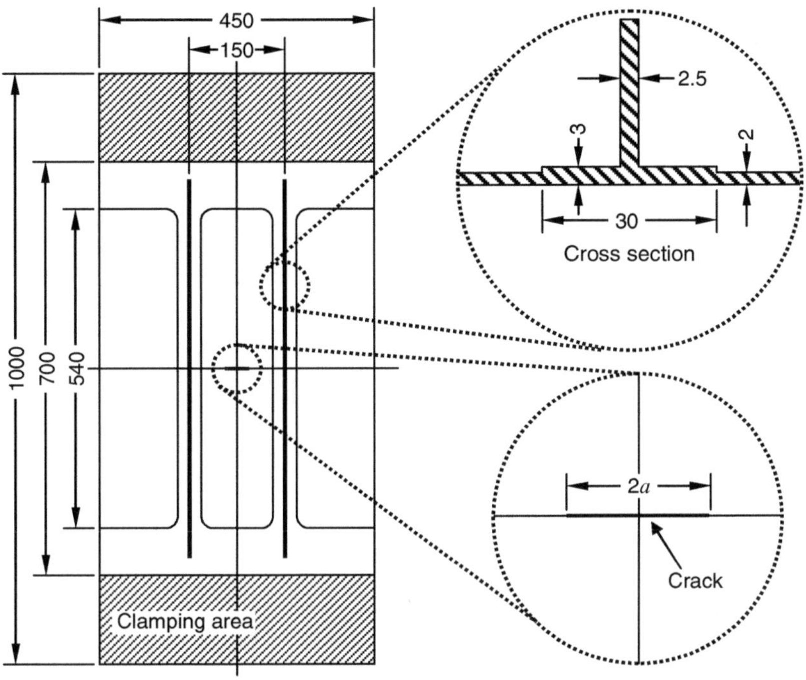

Fig. 4 Stiffened panels geometry

Two different manufacturing processes have been applied for the production of the stiffened panels: HSM and LBW. Using HSM stiffened panels are produced from machining of thick plates. By the application of HSM process it can be considered that no RS are introduced in the structure. The alternative LBW process applies two laser beams to weld stiffeners to the skin creating a T-joint. Nd YAG laser with a 3.5 KW power, 4 m/min welding speed, 600 μm fibre diameter and 150–200 mm focal length has been used. RS resulting from the LBW process are calculated by the numerical thermo-mechanical simulation of the LBW process described in [12]. The variation of the calculated RS field σ_x component (stress parallel to the weld line) along the panel width and at the centre of the stiffened panel (area of the crack) is presented in Fig. 5. The residual stress values at the top and the bottom of the skin are included in this diagram, while in the detail of Fig. 5 the stress variation at the stiffeners base area is presented. It may be observed that higher RS are developed at the top than at the bottom of the skin.

3.2 *Analysis of Stiffened Panels*

The validated FE model of the centrally cracked flat plate described in Sect. 2 is extended to the modelling of the cracked stiffened panel. The corresponding FE

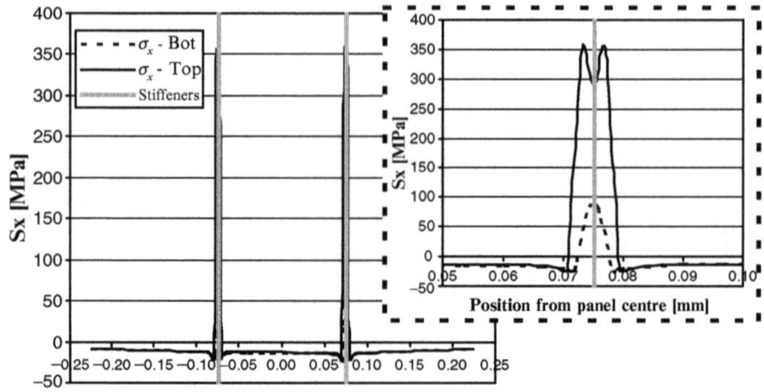

Fig. 5 Variation of the calculated RS field σ_x component along the panel width

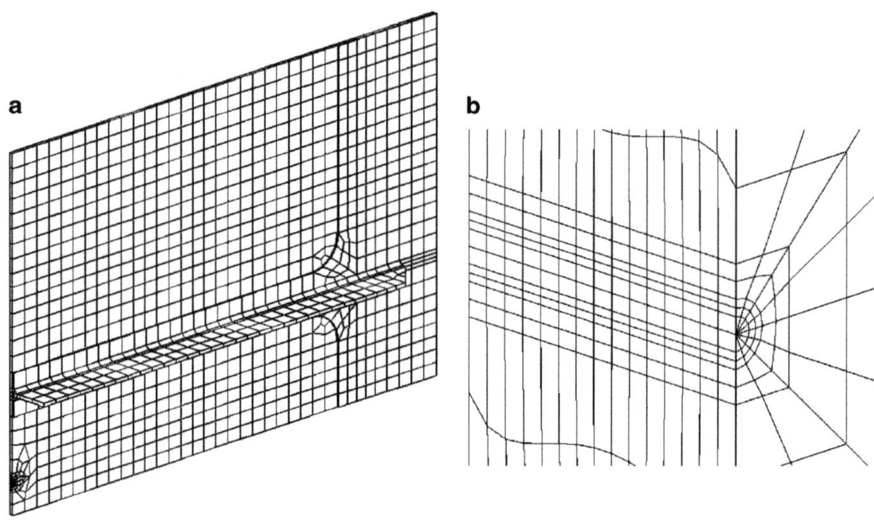

Fig. 6 FE model of cracked LBW panel

model is presented in Fig. 6a. The same model is used for the cases of HSM and LBW panels. As the panel has two axis of symmetry only one quarter of the geometry is modelled and proper boundary conditions are applied. The model consists of about 2,800 elements and 15,500 nodes. These numbers vary slightly depending on the size of the crack. Singular elements are used along with concentric rings of elements around the crack front forming a "spider-web" mesh pattern (Fig. 6b).

A remote tension fatigue load with $\sigma_{max} = 80$ MPa and $R = \sigma_{min}/\sigma_{max}$ ratio equal to 0.1 is applied at the panel. Additionally, for the case of LBW stiffened panel the residual stresses calculated by the numerical simulation of the thermo-mechanical process are applied at the element centroids. Due to the 3-D modelling of the plate, the developed bending moments and the 3-D nature of RS in the case of

LBW, a variation of the calculated SIF through-the thickness of the plate is observed. Again the average value K_{aver} of the through-the-thickness distribution is used.

The variation of calculated K_{aver} magnitudes for the maximum applied load is plotted against crack length in Fig. 7 for the cases of LBW and HSM panels. It is clear that residual stresses due to LBW lead to a reduction of the calculated SIF values. This is in qualitative agreement to experimental results of LBW panels, which exhibit lower fatigue crack propagation rates as compared to the respective in HSM panels (see Fig. 8) [13]. It may be attributed to the fact that, as it can be seen from the diagram in Fig. 5, RS are mainly compressive away of the stiffeners area, where most of the fatigue crack propagation takes place. However, no calculations concerning fatigue crack propagation in LBW panels could be made, as Forman law

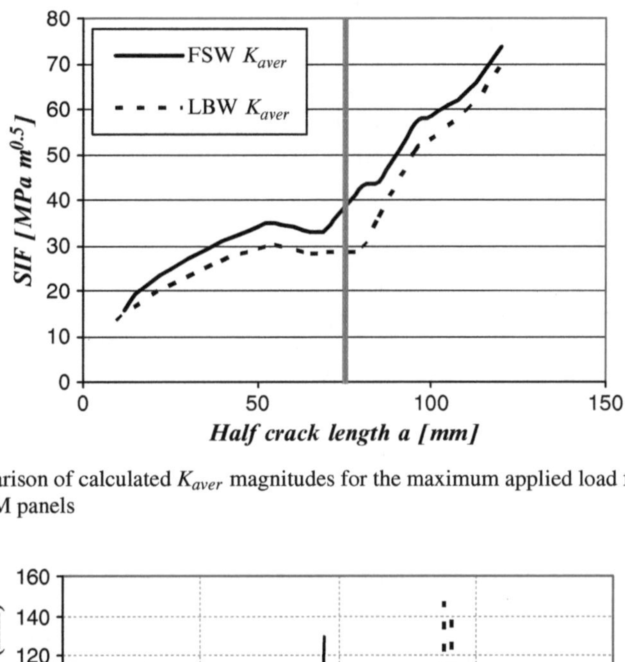

Fig. 7 Comparison of calculated K_{aver} magnitudes for the maximum applied load for the cases of LBW and HSM panels

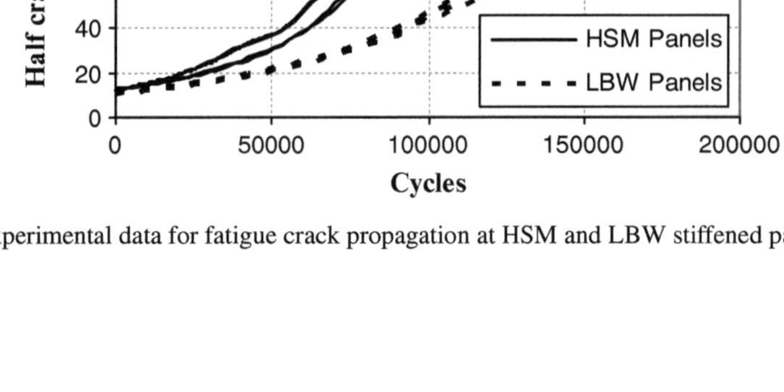

Fig. 8 Experimental data for fatigue crack propagation at HSM and LBW stiffened panels [13]

constants for the specific Al alloy (6,056) where not available; in this case Paris law is not applicable, as it neglects the effect of RS on the calculated fatigue crack propagation rate.

4 Conclusions

In the present work a numerical methodology has been developed, based on three-dimensional FEM, for the calculation of SIFs at cracks in structures under RS. The proposed methodology has been validated for the cases of a flat plate with RS and of a stiffened panel produced by HSM. From the application of the numerical methodology at RS affected panels the main conclusions are that through-the-thickness SIF variation is observed due to the transition from plane stress condition at the surface of the material to plane strain at the interior. The average value of through-the-thickness computed SIF values seems to be dominant for the fatigue crack propagation. Finally, residual stresses due to LBW lead to reduction of developed SIF for the crack configurations studied, where fatigue crack propagation area is mainly affected by compressive residual stresses.

Acknowledgement Part of this work was performed in the frame of the European Research Programme "Innovative Fatigue and Damage Tolerance Methods for the Application of New Structural Concepts" (DaToN). The financial support of the European Union under contract AST4-CT-2005-516053 is gratefully acknowledged. The experiments at stiffened panels were conducted by IDMEC, University of Pisa and Sheffield Hallam University in the frame of research activities of the above referred project.

References

1. Masubuchi K. Analysis of welded structures. Oxford: Pergamon Press, 1980.
2. Tsirkas SA, Papanikos P, Pericleous K, Strusevich N, Boitout F, Bergheau JM. Evaluation of distortions in laser welded shipbuilding parts using local-global finite element approach. Sci Technol Weld Joi 2003; 8:79–88.
3. Tsirkas SA, Papanikos P, Kermanidis Th. Numerical simulation of the laser welding process in butt-joint specimens. J Mater Process Technol 1995; 134(1):59–69.
4. Dalle Donne C, Biallas G, Ghidini T, Raimbeaux G. Effect of weld imperfections and residual stresses on the fatigue crack propagation in friction stir welded joints. In: Second International Conference on Friction Stir Welding, 26–28 June, 2000, Gothenburg, Sweden. Abington Hall: TWI 2000.
5. Cho SK, Yang YS, Son KJ, Kim JY. Fatigue strength in laser welding of the lap joint. Finite Elem Anal Des 2004; 40(9):1059–1070.
6. Paris P, Erdogan F. A critical analysis of crack propagation laws. J Basic Eng-Trans ASME 1963; 85:528–534.
7. Forman RG, Kearney VR, Engle RM. Numerical analysis of crack propagation in a cyclic-loaded structure. J Basic Eng-Trans ASME 1967;.89D:459.
8. Tada H, Paris PC. The stress intensity factor for a crack perpendicular to the welding bead. Int J Fracture 1982; 21(4):279–284.

9. Terada H. An analysis of the stress intensity factor of a crack perpendicular to the welding bead. Eng Fracture Mech 1976; 8(2):441–444.
10. Terada H, Nakajima T. Analysis of stress intensity factor of a crack approaching welding bead. Int J Fracture 1985; 27(2):83–90.
11. Tada H, Paris PC, Irwin G. The stress analysis of cracks handbook. New York, USA: The American Society of Mechanical Engineers, 2000.
12. Moraitis GA, Labeas GN. Residual stress and distortions calculation of laser beam welding for aluminum lap joints. J Mater Process Technol 2008; 198:260–269.
13. European Research Programme "Innovative Fatigue and Damage Tolerance Methods for the Application of New Structural Concepts" (DaToN). Contract AST4-CT-2005-516053.

BEM Solutions of Crack Problems in Gradient Elasticity

Gerasimos F. Karlis, Stephanos V. Tsinopoulos, Demosthenes Polyzos, and Dimitri E. Beskos

Abstract In this paper a boundary element method is used to solve gradient elastic fracture mechanics problems under static loading in three dimensions. A simplified version of Mindlin's Form II higher order gradient elastic theory is exploited and the Boundary Element Method (BEM), recently proposed by Polyzos et al. (2003) and Tsepoura et al. (2003), is utilized for the solution of Mode I gradient elastic problem. The Stress Intensity Factors (SIFs) are determined with the aid of a new special variable-order singularity discontinuous element applied at the tip of the crack. This element is designed to deal with fields that have different orders of singularities, as it is the case for displacement and traction fields in classical elastic case. Finally, a numerical example concerning the SIF calculation of a 3D mode I crack is presented.

Keywords Gradient elasticity · BEM · Fracture mechanics · Microstructure · 3D mode I crack

---

G.F. Karlis
Department of Mechanical and Aeronautical Engineering, University of Patras, GR-26500, Patras, Greece
e-mail: gkarlis@mech.upatras.gr

S.V. Tsinopoulos
Department of Mechanical Engineering, Technological and Educational Institute of Patras, GR-26334, Patras, Greece
e-mail: stsinop@teipat.gr

D. Polyzos (✉)
Department of Mechanical and Aeronautical Engineering, University of Patras, GR-26500, Patras, Greece and Institute of Chemical Engineering and High Temperature Chemical Process, Patras, Greece
e-mail: polyzos @mech.upatras.gr

D.E. Beskos
Department of Civil Engineering, University of Patras, GR-26500, Patras, Greece
e-mail: d.e.beskos@upatras.gr

S. Pantelakis, C. Rodopoulos (eds.), *Engineering Against Fracture: Proceedings of the 1st Conference*,

1 Introduction

It is well known that in classical elasticity all the fundamental quantities and material constants defined at any point of the analyzed elastic body are taken as mean values over very small volume elements the size of which must be sufficiently large in comparison with the material microstructure. Considering a very simple one dimensional example and taking Taylor expansions for displacements around the point of interest, Exadaktylos and Vardoulakis [1] explain that this assumption is possible only when displacements vary constantly or linearly throughout the aforementioned representative volume elements (RVEs). In cases where non-linear variations of displacements are observed, higher order Taylor expansion terms and thus higher order gradients of displacements should be taken into account. Making use of higher Taylor terms, however, some new internal length scale constants correlating the microscopic RVE with the macrostructure are introduced in the constitutive equations of the considered elastic continuum [2]. Thus, in fracture mechanics problems where near the tip of the crack abrupt changes of strains and stresses are observed, enhanced elastic theories that take into account higher order gradients of strains and stresses and introduce new internal length scale parameters to describe microstructural effects, should be applied.

In the framework of isotropic linear elasticity, the most general and comprehensive gradient elastic theory is the one proposed by Mindlin [3] . However in order to balance the dimensions of strains and higher order gradients of strains as well as to correlate the micro-strains with macro-strains, Mindlin utilized 18 new constants rendering thus his initial general theory very complicated from physical and mathematical point of view. In the same paper, considering long wavelengths and the same deformation for macro and micro structure Mindlin proposed 3 new simplified versions of his theory, known as Form I, II and III, where beyond the 2 Lamè constants other 5 ones are introduced instead of 16 employed in his initial model. The most important difference among those is the fact that Form II leads to a total stress tensor, which is symmetric as in the case of classical elasticity. This symmetry avoids the problems introduced by the non-symmetric stress tensors in Cosserat and couple stresses theories. In the middle of 1990s, Vardoulakis et al. [3] simplified further Mindlin's Form II gradient elastic theory, proposing, for the static case, the use of only one internal length scale parameter, instead of five introduced in the initial Form II theory. Adopting this theory, Polyzos et al. [4] and Tsepoura et al. [5] developed an advanced Boundary Element Method (BEM) for solving two dimensional (2D) and three dimensional (3D) static gradient elastic problems.

Very recently, Karlis et al. [6] addressed a numerical methodology, which combines the aforementioned BEM with special crack tip boundary elements for the numerical determination of the Stress Intensity Factor (SIF) in plane Mode I and Mixed Mode (I & II) fracture mechanics gradient elastic problems. Adopting the idea of variable-order singularity boundary elements around the tip of the crack for the evaluation of the corresponding SIFs (Lim et al. [7]; Zhou et al. [8]), a new special variable-order singularity discontinuous element was proposed for the treatment of singular fields around the tip of the crack.

The present paper can be considered as an extension of the above work to three dimensions and is organized as follows: the constitutive equations and the classical as well as the non-classical boundary conditions of the considered gradient elastic theory, which is a simplified version of Mindlin's Form II gradient elastic theory, are presented in the next section. The 3D special variable-order discontinuous elements are illustrated in the section after next. The integral representation, the numerical implementation of the problem and the solution procedure of the utilized BEM code and the SIF calculation are explained in brief in the fourth section. Finally, a 3D mode-I crack problem is solved and the obtained crack profiles, tractions and the corresponding SIFs are presented and discussed.

2 Simplified Form II Gradient Elastic Theory

Mindlin in the Form II version of his gradient elastic theory [2] considered that the potential energy density $\hat{W}$ is a quadratic form of the strains and the gradient of strains, i.e.

$$\begin{aligned}\hat{W} = \frac{1}{2}\lambda\varepsilon_{ii}\varepsilon_{jj} + \mu\varepsilon_{ij}\varepsilon_{ij} + \hat{a}_1\,\hat{\kappa}_{iik}\,\hat{\kappa}_{kjj} + \hat{a}_2\,\hat{\kappa}_{ijj}\,\hat{\kappa}_{ikk}\\ + \hat{a}_3\,\hat{\kappa}_{iik}\,\hat{\kappa}_{jjk} + \hat{a}_4\,\hat{\kappa}_{ijk}\,\hat{\kappa}_{ijk} + \hat{a}_5\,\hat{\kappa}_{ijk}\,\hat{\kappa}_{kji}\end{aligned} \tag{1}$$

where $\varepsilon_{ij} = (\partial_i u_j + \partial_j u_i)/2$, $\hat{\kappa}_{ijk} = \partial_i\varepsilon_{jk} = (\partial_i\partial_j u_k + \partial_i\partial_k u_j)/2 = \hat{\kappa}_{ikj}$ and λ, μ and $\hat{a}_1 \div \hat{a}_5$ material constants. It should be noticed that the constants λ, μ are not the same with the corresponding Lamè ones of the classical elasticity.

Strains ε_{ij} and gradients of strains $\hat{\kappa}_{ijk}$ are dual in energy with the Cauchy and double stresses, respectively, defined as

$$\hat{\tau}_{ij} = \frac{\partial\hat{W}}{\partial e_{ij}} = \hat{\tau}_{ji} \quad \text{and} \quad \hat{\mu}_{ijk} = \frac{\partial\hat{W}}{\partial\,\hat{\kappa}_{ijk}} = \hat{\mu}_{ikj} \tag{2}$$

which implies that

$$\hat{\tau}_{pq} = 2\mu e_{pq} + \lambda e_{ii}\delta_{pq} \tag{3}$$

and

$$\begin{aligned}\hat{\mu}_{pqr} = \frac{1}{2}\hat{a}_1\left[\hat{\kappa}_{rii}d_{pq} + 2\hat{\kappa}_{iip}d_{qr} + \hat{\kappa}_{qii}d_{rp}\right] + 2\hat{a}_2\,\hat{\kappa}_{pii}d_{qr}\\ + \hat{a}_3(\hat{\kappa}_{iir}d_{pq} + \hat{\kappa}_{iiq}d_{pr}) + 2\hat{a}_4\,\hat{\kappa}_{pqr} + \hat{a}_5\left(\hat{\kappa}_{rpq} + \hat{\kappa}_{qrp}\right)\end{aligned} \tag{4}$$

The first subscript of the double stress tensor μ_{ikj} indicates the normal vector on the surface on which the double stresses act, while the other two have the same significance as the corresponding ones of the classical stress tensor τ_{ij}.

Ignoring body forces, considering smooth boundaries and taking the variation of Eq. (1), one obtains the following equilibrium equation

$$\partial_j\sigma_{jk} = \partial_j(\hat{\tau}_{jk} - \partial_i\hat{\mu}_{ijk}) = 0 \tag{5}$$

accompanied by the boundary conditions

$$u_k = \bar{u} \text{ and/or } \hat{P}_k = \bar{p} \tag{6}$$

$$n_l \partial_l u_k = \bar{q} \text{ and/or } \hat{R}_k = \bar{R} \tag{7}$$

where $\sigma_{jk}, \hat{\tau}_{jk}, -\partial_i \hat{\mu}_{ijk}$ represent the symmetric total, Cauchy and relative stress tensors, respectively, $\bar{u}, \bar{p}, \bar{q}, \bar{R}$ are prescribed vectors, n_k is the unit vector normal to the boundary and $\hat{P}_k, \hat{R}_k$ are the traction and double traction vectors, respectively, having the form

$$\hat{P}_k = n_j \hat{\tau}_{jk} - n_i n_j D \hat{\mu}_{ijk} - (n_j D_i + n_i D_j) \hat{\mu}_{ijk} + (n_i n_j D_l n_l - D_j n_i) \hat{\mu}_{ijk} \tag{8}$$

$$\hat{R}_k = n_i n_j \hat{\mu}_{ijk} \tag{9}$$

where $D_i = \partial_i - n_i n_k \partial_k$ and $D = n_l \partial_l$.

Inserting (3) and (4) into (5) one obtains the equilibrium equation of a Form II gradient elastic material written in terms of displacements as

$$(\lambda + 2\mu)(1 - \hat{l}_1^2 \nabla^2)\nabla\nabla \cdot \mathbf{u} - \mu(1 - \hat{l}_2^2 \nabla^2)\nabla \times \nabla \times \mathbf{u} = 0 \tag{10}$$

where $\hat{l}_1^2 = 2(\alpha_1 + \alpha_2 + \alpha_3 + \alpha_4 + \alpha_5)/(\lambda + 2\mu)$ and $\hat{l}_2^2 = (\alpha_3 + \alpha_4 + \alpha_5)/2\mu$.

Although very elegant, the use of the Form II gradient elastic theory for the solution of real problems is discouraging since seven new material constants, i.e. λ, μ, $a_1 \div a_5$ have to be determined. If the constants $a_1 \div a_5$ take the values $\hat{a}_1 = \hat{a}_3 = \hat{a}_5 = 0$, $\hat{a}_2 = (\lambda/2) g^2$, $\hat{a}_4 = \mu g^2$, then the constitutive Eqs. (3, 4) as well as the equilibrium Eq. (10) can be written as

$$\hat{\tau} = \mu(\nabla\mathbf{u} + \mathbf{u}\nabla) + \lambda(\nabla \cdot \mathbf{u})\mathbf{I} \tag{11}$$

$$\hat{\mu} = g^2 \nabla \hat{\tau} \tag{12}$$

$$(\lambda + 2\mu)\nabla\nabla \cdot \mathbf{u} - \mu\nabla \times \nabla \times \mathbf{u} - g^2 \nabla^2 \left[(\lambda + 2\mu)\nabla\nabla \cdot \mathbf{u} - \mu\nabla \times \nabla \times \mathbf{u}\right] = 0 \tag{13}$$

where λ, μ are the classical Lamè constants and g^2 is the volumetric strain gradient energy coefficient or simply the gradient coefficient, which is introduced to balance the dimensions of strains and strain gradients and moreover to relate the microstructure with the macrostructure representing a characteristic length of the material. As it is stated in [3], the gradient elastic theory described by Eqs. (11–13) and accompanied by the classical and non-classical boundary conditions (6) and (7), respectively, is the simplest possible special case of Mindlin's Form II theory. As it has already been mentioned, the present work exploits this simple enhanced elastic theory, which from now will be referred to as Mindlin's simplified Form II gradient elastic theory.

3 Special Elements with Variable Order of Singularity

As it is illustrated in [6], near the crack front, displacements **u**, normal derivative of **u**, double tractions **R** and tractions **P** vary as $r^{3/2}$, $r^{1/2}$, $r^{-1/2}$ and $r^{-3/2}$ respectively, with r being the distance from the crack front. Adopting the idea of using variable-order elements [7, 8] a quadrilateral, eight-nodded element with variable-order singularity is proposed for the treatment of the fields around the crack front.

In order to deal with all these fields, the interpolation functions of the proposed element are considered to be not only functions of the local variables ξ_1 and ξ_2, but also functions of the distance of the point from the crack front. In addition, two extra parameters have been defined to describe the field-specific behavior.

The elements' functional nodes are the same as these of the corresponding classical ones and all their sides are allowed to be continuous or discontinuous, with the exception of the side, where the crack tip or crack front resides, for which discontinuity is compulsory.

Since the crack front is allowed to reside on any of the elements' sides, a renumbering of the local nodes is applied to address all the possible crack locations. This local renumbering is illustrated with an example in Fig. 1 and ensures that the crack front will always reside on the first side of the element.

The field of interest **F** near the crack front can be expressed in terms of the asymptotic solution as

$$\mathbf{F}\left(\xi'_1, r\right) = N^i\left(\xi'_1, r\right)\mathbf{F}^i, i = 1, ..., 8 \tag{14}$$

where r is the distance of the current point from the crack front, ξ'_1 is explained in Fig. 1, the symbol **F** represents any of the fields **u**, **q**, **R** and **P** and the shape functions N^i have the form

$$\begin{aligned} N^i\left(\xi'_1, r\right) = & \left(e^i_1 + e^i_2\xi'_1 + e^i_3\xi'^2_1\right) r^{\lambda_1} \\ & + \left(e^i_4 + e^i_5\xi_1 + e^i_6\xi'^2_1\right) r^{\lambda_2} + e^i_7 + e^i_8\xi'_1 + e^i_9\xi'^2_1 \end{aligned} \tag{15}$$

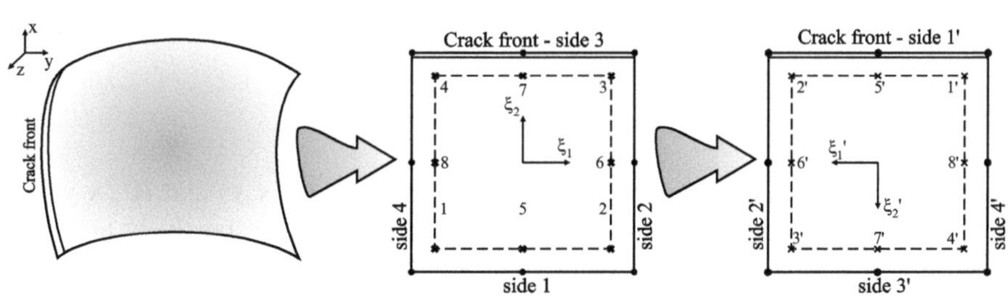

Fig. 1 Transformation from the real 3D space to the parametric representation of the element and nodal renumbering for the case of a fully discontinuous element

Table 1 Values of λ_1, λ_2 in Eq. (15)

F	λ_1	λ_2
u	3/2	1
q	1/2	1
R	−1/2	1
P	−3/2	−1/2

where λ_1, λ_2 take the values of Table 1 and the constants e^i_j can be determined by the well-known δ-property of shape functions

$$N^i\left(\xi'^{j}_{1,r^j}\right) = \delta_{ij},\ \ i,j = 1,...,8 \tag{16}$$

with δ_{ij} being the Kronecker delta, $r^j = \left|\mathbf{x}\left(\xi'^{j}_1, \xi'^{j}_2\right) - \mathbf{y}\left(\xi'^{j}_1, -1\right)\right|$ and $\left(\xi'^{j}_1, \xi'^{j}_2\right)$ the local coordinates of the j functional node. However, since the element is eight noded and only eight equations arise for each functional node, one of the e^i_j constants cannot be calculated and must be omitted. In the present work the constant e^i_9 was omitted, since as $r \to 0$, i.e. near the crack front, the terms r^{λ_1} and r^{λ_2} become dominant in the traction and double traction fields and their coefficients are useful for the calculation of the SIFs, as will be shown in the next section.

4 BEM Procedure and SIF Calculation

Considering a gradient elastic material of volume V surrounded by a smooth boundary S and characterized by two Lamè constants λ, μ and a gradient coefficient g^2, Polyzos et al. [4] showed that for any static gradient elastic boundary value problem its integral representation reads

$$\begin{aligned}\tilde{\mathbf{c}}(\mathbf{x}) \cdot \mathbf{u}(\mathbf{x}) + \int_S \{\tilde{\mathbf{P}}^*(\mathbf{x},\mathbf{y}) \cdot \mathbf{u}(\mathbf{y}) - \tilde{\mathbf{u}}^*(\mathbf{x},\mathbf{y}) \cdot \mathbf{P}(\mathbf{y})\} dS_\mathrm{y} \\ = \int_S \left\{ \frac{\partial \tilde{\mathbf{u}}^*(\mathbf{x},\mathbf{y})}{\partial n_\mathrm{y}} \cdot \mathbf{R}(\mathbf{y}) - \tilde{\mathbf{R}}^*(\mathbf{x},\mathbf{y}) \cdot \mathbf{q}(\mathbf{y}) \right\} dS_\mathrm{y}\end{aligned} \tag{17}$$

where $\tilde{\mathbf{u}}^*(\mathbf{x},\mathbf{y})$ is the fundamental solution given in [4], $\tilde{\mathbf{P}}^*(\mathbf{x},\mathbf{y})$ and $\mathbf{R}^*(\mathbf{x},\mathbf{y})$ are the fundamental traction and double stress traction tensors, respectively and $\tilde{\mathbf{c}}(\mathbf{x})$ is the well-known jump tensor being equal to $(1/2)\tilde{\mathbf{I}}$ for $\mathbf{x} \in S$ and equal to $\tilde{\mathbf{I}}$ if $\mathbf{x} \in V \cap S$. Observing Eq. (17) one can see that it contains four unknown fields $\mathbf{u}(\mathbf{x})$, $\mathbf{P}(\mathbf{x})$, $\mathbf{R}(\mathbf{x})$ and $\mathbf{q}(\mathbf{x}) = \partial \mathbf{u}/\partial n$. Thus, the evaluation of the unknown fields requires the existence of one more integral equation. This integral equation is obtained by applying the operator $\partial/\partial n_x$ on Eq. (17), i.e.

$$\begin{aligned}&\mathbf{c}(\mathbf{x})\cdot\frac{\partial\mathbf{u}(\mathbf{x})}{\partial n_x}+\int_S\left\{\frac{\partial\tilde{\mathbf{P}}^*(\mathbf{x},\mathbf{y})}{\partial n_x}\cdot\mathbf{u}(\mathbf{y})-\frac{\partial\tilde{\mathbf{u}}^*(\mathbf{x},\mathbf{y})}{\partial n_x}\cdot\mathbf{P}(\mathbf{y})\right\}dS_{\mathrm{y}}\\&=\int_S\left\{\frac{\partial^2\tilde{\mathbf{u}}^*(\mathbf{x},\mathbf{y})}{\partial n_x\partial n_y}\cdot\mathbf{R}(\mathbf{y})-\frac{\partial\tilde{\mathbf{R}}^*(\mathbf{x},\mathbf{y})}{\partial n_x}\cdot\mathbf{q}(\mathbf{y})\right\}dS_{\mathrm{y}}\end{aligned}\tag{18}$$

All the kernels involved in Eqs. (17 and 18) are given explicitly in [4].

Integral Eqs. (17) and (18) accompanied by the classical and non-classical boundary conditions (6) and (7) form the integral representation of the general gradient elastic boundary value problem.

The goal of the BEM is to solve numerically the just described well-posed boundary value problem. To this end the global boundary S is discretized into quadratic, continuous and discontinuous isoparametric boundary elements, while special variable-order singularity, discontinuous elements are placed on both sides of the crack tip or crack front as it is illustrated in Fig. 2.

Then, for a node k the integral Eqs. (17 and 18) are written as

$$\frac{1}{2}\mathbf{u}^k+\sum_{\beta=1}^{M}\tilde{\mathbf{H}}^k_\beta\cdot\mathbf{u}^\beta+\sum_{\beta=1}^{M}\tilde{\mathbf{H}}^k_\beta\cdot\mathbf{q}^\beta=\sum_{\beta=1}^{M}\tilde{\mathbf{G}}^k_\beta\cdot\mathbf{P}^\beta+\sum_{\beta=1}^{M}\tilde{\mathbf{L}}^k_\beta\cdot\mathbf{R}^\beta\tag{19}$$

$$\frac{1}{2}\mathbf{q}^k+\sum_{\beta=1}^{M}\tilde{\mathbf{S}}^k_\beta\cdot\mathbf{u}^\beta+\sum_{\beta=1}^{M}\tilde{\mathbf{T}}^k_\beta\cdot\mathbf{q}^\beta=\sum_{\beta=1}^{M}\tilde{\mathbf{V}}^k_\beta\cdot\mathbf{P}^\beta+\sum_{\beta=1}^{M}\tilde{\mathbf{W}}^k_\beta\cdot\mathbf{R}^\beta\tag{20}$$

where M is the total number of nodes. Explicit expressions for $\tilde{\mathbf{H}}^k_\beta, \tilde{\mathbf{K}}^k_\beta, \tilde{\mathbf{G}}^k_\beta, \tilde{\mathbf{L}}^k_\beta, \mathbf{S}^k_\beta$, $\tilde{\mathbf{T}}^k_\beta, \tilde{\mathbf{V}}^k_\beta$ and $\tilde{\mathbf{W}}^k_\beta$ are given in [5].

Collocating Eqs. (19 and 20) at all nodal points M and applying the boundary conditions one produces the final linear system of algebraic equations of the form $\tilde{\mathbf{A}}\cdot\mathbf{X}=\mathbf{B}$, where the vectors $\mathbf{X}$ and $\mathbf{B}$ contain all the unknown and known nodal components of the boundary fields, respectively. The singular and hypersingular integrals involved, are evaluated with high accuracy by applying a methodology for direct treatment of CPV and hypersingular integrals explained

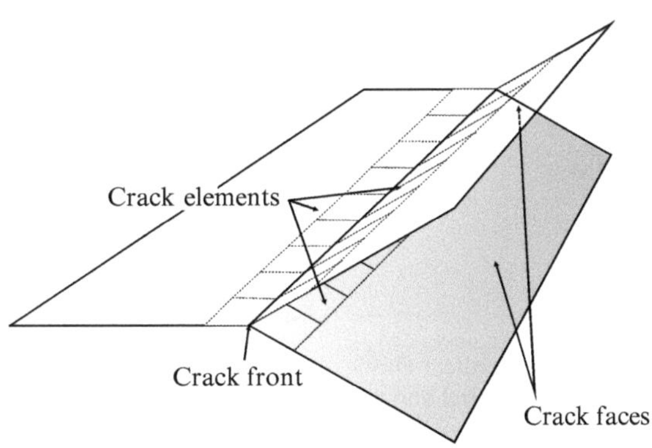

Fig. 2 Position of the variable order of singularity elements

in [5]. Near to the crack front, an extra singularity due to the singular behavior of the interpolation functions should be taken into account. The numerical treatment of all possible singular integrals defined on the special elements is illustrated in [6]. Finally, the linear system is solved via a typical LU-decomposition algorithm and the vector **X** comprising all the unknown nodal values of **u**, **P**, **R**, **q** is evaluated.

Once the boundary value problem has been solved, the calculation of SIFs is accomplished via the nodal traction values of the special elements.

Approaching the crack front ($r \to 0$), the traction **P**, in view of Eqs. (14, 15) admits a representation of the form

$$\mathbf{P} = \frac{\mathbf{K}_1(\mathbf{P}_1,\ldots,\mathbf{P}_8)}{\sqrt{2\pi}} \lim_{r\to 0} r^{-3/2} + \frac{\mathbf{K}_2(\mathbf{P}_1,\ldots,\mathbf{P}_8)}{\sqrt{2\pi}} \lim_{r\to 0} r^{-1/2} + \mathbf{C}(\mathbf{P}_1,\ldots,\mathbf{P}_8) \quad (21)$$

Taking into account relation (15) the SIFs are obtained by

$$\begin{aligned} \mathbf{K}_1(\xi_1) &= \sqrt{2\pi}\left(\mathbf{D}_1 + \xi_1\mathbf{D}_2 + \xi_1^2\mathbf{D}_3\right) \\ \mathbf{K}_2(\xi_1) &= \sqrt{2\pi}\left(\mathbf{D}_4 + \xi_1\mathbf{D}_5 + \xi_1^2\mathbf{D}_6\right) \end{aligned} \quad (22)$$

with $\mathbf{D}_i = e_i^j \mathbf{P}_j$, $i = 1,\ldots,6$ and $j = 1,\ldots,8$ for three dimensions and e_i^j constants taken from the solution of the system of equations represented by Eq. (16).

5 3D Mode I Crack Problem: Numerical Results

The problem of a gradient elastic cube containing a central horizontal line crack (Fig. 3) and subjected to an external load being perpendicular to the crack face is solved in this section. The cube has rounded corners and edges with very small radii of curvature, in order to have a smooth boundary. The edges of the cube have length $16a$ and the length of the crack is $2a$. The Poisson ratio and Young's modulus were chosen to be $\nu = 0.2$ and $E = 210\ GPa$, respectively. The gradient elastic coefficient g takes values 0.5, 0.3, 0.1 and 0.01. Due to the existing symmetries, only one eighth

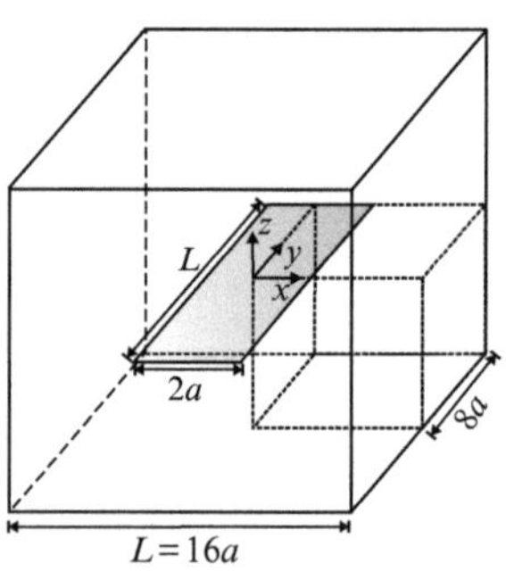

Fig. 3 The gradient elastic cube with a central horizontal crack

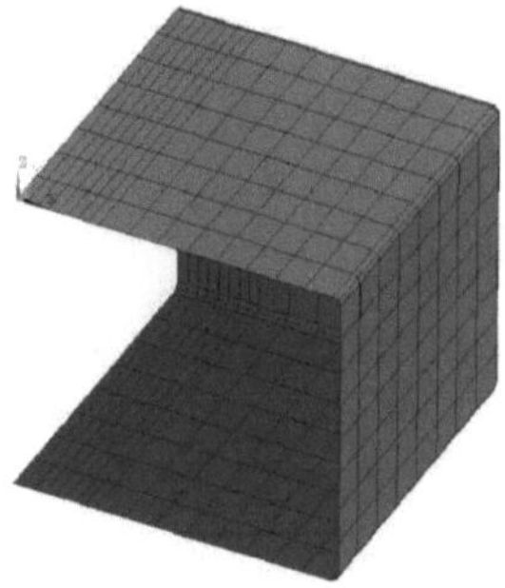

Fig. 4 The discretized domain (one eighth of the cube)

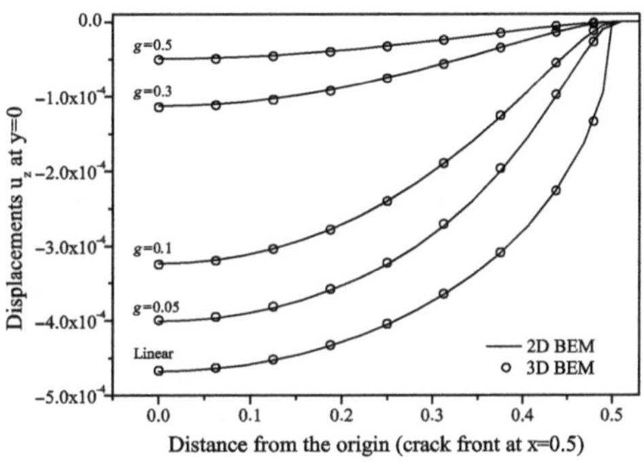

Fig. 5 Shape of mode I crack for the linear elastic case and for different values of the gradient elastic coefficient g

of the domain has been discretized. The crack face was discretized using a mesh of 4×8 elements on both sides (face and interface). From these, eight elements of each side, are special elements (Fig. 4).

In Fig. 5 the crack profile across the axis of symmetry is presented for the linear elastic case as well as for the gradient elastic case. The change in the shape of the crack profile and how the crack stiffens as g increases is apparent. An additional observation is that the profile of the crack undergoes a change in its curvature approaching the tip, an effect that becomes more pronounced as g increases. This result is consistent with Barenblatt's cohesive zone theory [9], which takes into account the interatomic forces appearing on the crack tip. These forces appear as compressive stresses that have the tendency to resist the opening of the crack. It is important to note that this kind of behavior appears without considering any other forces than those implied by Mindlin's simplified Form II gradient elastic theory.

The SIFs K_1 and K_2, explained in the previous section are calculated and plotted in Fig. 6 with respect to the gradient elastic material constant g. The obtained results are compatible with those of [6] taken for 2D cracks, with the $(K_I)_1$ SIF tending to zero and $(K_I)_2$ SIF to the linear elastic solution as g approaches zero. Finally, as g increases, the $(K_I)_1$ SIF becomes more significant and the $(K_I)_2$ SIF becomes more difficult to calculate, as it is no more the dominant term.

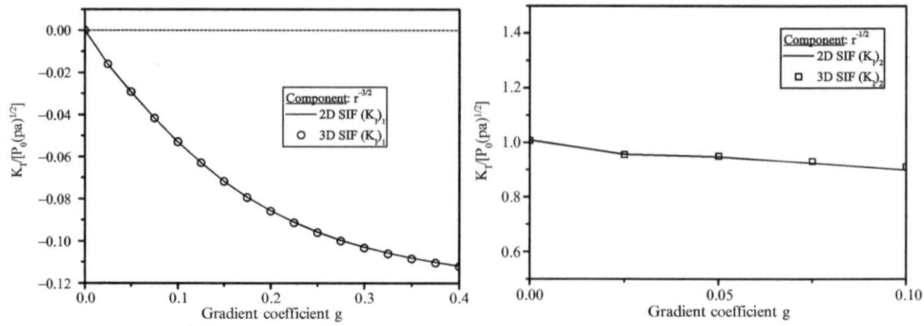

Fig. 6 The SIFs $(K_I)_1$ and $(K_I)_2$ for the mode I 3D crack corresponding to $r^{-3/2}$ and $r^{-1/2}$ terms

6 Conclusions

A displacement based BEM was employed for the solution of a 3D mode-I crack problems characterized by a linear and isotropic gradient elastic material. The Mindlin's simplified Form-II gradient elastic theory was adopted and a corresponding 3D boundary element methodology was employed. A new eight-noded discontinuous boundary element of variable singularity has been developed and used for the determination of the mode-I SIFs directly via the nodal traction values and after the determination of boundary tractions and displacements.

All the results concerning the behavior of the considered fields near the crack tip or front were obtained directly in the context of Mindlin's simplified Form-II gradient elastic theory, without considering extra conditions. Stresses and displacements around the crack tip were calculated with high accuracy and showed to be more physically acceptable than those of the case of classical elasticity.

References

1. Exadaktylos GE, Vardoulakis I (2001) Microstructure in linear elasticity and scale effects: a reconsideration of basic rock mechanics and rock fracture mechanics. Technophysics 335: 81–109.
2. Mindlin RD (1964) Microstructure in linear elasticity. Arch Rat Mech Anal 10:51–78.
3. Vardoulakis I, Exadaktylos GE, Aifantis ED (1996) Gradient elasticity with surface energy: mode III crack problem. Int J Solid Struct 33:4531–4539.
4. Polyzos D, Tsepoura KG, Tsinopoulos SV, Beskos DE (2003) A boundary element method for solving 2-D and 3-D static gradient elastic problems. Part I: integral formulation. Comp Meth Appl Mech Eng 192(26–27):2845–2873.
5. Tsepoura KG, Tsinopoulos SV, Polyzos D, Beskos DE (2003) A boundary element method for solving 2-D and 3-D static gradient elastic problems. Part II: numerical implementation. Comp Meth Appl Mech Eng 192(26–27):2875–2907.
6. Karlis GF, Tsinopoulos SV, Polyzos D, Beskos DE (2007) Boundary element analysis of mode I and mixed mode (I and II) crack problems of 2-D gradient elasticity. Comput Meth Appl Mech Eng 196:5092–5103.

7. Lim KM, Lee KH, Tay AAO, Zhou W (2002) A new variable-order singular boundary element for two-dimensional stress analysis. Int J Numer Meth Eng 55:293–316.
8. Zhou W, Lim KM, Lee KH, Tay AAO (2005) A new variable-order singular boundary element for calculating stress intensity factors in three-dimensional elasticity problems. Int J Solid Struct 42:159–185.
9. Barenblatt GI (1962) Mathematical theory of equilibrium cracks in brittle fracture. Adv Appl Mech 7:55–129.

Fracture Analysis of Medium Density Polyethylene

Ivica Skozrit and Zdenko Tonković

Abstract The paper deals with the application of the reference stress method (RSM) to estimate the J and C integrals of cracked thick-walled metal as well as medium density polyethylene (MDPE) pipes. Unlike the existing solutions, the newly developed analytical approximations of the plastic limit pressure and J-integral are applicable to a wide range of crack dimensions. Based on the experimental data from literature and analogy between plasticity and creep, the paper discusses a method used to develop the efficient computational strategy for modeling creep fracture mechanisms by slow crack growth in a MDPE pipes.

Keywords Finite element analysis · Polyethylene · Creep · C^*-integral

1 Introduction

The increasing application of polymeric materials especial polyethylenes as structural materials demands new methodologies in order to assess the material capability to withstand loads. The use of MDPE pipes for water and gas distribution is one of the most common examples. An accurate modeling of fracture and viscoelastic material responses of such structures, represent a key for the prediction of structural integrity.

Since the polyethylene structures are mostly subjected to creep loadings, the present paper is concerned with the numerical modeling of creep fracture mechanisms by slow crack growth in MDPE pipes [1]. The failure assessment philosophy for polymers is similar to philosophies for metals [2]. Moreover, the J-integral

I. Skozrit (✉) and Z. Tonković
Faculty of Mechanical Engineering and Naval Architecture, University of Zagreb, I. Lučića 5, HR-10000 Zagreb, Croatia
e-mail: ivica.skozrit@fsb.hr, ztonkov@fsb.hr

S. Pantelakis, C. Rodopoulos (eds.), *Engineering Against Fracture: Proceedings of the 1st Conference*,

equations can be used in principle to estimate the C-integral, by replacing strain with the strain rate [3]. In contrast to the internal axial semi-elliptical surface cracks [4,5], a very limited number of studies have been reported in the area dealing with the determination of the J-integral and plastic limit load for metal pipes with external axial semi-elliptical surface cracks. In accordance with this, the first part of the present contribution is concerned with the structural integrity of the metal thick-walled pipes subjected to internal pressure, having external axial surface cracks at the external wall.

Based on the experimental data by Ben Hadj Hamouda et al. [6, 7] and analogy between plasticity and creep, the second part of the paper discusses a method used to develop the efficient computational strategy for modeling creep fracture mechanisms by slow crack growth in a MDPE pipes.

2 Reference Stress Based J Estimation for Surface Cracked Pipe

Finite element (FE) analysis has been performed to evaluate J-integral and plastic limit pressure for a thick-walled pipe with an external axial surface crack subjected to internal pressure. The pipe geometry and loading are shown in Fig. 1. R_i and R_o are the inner and the outer radius, while t denotes the wall thickness. The external surface crack is assumed to have a semi-elliptical shape described by a length $2c$ and depth a. Geometrically, an axial surface-cracked pipe is characterized by three non-dimensional parameters, i.e. R_i/t, c/a and a/t. As evident from Fig. 1, the internal pressure p is applied as a distributed load to the inner surface, together with an axial tension force F equivalent to the internal pressure applied at the end of the pipe to simulate the closed end. A parametric study is performed, in which an inner radius-to-thickness ratio R_i/t takes two values of 4 and 10. Thereat, the wall thickness of the pipe t is held constant at a value of 1.625 mm. These values are chosen to represent a standard thick-walled pipes geometry (mean radius-to-thickness ratio R_m/t typically less than 10). Four different half crack length-to-crack depth ratios of $c/a = 5$, 10, 15 and 20, and crack depth-to-thickness ratios of $a/t = 0.2$, 0.4, 0.6 and 0.8 are considered.

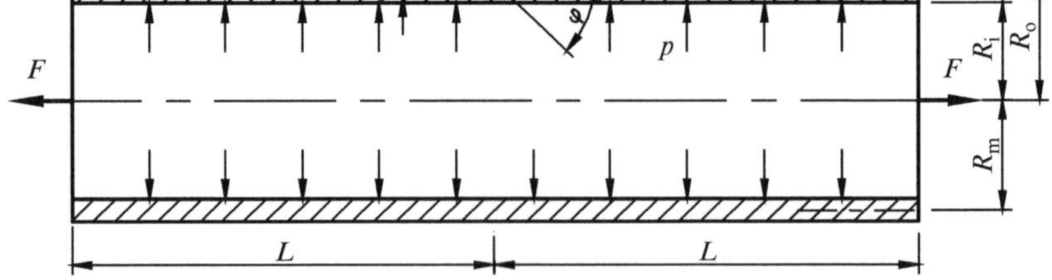

Fig. 1 Geometry and dimensions of a pipe subjected to internal pressure

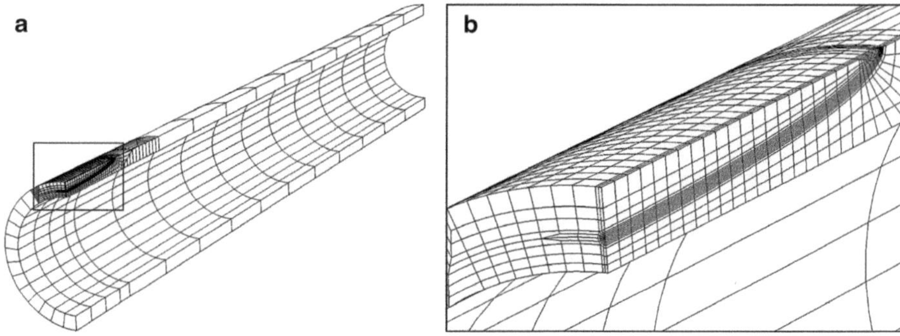

Fig. 2 Typical FE mesh for a pipe: **a** whole mesh; **b** crack-tip mesh

The finite element analysis is performed by using the commercial finite element package ABAQUS/Standard [8]. Twenty-noded isoparametric solid elements with reduced integration (C3D20R) are used to model the pipe. Due to symmetry, only a quarter of the pipe is modeled as shown in Fig. 2.

In the structural integrity assessment procedures the values of the elastoplastic J_{ep}-integral are obtained using the deformation theory of plasticity where the stress–strain response in the finite element analyses is described by the well-known Ramberg–Osgood model

$$\varepsilon/\varepsilon_0 = \sigma/\sigma_y + \alpha\left(\sigma/\sigma_y\right)^n. \tag{1}$$

In Eq. (1), σ_y is the yield stress, ε_0 denotes the associated reference strain $\varepsilon_0 = \sigma_y/E$ and E is the Young's modulus. The values α and n represent the parameters fitting the experimentally obtained curve. With the method developed by the General Electric/Electric Power Research Institute (GE/EPRI), the total crack driving force J_{ep} can be split into elastic, J_{el}, and plastic, J_{pl}, parts, as

$$J_{ep} = J_{el} + J_{pl}. \tag{2}$$

In the present investigation a total of 160 cases are considered to determine the J_{ep}-integral for external axial semi-elliptical surface cracks in thick pipes. Herein values of the strain hardening index n are systematically varied: $n = 3, 5, 7$ and 10, while parameter α is fixed to $\alpha = 1$. In addition, linear analyses with Poisson's ratio $\nu = 0.3$ are performed to determine elastic component of J_{ep}, which is formulated according to [9]

$$J_{el}(\varphi) = \left(p \cdot R_m/t\right)^2 \pi \cdot a \cdot F^2\left(R_m/t, a/t, c/a, \varphi\right), \tag{3}$$

where $F(R_m/t,\ a/t,\ c/a,\ \varphi)$ is a dimensionless function depending on the pipe and crack geometry and φ is the angle defining the crack front position (see Fig. 1). The values of function $F\left(R_m/t, a/t, c/a, \varphi\right)$ obtained from the FE analysis for the

Table 1 Dimensionless function F for the stress intensity factor

R_m/t	4				10			
a/t	c/a				c/a			
	5	10	15	20	5	10	15	20
0.2	0.921	0.993	1.019	1.030	1.031	1.117	1.151	1.168
0.4	1.171	1.350	1.413	1.443	1.280	1.516	1.615	1.667
0.6	1.593	1.966	2.098	2.165	1.693	2.235	2.486	2.614
0.8	2.187	2.895	3.138	3.256	2.200	3.346	3.911	4.230

considered pipe geometry are tabulated in Table 1. As the J_{el} values reach the largest values at the deepest crack front location ($\varphi = \pi/2$), only the results at that location are given.

Following [4], the plastic part of J_{ep}-integral, J_{pl}, for semi-elliptical surface cracked pipes can be given by

$$J_{pl} = \alpha\sigma_y^2/E\,(t-a)\,h_1\,(R_m/t, a/t, c/a, \varphi)\,(p/p_L)^{n+1}, \qquad (4)$$

where h_1 is the dimensionless plastic influence function which depends on the crack and pipe geometry, while p_L is the plastic limit pressure that produces the plastic collapse of the surface cracked pipe for a rigid-plastic, non-hardening material with yield stress σ_y. Based on the FE results, the empirical expression for the estimation of the plastic limit pressure in terms of non-dimensional crack configuration parameters as follows:

$$p_L = p_0 \cdot P\,(a/t, \rho), \qquad (5)$$

where p_0 is the limit pressure of an un-cracked thick-walled pipe based on the Von Mises yield criterion, $p_0 = 2/\sqrt{3}\sigma_y \ln(R_o/R_i)$, and ρ is normalized crack length defined as $\rho = c/\sqrt{R_m t}$. In Eq. (5), P is the dimensionless parameter which is a function of crack configuration parameters a/t and ρ, as follows:

$$\begin{aligned} &P = 1 + P_1\,(a/t) + P_2\,(a/t)^2, \\ \text{where } &P_1 = 0.135312 - 0.351517\cdot\rho + 0.067173\cdot\rho^2 - 0.004954\cdot\rho^3, \\ &P_2 = -0.123488 - 0.011068\cdot\rho - 0.009342\cdot\rho^2 + 0.001921\cdot\rho^3. \end{aligned} \qquad (6)$$

For the considered ranges, this empirical relation predicts the limit pressure which differs by less than 5% from values obtained by the finite element computation. Recently, Kim et al. [5] have proposed a similar equation for the plastic limit pressure based on selected FE limit analyses using the parameters $R_m/t = 20$, $\rho = 0.5$, 1.0, 2.0 and 3.0. The plastic limit pressures for the considered pipe geometry obtained by the equation proposed by Kim et al. [5] are compared with the present FE results

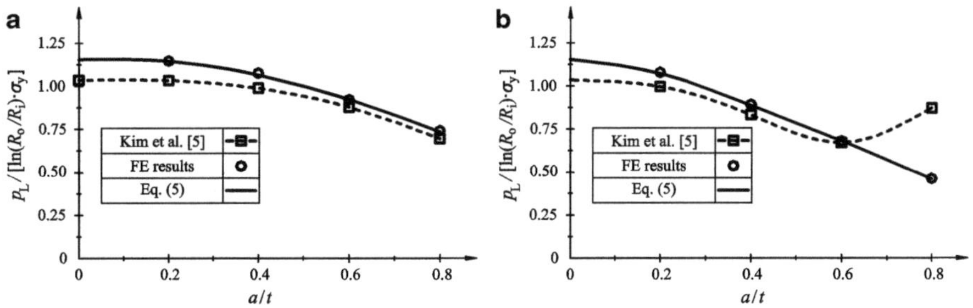

Fig. 3 A comparison of plastic limit pressure p_L between the present work and the published solutions by Kim et al. [5] ($R_i/t = 4$): **a** $c/a = 5$; **b** $c/a = 20$

and Eq. (5) in Fig. 3. As may be seen, the finite element results presented in this work for the pipes with rather thick walls ($R_i/t = 4$) and large crack lengths are generally different from those obtained by [5]. These differences could arise from the fact that solutions obtained by [5] are given only for limited ranges of pipe and crack geometries.

In order to obtain the values of the function $h_1\,(R_m/t, a/t, c/a, \varphi)$ from Eq. (4), the plastic part of J_{ep} integral is determined by

$$J_{pl,FE} = J_{FE} - J_{el}. \tag{7}$$

It is to note that the value of h_1 depends on the load magnitude. Since the power-law part of Eq. (1) describing material behaviour during the loading process, is dominant at sufficiently large loads, the values of h_1 at high loads should be taken from appropriate diagrams. The values of h_1 for the considered pipe geometry at the deepest crack front location ($\varphi = \pi/2$) are presented in Table 2.

As presented in literature, the GE/EPRI results can be very sensitive due to the fitting of the stress–strain data when use the Ramberg–Osgood relation. For application to general stress–strain laws, the J_{ep} estimation equation is now re-formulated using the RSM [10]:

$$J_{ep}/J_{el} = E\varepsilon_{ref}/\sigma_{ref} + 0.5\left(\sigma_{ref}/\sigma_y\right)^2 \cdot \left(\sigma_{ref}/E\varepsilon_{ref}\right), \tag{8}$$

where σ_{ref} denotes the reference stress, $\sigma_{ref} = (p/p_{ref})\sigma_y$ and ε_{ref} is reference strain at $\sigma = \sigma_{ref}$, determined from the true-stress strain data, while p_{ref} stand for the reference pressure.

Based on the FE results, the empirical expression for the estimation of the reference pressure p_{ref} in terms of non-dimensional crack configuration parameters a/t and ρ is derived

$$p_{ref} = \left(2/\sqrt{3}\right) \cdot \sigma_y \cdot \ln\left(R_o/R_i\right)\left[1 + A_1\,(a/t) + A_2\,(a/t)^2\right],$$

Table 2 Values of the plastic influence h_1-functions

R_m/t	c/a	a/t	n			
			3	5	7	10
4	5	0.2	1.358	1.737	2.020	2.370
		0.4	5.563	7.228	8.357	9.482
		0.6	15.943	18.571	19.581	20.009
		0.8	37.093	38.727	38.132	34.822
	10	0.2	1.570	2.075	2.449	2.868
		0.4	6.556	8.604	9.708	10.356
		0.6	16.592	17.815	17.681	17.112
		0.8	29.855	27.973	25.438	21.583
	15	0.2	1.581	2.102	2.483	2.879
		0.4	6.312	8.256	9.146	9.504
		0.6	14.925	15.394	14.875	14.002
		0.8	27.514	27.749	27.309	25.622
	20	0.2	1.536	2.033	2.392	2.743
		0.4	5.958	7.772	8.554	8.803
		0.6	14.133	14.476	14.087	13.490
		0.8	22.731	22.968	22.615	21.040
10	5	0.2	1.521	1.872	2.161	2.540
		0.4	6.424	8.254	9.739	11.386
		0.6	19.903	23.354	25.421	26.766
		0.8	51.712	57.183	60.255	59.333
	10	0.2	1.803	2.299	2.688	3.174
		0.4	8.395	11.071	12.818	14.223
		0.6	26.156	29.434	31.093	32.007
		0.8	57.742	58.603	60.943	60.396
	15	0.2	1.865	2.405	2.819	3.324
		0.4	8.678	11.579	13.262	14.492
		0.6	25.213	27.009	27.327	27.030
		0.8	47.198	45.301	45.762	43.527
	20	0.2	1.850	2.388	2.792	3.272
		0.4	8.387	11.241	12.750	13.676
		0.6	23.587	24.393	24.021	23.255
		0.8	43.821	42.614	44.701	45.104

$$\text{where } A_1 = -0.92351 - 0.18551 \cdot \rho + 0.058677 \cdot \rho^2 - 0.00539 \cdot \rho^3, \quad A_2 = 1.066525 - 0.16797 \cdot \rho - 0.00265 \cdot \rho^2 + 0.002479 \cdot \rho^3. \tag{9}$$

The proposed Eq. (8) is compared with three-dimensional finite element results using actual experimental uniaxial stress–strain data of austenitic steel 08X18H10T [11] and incremental plasticity as well as GE/EPRI solutions. As may be observed from Fig. 4, excellent agreement of the two RSM and FE solutions is exhibited. On the other hand, the results obtained by the GE/EPRI deviate significantly.

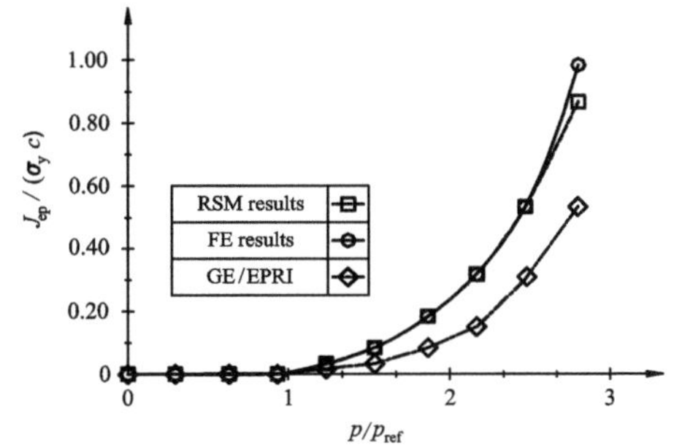

Fig. 4 Comparison of FE and GE/EPRI J results with the proposed reference stress based J estimations $R_m/t = 4.8$, $c/a = 2.08$, $a/t = 0.8$

3 *C*-Integral Estimation for MDPE

3.1 Analogy Between Plasticity and Creep

In the previous section, an empirical expression for the estimation of the J-integral of cracked thick-walled pipes is proposed, based on the RSM. As presented in literature, at the sufficiently large load, Eq. (8) can be approximated as

$$J_{ep} = J_{el}\left(E\varepsilon_{ref}/\sigma_{ref}\right). \tag{10}$$

Based on the analogy between plasticity and creep, the C^*-integral for steady-state creep conditions is formulated using the RSM introduced by Ainsworth [10]

$$C^* = J_{el}\left(E \cdot \dot{\varepsilon}_c/\sigma_{ref}\right), \tag{11}$$

where $\dot{\varepsilon}_c$ is the creep strain rate at $\sigma = \sigma_{ref}$, determined from the actual creep-deformation data.

3.2 Numerical Modeling

The nonlinear time-dependent fracture behavior of MDPE has been investigated recently by Ben Hadj Hamouda et al. [6,7]. Creep cracking test has been performed on axisymmetrically cracked specimen denoted as full notched crack tensile (FNCT) and shown in Fig. 5. As reported in [6,7], the notch is made using a fresh razor blade with the crack tip radius approximately of 10 μm.

The experimental creep displacement δ versus time t diagram is obtained for a FNCT specimen tested at 60°C under a net stress σ_{net} (load divided by the initial minimal cross-section) of 8 MPa (Fig. 6). As may be observed from Fig. 6, three stages can be distinguished in diagram. The primary creep stage (Stage I) is characterized by continuously decreasing of the displacement rate. During the Stage II constant displacement rate is obtained that corresponds to the stationary creep, while

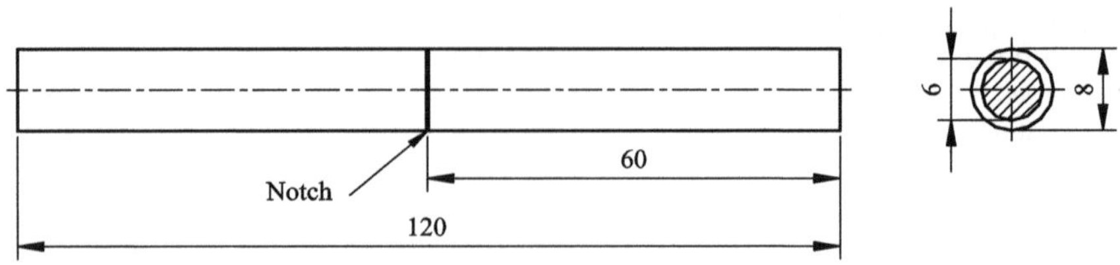

Fig. 5 Full notched crack tensile (FNCT) specimen

Fig. 6 Variation of creep displacement with time

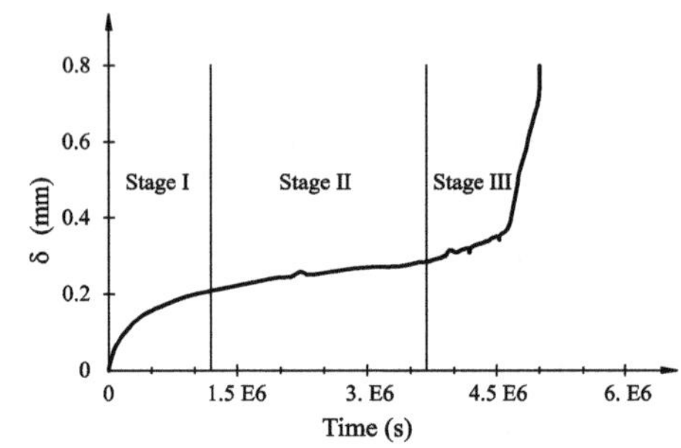

Fig. 7 True stress–strain curve [6, 7]. for a FNCT specimen at 60°C with $\sigma_{net} = 8\,\text{MPa}$ [6, 7]

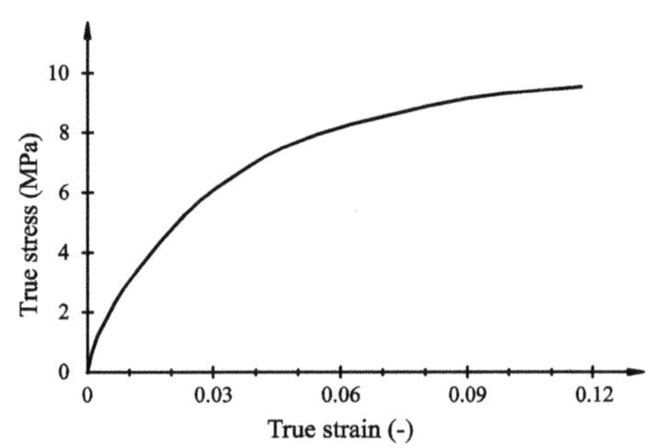

during Stage III crack accelerates and propagates through the remaining ligament until the ultimate fracture (tertiary creep).

Based on the above experimental results, the primary–secondary creep constitutive law is proposed as follows

$$\dot{\varepsilon}_c = B_1 p t^{(p-1)} \sigma^{n_1} + B_2 \cdot \sigma^{n_2}, \tag{12}$$

where σ and ε_c are the stress and creep strain, while B_1, B_2, p, n_1 and n_2 stand for the material coefficients ($B_1 = 5.73 \cdot 10^{-9}$, $B_2 = 1.13 \cdot 10^{-12}$, $p_1 = 0.41$, $n_1 = 5.81$, $n_2 = 6.1$). In Eq. (12) dot designates derivative with time t. The first term on the right-hand side of Eq. (12) represents the primary creep and the second term is the secondary portions of creep deformation. Fig. 7 displays the nonlinear elastic behavior of the considered MDPE in the form of the true stress–strain curve.

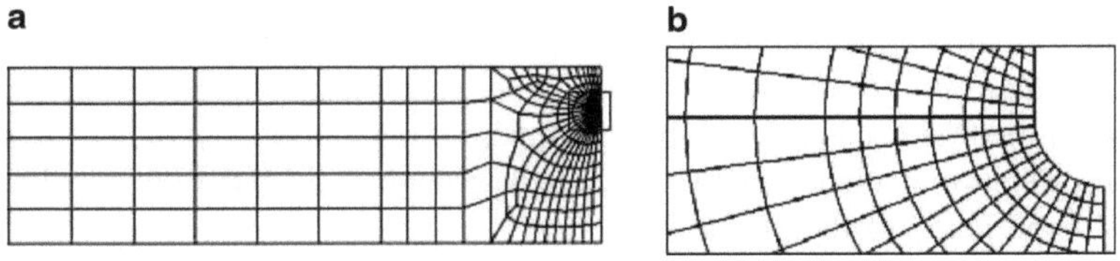

Fig. 8 Finite element mesh for FNCT specimen: **a** whole mesh; **b** notch-tip mesh

The main intention of the present work is to derive efficient computational strategy for modeling creep fracture mechanisms by slow crack growth in a MDPE, employing described material model. For power-law metal creep ABAQUS/Standard [8] provides an in-built routine for calculation of the C-integral. However, for generalized creep law, such as mentioned creep law for MDPE, calculation of the C-integral requires a user subroutine CREEP or UMAT.

Currently we are working on the development of the efficient numerical algorithm for integration of MDPE constitutive model. The computational strategy is based on the time hardening integration approach. In the next phase, the derived algorithm will be implemented at the material point level of the availabel finite elements in the code ABAQUS [8] by using a user subroutine UMAT.

A typical finite element mesh for FNCT specimen which will be applied in the analysis is shown in Fig. 8a. The mesh refinement in the vicinity of the blunted notch-tip is depicted in Fig. 8b. Employing symmetry, one quarter of the FNCT specimen section is modeled. For the model discretization, the eight-noded axisymmetric solid elements with reduced integration (CAX8R) are applied. The large strain FE model will be employed by invoking the NLGEOM option within ABAQUS. A tension load will be first applied to the FE model using a nonlinear elastic calculation at time $t = 0$. The load will be then held constant and subsequent time-dependent creep analyses will be performed. Next, the accuracy of the computational procedure will be tested by comparing the computed results with the real experimental data.

In the last phase, systematic detailed non-linear finite element analyses will be carried out to determine the C-integral as a function of time for a thick-walled MDPE pipe with an external axial surface crack subjected to internal pressure. In order to achieve high numerical efficiency combined explicit and implicit integration schemes will be used for the creep calculations. The values of the C-integral and time to fracture will be compared with the published empirical solutions [6, 7].

4 Conclusions

The application of the reference stress method to estimate the J-integral of cracked thick-walled metal pipes is presented. It is shown that, compared to existing solutions, the newly developed analytical approximations of the plastic limit pressure and J-integral are applicable to a wide range of crack dimensions. Based on the experimental data from literature and analogy between plasticity and creep, the paper discusses a method used to develop the efficient computational strategy for modeling creep fracture mechanisms by slow crack growth in a MDPE pipes which will be considered in the further work.

References

1. Lu X. and Brown N., 1990. The transition from ductile to slow crack growth failure in a copolymer of polyethylene. Journal of Materials Science 25 411–416.
2. Krempl E. and Kahn F., 2003. Rate (time)-dependent deformation behavior: an overview of some properties of metals and solid polymers. International Journal of Plasticity 19, 7, 1069–1095.
3. Kim Y.-J., Kim J.-S., Huh N.-S. and Kim Y.-J., 2002. Engineering C-integral estimates for generalised creep behaviour and finite element validation. International Journal of Pressure Vessels and Piping 79, 427–443.
4. Kim Y.-J., Kim J.-S., Park Y.-J. and Kim Y.-J., 2004. Elastic-plastic fracture mechanics method for finite internal axial surface cracks in cylinders. Engineering Fracture Mechanics 71, 925–944.
5. Kim Y.-J., Shim D.-J., Nikbin K., Kim Y.-J., Hwang S.-S. and Kim J.-S., 2003 Finite element based plastic limit loads for cylinders with part-through surface cracks under combined loading. International Journal of Pressure Vessels and Piping 80, 527–540.
6. Hamouda H.B.H., Laiarinandrasana L. and Piques R., 2007. Fracture mechanics global approach concepts applied to creep slow crack growth in a medium density polyethylene (MDPE). Engineering Fracture Mechanics 74, 2187–2204.
7. Hamouda H.B.H., Laiarinandrasana L. and Piques R., 2007. Viscoplastic behaviour of a medium density polyethylene (MDPE): Constitutive equations based on double nonlinear deformation model. International Journal of Plasticity 23, 1307–1327.
8. ABAQUS/Standard, 2006. User's guide and theoretical manual, Version 6.6, Hibbitt, Karlsson & Serensen.
9. Raju I.S. and Newman J.C., 1982. Stress-intensity factors for internal and external surface cracks in cylindrical vessels. Journal of Pressure Vessel Technology 104, 293–298.
10. Ainsworth R.A., 1984. The assessment of defects in structures of strain hardening material. Engineering Fracture Mechanics 19, 633–642.
11. Tonković Z., Skozrit I. and Alfirević J., 2008. Influence of flow stress choice on the plastic collapse estimation of axially cracked steam generator tubes. Nuclear Engineering and Design 238, 1762–1770, doi:10.1016/j.nucengdes.2008.01.008

Session X
Structural Analysis (Metals)

Buckling Evaluation in Case of Complicated Stress Condition

Ausra Zilinskaite and Antanas Ziliukas

Abstract Probability analysis of buckling stresses in this study allows to present features of biaxial and triaxial stress condition compared to linear stress condition. Absolute characteristics of buckling in case of triaxial stress condition compared to linear stress condition are described in function with one maximum value. This study shows that biaxial and triaxial stress condition is always more dangerous compared to monoaxial. Stress condition is characterized by tensor invariants. Therefore, buckling stresses must be measured under invariants of stress condition, and the criterion of buckling in case of complicated stress condition is obtained measuring these stresses and geometrical parameters of the construction.

Suggested criterion is verified experimentally and compared to analytical calculations. Criterion of buckling in case of complicated stress condition requires having yield stress of the material, performing single experiment by measuring critical stresses of constructional element with given geometrical values and after calculating constants can be applied to any cases of complicated stress condition not exceeding the yield stress of material.

Keywords Buckling · Probability · Stress · Thin-walles construction

1 Introduction

Constructional elements with cross-section values significantly less than length often loss stability under influence of axial forces and other complicated loads. Stress condition, as well as geometrical values, influence buckling. Measuring geometrical values makes no trouble, on the contrary, one for stress condition, especially in case of complicated loading, still lacks investigation. Deterministic solutions for

A. Zilinskaite and A. Ziliukas (✉)
Kaunas University of Technology, Kestucio St. 27, 44312 Kaunas, Lithuania
e-mail: ausrazilinskaite@takas.lt, antanas.ziliukas@ktu.lt

S. Pantelakis, C. Rodopoulos (eds.), *Engineering Against Fracture: Proceedings of the 1st Conference*,

buckling are given for cases with one type loads [1–8]. In case of complicated loading with no analytical solutions solving the problem is more simple with probabilistical explanation of the process.

2 Probability Analysis for Buckling Stresses

Describing the process of buckling in probability it can be stated that in case of triaxial stress condition probability P123 with principal stresses σ_K can be given as function

$$F[P(A1),\ P(A2),\ P(A3)] = P123 \tag{1}$$

where: $P(A_1)$ – buckling probability of principal stress σ_1, $P(A_2)$ – one of σ_2 and $P(A_3)$ one of σ_3.

According to theory of probabilities the whole of finished sequence of events A_1, $A_2, \ldots A_n$ is event $(A_1 \cup A_2 \cup A_3 \ldots . \cup A_n)$, that has a probability $P(A_1 \cup A_2 \cup A_3 \ldots . \cup A_n)$ and in case of triaxial stress condition

$$\begin{aligned} &P(A1 \cup A2 \cup A3) = P(A1) + P(A2) + P(A3) - P(A1A2) \\ &- P(A2A3) - P(A3A1) + P(A1A2A3) = P123 = Pc \end{aligned} \tag{2}$$

where P_c – probability of critical buckling.

Evaluating that

$$\begin{aligned} \sigma_1 &= \sigma_{cr,0}\ P(A_1); \\ \sigma_2 &= \sigma_{cr,0}\ P(A_2); \\ \sigma_3 &= \sigma_{cr,0}\ P(A_3), \end{aligned} \tag{3}$$

where $\sigma_{cr,0}$ – critical buckling stresses.

It can be put that

$$\begin{aligned} &P(A_1) + P(A_2) + P(A_3) = F_1(\sigma_1 + \sigma_2 + \sigma_3) = F_1(I_1); \\ &P(A_1)P(A_2) + P(A_2)P(A_3) + P(A_3)P(A_1) = F_2(\sigma_1\sigma_2 \\ &+ \sigma_2\sigma_3 + \sigma_3\sigma_1) = F_2(I_2); \\ &P(A_1)P(A_2)P(A_3) = F_3(\sigma_1\sigma_2\sigma_3) = F_3(I_3), \end{aligned} \tag{4}$$

where I_1, I_2, I_3 – first, second and third invariant of stress condition.

Thus, critical buckling probability

$$P_{123} = P_c = F_1(I_1) - F_2(I_2) + F_3(I_3). \tag{5}$$

We will take probability P_{123} for buckling criterion. If two uniform constructional elements in case of different complicated stress condition have same buckling probability $P^{I}_{123} = P^{II}_{123}$, these stress conditions are uniformly dangerous. Consequently, term of buckling equivalence can be formulated: complicated stress condition equals to linear stress condition if buckling probability are uniform

$$P = P_c \tag{6}$$

where P – buckling probability in case of linear stress condition.

Taking on account Eqs. (5, 6) can be given as

$$P = P_{ekv} = P_{ijk} = P_c \tag{7}$$

Consequently, term of statistical equivalence Eq. (6) in case of linear stress condition can be followed when P(A) reaches value $P = P_{ijk} = P_{ekv}$

Criterion Eq. (7) is summarized and sufficient as perfectly describing the term of buckling.

$$P_{ekv} = P_{123} = P_c \leq P_{all} \tag{8}$$

where P_{all}– allowed probability.

Now, comparative term of buckling in case of linear biaxial and triaxial stress condition is to be analyzed using analogical research in the area of metal fatigue [9]. Probability $P(A_1)$ in case of linear stress condition and $P(A_1) = P(A_2) = P(A_3)$ in case of triaxial stress condition are given on x-axis, and probability $P(A_1)$, P_{12} and P_{123} on y-axis respectively (Fig. 1). Buckling probability criteria are presented for three cases with σ_1, $\sigma_1 = \sigma_2$ or $\sigma_1 = \sigma_2 = \sigma_3$. In case of linear stress condition, probability $P(A_1)$ is represented as diagonal (Fig. 1, line 1). In case of biaxial stress condition, probability of buckling cannot exceed $P_{12\max}$, obtained this way

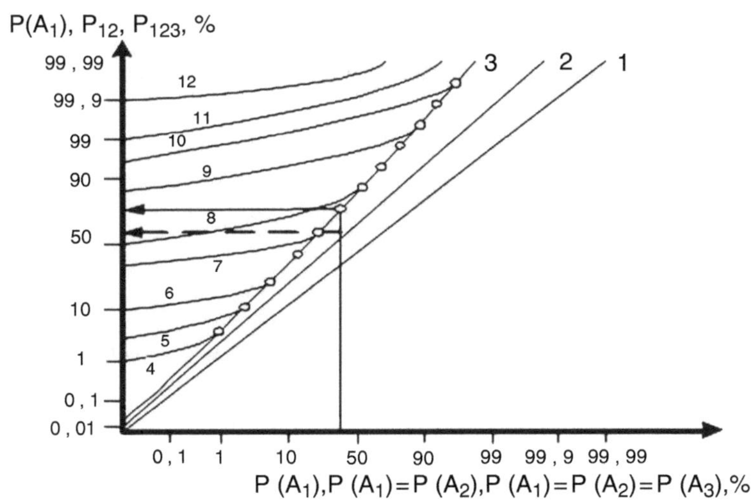

Fig. 1 Critical values of probable buckling criterion in case of linear, biaxial, and triaxial stress condition: 1 – $P(A_1)$; 2 – P_{12}max; 3 – P_{123}max; 4–12 – P123 with P(A1) = 0.01; 0.05; 0.1; 0.3; 0.5; 0.8; 0.95; 0.99; 0.999 [9]

(Fig. 1, line 2). From Eq. (2), it can be said that with σ_1, σ_2, probability of buckling P_{12} always exceeds the one in case with only σ_1 and is always the biggest when $\sigma_1 = \sigma_2$. In this case, $P(A_1) = P(A_2) = P(A)$.

Probability of buckling $P_{12\max}$ can be calculated as

$$P_{12\max} = 2P(A) - [P(A)]^2 = P(A)[2 - P(A)] \tag{9}$$

Line family 4–12 (Fig. 1) obtained from Eq. (2). These are lines of probabilities P_{123}, with fixed $P(A_1)$ as $P(A_2) = P(A_3) \leq P(A_1)$.

For example, line P_{123} and probability of buckling cannot exceed the value 0.9. More, each line of the family 4–12 cannot continue further than line 3, i.e. points represent cases with $P(A_1) = P(A_2) = P(A_3)$. Thus, line $P_{123\max}$ is analogous to line $P_{12\max}$.

In critical case, Eq. (2) with $P(A_1) = P(A_2) = P(A_3) = P(A)$ can be given as

$$P_{123\max} = 3P(A) - 3[P(A)]^2 + [P(A)]^3 \tag{10}$$

Ordinate differences of lines $P(A_1)$, $P_{12\max}$, and $P_{123\max}$ show quantitative dangers of biaxial and triaxial stress condition with equal principal stresses, i.e. $P(A_1) = P(A_2) = P(A_3)$.

For example, with $P(A_1) = P(A_2) = 0.4$ probability of buckling increases to $P_{12\max} = 0.64$, and with triaxial stress condition and $P(A_1) = P(A_2) = P(A_3) = 0.4$, then $P_{123\max} = 0.784$ (cases shown in Fig. 1, arrows).

Therefore, it is obtained that in case of linear (axial) stress condition $P(A_1) = 0.4$, probability of biaxial buckling changes in $(0.64 - 0.4)/0.4 \cdot 100\% = 60\%$, and with triaxial buckling $-(0.784 - 0.64)/0.64 \cdot 100\% = 22.5\%$ compared to biaxial. This shows that less buckling danger is presented by change from biaxial to triaxial stress condition (22.5%) than from monoaxial to biaxial (60%).

Buckling danger comparing one stress condition with another can be characterized in such ratios:

$$\begin{aligned}\frac{P_{12\max} - P(A_1)}{P(A_1)} \cdot 100\% &= P_{12/1}\\ \frac{P_{123\max} - P(A_1)}{P(A_1)} \cdot 100\% &= P_{123/1}\\ \frac{P_{123\max} - P_{12\max}}{P_{12\max}} \cdot 100\% &= P_{123/12}\end{aligned} \tag{11}$$

where $P_{12/1}$, $P_{123/1}$, $P_{123/12}$ – probabilities compared to one another

Absolute values can be put as

$$\begin{aligned}P_{12\max} - P(A_1) &= \Delta P_{12-1},\\ P_{123\max} - P(A_1) &= \Delta P_{123-1},\\ P_{123\max} - P_{12\max} &= \Delta P_{123-12}.\end{aligned} \tag{12}$$

where ΔP_{12-1}, ΔP_{123-1}, ΔP_{123-12} – differences in probabilities.

Analysis of stress condition probability has shown that:

1. Compared to linear stress condition, dangerous is biaxial stress condition with $P(A_1) = P(A_2) = 0.5$ and $\Delta P_{12-1} = 0.25$. With $P(A_1) = P(A_2) < 0.5$ and $P(A_1) = P(A_2) > 0.5$. Danger of biaxial stress condition also decreases with $P(A_1) = P(A_2) = 0.1$ or $P(A_1) = P(A_2) = 0.9$, and value ΔP_{12-1} decreases to 0.09.
2. Compared to linear stress condition, dangerous is triaxial stress condition with $P(A_1) = P(A_2) = P(A_3) = 0.40$. In this case, $\Delta P_{123-1} = 0.384$. All other stress combinations of σ_1, σ_2, σ_3 give less values ΔP_{123-1}.
3. Triaxial stress condition is more dangerous than biaxial one compared to linear stress condition with $P(A_1) = P(A_2) = P(A_3) < 0.95$. With $P(A_1) = P(A_2) = P(A_3) > 0.95$ biaxial and triaxial stress conditions are equally dangerous compared to linear stress condition (differences less than 5%).
4. Compared to biaxial stress condition, most dangerous is triaxial stress condition with $P(A_1) = P(A_2) = P(A_3) = 0.35$; in this case, $\Delta P_{123-12} = 0.15$. In other cases, danger of triaxial stress condition decreases compared to biaxial one.
5. With $P(A_1) = P(A_2) = P(A_3) < 0.05$ triaxial stress condition makes practically equal danger compared to biaxial, as well as biaxial compared to linear stress condition. With $P(A_1) = P(A_2) = P(A_3) > 0.05$, biaxial stress condition significantly increases the probability of buckling compared to linear stress condition than in the case of triaxial and biaxial stress condition.
6. Triaxial and biaxial stress condition compared to linear one are less dangerous with increasing probabilities $P(A_1) = P(A_2) = P(A_3)$. With $P(A_1) = P(A_2) = P(A_3) > 0.95$, biaxial and triaxial stress condition are almost equally dangerous compared to linear stress condition.
7. The less are probabilities $P(A_1) = P(A_2) = P(A_3)$ in case of linear stress condition, the more influence they have on P_{12} and P_{123}.
8. When $P(A_1) > 0.95$, stresses σ_2 and σ_3 can be almost unmeasured analyzing triaxial stress condition as P_{123} has influence of 5%.
9. Biaxial and triaxial stress conditions are always more dangerous compared to monoaxial than triaxial compared to biaxial.

3 Buckling Criterion in Case of Complicated Stress Condition

As shown in analysis of probable buckling stresses, in case of complicated stress condition, buckling criterion depends on measuring right stress condition. This can be obtained by giving equivalent stress as

$$\sigma_{ekv} = \sigma_c = \varphi(\sigma_i, \sigma_{ci}, m_j, ..) \tag{13}$$

and evaluating former conclusion that probability of buckling depends on invariants I_1, I_2, I_3 of stress condition and while measuring unlinearity it can be stated that equivalent stress depends on invariants I_1 and I_2 according to quadratic dependence [10]

$$m_1(\sigma_i)^2 + m_2(\sigma_o)^2 = \sigma_c^2. \tag{14}$$

where: σ_c-critical buckling stress in case of simple loading; m_1, m_2- material constants;

$$\sigma_i = \frac{1}{\sqrt{2}}\sqrt{(\sigma_1-\sigma_2)^2+(\sigma_2-\sigma_3)^2+(\sigma_3-\sigma_1)^2};$$

$$\sigma_o = \frac{\sigma_1+\sigma_2+\sigma_3}{3}.$$

In case of bending and buckling with active normal stress σ and shear stress τ Eq. (14) can be given as

$$m_1(\sigma_1^2+3\tau^2)+m_2\left(\frac{\sigma_1}{3}\right)^2 = \sigma_c^2. \tag{15}$$

where: $\sigma_1 = \frac{M_x}{W_x}$ (M_x – bending moment, W_x – axial section modulus);
$\tau = \frac{M_{tor}}{W_{tor}}$ (M_{tor} – torsion moment; W_{tor} – section modulus of torsion).

Constants can be obtained from critical conditions:

1. When stresses reach the limit of yielding and $\tau = 0$, then Eq. (15) gives the following:

$$\begin{aligned} &m_1\sigma_1^2 + m_2\frac{\sigma_1^2}{9} = \sigma_Y^2 \text{ with } \sigma_1 = \sigma_Y; \\ &m_1 + \frac{m_2}{9} = 1. \end{aligned} \tag{16}$$

where: σ_Y – yields stress.

2. During the buckling from one test with simply supported beam loaded with bending force and moment of torsion at one end, from Eq. (15) we get the following:

$$\frac{m_1F_Y^2L^4}{D^2}\left(\sigma_1^2+3\tau^2\right)+\frac{9(1-m_1)F_Y^2L^4}{D^2}\left(\frac{\sigma_1}{3}\right)^2 = \sigma_Y^2 \tag{17}$$

where: $D = F_cL^2$

F_Y – force matching the limit of yielding;
Equation (17) gives the constant m_1, and Eq. (16) gives m_2.

4 Experimental

Experiments were made for checking buckling criterion in case of complicated stress condition. Steel gird of various lengths was chosen. This constructional element was strengthened in one end and loaded with transversal force and buckling moment in free end.

In case of simple loading with $\tau = 0$, critical force F_c was calculated under the formula [3]

$$F_c = \frac{4{,}0126}{L^2}\sqrt{EI_yGI_{tor}} \tag{18}$$

where: E – modulus of elasticity; G – shear modulus; I_y – axial moment of inertia; I_{tor} – torsional moment of inertia

This case was tested experimentally.

Results are given in Table 1.

In case of complicated various loading (bending and torsion) critical force F_c and critical moment of torsion $M_{c,tor}$ can be obtained from Eq. (15).

In case of complicated loading results are given in Table 2.

Theoretical and experimental results are presented graphically in Fig. 2. The constant gives the following: $m_1 = 0.102$ and $m_2 = 8.079$.

Table 1 Experimental and theoretical critical buckling stress in case of simple loading

Cross-section values (mm)	Length (m)	$\sigma_{1\,(\text{experimental})}$ (MPa)	$\sigma_{1\,(\text{theoretical})}$ (MPa)	Variation ratio
	1	244.80249.60259.20256.80245.76	245.66	1.16
50 × 5	1.5	162.00165.60163.44167.04172.80	163.77	1.12
	2	128.64122.88126.72134.40130.56	122.83	1.49

Table 2 Experimental critical buckling stress in case of complicated loading

Cross-section values (mm)	Length (m)	σ_1 (experimental) (MPa)					Variation ratio
		With $\tau = 128$ MPa					
	1	216.00	223.20	235.20	230.40	225.60	1.44
50 × 5	1.5	147.60	140.40	151.20	144.00	152.64	1.54
	2	96.00	105.60	103.68	100.32	102.72	1.62
		With $\tau = 64$ MPa					
	1	240.00	230.40	238.56	243.12	249.60	1.30
50 × 5	1.5	149.76	162.00	158.40	159.84	151.20	1.56
	2	110.40	115.20	119.04	117.12	120.00	1.46
		With $\tau = 32$ MPa					
	1	261.60	259.20	254.40	249.60	245.04	1.20
50 × 5	1.5	162.72	164.16	151.20	162.00	158.40	1.46
	2	124.80	121.92	129.60	120.00	127.68	1.42

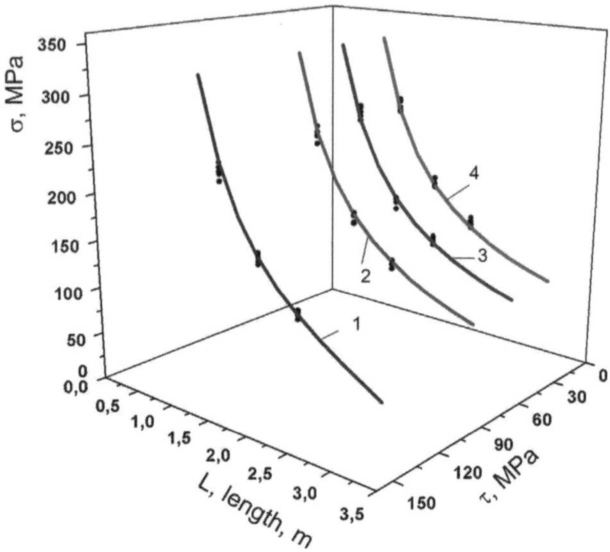

Fig. 2 Diagram of buckling stresses: straight lines – theoretical; ∗ – experimental data; 1 – $\tau = 128\,\text{MPa}$; 2 – $\tau = 64\,\text{MPa}$; 3 – $\tau = 32\,\text{MPa}$; 4 – $\tau = 0\,\text{MPa}$

5 Conclusions

1. Deterministic buckling solutions show that they can be applied in case of one type loads, and in case of complicated loading it is better to analyze probabilistical processes.
2. Probabilistical analysis shows that in all cases biaxial and triaxial stress condition are more dangerous compared to monoaxial one.
3. Buckling chances are related to invariants of stress condition, and buckling criterion must be described under equivalent stress.
4. Referring to probabilistical analysis equivalent buckling stress depends on stress condition invariants unlinearly and that gives unlinear buckling criterion.
5. Buckling criterion obtained from one complicated test results and material yield stress limit allows us to foresee various cases of buckling.

References

1. BS590 (2000) Structural Uses of Steelwork in Building. British Standards Institution, London
2. Eurocode No3 (1990) Design of Steel Structures. Part 1- General Rules and Rules for Buildings Eurocode 3 Editorial Group. Commission of European Communities, Brussels
3. Timoshenko SP, Gere JM (1961) Theory of Elastic Stability. 2nd Ed. McGraw-Hill, New York
4. Galambos TV (1998) Guide to Stability design Criteria for Metal Structures. 5th Ed. Wiley, New York
5. Chen WF, Atsuta T (1977) Theory of Beam – Columns. McGraw-Hill, New York

6. Lindner J (2000) Stability of structural members. Journal of Constructional Steel Research 55:29–44
7. AISC (1993) Load and Resistance Factor Design Specification for Structural Steel Buildings. AISC, Chicago, IL
8. Wang CM, Wang CY, Reddy JN (2005) Exact Solutions for Buckling of Structural Members. CRC Series in Computational Mechanics and Applied Analysis, CRC Press LLC, Boca Raton, FL
9. Troshenko VT, Sosnovsky LA (1979) Statistical theory of fatigue fracture in the case of complicated stress condition (in Russian). Strength of Materials 7:3–11
10. Ziliukas A (2006) Strength criterion of unidirectional composite. Indian Journal of Engineering and Materials Sciences 13:520–524

Comparison of Two Currently used and One Proposed Seismic Design Methods for Steel Structures

Theodore L. Karavasilis, Nikitas Bazeos, and Dimitri E. Beskos

Abstract The forced-based, the displacement-based and the hybrid seismic design methods as applied to plane steel frames are briefly presented and critically compared. The forced-based seismic design method forms the basis of almost all the current seismic design codes, while the displacement-based one is a rather new seismic design method already adopted by some seismic design codes. The hybrid seismic design method is a new method which appropriately combines the best elements of both the force and the displacement-based methods and affects considerable improvements upon these elements. Advantages and disadvantages of these three seismic design methods are presented as derived first on the basis of their description and on the basis of the results of their application to three plane steel frames as compared to those coming out of inelastic time-history analyses involving eight different earthquakes.

Keywords Seismic design · Plane steel frames · Force-based design · Displacement-based design · Hybrid force/displacement-based design · Seismic codes · Performance-based design

1 Introduction

The traditional procedure for seismic design of building structures has been generally termed the force-based design (FBD) method. The implementation of FBD in seismic codes, e.g. EC8, 2004 does not clearly define and employ the terms performance objective, performance level or limit state and expected level of seismic

T.L. Karavasilisa
ATLSS Centre, Lehigh University, Bethlehem, PA, USA
e-mail: thk207@Lehigh.EDU

N. Bazeos (✉) and D.E. Beskos
University of Patras, Department of Civil Engineering, Patras, Greece
e-mail: {n.bazeos, d.e.beskos}@upatras.gr

S. Pantelakis, C. Rodopoulos (eds.), *Engineering Against Fracture: Proceedings of the 1st Conference*,

motion or seismic intensity. In contrary, the FBD method implements performance objectives in a very approximate manner, actually through the use of one performance level, that is the ultimate limit state (ULS) and one seismic intensity for 475-year period ground motions. The FBD method uses the importance factor for different seismic intensities and the reduction factor ν for different limit states. The FBD method usually overestimates the inelastic displacements.

The design procedure would be more rational if the performance of the structure was quantified through a target value of deformation treated as an input variable in that design procedure. This target value of deformation can be assigned to different performance objectives and is the starting point for the development of the direct displacement-based seismic design (DDBD) method [1, 2]. This design method is being advocated as a promising method for the next generation of seismic codes and has already been adopted by seismic design provisions in the United States [3] as one of the proposed displacement-based seismic design methods for the performance-based seismic design.

The third seismic design method presented and compared in this work was originally proposed in a sketchy manner by Bazeos and Beskos [4] and further developed in detail by Karavasilis [5]. This method is called the hybrid force/displacement-based seismic design (HFD) method as it combines the best elements from both the force-based and the displacement-based methods in order to produce an effective design scheme. The starting point in the HFD method is the maximum allowable roof displacement of the MDOF structure computed through a new simple expression proposed by Karavasilis et al. [6], which takes into account structure properties and seismic excitation characteristics. Then, a new relation is used for the calculation of the behaviour factor [6], which is used in a similar way as in the FBD method.

The objective of this article is to critically compare three seismic design methods, the two well established force-based and displacement-based methods, and the new hybrid method.

2 Force Based Design (FBD)

The FBD method relies on the ability of the behaviour factor q to estimate both strength and displacement demands. According to EC8 seismic code, the following steps shall be followed for the design of a building in terms of the FBD method:

2.1 Definition of the Performance Level of the Building

EC8 seismic code identifies two levels of performance: (1) no-(local)-collapse requirement which is associated with the ultimate limit state (ULS) of the structure subjected to destructive earthquakes; (2) damage limitation requirement which is associated with the serviceability limit state (SLS) of the structure for frequent seismic action.

2.2 Design at the Ultimate Limit State (ULS)

Current design practice implies that the structure is first designed at the ULS and subsequently checked at SLS. Thus, firstly the design seismic action is determined by adopting an appropriate elastic response spectrum with damping corresponding to the construction site and then the behaviour factor q is computed. According to EC8, for high ductility class of buildings, the maximum allowable behaviour factor q is given by the equation

$$q = 5\frac{a_u}{a_1} \tag{1}$$

where a_u is the value by which the horizontal seismic design action is multiplied in order to form plastic hinges in a number of sections sufficient for the development of overall structural instability and a_1 is the value by which the horizontal seismic design action is multiplied in order to first reach the plastic resistance in any member in the structure.

2.3 Computation of Inelastic Displacements and Interstorey Drifts

The inelastic displacement $u_{\mathrm{inel,i}}$ and maximum interstorey drift ratio $\mathrm{IDR}_{\mathrm{max,i}}$ in storey i of a multi-degree of freedom (MDOF) structure are computed by the equations

$$u_{\mathrm{inel,i}} = q \cdot u_{\mathrm{d,i}} \text{ and } IDR_{\mathrm{max},i} = \frac{q \cdot (u_{\mathrm{d,i}} - u_{\mathrm{d,i-1}})}{h_i} \tag{2}$$

respectively, where $u_{\mathrm{d,i}}$ is the elastic displacement of storey i calculated in step 2 and h_i is the storey height.

3 Direct Displacement Based Design (DDBD)

The direct displacement-based design method (DDBD) is probably the most well known displacement-based design method and is adopted by modern design codes. The following steps shall be followed for the design of a building in terms of the DDBD method:

3.1 Definition of the Performance Level of the Building

Performance-based seismic engineering adopts different design levels to meet different 'performance levels' at different seismic 'hazard levels' [7]. In the DDBD method the hazard levels are identified by the appropriate elastic displacement response spectrum and the performance levels are determined by the maximum IDR values defined according to a modern design code, e.g., SEAOC [3].

3.2 Computation of the Target Displacement of the SDOF

The target displacement $u_{eff,t}$ of the equivalent substitute single-degree of freedom (SDOF) system to the MDOF given structure is computed by the equation

$$u_{\mathrm{eff,t}} = \frac{\sum_{i=1}^{n} m_{\mathrm{i}} u_{\mathrm{i}}^2}{\sum_{i=1}^{n} m_{\mathrm{i}} u_{\mathrm{i}}} \tag{3}$$

where n is the number of stories and m_i the mass and u_i the maximum displacement of storey i, respectively. The maximum displacement profile is an important issue in the DDBD, because it correlates the damage index (IDR_{max}), as defined for a specific performance level, with the maximum displacement of the stories. For the computation of the maximum displacement profile, one can use the profiles suggested by Loeding et al. [1] for regular reinforced concrete frames in the elastic range.

3.3 Computation of the Yield Displacement of the SDOF

The yield displacement $u_{eff,y}$ of the equivalent substitute SDOF system is computed by Eq. (2) for damage index (IDR_y) corresponding to the yielding of the frame.

3.4 Computation of the Equivalent Ductility and Equivalent Damping of the SDOF

The ductility μ_{eff} of the equivalent substitute SDOF system is calculated through the expression

$$\mu_{\mathrm{eff}} = \frac{u_{\mathrm{eff,t}}}{u_{\mathrm{eff,y}}} \tag{4}$$

where the $u_{eff,t}$ and $u_{eff,y}$ are the maximum and yield displacement, respectively, of the SDOF substitute system.

The effective damping associated with a specific level of ductility can be calculated by using the expression proposed by Borzi et al. [8]

$$\xi_{\mathrm{eff}} = a \cdot \left(1 - \frac{1}{\mu_{\mathrm{eff}}}\right) + \xi_v \tag{5}$$

where α is a coefficient depending upon the ductility and ξ_v stands for the viscous part of the damping, usually taken to be 5%.

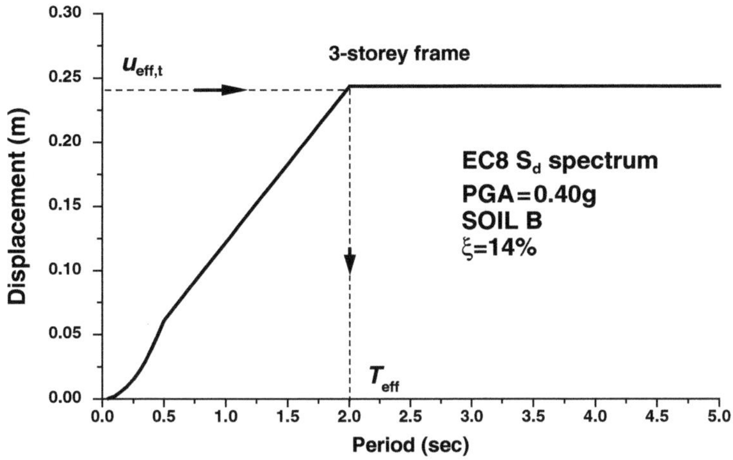

Fig. 1 Elastic displacement response spectrum

3.5 Computation of the Equivalent Period of the SDOF

With a target displacement $u_{eff,t}$ and an effective damping ξ_{eff} calculated in the previous steps, the effective structural period T_{eff} can be directly obtained from the elastic displacement response spectrum (Fig. 1).

3.6 Computation of the Equivalent Stiffness and Base Shear of the SDOF

The effective stiffness K_{eff} is calculated by using the simple relation of structural dynamics

$$K_{eff} = \frac{4 \cdot \pi^2}{T_{eff}^2} \cdot M_{eff} \tag{6}$$

where M_{eff} is the mass of the *SDOF* structure given by

$$M_{eff} = \sum_{i=1}^{n} m_i \frac{u_i}{u_{eff,t}} \tag{7}$$

Finally, the required shear strength V_{eff} of the SDOF system can be computed by the equation

$$V_{eff} = K_{eff} \cdot u_{eff,t} \tag{8}$$

3.7 Design of the MDOF Structure

The strength of the structure refers to the maximum inelastic base shear which the structure should have in order to achieve the requirements of the performance level under consideration (step 1 of the DDBD method). This base shear V_{eff} should be divided by the overstrength factor Ω in order to produce the design one, V_d, i.e.,

$$V_d = \frac{V_{eff}}{\Omega} \tag{9}$$

The overstrength factor Ω depends on the number and the sequence of development of plastic hinges in the structure and on the geometric and mechanical characteristics of the structure. One may rely on the recommendation of EC8 and take $\Omega = 1.3$.

4 Hybrid Force/Displacement Based Design (HFD)

The current version of the newly proposed hybrid force/displacement (HFD) seismic design method is valid for plane steel frames (moment resisting, x-braced, with setbacks or with mass irregularities) [5]. The following steps shall be followed for the design of a plane steel moment resisting frame in terms of the HFD method.

4.1 Definition of the Performance Level of the Building

In the HFD method the seismic hazard levels are identified with the appropriate elastic acceleration response spectrum, as it is defined in EC8 and the performance levels are determined by the maximum IDR values.

4.2 Computation of the Maximum Allowable Roof Displacement of the Building

The maximum allowable roof displacement $u_{r,max}$ of the building can be computed on the basis of the maximum IDR value, as defined in the previous step, with the aid of the expression

$$u_{r,max,IDR} = \beta \cdot IDR_{max} \cdot H \tag{10}$$

where H is the building height from its base and β is a coefficient depending on building properties and seismic excitation characteristics, which can be calculated through the equation [5]

$$\beta = 1.0 - 0.193 \cdot (n_s - 1.0)^{0.54} \cdot \rho^{0.144} \cdot \alpha^{-0.19} \tag{11}$$

In the above, n_s is the number of stories, ρ is the stiffness ratio of the frame calculated for the storey closest to the mid-height of the frame via the expression

$$\rho = \frac{\Sigma (I/l)_b}{\Sigma (I/l)_c} \tag{12}$$

with I and l being the second moment of inertia and length of the steel member (column c or beam b), respectively and α is the ratio defined as

$$\alpha = \frac{M_{RC,1,av}}{M_{RB,av}} \tag{13}$$

where $M_{RC,1,av}$ is the average of the plastic moments of resistance of the columns of the first storey and $M_{RB,av}$ is the average of the plastic moments of resistance of the beams of all the stories of the frame.

4.3 Calculation of the Required Strength and Design of Structure

First the maximum allowable roof displacement ductility factor μ_δ is computed by the expression

$$\mu_\delta = \frac{u_{r,max,IDR}}{u_{r,max,y}} \tag{14}$$

where $u_{r,max,IDR}$ and $u_{r,max,y}$ are defined in the previous step. Then, the behaviour factor q can be calculated by the equations [5]

$$q = 1 + 1.39 \cdot (\mu_\delta - 1) \text{ for } \mu_\delta \leq 5.8 \tag{15a}$$

$$q = 1 + 8.84 \cdot (\mu_\delta^{0.32} - 1) \text{ for } \mu_\delta > 5.8 \tag{15b}$$

The behaviour factor q is used in a similar way as in FBD.

5 Comparison of the Methods Through a Design Example

Consider a S275 plane steel moment resisting frame with three storeys and two bays. All bay widths are assumed equal to 7 m and all storey heights equal to 3.5 m. The gravity load on beams is equal to 30 kN/m and the viscous damping ratio ξ is equal to 3%. The expected ground motion is represented by the design elastic spectrum of the EC8 seismic code with peak ground acceleration equal to 0.4 g and a soil class B. The frame is designed according to EC3, 1992 with the aid of the commercial analysis and design software package SAP2000. HEB profiles are used for the columns and IPE profiles for the beams. This frame was seismically designed by the three methods described previously and comparisons were made with respect to their accuracy, efficiency and degree of conservatism.

5.1 Force Based Seismic Design (FBD)

The value of q is selected equal to 6.5 according to Eq. (1) since the ratio α_u/α_1 for moment resisting frames with more than one bay is equal to 1.3 (EC8). Elastic modal analysis and design leads to the optimum sections HEB240 for columns and IPE330 for beams. The maximum roof displacement $u_{d,r} = 3.4$ cm, while the maximum interstorey drift ratio occurs at the second floor and is equal to 0.46%. Thus, their maximum inelastic counterparts are $u_{inel,r} = 6.5*3.4 = 22.1$ cm and $IDR_{max} = 6.5*0.0046 = 2.9\%$, respectively. Finally, the base shear is $V_b = 130$ kN.

5.2 Displacement Based Seismic Design (DDBD)

For comparison purposes with the FBD method described previously, the value of IDR_{max} computed in the last step of the FBD method is chosen here as the appropriate damage level, i.e. $IDR_{max} = 2.9\%$. By using Eq. (3) and the displacement profile proposed by Loeding et al. [1], the target displacement $u_{eff,t}$ of the SDOF system is computed equal to 24.0 cm. Similarly, using Eq. (3) with interstorey drift ratio at yielding of the equivalent elasto-plastic SDOF system, i.e., $IDR_y = 1.0\%$, the yield displacement $u_{eff,y}$ of the SDOF system is calculated equal to 8.1 cm. Then, the ductility μ_{eff} and the effective damping ξ_{eff} are evaluated from Eqs. (4) and (5) and they are equal to 2.96% and 14.3%, respectively. By entering the displacement response spectrum (Fig. 1) with a damping value equal to 14.3% and a displacement equal to 24.0 cm, the effective period T_{eff} is found equal to 2.00 s. The effective mass is calculated through Eq. (7) and consequently the effective stiffness is found through Eq. (6) to be equal to 1,071 kN/m. The required strength in terms of the base shear is found through Eq. (8) to be equal to 257 kN. Thus, the design base shear V_d is computed according to Eq. (9) with a value of Ω equal to 1.3 and found to be equal to 197.7 kN. This design base shear is distributed linearly along the height of the frame and an elastic analysis and design is performed, which leads to the optimum sections HEB280 for columns and IPE360 for beams. The maximum roof displacement is estimated from the displacement profile proposed by Loeding et al. [1] to be equal to $u_{d,r} = 30.0$ cm.

5.3 Hybrid Force/Displacement Seismic Design (HFD)

For comparison purposes with the previous methods, a value of IDR_{max} equal to 2.9% is chosen. The maximum allowable roof displacement $u_{r,max}$ of the building is then computed by using Eqs. (10) and (11), with $H = 10.5$ m, $n_s = 3$, $\rho = 0.5$ and $\alpha = 1.3$ and found to be equal to 23 cm. The maximum roof displacement $u_{r,max,y}$ can be calculated from Eq. (10) for IDR_{max} equal to IDR_y. Also $u_{r,max,y} = IDR_y^*H$, since

Table 1 Comparison results

	FBD	DDBD	HFD	Nonlinear	Nonlinear
Sections	HEB240-IPE330	HEB280-IPE360	HEB240-IPE330	HEB240-IPE330	HEB280-IPE360
$IDR_{max}(-)$	2.9%	2.9%	2.9%	3.0%	2.4%
$u_{r,max}(cm)$	22.1	30.0	23.0	24.0	19.0
$V_b(kN)$	130	197	108	–	–

the displacement profiles are linear in the elastic response of the frame and IDR_y is equal to 0.4% for S275 [6]. Thus one obtains $u_{r,max,y} = 4.20$ cm. The maximum allowable roof displacement ductility factor μ_δ as computed by Eq. (14), is obtained equal to 5.48, while the behaviour factor q as evaluated by Eq. (15a) is found to be equal to 7.20. The elastic response spectrum of EC8 is divided by q and an elastic modal analysis is performed which yields the optimum sections HEB240 for columns and IPE330 for beams. These sections are the same with those of the FBD method. The fundamental natural period of the frame is found to be equal to 1.20 s and the base shear is found to be equal to 108 kN. Finally, the values of ρ, α and $u_{r,max,y}$ are computed, compared to those initially assumed and found to be very close to them.

Table 1 summarizes the results of the three seismic design methods. In order to compare the three seismic design methods, nonlinear time history analyses of the designed frames are performed using the well-known program DRAIN-2DX. Eight semi-artificial accelerograms compatible with the EC8 spectrum were generated via a deterministic approach. The results, shown in Table 1, reveal that the proposed method along with the FBD method yield more economical sections than the DDBD method, but the proposed method seems to predict more accurately the maximum roof displacement, $u_{r,max}$, compared to the two other methods.

References

1. Loeding S, Kowalsky MJ, Priestley MJN. Direct Displacement-Based Design of Reinforced Concrete Building Frames. Report no. SSRP-98/08, University of California, San Diego, CA, 1998.
2. Priestley MJN, Calvi GM, Kowalsky MJ. Displacement-Based Seismic Design of Concrete Structures. Proceedings of 6th National Conference on Earthquake Engineering, Turkey, 2007.
3. SEAOC (Structural Engineers Association of California). Recommended Lateral Force Requirements and Commentary. 7th ed., Seismology Committee, Sacramento, CA, 1999.
4. Bazeos N, Beskos DE. New Seismic Design Method for Building Framed Structures. In: Proceedings of International Conference on Computational and Experimental Engineering and Sciences, Atluri SN, Beskos DE, Polyzos D (eds), TechScience Press, Corfu, Greece, 2003.
5. Karavasilis TL. New seismic design method for steel frames. Ph.D. thesis, Department of Civil Engineering, University of Patras, Patras, Greece, 2007 (in Greek).

6. Karavasilis TL, Bazeos N, Beskos DE. Drift and Ductility Estimates in Regular Steel MRF subjected to ordinary Ground Motions: A Design-Oriented Approach, *Earthquake Spectra*, Vol. 24, No. 2, pp. 431–451, 2008.
7. Fardis MN. Code Developments in Earthquake Engineering. Proceedings of the 12th European Conference on Earthquake Engineering, London, United Kingdom, 2002.
8. Borzi B, Calvi GM, Elnashai AS, Faccioli E, Bommer JJ. Inelastic Spectra for Displacement-Based Design. Soil Dynamics and Earthquake Engineering 2001, 21:47–61.

Experimental and Numerical Investigation of Failure Pressure of Valve Housing

Ivica Galić, Zdenko Tonković, and Krešimir Vučković

Abstract This paper presents an experimental and a finite element analysis which are conducted to determine the allowable and failure pressures in unflawed valve housing (designation DN100 PN40). Extensive three-dimensional non-linear finite element analyses are performed to obtain the plastic limit, collapse and instability loads for valve housing under internal pressure and to predict the burst pressure obtained by the experimental tests. Finite element results show that the failure pressures obtained by limit load analyses give conservative results, except those obtained by using ultimate tensile stress as the flow stress. It is shown that the plastic collapse pressure results obtained by a real strain hardening material model agree well with the experimental results.

Keywords Burst pressure · Experiments · Finite element analysis · Plastic collapse pressure · Valve housing

1 Introduction

In the process industry, valves are used as important elements of process control with a function to influence the process parameters by increasing or decreasing the flow or pressure, or to influence indirectly the temperature of the flowing medium. The main valve element carrying the internal pressure is the valve housing, which is often exposed to high temperatures due to the process requirements. In most cases, housings are complex in form and are made by casting. It is the main reason why

I. Galić (✉)
ATM Žitnjak bb, 10000 Zagreb, Croatia
e-mail: ivica.galic@atm.hr

Z. Tonković and K. Vučković
Faculty of Mechanical Engineering and Naval Architecture, University of Zagreb,
I. Lučića 5, 10000 Zagreb, Croatia

S. Pantelakis, C. Rodopoulos (eds.), *Engineering Against Fracture: Proceedings of the 1st Conference*,

the valve wall often contains defects like casting cavities, slag inclusions, pores or cracks. In the life cycle of valve housing, such defects can develop or new ones can originate during operation and grow to a critical size at which the structure fails as shown in [9]. In the design phase, the calculation of the valve wall thickness is carried out in accordance with standards by means of which the complex three-dimensional valve geometry is approximated with the two-dimension valve geometry and the equalization of the pressured surfaces with the supporting surfaces is done [2].

For safe exploitation, the allowable pressure up to which the valve can be pressurized needs to be known. In order to define the allowable load of the pressure vessel components according to the limit design method [3], it is necessary to determine the plastic collapse load due to the static load. Therefore, the plastic collapse load can be predicted by the limit load analysis which defines two additional characteristic loads, i.e. the plastic yield and the plastic instability load. The plastic yield load is determined on the basis of limit analysis by the application of the small strain theory and elastic–perfectly plastic material assumptions [4]. Unlike the plastic yield load, the determination of the plastic collapse load and the plastic instability load is based on the large strain theory, emphasizing the effect of strain hardening [5–7]. In that case, the plastic collapse load is not a load necessary to cause the physical collapse of the structure; it is rather a load under which significant plastic strain occurs. The commonest methods for the determination of the plastic collapse load used in literature are the twice-elastic-slope (TES) method [3] and the tangent intersection (TI) method [1]. The TES method defines the collapse load as the load at the intersection of a line drawn from the origin of the load-deformation curve, at twice the slope of the linear part of the load-deformation curve. For the determination of the collapse load by means of the TI method, it is necessary to draw the tangent line to the linear and nonlinear parts of the load-deformation curve. The load at the intersection of tangent lines defines the plastic collapse load. In contrast to the TES method, which gives consistent results, the TI method is sensitive to the selection of the point in which the tangent line is drawn to the nonlinear part of the curve. In contrast to the plastic collapse load, the plastic instability load can be defined as the maximum load which the structural element made of a work-hardening material could withstand [8].

The aim of this paper is to determine the allowable and failure pressures of the valve housing marked DN100 PN40. The finite element (FE) analysis is conducted along with the experimental research. The scope of the experimental research is shown in the second section of the present contribution. Three-dimensional limit load finite element analyses have been performed in the third section. Therein, the experimentally determined plastic collapse pressure and the burst pressure have been compared with the results obtained using the finite element method (FEM). In the last section, the allowable housing pressure is determined and compared with the solution obtained by the method proposed in [2].

2 Experiments

The housing on which the experiment was conducted was made by sand casting. Its complex structure with basic dimensions is shown in Fig. 1a. As may be observed, the housing consists of two lateral flanges and one upper and one lower flange. Bonnets are connected to the upper and the lower flange, while the valve is connected to the pipeline through the lateral flanges.

The valve housing material is steel casting GS-C25 (according to DIN) with the yield stress $\sigma_y = 250\,\text{MPa}$, the ultimate stress $\sigma_u = 475\,\text{MPa}$ and the modulus $E = 205\,\text{GPa}$. For the experimental purpose, the lateral flanges were removed and made into test specimens for the tensile test. Material testing was done at room temperature. For the large strain FE analysis, the true stress and strain values are determined [8]. Lateral flanges have an irregular bore which cannot be sealed in a satisfying manner. Besides, the flanges cannot endure high internal pressures which pressurize the housing during the experiment. That is the reason why the side flanges were removed and instead the flanges connected with screws were mounted. The upper and lower flanges have regular bores which can be sealed for high internal pressures and blind flanges were mounted on them. One blind flange contains the entrance for the pressurizing medium and the other flange is connected to the measuring range gage. Additionally, strain gages were pasted to the housing as shown in Fig. 1b. For experimental purposes, foil strain gages with two perpendicular measuring grids were used. The strain gages were bonded at four different points and their measuring grids were oriented in the axial and the circular direction in accordance with the geometry of the observed housing area. For the experiment control, pairs of strain gages at positions 1 and 3 and the ones at positions 2 and 4 were positioned in a way that they should give the same data.

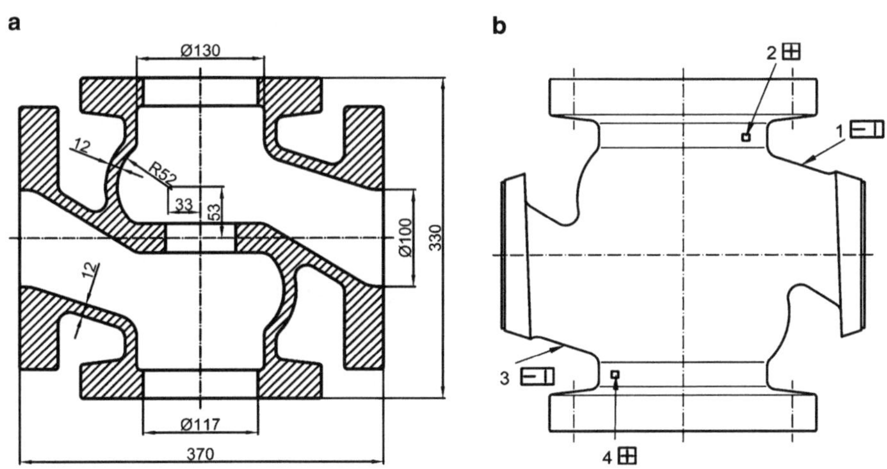

Fig. 1 **a** Layout and dimensions of the valve housing; **b** location of installed strain gages

The strain gages at positions 1 and 3 have the measuring grids placed one above the other, while the strain gages at positions 2 and 4 have the measuring grids one beside the other. Two additional experiments, in which the same strain gages were used, were conducted. However, their types, concerning the measuring grid placement, were switched between the first strain gage pair at positions 1 and 3 and the second one at positions 2 and 4. All strain gages used in the experiment were calibrated to the nominal resistance of $120 \pm 0.1\,\Omega$.

The valve housing was pressurized by a pump and the pressurizing medium is water with the temperature of about 15°C. For pressure measurement, a moderate measuring range gage with the measuring scope of up to 100 MPa was used. As control, two moderate manometers with the measuring scope of up to 100 MPa were used. Before the measurement started, the valve housing had been vented and pressurized two times to 2 MPa and then depressurized to the atmosphere pressure for the strain gage response control.

The analyzed housing is used for the nominal internal pressure of up to 4.0 MPa. However, much higher pressures are needed for the housing failure. At such pressures, large strains occur on the housing surfaces, which cannot be measured by strain gages as their accuracy decreases. On account of that, the experiment results include only the data up to 35.0 MPa.

The experimental results are shown in Fig. 2 giving the pressure versus strain diagrams of the gages. Figure 2a shows the experimental results for the gages 1 and 3 and Fig. 2b for 2 and 4. The results for the gage pairs 1 and 3 and 2 and 4 deviate insignificantly. For easier overview, Fig. 2a, b give only the results for gages 1 and 2. The cause of deviations is the imperfection of the position of attached gages on the housing.

Based on the experimental results, the plastic collapse pressure for the housing is determined separately for the position of gages 1 and 3 and for 2 and 4. Two methods described in the introduction of the paper were used, i.e. the twice-elastic-slope (TES) method [3], and the tangent intersection (TI) method [1]. In the internal pressure-strain diagrams, the larger component, i.e. the circular one, was used for the strain. Figure 3 presents the results where $p_{\mathrm{C,exp}}^{\mathrm{TES}}$ is the plastic collapse pressure determined by the TES method and $p_{\mathrm{C,exp}}^{\mathrm{TI}}$ is the plastic collapse pressure obtained by the TI method.

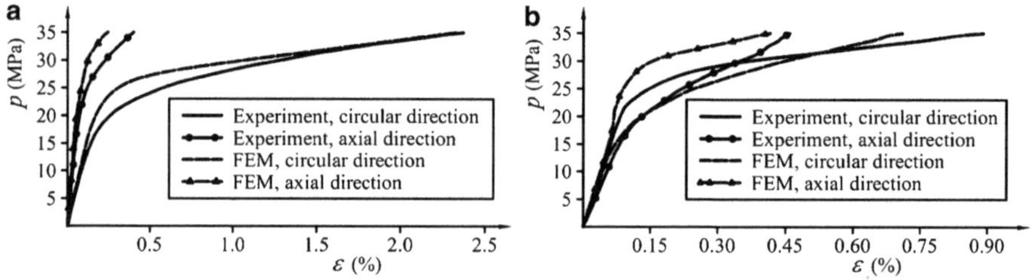

Fig. 2 The comparison of experimental and FE results for the strain gages: **a** 1 and 3; **b** 2 and 4

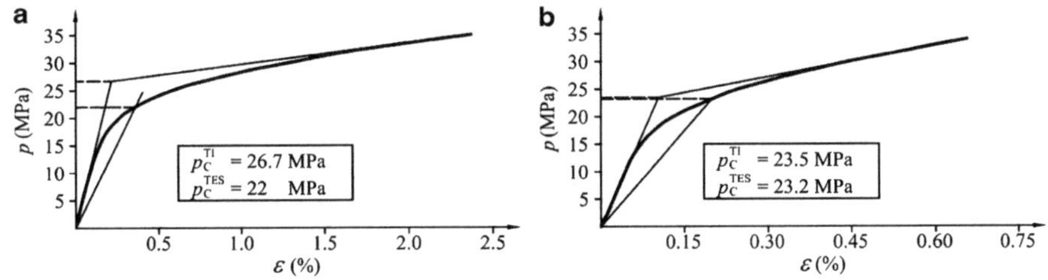

Fig. 3 Plastic collapse pressure based on the experimental results for the strain gages: **a** 1 and 3; **b** 2 and 4

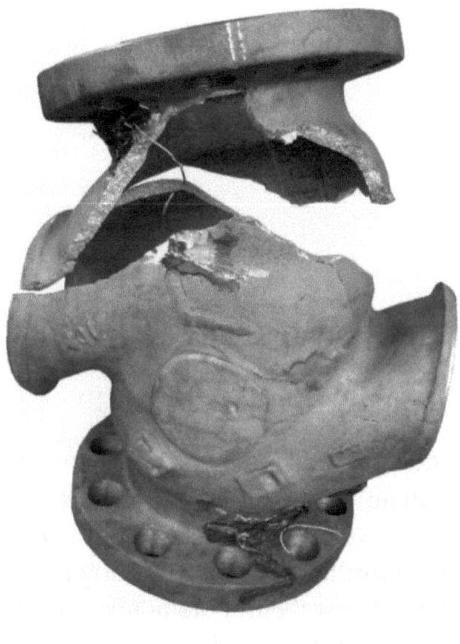

Fig. 4 Rupture of the valve housing

Results up to the internal pressure of 35.0 MPa have been shown in previous figures. As said earlier, after that pressure value due to the occurrence of large strains, the strain gages slide with respect to the valve housing wall so that there was no point in showing the results for pressures above that value. At further pressure increase, the gages stopped giving signals one by one until, at the pressure of 48.2 MPa, none of them gave any response. The final housing rupture occurred at the pressure of 69.5 MPa. The rupture of the valve housing is shown in Fig. 4.

3 Finite Element Analysis

The finite element analysis is performed by using the commercial finite element package ABAQUS/Standard (2007) [9]. Only one half of the housing is modeled for symmetry purposes. Since this is a case of a very complex geometry, a free-meshing

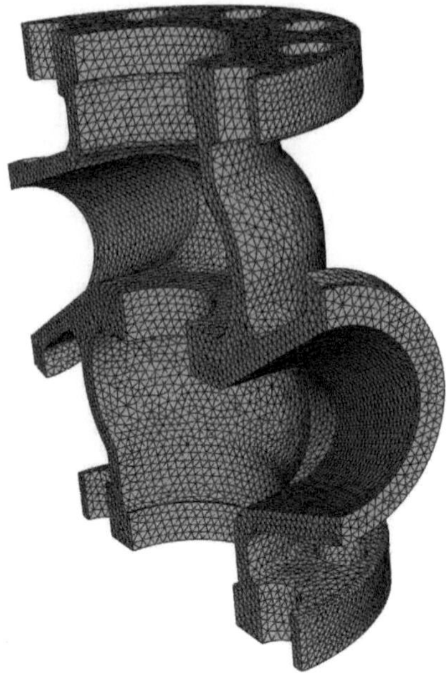

Fig. 5 Typical FE mesh

algorithm was applied. For the model discretization, a modified quadratic tetrahedral element (C3D10M) is applied as adequate for both small and large deformations. Several different meshes were analyzed and a typical mesh, for which solutions are given further in the text, is shown in Fig. 5 in two different views. The mesh contains 168,879 tetrahedral elements and 425,440 nodes.

Boundary conditions are defined by restraining displacements in all nodes in the symmetry plane in the direction vertical to the plane. Additionally, the displacement of nodes leaning on the lateral flanges in the circular direction is restrained. The housing load is the internal pressure p, which was applied as a distributed load to the internal walls of the finite element model. The pressure affecting the upper and the lower flange is calculated into the screw force and given as the surface traction, upwards for the upper flange and downwards for the lower flange, and is assigned to the bores in which the screws are mounted.

To compute the plastic yield, collapse and instability pressures for which the collapse of the valve occurs, an incremental load–displacement analysis is performed by using the modified arc-length Riks method available in the ABAQUS program. Non-linear material behavior is modeled using the incremental plasticity with the von Mises yield function, together with the associated flow rule and isotropic hardening. The large strain FE model is employed by invoking the NLGEOM option within ABAQUS.

4 Finite Element Results and Comparison with the Experiment

With the incremental increase of the internal pressure, the housing deformation occurs so that at the pressure of about 11.0 MPa the first areas on the model appear, in which the equivalent stresses overcome the yield stress. With a further pressure increase, larger elastic-plastic strains occur on the model. At the pressure of about 70.0 MPa, on a large part of the housing the Von Mises stresses reach the value of the material ultimate stress. For a comparison of experimental and FE results, at the positions where strain gages were attached in the real model (Fig. 1b), the values of relevant strains for pressure increments were read in the nodes of the numerical model finite elements. In Fig. 2a, b, a comparative illustration of experimental and FE results is given in pressure versus strain diagrams. Figure 2a shows the results for the strain gages at positions 1 and 3, while Fig. 2b shows the results for the strain gages at positions 2 and 4. As the experimental results include only the pressure values up to 35.0 MPa, the FE results are also given up to that pressure range. As may be observed from Fig. 2a, the finite element results correspond well with the experimental results for the strain gages at positions 1 and 3, both in the axial and in the circular direction. However, the results from Fig. 2b for the strain gages at positions 2 and 4 do not correspond that well. The cause of that is that the strain gages at positions 2 and 4 were placed in the area of an abrupt geometry change. During their pasting, a small deviation from the ideally circular and axial directions of that area of the housing geometry was introduced.

In the same way as described earlier, the plastic collapse pressure based on the FE analysis was determined for the area in which the strain gage was attached. Figure 6 shows the results where $p_{\mathrm{C,FEM}}^{\mathrm{TES}}$ is the plastic collapse pressure determined by the twice-elastic-slope method and $p_{\mathrm{C,FEM}}^{\mathrm{TI}}$ is the plastic collapse pressure obtained by the tangent intersection method. In addition to that, areas on the housing with critical stresses and strains were defined based on the FE results. Since these areas are very close to the position of strain gages 2 and 4, their internal pressure-deformation diagrams and also the plastic collapse pressure correspond to those shown in Fig. 6b. From the presented results it is obvious that the plastic collapse pressure values determined experimentally are lower than those determined by the FE analysis. The

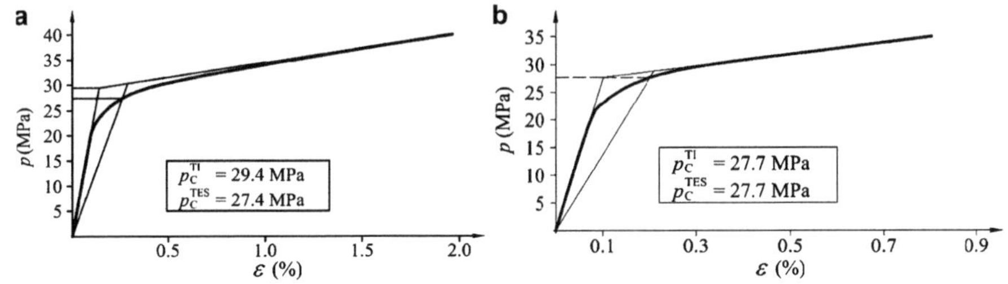

Fig. 6 Plastic collapse pressure based on the FE results for the strain gages: **a** 1 and 3; **b** 2 and 4

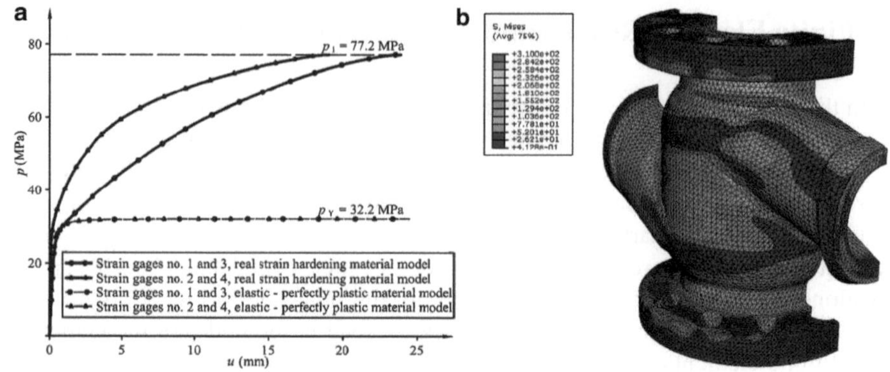

Fig. 7 **a** Variation of pressure with radial displacement of point at the position of the strain gages 1 and 3 as well as 2 and 4; **b** Von Mises stress distribution for the internal pressure $p = 18\,\text{MPa}$

mean value of the plastic collapse pressure determined experimentally is 23.9 MPa, and the one determined numerically 28.1 MPa, which is a satisfactory correspondence. Slightly better result correspondence is obtained by means of the *tangent intersection* method in comparison with the *twice-elastic-slope* method.

Figure 7a shows load-displacement diagrams for the numerical model finite element nodes at the positions 1 and 3, and 2 and 4, of strain gages. The radial components of displacement were taken into consideration as displacements in accordance with the geometry of the observed housing area. From the load-displacement curves, the plastic instability pressure p_{I} is obtained as the highest point on the curve, i.e. the maximum pressure that could be attained. In the FE analyses it is assumed that the material will fail in a ductile manner so that the plastic instability pressure p_{I} represents the failure pressure p_{F} that causes the plastic collapse of the housing. In such a manner, the numerically obtained failure pressure of 77.2 MPa corresponds well with the experimentally determined burst pressure of 69.5 MPa.

The approximate result for the failure pressure p_{F} can be obtained without the consideration of the real material hardening and geometrical nonlinearity. In some Failure Assessment Diagram (FAD) methods [4], the plastic yield load of a structural component is used to define a parameter L_{r} that measures the probability of a ductile failure, where a limit value of the plastic collapse or a cut-off on the Failure Assessment Diagram $L_{\mathrm{r}}^{\max}$ is defined as follows:

$$L_{\mathrm{r}}^{\max} = F_{\mathrm{F}}/F_{\mathrm{Y}} = \sigma_{\mathrm{f}}/\sigma_{\mathrm{y}} \tag{1}$$

where F_{Y} denotes the plastic yield load, F_{F} is the ductile failure load, while σ_{f} stands for the flow stress. As described in the introduction, the definition of the plastic yield load F_{Y}, according to classical limit load analysis, is an idealized one. It ignores geometric nonlinearity and strain hardening effects since it is built on the small deformation theory and assumes an elastic-perfectly plastic material behavior.

Figure 7a, b show the internal pressure-displacement curves for the determination of the instability pressure p_I, and also the curves for the determination of the plastic yield pressure p_Y. The diagram shows that for the analyzed housing the plastic yield pressure amounts to 32.2 MPa. As can be usually found in literature, to take the strain hardening effectiveness into account, the flow stress can be introduced into the limit load analysis as the material yield stress. It is generally known that the flow stress has a value between the yield and ultimate tensile strength $(\sigma_y \leq \sigma_f \leq \sigma_u)$ and it is usually defined as follows:

$$\sigma_f = k\left(\sigma_y + \sigma_u\right) \tag{2}$$

where k is the flow stress factor. The flow stress is most often evaluated as a simple average between the yield and tensile stress ($k = 0.5$, recommended for most assessments), but there are also other definitions of the flow stress, as described in [8]. In accordance with the expression (1), the failure pressure is

$$p_F = p_Y \sigma_f / \sigma_y. \tag{3}$$

If the term σ_f is inserted into the previous expression for the mean value between the yield stress and the ultimate tensile stress ($k = 0.5$), a rather conservative solution for the failure pressure ($p_F = 46.7$ MPa) is obtained. But if the ultimate tensile stress is assumed for the flow stress ($\sigma_f = \sigma_u$), the obtained failure pressure in the amount of 61.2 MPa corresponds well to the experimentally determined burst pressure of 69.5 MPa.

5 Allowable Pressure

In accordance with [3], the allowable pressure $p_{allowable}$ is determined from the plastic collapse pressure p_C through the expression $p_{allowable} = 2/3 p_C$. If the mean value of the plastic collapse pressure determined experimentally ($p_{C,exp} = 23.9$ MPa) is inserted into the previous expression, it follows that the valve housing allowable pressure is 15.9 MPa. On the other hand, if the FEA results ($p_{C,FEM} = 28.1$ MPa) are applied, the allowable pressure is 18.7 MPa. Figure 7b shows the Von Mises stress distribution for the internal pressure $p = 18.0$ MPa which is close to the allowable pressure. It can be noted in Fig. 7b that under that kind of load, the local plastic region starts occurring only at small areas with significant stress concentration. Since the housing is intended for the pressure of up to 4.0 MPa, it is completely in the elastic area for such a load.

In common engineering practice, the calculation of the valve wall thickness is done in accordance with the EN 12516-2 standard. Maximum allowable pressure for the right and left housing sides is also determined in accordance with [2] as following: $p^R = 17.6$ MPa and $p^L = 19.7$ MPa. The conclusion is that the right housing side is weaker and the maximum allowable pressure according to the EN12516-2 standard is 17.6 MPa. If results are compared, it is clear that the allowable pressure values determined by the application of the EN12516-2 standard are close to the mean of the FE results and the experimental results obtained by the application of the limit design method [3].

6 Conclusion

Extensive experimental and non-linear finite element analyses were performed in order to obtain the allowable and failure pressures for the valve housing. Experimental analyses were conducted on three different housings and the results are shown in pressure versus strain diagrams of the gages which were pasted to the critical places on the structure. The gage position was determined based on the previous finite element analysis (FEA). At the internal pressure of 35.0 MPa, large strains on the housing occur and the gages slide with respect to the housing wall. As a consequence, the experimental results are shown up to that value.

In the finite element analysis, the real material hardening and geometrical nonlinearity were modeled. The conclusion was that the experimental and FE results correspond well. Differences exist only in the points of abrupt geometry change, where, during the placement of gages, a small deviation from the ideally circular and the axial direction of that part of the housing geometry was introduced. Based on the experimental and numerical results, the housing plastic collapse pressure was determined by means of two methods: the twice-elastic-slope and the tangent intersection. In the result comparison it was noticed that the plastic collapse pressure values determined experimentally were lower than the ones obtained by the FE analysis and that a better result correspondence was obtained by the application of the tangent intersection method than by the twice-elastic-slope method.

By the application of experimental and FE results for the plastic collapse pressure, in accordance with the limit design method, the allowable pressure for the housing was derived and compared with the one obtained by the application of the EN 12516-2 standard. The conclusion was that the allowable pressure determined by the application of the EN 12516-2 standard is close to the mean of the FE results and the experimental results obtained by the application of the limit design method.

In the FE analyses it is assumed that the material will fail in a ductile manner. In accordance with that, from the load-displacement diagram obtained by the FE analysis, the failure pressure, i.e. the plastic instability pressure, was determined as the highest point on the curve. The numerically obtained failure pressure in the value of 77.2 MPa corresponds well to the experimentally determined burst pressure of 69.5 MPa.

Additionally, it is shown that the results for the failure pressure can be obtained by a simpler FE analysis without taking into consideration the material hardening and geometrical nonlinearity. Thereby, FE results show that the failure pressures obtained by the limit load analyses give conservative results, except those obtained by using the ultimate tensile stress as the flow stress.

References

1. FKM Guideline, 2005. Fracture Mechanics Proof of Strength for Engineering Components, Editor: Forschungskuratorium Maschinenbau (FKM), VDMA Verlag GmbH, Frankfurt/Main.
2. EN 12516-2, 2004. Industrial valves Shell design strength Part 2: Calculation method for steel valve shells, ICS 23.060.01.
3. ASME, 1995. Boiler and pressure vessel code section III and VIII, New York: American Society for Mechanical Engineers.
4. SINTAP, 1999. Structural Integrity Assessment Procedure for European Industry. Final Procedure.
5. Sang, Z.F, Xue, L.P., Lin, Y.J. and Widera, G.E.O., 2002. Limit and burst pressures for a cylindrical shell intersection with intermediate diameter ratio. Int. J. Pressure Vessels Piping 79, 341–349.
6. Li, H. and Mackenzie, D., 2005. Characterizing gross plastic deformation in design by analysis. Int. J. Pressure Vessels Piping 82, 777–786.
7. Chattopadhyay, J. and Tomar, A.K.S., 2006. New plastic collapse moment equations of defect-free and through wall circumferentially cracked elbows subjected to combined internal pressure and in-plane bending moment. Engineering Fracture Mechanics 73, 829–854.
8. 8. Tonković, Z., Skozrit, I. and Alfirević, I., 2008. Influence of flow stress choice on the plastic collapse estimation of axially cracked steam generator tubes. Nuclear Engineering and Design 238, 1762–1770.
9. Hibbitt, Karlsson & Serensen, Inc., 2007. ABAQUS/Standard. User's guide and theoretical manual, Version 6.7.

Fracture Analysis of a Bolted Joint of a Large Pump Frame of a Desalination Plant

Luigi Carassale and Cesare Mario Rizzo

Abstract The present paper deals with the fracture analysis of the frame of a very large pump. The complex geometry of the structure as well as the loading conditions were responsible of premature fracture failures localized in the neighborhood of the bolted connection between the column pipe and the basement plate.

Retrofitting solutions and appropriate countermeasures were proposed after long lasting hydrodynamic and structural analyses that needed to account for very strict operational and time constraints.

The fracture strength as well as the vibration behavior of the structure were considered taking advantage of both numerical FE calculations and experimental in-situ measurements. The parallel numerical and experimental analyses leaded to a set of modifications of the structure hereunder described in detail.

The integration among experimental and theoretical results concurred to obtain a suitable solution for a quite complex case study because it allowed improving aspects related to both the strength and the loading conditions of the structure.

Keywords Pump · Fatigue · Fracture · Vibration · Finite elements · Measurements

1 Introduction

Desalination plants produce fresh water via partial distillation of seawater. The by-product of the process is high-concentrate seawater that is recirculated by pumps referred to as Brine Recirculation Pumps (BRP). BRPs are very large pumps

L. Carassale
Department of Civil Environmental and Architectural Engineering, University of Genova, Via Montallegro, 1 – 16145 Genova, Italy
e-mail: luigi.carassale@unige.it

C.M. Rizzo (✉)
Department of Naval Architecture and Marine Engineering, University of Genova, Via Montallegro, 1 – 16145 Genova, Italy
e-mail: cesare.rizzo@unige.it

S. Pantelakis, C. Rodopoulos (eds.), *Engineering Against Fracture: Proceedings of the 1st Conference*,

vertically mounted in barrels that form their intake wells. Due to economical reasons, the size of the distillers is constantly increasing and, as a consequence, the size of BRPs is larger and larger. This trend often produces the occurrence of unpredicted phenomena related to the change of scale causing reliability concerns for plant constructors and managers, as well as the need to realize retrofitting solutions. BRPs are usually realized in such a way that the pump body is suspended below a basement plate through a hydraulic connection referred to as column pipe. Above the basement plate, the pump is connected to the discharge pipe line by an elbow. The pump can be idealized as a vertical cantilever beam subject to horizontal vibrations; the clamping of its upper part is provided by the stiffness of basement plate and of the discharge elbow. The connection between the column pipe and the basement plate, realized by a bolted joint, is a high-stressed area and has a large number of various crack raisers due to its complex geometry and the presence of flanges, bolts, brackets. The structural components are heavily stressed by a time-constant stress component, mainly due to the pump head, and by a fluctuating stress component related to vibrations and responsible for fatigue damage accumulation. Vibrations are generated by the deterministic response to the shaft unbalance, by the random response to the turbulence incoming in the suction line and by the response to intermittent or non-synchronous excitation components related to fluid instabilities as well as cavitations. BRPs have usually natural frequencies lower than their shaft frequency. This implies that the deterministic response to the pump unbalance has usually irrelevant effects; however its interactions with other form of excitation may produce significant consequences.

2 Description of the Case-Study

The present paper describes a case study related to a set of large BRPs that experienced premature fracture failures localized in the neighborhood of the connection between the column pipe and the basement plate (Fig. 1). The problem was investigated by carrying out several numerical analyses through Finite Element (FE) models and various in situ acceleration and strain measurements of the structural response as well as of hydraulic parameters such as the pressure in several points of the pump and the concentration of vapor bubbles entrapped in the suction line. All the measurements have been repeated in all the relevant working regimes of the plant, before, during and after retrofitting works.

Issues about the reliability of the pumps were raised as some bolts of the flange were found broken during an early inspection after a few weeks of working. Together with such failures, some cracks at the toes of the brackets connecting the upper part of column pipe to the basement plate (Fig. 1) were found. It was unclear at that time if the failure of bolts was the cause of cracks at bracket toes or vice versa. Furthermore, causes of the loads generating such failures were unknown.

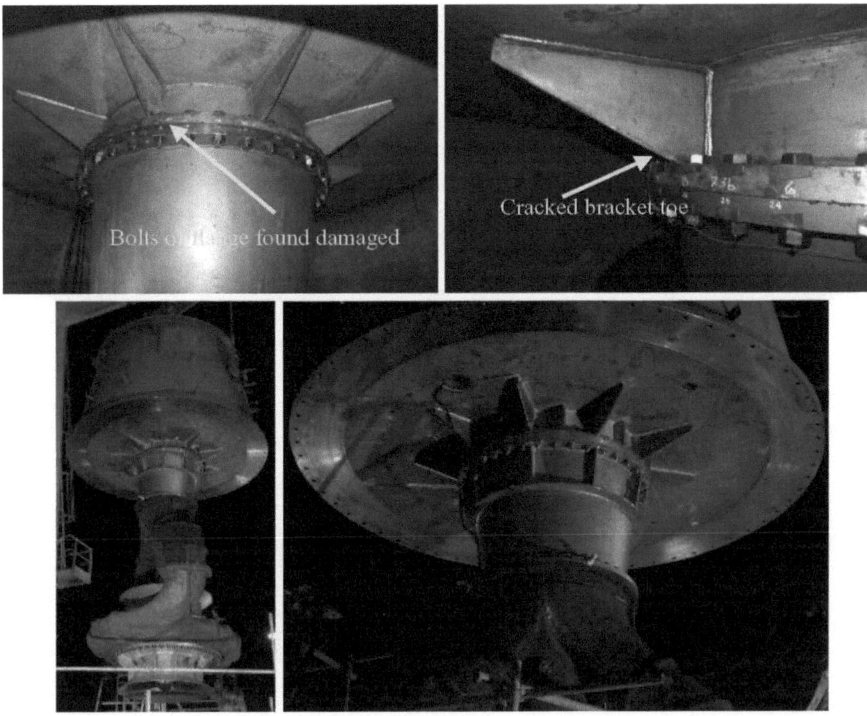

Fig. 1 BRP flange as built (*top*) and BRP during re-assembling after retrofitting works (*bottom*)

3 Investigation Activity

As a first step of the investigation, vibration measurements were carried out in order to identify the natural first frequencies of the pump frame, verify the gap between them and the shaft rotational speed and evaluate the amplitude of vibration in some specific points of the pump. For that purpose, one pump was instrumented by four accelerometers installed on the casing and on the column pipe, 14 strain gauges to monitor the deformation of the discharge elbow and of the basement plate and two position probes installed to measure the displacement of the discharge line with respect to its flexible support. The signals deriving from all the sensors were simultaneously sampled and recorded for several days, spanning all the relevant working conditions of the plants (in terms of flow-rate and suction pressure). The data-interpretation included the analysis of some single-time statistics and of the spectral properties of the acquired signals by applying standard and advanced statistical estimators [?]. A particular attention was focused on the study of the first two vibration modes corresponding to the deflection of the pump frame in two planes parallel and orthogonal to the discharge axis.

Figure 2 shows a typical power spectral density function (PSD) estimated from the acceleration measurements; the frequency is normalized by the shaft rotational

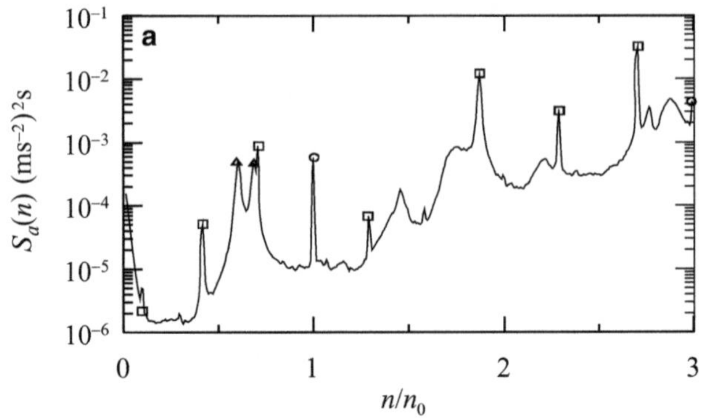

Fig. 2 Typical PSD of the acceleration measurements: peaks at synchronous (O) and non-synchronous (□) frequencies; resonances (Δ)

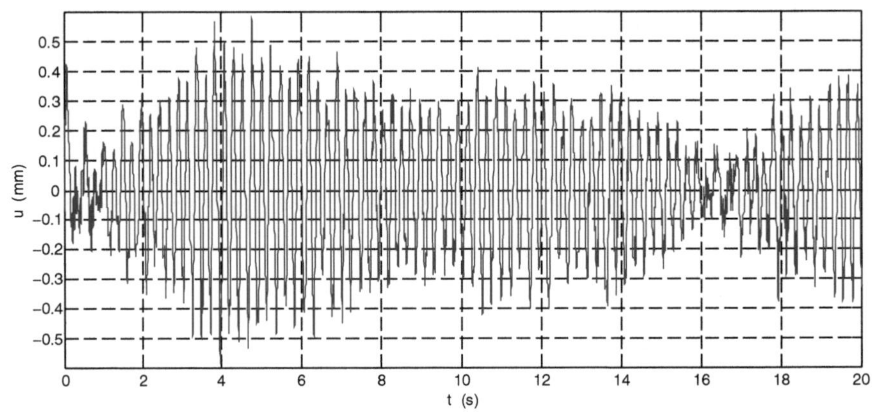

Fig. 3 Example of horizontal displacement time-series estimated for the lower flange of the column pipe from acceleration measurements

speed n_0. Several peaks can be clearly distinguished either at synchronous (symbol o) and non-synchronous frequencies (symbol □). Two non-synchronous peaks have been interpreted as the trace of the mechanical resonances ($n/n_0 = 0.60$ and 0.69, symbol Δ), while the other ones have been related to fluid-induced excitations since they were clearly found as dependent on the flow rate. In particular, the two spectral peaks on the left and on the right hand side of the resonances were observed to shift in such a way to become resonant in some specific working condition, producing large amplitude vibrations.

A characteristic amplitude of displacement was evaluated both for resonant and non-resonant regimes averaging the envelope of the displacement obtained by integrating the measured acceleration between 2 and 200 Hz (Fig. 3).

Owing the results of mentioned experimental campaigns, part of the structure was analysed by a FE numerical model calibrated in such a way to satisfactorily match the amplitude of displacements and strains evaluated for all the measuring points. The main aims of the structural analyses were then the following:

1. To identify the most stressed locations in the area where damages had been found and to evaluate the stress field in the neighborhood such locations
2. To estimate the axial tension stress acting on bolts joining the flanges and, bending stress on bolts, if any
3. To obtain the acting stress range at damaged locations for subsequent fatigue strength assessments.

4 FE Modeling of the as Built Structure

The Ansys 8.0 software has been used for modeling the original structure [?]. The FE model is built up by approximately 43,000 nodes and 39,000 elements using beam, shell and solid elements, according to the aims of the study: spar elements (LINK8 in the Ansys manual) have been used for the simulation of the tension force of the bolts, shell elements (SHELL63) have been used for the whole model except for the flanges, built up by brick elements (SOLID45). Figure 4 shows the model:

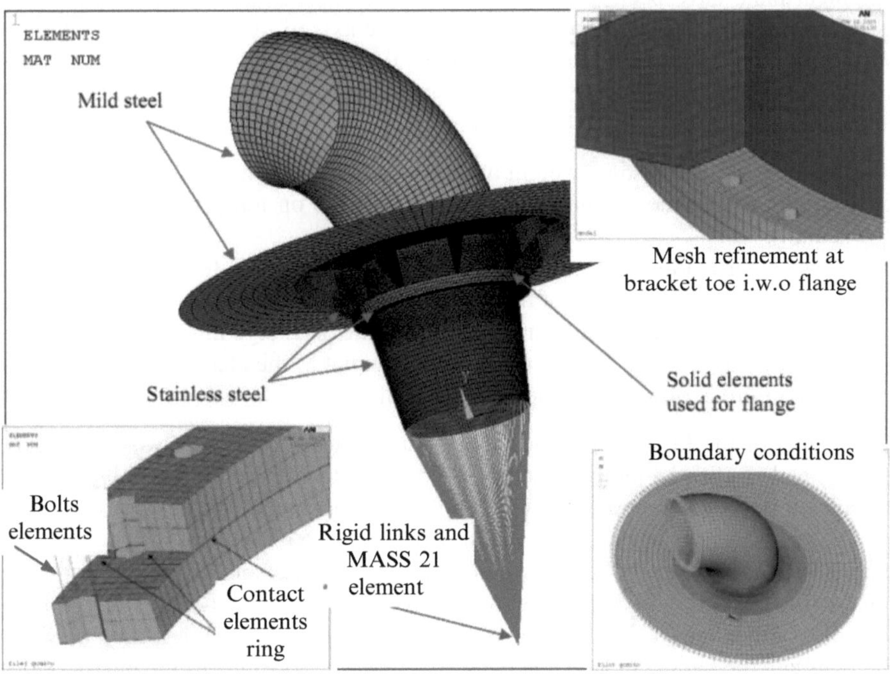

Fig. 4 As built frame FE model: boundary conditions, bolts and flange contact simulation, and mesh refinement on one bracket toe

the part of the pump frame below the column pipe was conventionally modeled by very stiff spar elements linked to a mass element (MASS21), simulating the weight of the non-modeled parts. The location of the mass has been selected as to obtain an easy calibration with the experimental estimate of displacements by accelerometers.

The mesh has been refined approaching the areas of interest (i.e. brackets toes at the basement plate and at the flange as shown in Fig. 4) obtaining an elements size of 4 mm along bracket edges, as required by current IIW standards for stress analyses of hot spots type B using relatively coarse meshes [1, 2]. The flange has been modeled by solid elements, joined to the shell ones of the pipe wall by the corresponding nodes. Two layers of element are used in the thickness of flanges and bolts' holes are schematized as hexagonal holes, accounting for the shape of bolts' heads and nuts. Non-linear link elements (LINK10) were used to connect the upper and lower flanges of the bolted joint in such a way to allow possible openings. Such elements are in practice very short beam elements linking the nodes of the opposite sides of the flange in such a way to offer a compression-only elastic reaction in vertical direction. The horizontal degrees of freedom of the nodes on opposite sides of the flange are coupled. The pre-tension in bolts has been accounted for by imposing a suitable initial strain whose value was selected according to manufacturer specifications. Bolts were modelled by six conventional beam elements.

The boundary conditions and the representative loading cases were calibrated by the results of the in-field measurements. All the DOFs of nodes on the edge of the basement plate have been fixed while the DOFs of nodes at the extremity of the discharge elbow have been constrained along the axis of the pipe by a spring with stiffness $k = 10$ kN/mm and the remaining DOF fully fixed.

The following loading cases have been run, accounting for the non linearity due to the contact elements of the bolted flange:

1. Vertical loads due to own weight and pressure effects (internal pressure on wall of column pipe due to pumped fluid and pressure on top-plate due to internal vacuum).
2. Loads as per loading case 1 and an incremental horizontal force applied on the mass element at the bottom of the model; the force was applied along two directions parallel and perpendicular to the axis of the discharge line and calibrated in such a way to produce a static displacement equal to the characteristic displacement amplitude observed in the experimental results.

The results of the FE demonstrated that the opening of the flange was significant (Fig. 5) and confirmed the presence of a stress fluctuation acting on bolts. Also, a significant stress concentration at the flange-side bracket toe was observed.

The stress acting on bolts was obtained from the axial tension of the six beam elements connected to the nodes around the holes of the flange. It is interesting to analyse the bolt stress in connection with the ones of the contact elements in way of the holes of the flange (e.g. as in Fig. 6), in order to highlight the interaction between the bolts and the flange. Indeed, the flange deformation (opening of the internal side) induces the bending of the bolts. After several runs of the FE analyses, the following behavior is noted:

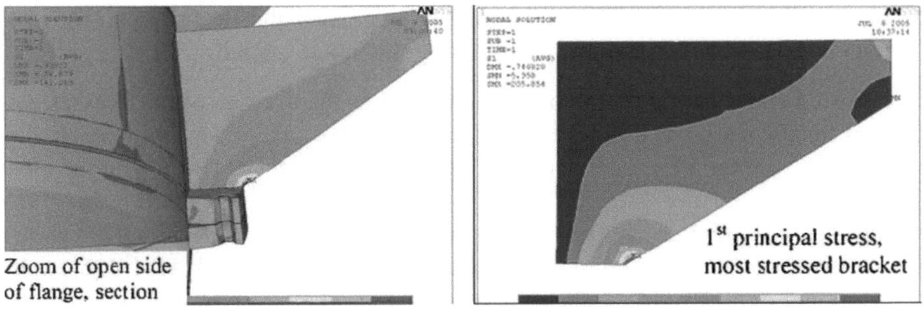

Fig. 5 FEM results at critical hot spots of the as built structure

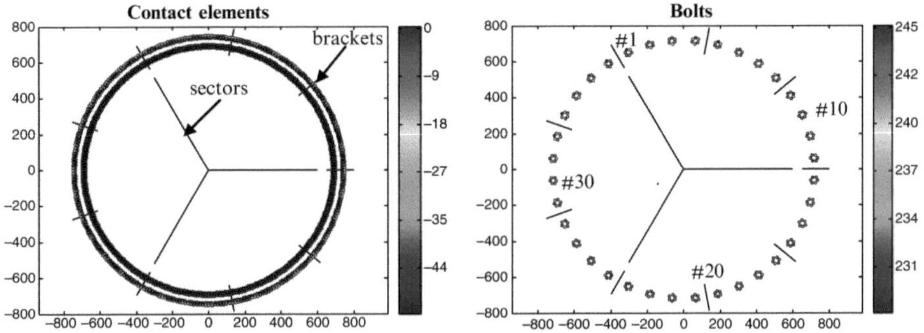

Fig. 6 Example of plots of stresses in contact elements and bolts elements

- The stress–displacement relationship for bolts is non linear with a highly increasing rate towards high stress values.
- Some of the contact area between the two flanges is open even with vertical loads only.
- Uncertainties in displacements estimation can significantly affect the evaluation of the stress range value; the definition of loading cases should accounts for the extreme conditions measured by accelerometers as well as for the time history in estimates of fatigue damage accumulation.

5 FE Modeling of Possible Retrofitting Solutions

Since the opening of the flange was believed as a major cause of bolts failure, three possible retrofitting solutions were tested by the FE model described above:

1. Substitution of the original AISI 316L bolts with duplex bolts in order to increase their preload, Nb, and reduce the opening of the flange
2. Substitution of the original bolts with duplex reduced shank bolts having a profile specifically designed to increase their fatigue strength

3. Introduction of additional webs aimed at increasing the flectional stiffness of both upper and lower flanges, thus limiting the flange opening

The tension stress acting on the bolts is obtained as the average of the stresses calculated for the six beam elements simulating each bolt. However, bolts result also in bending from the FE model and therefore the maximum stress value on the most stressed side of the most stressed bolt is estimated by the most stressed beam of the six of each bolt, namely the one on the internal side of the pipe. Figure 7 shows a comparison of the maximum stress calculated for the original AISI 316L bolts (Fig. 7a) and duplex reduced shank bolts (Fig. 7b). The stress is shown as a function of the horizontal displacement of the lower flange of the column pipe. It can be observed that the average stress (with zero displacement) for the duplex reduced shank bolts is higher that for the AISI 316L ones, due to the higher preload and the reduced cross-section; on the contrary, the related acting stress fluctuation results lower. Table 1 summarizes the average stress and the stress range calculated for ±0.5 mm displacement amplitude for the three types of considered bolts.

An estimate of the fatigue life of the bolts was supported by experimental testing of a few bolts based on the stress range evaluated by FE analyses. Figure 8 shows

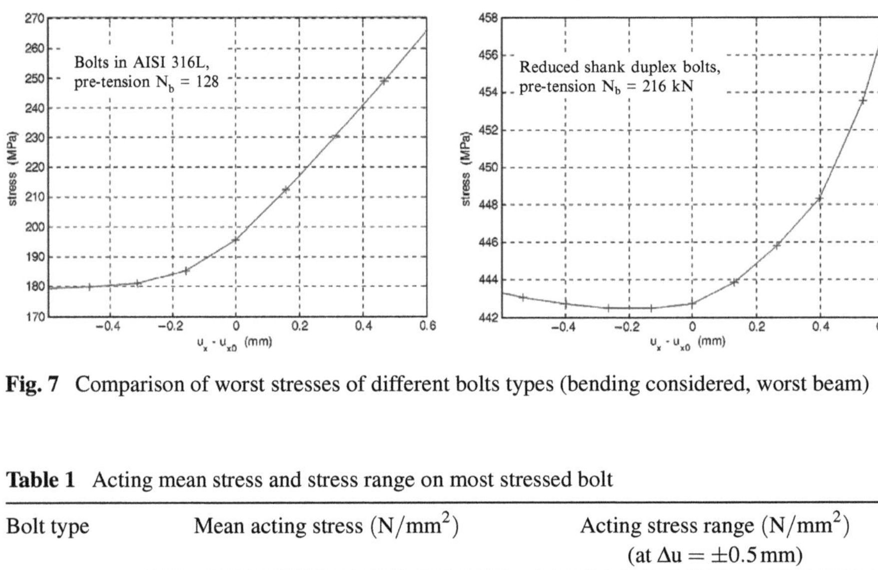

Fig. 7 Comparison of worst stresses of different bolts types (bending considered, worst beam)

Table 1 Acting mean stress and stress range on most stressed bolt

Bolt type	Mean acting stress (N/mm^2)		Acting stress range (N/mm^2) (at $\Delta u = \pm 0.5$ mm)	
	Average value in the cross section	*Maximum value at bolt external side (bending)*	*Average value in the cross section*	*Maximum value at bolt external side (bending)*
AISI 316L	219	259	69	119
Duplex	298	314	29	55
Duplex, red. shank	462	481	36	66

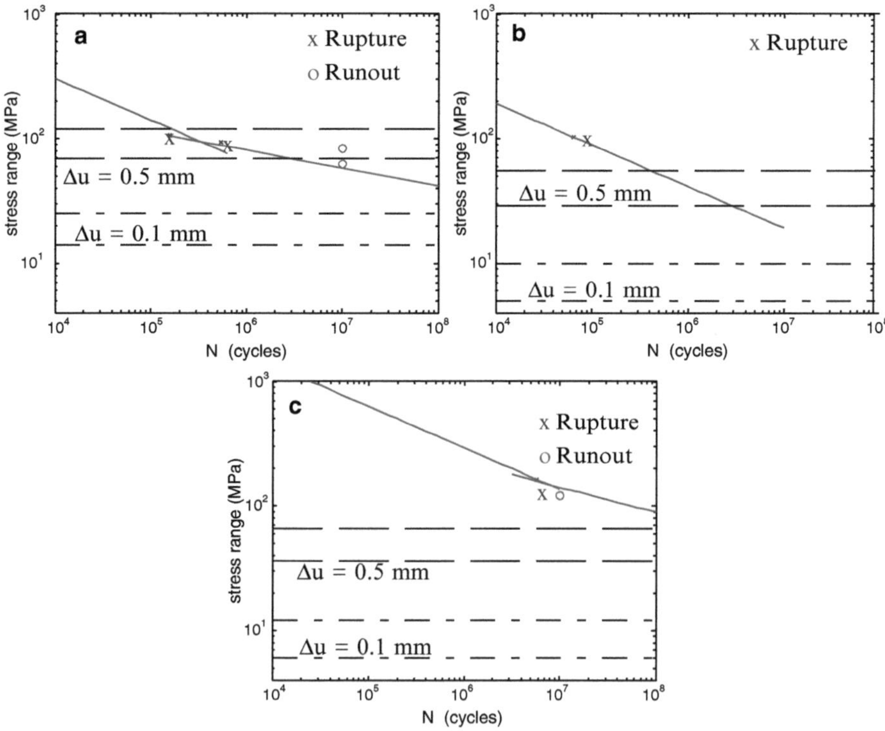

Fig. 8 Results of fatigue tests for AISI 316L bolts **a**, duplex bolts **b** and reduced shank bolts **c**; acting stress range (mean and maximum in the bolt cross-section) corresponding to $\Delta u = 0.5$ mm (dashed lines) and $\Delta u = 0.1$ mm (dash-dot line) reported for comparison purposes

the results of such experiment carried out on AISI 316L bolts (Fig. 8a), duplex bolts (Fig. 8b) and reduced shank duplex bolts (Fig. 8c). The experimental data have been interpolated by lines with slope 3 and 5 and have been compared with the maximum (most stressed element of six beams simulating a bolt) and the mean (average of beam elements) acting stress range evaluated by the FE analysis for the amplitude of displacement $\Delta u = 0.5$ mm and $\Delta u = 0.1$ mm. Due to the very limited time available, the experimentation was limited to few endurance tests; however, it is observed that the reduced shank duplex bold provide the best fatigue strength with respect to their stress range evaluated by the FE analyses.

Once the most efficient type of bolt was selected according to the above considerations, some structural modifications aimed at increasing the flectional stiffness of the flanges (in particular the lower one) were implemented and tested through the FE model described above. Figure 9 shows two of the many analyzed designs. The first solution (A) is obtained installing on the existing column pipe nine brackets with a 30° chamfered end. Variations of this solution where obtained by profiling the brackets end in order to minimize the stress concentration of the column pipe (B and C). A further solution (D) is obtained connecting the lower extremities of

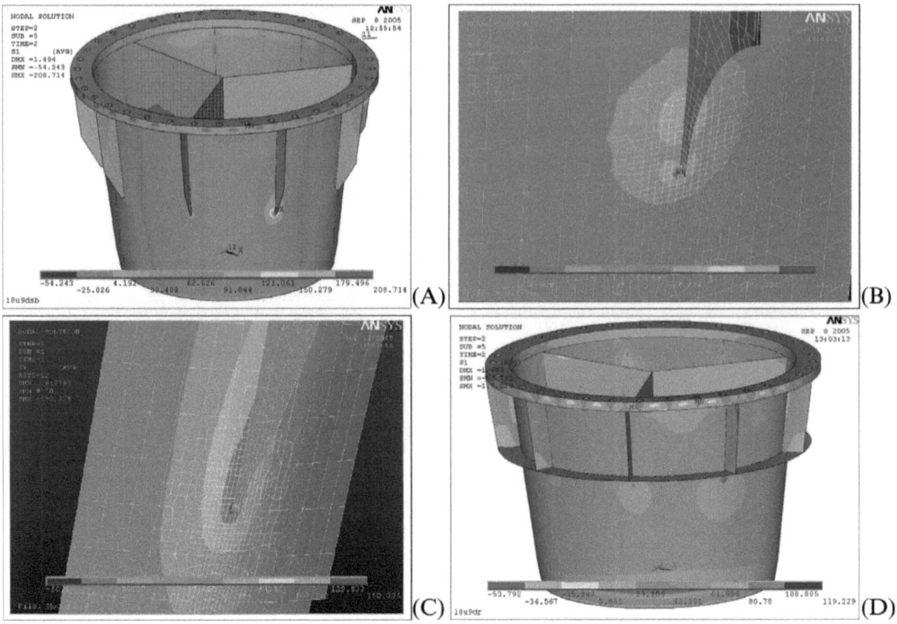

Fig. 9 Two modifications aimed at increasing the flectional stiffness of the flange

Table 2 Fatigue life estimates of some different solutions for lower brackets toes

Lower flange bracket, lower end toe detail	FAT class	Estimated life[a]
Solution A: non rounded bracket toe (30° chamfer, 20 mm step)	90	5.368E + 05
Solution B: soft bracket toe (R = 150 mm)	90	2.875E + 06
Solution C: sloped and soft bracket toe ($\phi = 15°$, R = 100 mm)	90	1.561E + 06
Solution D with annular ring (non load carrying fillet weld)	100	4.926E + 06

[a]Survival probability of SN curves: 97.7%, misalignment not considered, welding seam not modeled by FE, hot spot stress approach type A, using fine mesh according to ref. 1. Extrapolation of hot spot stresses using stress at nodes i.w.o. 0.4t and 1.0t from hot spot.

the brackets by a ring introduced to provide a constraint such to equilibrate the deflection transmitted from the flange. Also, the brackets of the upper flange were duplicated adding nine small rectangular brackets. Indeed, the lower toes of the additional brackets installed on the lower column pipe introduce additional crack raisers in the structure that have been verified versus a possible fatigue failure as shown in Table 2, according to [1] and [2].

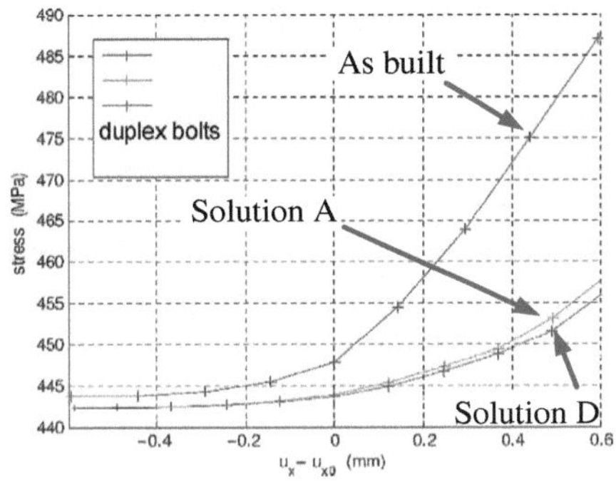

Fig. 10 Mean stress in bolts vs displacement of as built and modified flange brackets

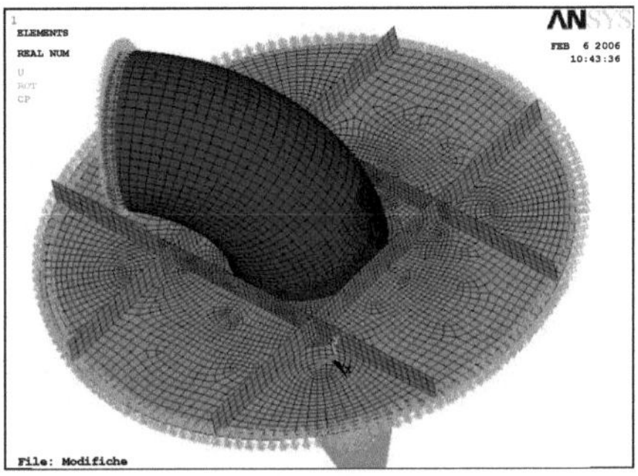

Fig. 11 Additional stiffeners on the top of the basement plate

It should be noted that the fatigue class of the annular ring detail is higher; moreover the FE model estimates a lower acting stress range. The bolts behavior after the structural modification was assessed as well, resulting in an additional improvement (Fig. 10).

A further retrofit work was considered in order to avoid the occurrence of the resonant response highlighted by measurements: the global stiffness was increased by welding four stiffeners on the top of the basement plate (Fig. 11). The stiffness increment of the natural frequencies due to these modifications was estimated by some static FE analyses calibrated with experimental data.

6 Refinements of Retrofitting Works

After the adoption of the new bolts type and the modification of the flange according to solution D described Table 2 and Fig. 9, a few cracks in way of the toes of the upper brackets were found after approximately 1 year exercise. It was questioned whether such cracks were due to an excessive acting stress or rather to local fabrication defects also produced during the numerous reparations made during the first month of work. In order to eliminate such problems, different geometries of the upper brackets were thereafter implemented in the FE model and tested imposing a reference displacement.

It is well known that the fatigue strength of a structure can be significantly enhanced by reducing the stress concentration at notches (at bracket toes in the present case), as well as by improving the quality of the details defining a suitable procedure for the in-situ implementation of the modifications. Obviously, the two mentioned aspects are strictly related and some iteration between the design of the profile and the definition of the modification procedure is generally necessary to achieve the convergence to a satisfactory solution. In practice, the final design of the bracket profile has to be as simple as possible to be realized by in-situ operations in order to limit fabrication defects.

The final aim of the calculations is to assess the stress concentrations at the toe of the brackets and to compare possible different structural modifications. Four possible solutions have been considered (Fig. 12) and have been tested by FE analyses. Even if a more refined mesh would be necessary in order to correctly estimate the stress values at the notch extrapolating the stress values from adjacent nodes according to [1] and [2], the trend of the stresses approaching the singularity point is significant for a quick comparison of different solutions, matching the time constraints of plant overhaul. The results presented in Fig. 12 can be summarized as:

- Solution 2 provides a significant reduction of the stress field (both mean value and fluctuation) in the neighbourhood of the lower toe, while does not produce any advantage (or is negative) concerning the stress field near the upper toe.

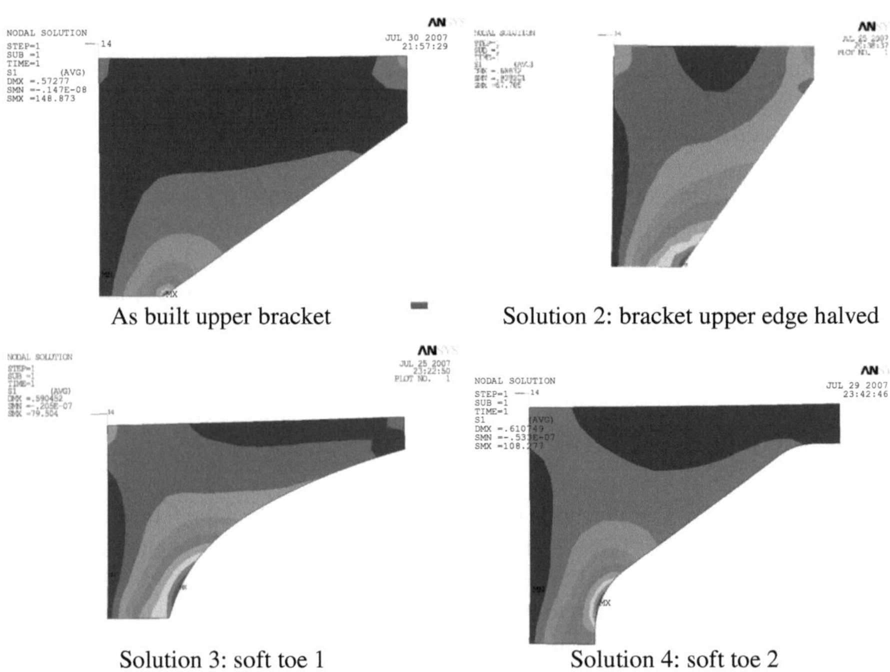

Fig. 12 Different proposed upper bracket profiles, first principal stresses

Table 3 Reduction of stiffness and natural frequencies with respect to present configuration

Bracket profile	Reduction of stiffness (%)	Reduction of natural frequencies (%)
Solution 2	11	5
Solution 3	8	4
Solution 4	9	4

- Solution 3 has a performance similar to Solution 2 at the lower toe, but the stress transferred to the plate at the upper toe is significantly reduced with respect to the "as built" configuration.
- Solution 4 is similar to Solution 3, but produces a lower stress concentration at the lower toe, moving the stress peak away from the weld notch; the maximum stress occurs in the parent material quite far from the heat affected zone. This is certainly beneficial provided that the bracket edge is correctly cut and cold worked. Moreover, in situ implementation of this solution seems easier.

According to the FE model results, Solution 4 appears as the best candidate for an effective reduction of the stress concentration in the neighbourhood of the hot spots where cracks were found and for the reduction of the force transferred to the basement plate. Before the actual implementation of Solution 4, the FE model was used to predict the stiffness reduction due to the modification and the consequent reduction of the first natural frequencies. Table 3 summarizes the results, appearing appropriate to avoid resonance problems.

7 Conclusions

For the time being, the plant has been working since the last major overhauling and no notice of further failures. The upper brackets were modified as shown by Fig. 13, i.e. basically following the indications provided by Solution 4, but cutting the edge of the bracket along smoothed curved path in order to achieve a more locally flexible connection between the column pipe and the basement plate of the structure and, at the same time, moving the hot spot away from the weld bracket toe.

Several lessons have been learned from this study; the main ones can be summarized as:

- Experimental and numerical studies carried out in parallel were crucial to obtain right and proper solutions within strict time and economical limits imposed by the scheduled overhauls of the plant and accounting for operational constraints due to lack of workshop facilities at the plant location.
- Countermeasures regarding both fracture strength of the structure and vibration levels were adopted, checking several aspects of the structural behavior by means of a rather complex well calibrated FE model, built up on purpose.

Fig. 13 Final solution implemented at connection between column pipe and basement plate

- Cooperation was essential managing a quite large team with several different skills working towards a unique goal.

Acknlowedgments Some of the described retrofitting actions, in particular the design of the reduced shank bolts, were proposed by the pump manufacturer, while the fatigue tests on bolts have been carried out at Istituto Scientifico Breda (Milan, Italy). Their valuable work and the professionalism of the plant constructors are gratefully acknowledged. The support by Ms C. Calderini e Mr G. Notaro in FE calculations is also acknowledged.

References

1. Hobbacher A. et al. (1996), Recommendations for Fatigue Design of Welded Joints and Components, International Institute of Welding (IIW), doc. XIII-1539-96/XV-845-96, Paris, France, July 1996.
2. Niemi E., Fricke W. and Maddox S.J. (2004), Structural hot-spot stress approach to fatigue analysis of welded components - Designer's guide, International Institute of Welding (IIW), doc. XIII-1819-00/XV- 1090-01, Paris, France, May 2004.

The Accurate Prediction of the Thermal Response of Welded Structures Based on the Finite Element Method: Myth or Reality?

Dimitrios Karalis, Vasilios Papazoglou, and Dimitrios Pantelis

Abstract The thermal and metallurgical response of thin welded plates made of Ck45 steel during shielded metal arc welding is investigated through numerical results using the finite element method (FEM) and by in situ experimental measurements.

The aim of this paper is to show that the typical and most common investigation strategy followed for the solution of the heat transfer problem of welded panels based on FEM investigations (prior to the commencement of real production welding) does provide engineers with an efficient temperature – prediction tool if autonomously used. The experimental welding verification is, however, deemed essential for further analysis.

The paper places emphasis on the difficulties arising when attempting to deal with the problem from the point of view of pure prediction. Detailed numerical thermal modeling, including most of the thermal and metallurgical parameters that are more or less involved in real welding, does not necessarily provide accurate results as far as the transient temperature distribution is concerned due to the unknowns involved. Typical unknown parameters such as the material model, the net heat input and the amount of thermal energy released to the surroundings by convection and/or radiation turn these numerical prediction tools into case sensitive and variable systems that should be used very cautiously for direct accurate welding thermal response predictions.

In the present paper the deviation between the predicted thermal results obtained using detailed Finite Element Modeling and those obtained by means of experimental measurements is clearly demonstrated. This fact implies the weakness of the numerical FEM strategies to accurately predict the actual thermal response during welding with respect to the modeling parameters that are unknown. An adaptation procedure on the basis of the weld metal shape is thus proposed. The adapted

D. Karalis, V. Papazoglou (✉), and D. Pantelis
Shipbuilding Technology Laboratory, School of Naval Architecture and Marine Engineering, National Technical University of Athens, 9 Heroon Polytechniou Avenue, Zografou, Athens GR-157 73, Greece
e-mail: papazog@deslab.ntua.gr, pantelis@central.ntua.gr

S. Pantelakis, C. Rodopoulos (eds.), *Engineering Against Fracture: Proceedings of the 1st Conference*,

model is then used for a thorough investigation of the thermal response analysis. These adapted thermal numerical results, later used for post-mechanical analysis, are expected to more accurately address the problem of the transient and residual stresses calculations, stresses which greatly affect the fracture behavior of a welded structure.

Keywords SMAW · FEA thermal modeling · Experimental measurement · Ck45 steel · Metallurgical study

1 Introduction

The aim of welding simulations is to analyze the evolution of the manufacturing process. Simulations are more significant if they aim at accurately predicting the magnitudes of interest, such as temperatures or deformations, prior to the commencement of the manufacturing process itself. This would provide an efficient tool for the engineers during the design process of welded steel structures.

In general, interest has been focused on the transient and steady state thermal and mechanical response of the welded system. If the numerical simulations are carried out prior to the experimental or manufacturing process, the simulation models are treated as prediction models. They aim at calculating the thermo-mechanical response without any real input from the exact fabrication process, with the exception of the welding parameters, welding setup, materials, initial clamps or tacks and probably some very simple trial tests. In this case, accurate prediction in terms of welding is a very complex and triggering task but its value is of great importance in case of ship hull structural members, like stiffened panels, corrugated bulkheads or stiffened curved side shells.

On the other hand, if the numerical simulations are carried out after the welding has been performed, the opportunity is provided for comparison of results and model adaptation. Comparisons between numerical and experimental results can lead to a process of numerical adaptation depending on the magnitude of results deviation. This process aims at improving the response of the system that is been modeled by the numerical simulations to follow. The drawback is that welding has already been completed and its steady state response, like residual deformations or final microstructure, if being of poor quality, has not been predicted on time by the simulation. The corrections and modifications undertaken in numerical analysis are included in the so called model-calibration or model-adaptation procedure. In this case, simulations lose their pure prediction character and are turned into analysis simulations, focusing on the thorough examination and optimization of the real process.

The thermal and mechanical response of steel plates during welding has been investigated by several authors during the last three decades. Most of the investigations are focused on either or both the thermal and the mechanical part of the structural response through a combination of experimental and numerical simulations. The numerical part of the investigation still remains under high interest due to its high complexity and due to the uncertainty in predicting the actual structural response prior the welding itself. An extensive review has been carried out by Radaj [1] and more recently by Lindgren [2, 3].

As far as the thermal problem during welding is concerned, it consists of three different stages: (a) the solution of the transient heat transfer problem based on numerical simulation, (b) the experimental measurement of transient temperatures or micro-structural characterization of the welded parts, and, if necessary, (c) the adaptation of the initial simulation model in order to attain convergence between numerical and experimental results. Adaptation aims at compensating for the unknown parameters that are explicitly or implicitly involved during the numerical simulation. For example, the arc efficiency, the heat losses to the surrounding environment (by convection, radiation or metal to metal contacts), the analysis simplifications (e.g. the two-dimensional heat flow consideration), the thermal material properties at elevated temperatures, and the experimental or in situ welding set-up are some typical examples. In some cases, if the transient solid or liquid state transformations of the material are also taken into account in the numerical investigation, the analysis is turned into a micro-structural dependent analysis which may add to the uncertainty of the modeling due to the increased complexity.

Further investigation can be carried out as soon as the transient temperature distributions of the welded structure between numerical and experimental results are in good agreement. The numerical models developed can be further used for parametric analysis focusing on the optimization of the design process of the marine structure as far as the micro-structural evolution or maximum temperatures is concerned, or they can alternatively be used for the calculation of the transient mechanical response of the welded system. In this case, the probability to correctly calculate the steady state mechanical response of the steel structure is higher due to exact thermal excitation.

The aim of this paper is to propose an investigation strategy to be followed for the solution of the heat transfer problem of thin welded panels based on numerical investigation and to emphasize the difficulties arising when trying to deal with the problem from the point of view of pure prediction. The numerical solutions obtained are then compared to the experimental results and adaptation modifications are carried out in order to bring numerical and experimental results close to each other. The adapted model is then used for a thorough investigation.

2 Experimental Set-Up

The aim of the experimental procedure, except for the welding itself, is to provide as accurate as possible measurements of the transient thermal response of the welded system. These would be later used in order to compare them with temperature results obtained from the numerical analysis.

In general, measurements usually employed for this task are temperature and micro-hardness measurements, combined with results from standard metallography technique using optical microscopy. As far as the temperature measurements is concerned, the maximum austenitization temperatures ($T_{a\text{-}max}$) reached as close as possible to the weld metal and the cooling rate between 800°C and 500°C ($\Delta t_{8/5}$) in the heat affected zone are usually of great interest. The former accounts for the austenite grain size and hardenability of steel and the latter for the solid state transformations during cooling. The exact size of the weld pool and the final microstructural state of the material at the steady-state condition (that is when the material has completely cooled down) is also to be taken into account.

The experimental setup consisted of thin plates made of Ck45 steel. The size of each of the steel plates was $700 \times 200 \times 4\,\text{mm}^3$ as shown in Fig. 1. The plates were free from initial curvatures and other imperfections, except that they were slightly oxidized. The shielded metal arc welding technique and basic electrodes made by ESAB, type OK55.00, E7018 class, according to AWS-A5.1 were employed for the welding. The latter is a very low carbon electrode suitable for Ck45 steel welding. The chemical composition of base and weld metal are listed in Table 1. The welding parameters were tuned at $V = 26\,\text{V}$, $I = 145\,\text{A}$ and $v = 0.003\,\text{m/s}$ [4]. Prior to the commencement of the experiments, the plates were tack welded at $x = 0$, 0.125, 0.250, 0.450, 0.575 and 0.7 m and the electrodes were dried at 350°C for 2 h.

Temperature measurements were obtained using ANSI-. Chromel-Alumel thermocouples of 0.002 m tip diameter, that were theoretically capable for measuring up to 1370°C. The exact location of the thermocouples, derived from local measurement carried out after the completion of the experiments, is shown in Fig. 1. The thermocouple positions listed in Table 1 were selected after extensive experimental

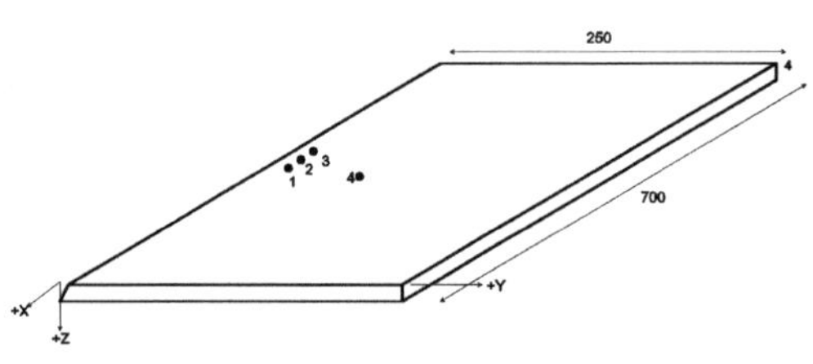

Thermocouple	Position (m)
No 1	x = +0.380, y = +0.004, z = +0.002
No 2	x = +0.360, y = +0.003, z = +0.002
No 3	x = +0.340, y = +0.005, z = +0.002
No 4	x = +0.360, y = +0.080, z = +0.002

Fig. 1 Plate dimensions and locations of thermocouples (No1 to No4)

Table 1 Chemical composition of Ck45 steel and weld metal

% w/w	(C)	(Si)	(Mn)	(S)	(P)
Ck45	0.42–0.5	0.15–0.35	0.5–0.8	0.035	0.035
Weld Metal	0.06	0.5	1.5	–	–

investigation based on the following criterion: the temperatures measured at the positions selected during all experiments should be higher than the transformation temperature of steel ($A_1 = 723°C$). The aim of the extensive experimental investigation was also to attain repeatability of the results due to the welding technique employed (SMAW).

3 Numerical Modelling

For the numerical simulation of the transient heat transfer response the finite element analysis method was used by developing numerical models using the commercial finite element software ABAQUS [5]. A two-dimensional analysis was set up referring to the cross section of the welded plates. Due to symmetry, only one half of the welded plate was modeled. Mesh density was tuned in order to attain convergence of thermal response, thus 1,575 linear four nodded plane quadrilaterals were used.

Heat input was modeled based on the double ellipsoid distribution of arc heat, as proposed by Goldak et al. [6–8]. This distribution is shown schematically in Fig. 2 and is described by Eqs. (1–5):

$$Q_f(x,y,z,t) = \frac{6\sqrt{3}r_f Q}{a_h b_h c_{hf} \pi\sqrt{\pi}} e^{-\frac{3(x-vt)^2}{c_{hf}^2}} e^{-\frac{3y^2}{a_h^2}} e^{-\frac{3z^2}{b_h^2}} \tag{1}$$

$$Q_r(x,y,z,t) = \frac{6\sqrt{3}r_b Q}{a_h b_h c_{hb} \pi\sqrt{\pi}} e^{-\frac{3(x-vt)^2}{c_{hb}^2}} e^{-\frac{3y^2}{a_h^2}} e^{-\frac{3z^2}{b_h^2}} \tag{2}$$

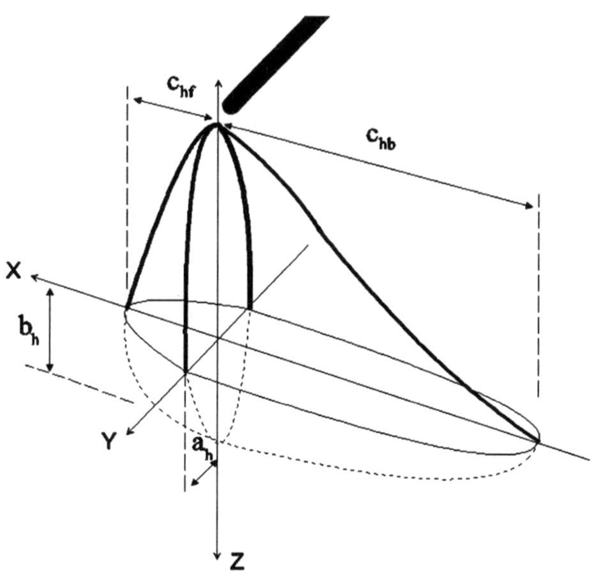

Fig. 2 The double ellipsoid heat distribution [6–8]

$$r_f = \frac{2c_{hf}}{c_{hf} + c_{hb}} \tag{3}$$

$$r_b = \frac{2c_{hb}}{c_{hf} + c_{hb}} \tag{4}$$

$$Q = \eta V I \tag{5}$$

in which v refers to the arc longitudinal velocity and η to the arc efficiency. For the modeling procedure, all elements that were involved within the double ellipsoid distribution were assigned an individual thermal load curve according to their co-ordinates (y, z) in the two-dimensional plane. The heat applied was considered to be distributed on the element surfaces [9], while the weld metal was activated at the melting temperature [2, 3]. For the estimation of double ellipsoid dimensions a simple "bead on plate" trial experiment was carried out with exactly the same welding parameters used for the welding experiments. This procedure allows the measurement of the double ellipsoid dimensions by simple means and without any additional cost prior to the commencement of the real weld; otherwise these have to be estimated according to the literature or based on semi-empirical methods. Similar procedures have been adopted by other investigators [10–12] for the estimation of the arc dimensions. The final values of the distribution dimension were tuned at $a_h = 0.006\,m$, $b_h = 0.004\,m$, $c_{hf} = 0.002\,m$ and $c_{hb} = 0.005\,m$. The arc efficiency was taken equal to $\eta = 0.75$, based on the values proposed in the literature [1]. All material and boundary non-linearities were included in the thermal part of the analysis. More specifically, temperature dependent thermal material properties [1–3,13] were assigned for the whole thermal analysis, taking into consideration that the final material microstructure was free of martensite that exhibits different thermal properties at room temperature compared to ferrite-pearlite which was the initial micro-structural state of the steel. This assumption was later checked through the cooling rates calculated by the software. The temperature and microstructural dependent thermo-physical properties of Ck45 steel employed for the numerical analysis are shown in Figs. 3–5. The value of heat conductivity above the liquidus temperature was set to be equal to $230\,W/m^\circ C$ in order to account for molten metal stir effects. Latent heat was also included in the analysis by specifying the solidus and liquidus temperature of the base and weld metal. Its value was set equal to $260\,kJ/kgr$. As far as the heat loss is concerned, radiation and heat convection to the surroundings was taken into account. The model used for the calculation of the heat loss was based on the equivalent convection model [14, 15] which is described by Eq. (6):

$$h = 24.1 \times 10^{-4} \times \varepsilon \times T^{1.61}\,(W/m^2 \cdot {}^\circ C) \tag{6}$$

in which ≪h≫ refers to the convection coefficient, ≪T≫ refers to the current surface temperature of the element and ≪ε≫ to the emissivity that was supposed to be equal to 0.8, a typical value for slightly oxidized steel plates. Finally, the initial temperature of the base metal was set to be 25°C referring to the room temperature.

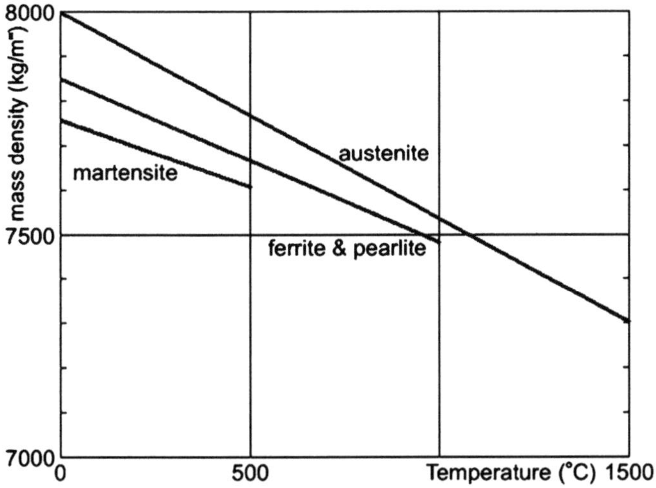

Fig. 3 Mass density dependence on temperature [1–3, 13]

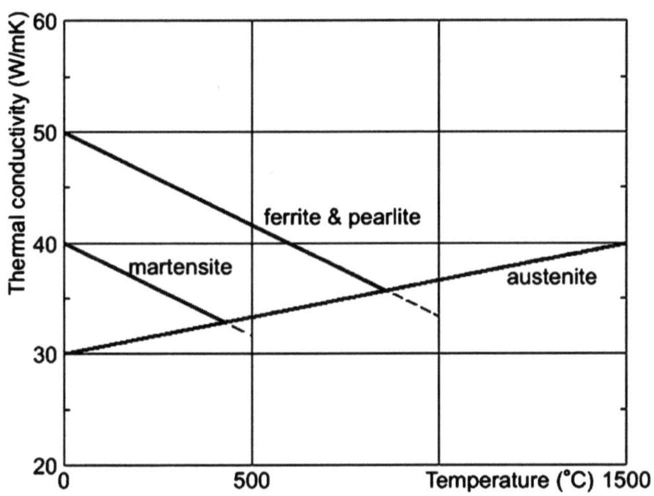

Fig. 4 Thermal conductivity dependence on temperature [1–3, 13]

4 Prediction Results and Discussion

A typical view of the cross section at x = 0.350 m of the welded plates is shown in Fig. 6 obtained my means of optical stereoscopy. In Fig. 7 the temperature distributions calculated using the finite element method are shown for the same cross section. In this figure, all the sectioned area included within the grey temperature isotherm refers to the melted zone (weld metal, dilution zone). The temperature

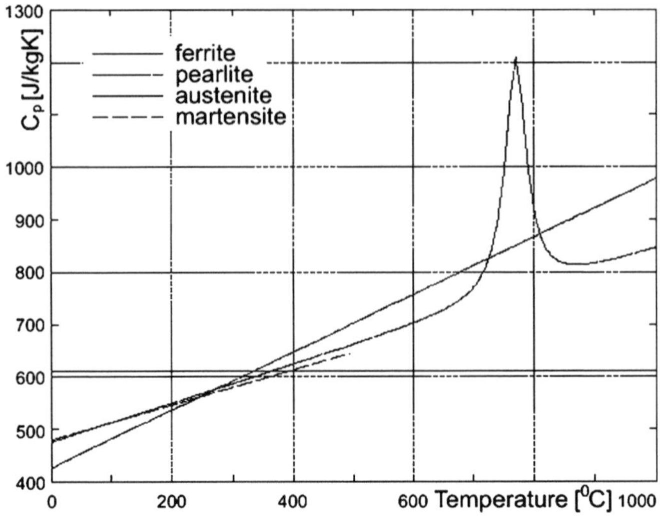

Fig. 5 Specific heat dependence on temperature [1–3, 13]

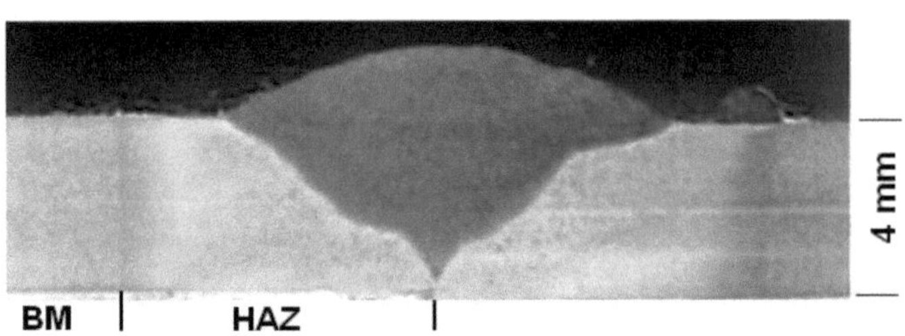

Fig. 6 Typical view of the weld cross section at x = 0.350 m (experimental results)

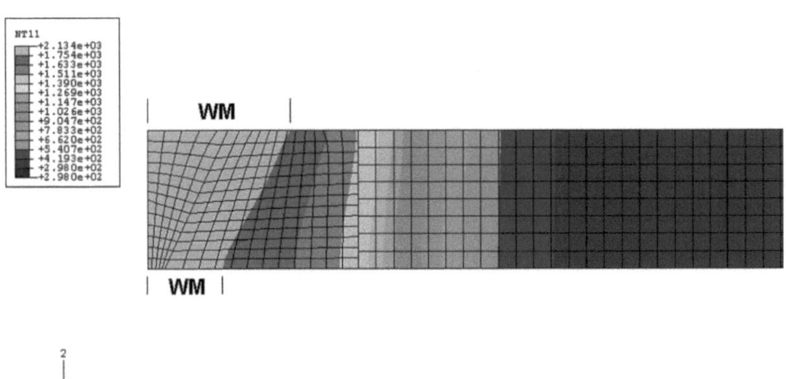

Fig. 7 Liquidus isotherm of the weld cross section, as predicted (units in K)

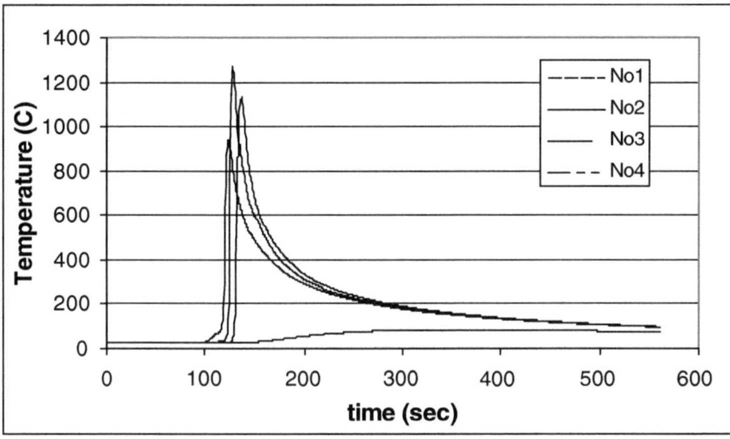

Fig. 8 Measured temperature distributions

results shown in Fig. 7 refer to the model developed before welding took place, aimed at predicting the temperature distributions and cooling rates. As far as the comparison between numerical and experimental results is concerned, a typical view of the measured temperature profiles at the locations shown in Fig. 1 is provided in Fig. 8. It is obvious that none of the thermocouples measured any temperature higher than 1,300°C. The overestimation of the liquidus isotherm is mainly focused at the bottom side of the plates ($z = 0.004\,m$ plane) in which the liquidus isotherm extends to a great distance form the axis of symmetry ($y = 0\,m$) compared to the actual weld metal size shown in Fig. 6. This observation is not valid for the top surface of the plates (plane $z = 0\,m$), on which the weld metal extension has been better predicted. On the other hand, the cooling rate $\Delta t_{8/5}$ calculated by the simulation and measured by experiment at the locations of thermocouples 1, 2, and 3 (see Fig. 1), was 21 and 22 s, respectively.

Prior to any further comparisons between measured and calculated temperatures, it has already been observed that the melted zone is greater in size in Fig. 7 than that of Fig. 6. This error may have been introduced due to several reasons such as modeling assumptions or modeling simplifications. As soon as the heat input was modeled as accurately as possible and time incrementation and meshing was considered to be of adequate density, any deviation between numerical and experimental observations, apart from the uncertain material properties close to the liquidus temperature, may be due to inaccurate estimation of the double ellipsoid dimensions, especially in the longitudinal direction, overestimation of the arc efficiency or inaccurate modeling of heat loss. The latter has also been discussed by Papazoglou et al. [16]. Notice that the two dimensional analysis formulation that was employed tends to over-predict temperature fields due to the fact that no heat flow is allowed in the longitudinal direction. Furthermore, the size of the hump that is generated on the upper surface of the weld metal during welding, as shown in Fig. 6, is not taken into account in the finite element model. The latter affects the accuracy of the calculated temperature distributions in both directions of the two-dimensional plane. Moreover, temperature distributions like these shown in Fig. 7 can not happen in real practice since, due to the gravity loads, the molten weld metal flows away, thus a large amount of thermal energy is removed. In this case, the welder would consciously adjust the arc speed in order to avoid melt or burn-through, a parameter that is considered to be constant during the simulation.

As a result, the unknown parameters involved in the analysis (e.g. heat input, arc efficiency, material properties, double ellipsoid dimensions, heat loss, size of the weld metal hump, variable welding speed) prevent the analyst to directly and systematically predict with high accuracy the temperature distributions during numerical welding simulation. The question that arises is whether the finite element results shown in Fig. 7 are sufficient for any further processing.

In the case of a general heat transfer investigation and analysis, the prediction of the temperature fields may be adequate regarding general thermal information. Alternatively, and aiming at compensating for the governing unknown parameters, further parametric analysis on the basis of these parameters may be carried out in order to predict the minimum and maximum size of the weld shape and heat affected zone. Thus it is expected that the results obtained during the pure prediction phase should be presented in a parametric way with respect to the governing unknown parameters. Typical results of this parametric analysis can look like these shown in Fig. 9 regarding e.g. the arc efficiency. Thus, if dealing with the problem on the basis of the aforementioned parametric analysis, the Finite Element Method is turned into an efficient prediction tool that can be autonomously used, but the prediction accuracy will obviously vary within a specific range, verified and confirmed by the end of the actual manufacturing itself.

As a result, it is derived that due to the unknowns involved in the thermal part of the numerical analysis, the systematic accurate prediction of the thermal response prior to the commencement of the actual welding itself and prior to any kind of numerical adaptations, tends to be more or less a myth since it is very difficult or even impossible to exactly estimate the actual values of all the unknowns involved. On the other hand the presentation of the thermal results on the basis of parametric analysis aiming at compensating for the unknowns is more than reasonable. However, the main governing parameters selected by the user depend on the analysis.

5 The Adaptation Procedure, Results and Discussion

In case of further processing aiming at a thorough micro-structural characterization of the heat affected zone and weld metal, the current model presented until now is considered insufficient. It is also doubtful whether the calculated temperature response is sufficient for further analysis of the mechanical response of the system, since the weld metal shape and size and the temperature distribution within the heat affected zone are considered as essential parameters for the mechanical response of thin steel plates. This topic has been also discussed by Camilleri et al. [17], Kuzminov [18], Vinokurov [19], and Radaj [1].

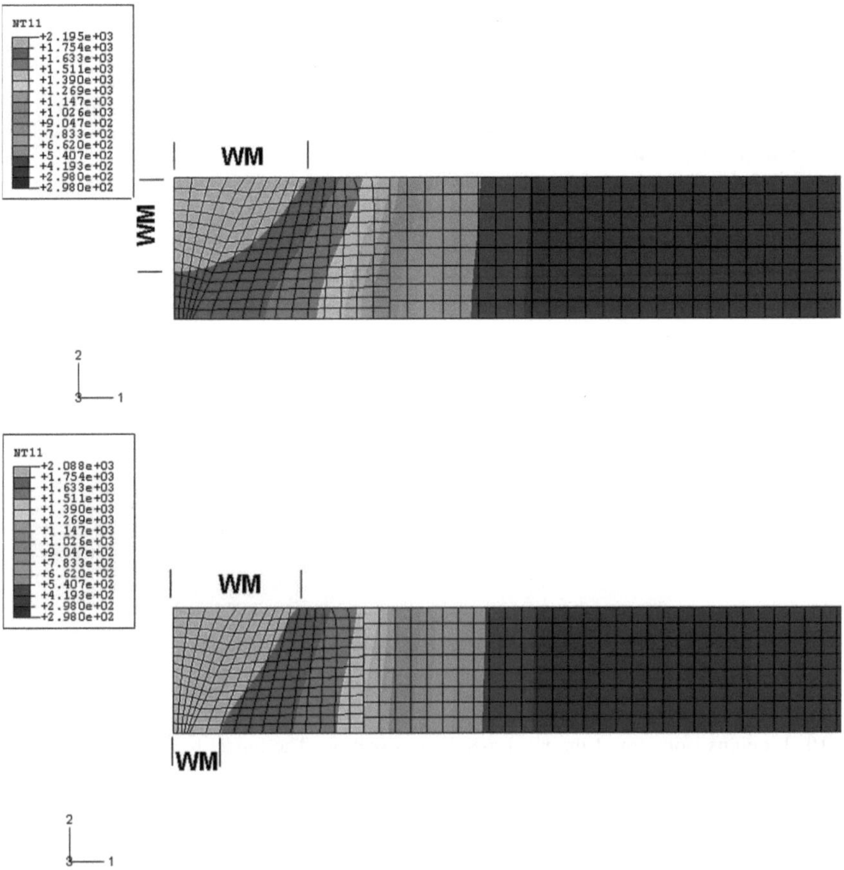

Fig. 9 Typical results regarding the weld metal shape and size on the basis of parametric analysis (units in K)

For the above reasons, an adaptation procedure was followed. During this adaptation, the surface convective loads applied on the bottom side of the plate were modified in order to account for the additional heat losses that were not taken into account [20], such as heat diffusion due to plate contact with the welding table and to compensate for the unknowns involved in the analysis. The value of the convection heat transfer coefficient was tuned until good agreement was attained between numerical and experimental results regarding the shape of the weld metal only. This procedure allows the adaptation of the numerical and experimental results but is strictly case-dependent and aims at the utilization of the specific thermal model for further investigation of similar welding set-ups only. It also allows the implicit consideration of all heat losses that are not taken into account during the initial

Table 2 Maximum temperatures and cooling rates $T_{a\text{-}max}$ (°C)/$\Delta t_{8/5}$ (s)

	Measured	FEA As predicted	FEA Adapted
Mid-plane position – y (m)			
0.003	1,280/22	1,500/21	1,338/18
0.004	1,140/22	1,402/21	1,264/18
0.005	941/22	1,292/21	1,156/18
0.080	83	107	87

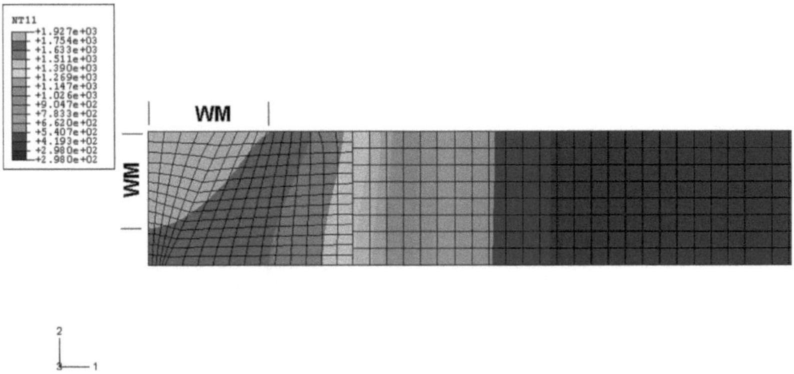

Fig. 10 Liquidus isotherm of the weld metal cross section. The model has been adapted to real measurements (units in K)

modeling procedure. Similar adaptation methods are discussed by the authors for non conventional welding techniques (solar welding) [21].

The temperature values of the initial and adapted finite element model, together with the measured temperatures are shown in Table 2. The extension of the weld metal at the same cross section ($x = 0.350$m) is shown in Fig. 10 for the adapted model. In this case, both weld metal size and shape, cooling rates and maximum temperatures are considered to be in better agreement with experimental results, but still not exact. Notice, that the adapted numerical model can not perfectly match the experimental results, which validates the assertion presented before, that the accurate prediction prior to the commencement of the actual welding itself and prior to any kind of numerical adaptations is unlikely to be realized on a scientific and clearly documented basis.

The heat affected zone shown in Fig. 11 extends up to $\Delta y = 0.0095$m from the axis of symmetry ($y = 0$m) and is considered to be in good agreement with the experimental results which extension was $\Delta y = 0.008$m. The cooling rate calculated at the far end of the heat affected zone ($y = 0.0095$m, $z = 0.002$m) was $\Delta t_{8/5} = 16.3$s which is a bit less compared to areas of the heat affected zone close to the weld metal, as shown in Table 2. This small difference of the cooling rate between the two areas of the heat affected zone leads to the assumption that a constant mean

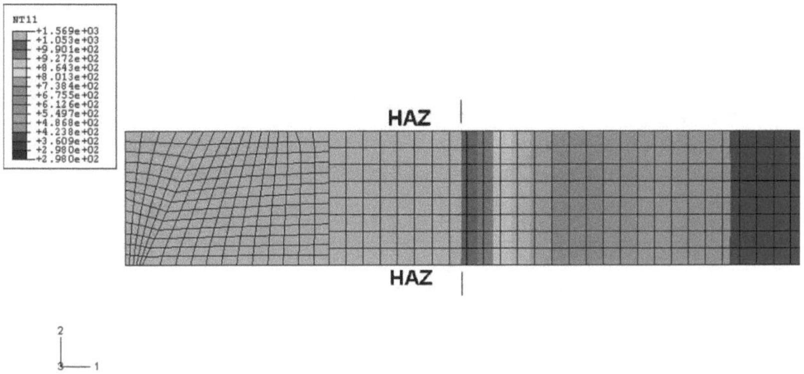

Fig. 11 Isotherm of 1,053 K(780°C) for the heat affected zone (model has been adapted to real measurements)

Fig. 12 Microstructure of the weld metal, heat affected zone and base metal

cooling rate value along the transverse axis would suffice for further micro-structural or mechanical modeling of the heat affected zone for the adopted welding.

The micro-structural state of the weld metal, heat affected zone and base metal are also shown in Fig. 12. The base metal consists of a mixture of ferrite and pearlite. The heat affected zone consists of a mixture of bainite, ferrite and pearlite. The weld metal consists of fine grained ferrite. Thorough investigation as far as the microstructure of low-alloy steel weld metals is concerned has been carried out by Bhadeshia et al. [22]. In his study, mixture of allotriomorphic ferrite, Widmanstätten ferrite side-plates, acicular ferrite and small quantities of microphases are mentioned as typical microstructure constituents in similar low carbon weld metals. Martensite was absent in the entire cross section of the weld, except for a few areas close to the boundary zone between weld metal and heat affected zone. The amount of martensite investigated in this area was treated as negligible by the authors.

The measured cooling rate between 800–500°C is plotted on the CCT diagram [13] as shown in Fig. 13, while the micro-hardness distribution along the transverse direction from the weld is shown in Fig. 14. From Fig. 13 it is derived, that the final micro-structural state of the heat affected zone would be a mixture of ferrite, pearlite and bainite with a mean hardness value between 353 and 254 HV.

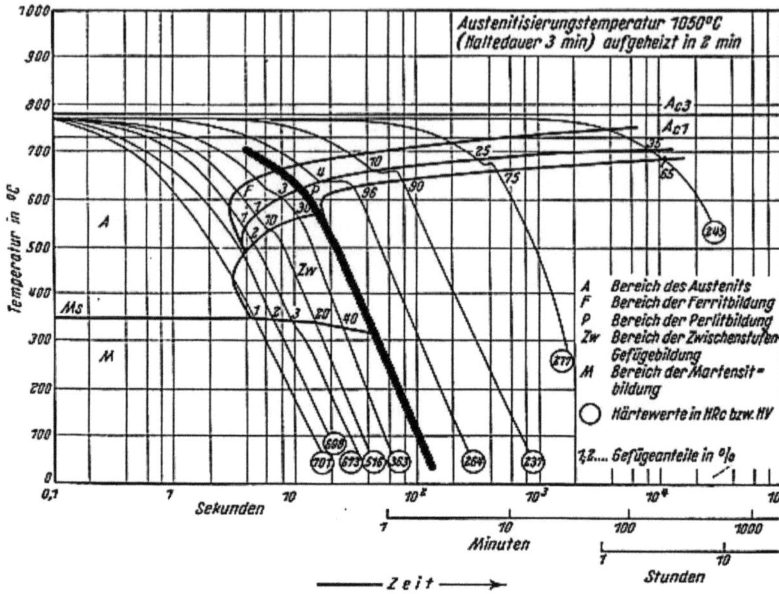

Fig. 13 Cooling curve plotted on the CCT diagram

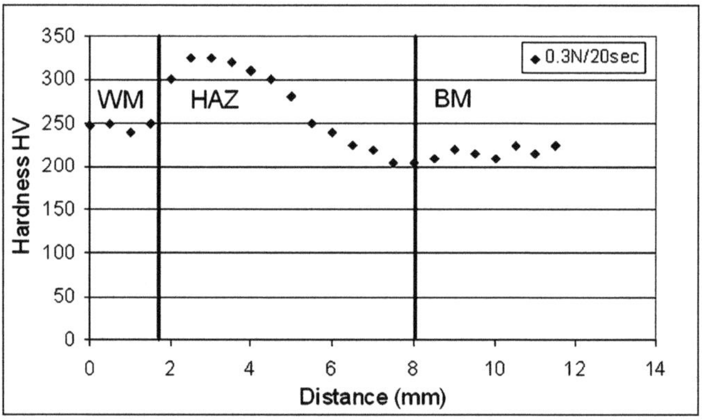

Fig. 14 Micro-hardness distribution along the transverse direction at mid-plane

This mean hardness is also proved by the experimental micro-hardness distribution shown in Fig. 14, in which the weld metal, the heat affected zone and the base metal have average hardness equal to 250, 300 and 220 HV, respectively. The maximum austenitization temperatures $T_{a\text{-}max}$ calculated by the numerical analysis along the heat affected zone from y = 0.007 m to y = 0.0095 m is shown in Fig. 15. The maximum temperatures reached (shown in Fig. 15 and Table 2) together with a mean value of cooling rate explain the micro-hardness profile within the heat affected zone obtained by the experimental measurements shown in Fig. 14 [1–3, 12].

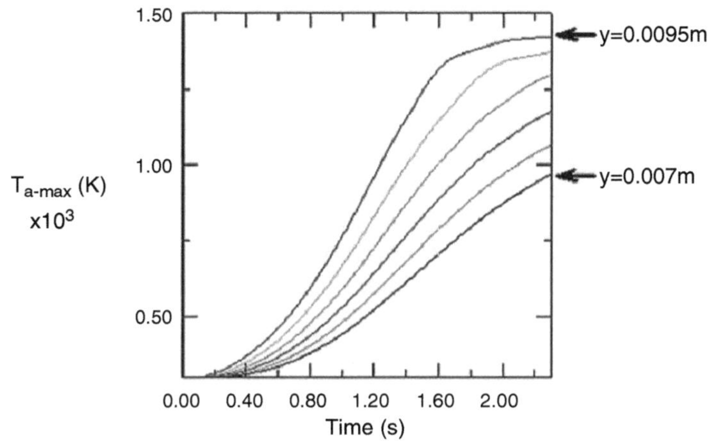

Fig. 15 Maximum austenitization temperature $T_{a\text{-}max}$ (K) calculated in the heat affected zone, from y = 0.007 m to y = 0.0095 m, at mid-plane (step of 0.0005 m)

Maximum austenitization temperatures and the mean cooling rate presented above can be further used for the exact description of the micro-structural dependence of the mechanical properties of the heat affected zone in a post-mechanical simulation.

6 Conclusions

The aim of the finite element analysis, as far as welding is concerned, is to predict the transient and steady state response of the welded system prior the welding itself. It aims at solving for temperatures and deformations, thus predicting the final structural quality providing a significant tool for the design of marine structures and their strength assessment.

Normally, a distinct deviation between numerical and experimental results is expected due to the unknowns involved and due to the increased complexity. Unknown material properties, arc efficiency, micro-structural dependency and the total heat loss are mentioned as typical common unknowns. A parametric analysis with respect to the governing unknown parameter(s) selected by the user may help better understand the range of this deviation. This distinct deviation generated during the early prediction phase (which turns the systematic accurate pure prediction into a myth) is not always clearly reported by investigators, since the numerical results published may already contain a kind of calibration or adaptation. Furthermore, the range of this deviation is treated as the decisive criterion for (a) further thorough investigation using the same numerical model, or (b) numerical model adaptation. The latter aims at adapting the numerical analysis to the real welding world, thus compensating for the governing unknowns. Attention should be paid whether the adaptation followed is general or case dependent. Thorough investigation of the process and analysis using the adapted model is then suggested.

As far as the specific study is concerned, a distinct deviation between numerical and experimental results was observed during the prediction phase as the weld metal predicted by simulation was larger in size and shape than in real welding. A case-dependent adaptation procedure on the basis of the weld metal shape and size was proposed in order to adapt numerical and experimental results. The latter is a cost effective adaptation procedure as it does not require any temperature measurements. Furthermore the weld metal size can easily be measured in situ at least for the single pass welds. The adapted model on the basis of the weld metal shape presented satisfactory results compared to the magnitudes calculated by the numerical simulation.

References

1. Radaj D (1992) Heat Effects of Welding. Springer-Verlag, Berlin, Heidelberg.
2. Lindgren LE (2001) Finite element modeling and simulation of welding part I: increased complexity. Journal of Thermal Stresses 24:141–192.
3. Lindgren LE (2001) Finite element modeling and simulation of welding part II: Improved material modeling. Journal of Thermal Stresses 24:195–231.
4. ESAB (1998) Welding Handbook. IIW.
5. ABAQUS 6.5 Documentation (2002) © Hibbitt, Karlsson & Sorensen, Inc.
6. Goldak JA, Chakravarti AP, Bibby M (1984) A new finite element model for welding heat sources. Metallurgical Transactions 15B:299–305.
7. Goldak J, Bibby M, Moore J, House R, Patel B (1986) Computer modeling of heat flows in welds. Metallurgical Transactions 17B:587–600.
8. J, Oddy A, McDill M, Chakravarti A (1986) Progress in computing residual stress and strain in welds. International Conference on Trends in Welding Research, ASM International, Gatlinburg, Tenn.
9. Bang IW, Son YP, Oh KH, Kim YP, Kim WS (2002) Numerical simulation of sleeve repair welding of in-service gas pipelines. Welding Journal 81:273s–282s.
10. Nguyen NT, Ohta A, Matsuoka K, Suzuki N, Maeda Y (1999) Analytical solutions for transient temperature of semi-infinite body subjected to 3-d moving heat sources. Welding Journal 78:265s–274s.
11. Nguyen NT, Mai YW, Simpson S, Ohta A (2004) Analytical approximate solution for the double ellipsoidal heat source in finite thick plate. Welding Journal 83:82s–93s.
12. Andersen LF (2000) Residual stresses and deformations in steel structures. Ph.D. thesis, Department of Naval Architecture and Offshore Engineering, DTU.
13. Geijselaers HJM (2000) Numerical simulation of stresses due to solid state transformations: the simulation of laser hardening. Ph.D. thesis, University of Twente.
14. Little GH, Kamtekar AG (1998) The effect of thermal properties and weld efficiency on transient temperatures during welding. Computers and Structures 68:157–165.
15. Wu CS, Yan F (2004) Numerical simulation of transient development and diminution of weld pool in GTAW. Modeling and Simulation in Science and Engineering 12:13–20.
16. Papazoglou VJ, Masubuchi K, Gonçalves E, Imakita A (1982) Residual stresses due to welding: Computer aided analysis of their formation and consequences. SNAME Transactions 90:365–390.
17. Camilleri D, Comlecki T, Gray TGF (2005) Computational prediction of out of plane welding distortion and experimental investigation. Journal of Strain Analysis 40:161–176.
18. Kuzminov AS (1956) Calculation principles of total deformation of ship hull structures. Works of Central Scientific Research Institute of Shipbuilding Industry, N9, Russia.
19. Vinokurov VA (1977) Welding distortion. British Library, Wetherby (Original: Svarochnye deformatsii I napryazheniya: methody ikh ustraneniya. Maschinostroeniu, Moscow 1968).

20. Hibbitt D, Marcal PV (1973) A numerical, thermo-mechanical model for the welding and subsequent loading of a fabricated structure. Computers & Structures 3:1145–1 174.
21. Karalis DG, Pantelis DI, Papazoglou VJ (2005) On the investigation of 7075 aluminum alloy welding using concentrated solar energy. Solar Energy Materials & Solar Cells 86:145–163.
22. Bhadeshia HKDH, Svensson LE, Gretoft B (1985) A model for the development of microstructure in low-alloy steel (Fe-Mn-Si-C) weld deposits. Acta Metallurgica 33:1271–1283.

Session XI
Theoretical Fracture Mechanics and Modeling II

Application of a Fracture Methodology for Studying the Mechanics that Govern Failure of Aluminum Structures

Konstantinos Panagiotis Galanis, Tomasz Wierzbicki, and Vassilios John Papazoglou

Abstract The existence of cracks besides the effect of structural degradation can cause partial or total failure of a structure. Usually, structures are designed to resist yielding, buckling and fatigue, but not fracture due to limited existing data on large scale models. As a result, robust methods and procedures to design structures against fracture have not been developed, although the industry is seeking to achieve more efficient concepts and designs with advanced safety and performance using optimized structural design. The rapidly increasing application of lightweight materials and thin-walled structures in several industries requires fundamental understanding of the mechanisms and mechanics of fracture that govern stiffened panels. A methodology consisting of the application of advanced fracture models, material calibration, and validation through component testing is provided that will increase the survivability envelope of new structures. This paper presents its application that will enable designers to evaluate critical areas within a structure with respect to crack initiation, propagation, optimum material usage, and computational cost. It is based on the effect of stiffening configurations on fracture of aluminum marine structures by studying the structural response of various stiffened plates, represented by small-scale compact tension (CT) and intermediate-scale specimens, which are compared with unstiffened plates. It is shown that mapping of crack patterns in stiffened plates is feasible. Numerical modeling and analyses of ductile fracture initiation and propagation on a precracked geometry using a commercial finite element

K.P. Galanis (✉) and T. Wierzbicki
Impact and Crashworthiness Laboratory, Mechanical Engineering Department,
Massachusetts Institute of Technology, 77 Massachusetts Avenue, Cambridge, 02139 MA, USA
e-mail: galanis@alum.mit.edu, wierz@mit.edu

V.J. Papazoglou
Shipbuilding Technology Laboratory, School of Naval Architecture and Marine Engineering,
National Technical University of Athens, 9, Heroon Polytechniou Ave.,
GR-157 73 Zografou Athens, Greece
e-mail: papazog@deslab.ntua.gr

S. Pantelakis, C. Rodopoulos (eds.), *Engineering Against Fracture: Proceedings of the 1st Conference*,

code (ABAQUS), taking into account the behavior of simple uncracked material, has been performed showing a very good agreement with small and intermediate scale tests.

Keywords Fracture · Aluminum · Propagation · Stiffener

1 Introduction

The examination of the shipbuilding industry was selected to serve as the basic case study. Modern vessels vary from conventional ones, which are built of mild steel and designed on the basis of experience, to front-end technology high speed craft made by sophisticated lightweight materials such as composites, high-strength steels and aluminum alloys. It is obvious that design procedures have to cover a wide range of applications and materials, and with the continuing development of faster vessels in highly optimized designs, the demand for research in this area will continue as the accidental loads and their consequences, due to lack of experience, are more or less unknown.

The approach used in this study is very general, whereas the material chosen is also applicable to a wide variety of other metallic structures, both fixed and floating. Consideration of additional types of loads and failure modes, plus some new examples would be required, however, to illustrate these other applications. Paik et al. [1,2] made a comparison of the various methods for ultimate limit state predictions of aluminum stiffened panels used for aerospace, marine and civil engineering applications and concluded that due to the different structural characteristics, material types (depending on structure types of each industry application) and the primary failure pattern, analyses using the same principles present different results.

The shipbuilding industry is seeking to achieve advanced and more efficient concepts and designs for ships and underwater vehicles with improved safety and performance using optimized structural design. To increase the ship's survivability, in-depth understanding and vast experience is required to study the methodology and dynamic effects of complex damage mechanisms on marine structures and ship systems. Damage prediction models should be created to enhance the ability of the ship to withstand all types of loads. Verification and validation of modeling and simulation is required for the models to be functional. The modeling could be either numerical or physical. All these will lead to the improvement of existing design procedures and criteria, which may lead to more affordable designs.

To pursue the objective of this study, an extended experimental program including coupon, small and intermediate scale tests was conducted examining the macroscopic effect of various parameters of stiffening mechanisms towards fracture.

Many civil and defense applications require a blast-resistant performance. The threat from a sudden overpressure can arise from a number of situations, ranging from accidents to attack by an enemy weapon. The impact of an explosive is one of the most threatening scenarios for a marine platform. A marine engineer has various options to reduce the vulnerability of the platform design, like structural arrangement and protection. One of the most important options is to increase the blast resistance of the hull. The methodology proposed provides a tool for engineers and designers to include blast resistance during design and evaluation of marine structures.

Fracture mechanics concerns the design and analysis of structures which contain cracks or flaws. On some size/scale all materials contain flaws, either microscopic, due to cracked inclusions, debonded fibers, etc., or macroscopic, due to corrosion, fatigue, welding flaws, etc. Thus, fracture mechanics is involved in any detailed design or safety assessment of a structure. As cracks can grow during service due to fatigue, fracture mechanics assessments are required throughout the life of a structure or component, not just at the start of the life cycle of a marine vessel. In which direction will a crack propagate when a stiffening mechanism is present? Which type of stiffener arrests a crack better? Which parameters affect the crack path? What should be done by a designer to improve the damage tolerance of a structure without penalty to the rest of the existing requirements? This paper fully or partially answers these questions.

2 Methodology Description

Crack propagation and, especially, arrest in marine aluminum panels has not been an area of in-depth research. Under the Tank Safety project, Wierzbicki et al. [3,4] presented extensive work of deformation and fracture of small-scale (1:50) longitudinally stiffened single and double hulls, Fig. 1.

The crack arrest phenomena in cylindrical containers and pipes have been extensively studied in relation to the pressure vessel technology by, e.g., Freund et al. [5]

Fig. 1 Existing work on damaged and fractured small-scale structural models subjected to localized transverse loading showing the crack propagation between and parallel to the stiffeners of stiffened panels for single (*left*) and double (*right*) plate structural configurations

and Zhuang and O'Donoghue [6]. Similar analyses, though, are not readily available for the case of the tearing fracture mode in plates, in which case the mechanics of the crack arrest process must be well understood.

At the dawn of the previous century the US Navy pursued research for the Advanced Double Hull (ADH) for naval combatants [7,8]. Tearing fracture in stiffened plates was considered up to now to be arrested for a while at the foot of the stiffener, and then be re-initiated on the other side of the stiffener; eventually the longitudinal stiffener would fracture as well. A test on a quarter model of splitting damage of a longitudinally stiffened, double hull, which was performed by Rodd [9], gives only a glimpse on this important phenomenon. It is still unknown, however, what aspect of the design (strength of the fillet weld, relative thickness of the base plate and stiffener, height of the stiffener or hull separation) will stop the crack, re-direct it or overcome the obstacle [10]. Recently, Paik and Thayamballi [11] have performed some preliminary work on the estimation of the strength of cracked aluminum unstiffened plates.

The application of the proposed methodology includes the definition of the problem, design of experiments, execution of an experimental program that builds up from coupon, small-scale to intermediate-scale (large-scale if economically feasible) testing and verification using numerical simulations.

2.1 Design of Experiments

A major challenge confronting researchers is how to efficiently allocate resources to an experimental effort in order to extract the maximum amount of information while minimizing the amount of required specimens. Design of experiments is a process of determining the functional relationships between independent experimental variables so that controlling factors can be predicted and stated with a statistical accuracy. Once a better understanding of the effects of the various parameters is achieved, the remaining resources can be directed more efficiently on those variables that can be controlled and that significantly influence the response.

The uniaxial tension test is one of the most widely used tests to characterize strength, ductility and other mechanical properties of a material. This type of test was selected because its analysis provides data about elastic and plastic deformation: modulus of elasticity or Young's modulus, Poisson's ratio, yield and tensile strengths, elongation and reduction in area, and strain-hardening characteristics.

2.2 Material Used

In all cases, an aluminum alloy relevant to the ISO 209-1 based AlMg6MnZr alloy of 4 mm thickness was selected which has a density of $2.65 * 10^3\,\mathrm{kg/m^3}$ and yield stress of 190 MPa. This aluminum-magnesium alloy is among the average strength heat non-hardenable alloys. Being heated by welding, alloys of this group do not reduce their strength properties in the heat-affected zone.

2.3 Specimen Selection and Experimental Setup

This research is concerned with the examination of fracture characteristics of a material with limited resources, and as a result, the experimental work has focused on the fundamentals of crack propagation in stiffened panels. The tests were, therefore, divided into three groups: (i) determination of material characteristics, (ii) small-scale tests of plates and panels, and (iii) intermediate-scale tests of structural configurations.

The first group consists of results obtained from six specimens that were tested in order to determine the material properties. Rectangular tensile thin sheet ("dog-bone") type specimens were used. As it will be shown later, there was a fairly good agreement between measured values, documentation provided and results obtained from numerical simulations.

The choice of test in the second group was based on the international standard ISO12135:2002 (Metallic materials – Unified method of test for the determination of quasi-static fracture toughness) [12], utilizing the CT specimen without fabrication of the precrack often used for fatigue tests, but only the notch. A series of fracture tests on small scale aluminum alloy sheets and panels were performed to examine and characterize the process by which cracks propagate and coalesce in stiffened configurations.

Special test fixtures were designed to withstand the tension loads applied to the panel specimens in a ±250 kN capacity MTS servo hydraulic universal testing machine. Thirteen single sheets of bare marine aluminum alloy, approximately 120 mm high, 125 mm wide, and 4 mm thick were fabricated with artificial through-cracks oriented horizontally at mid-height. Additionally, nine tests of marine aluminum alloy panels in several geometries, stiffener types, stiffener configurations, and displacement rates were conducted.

In the third group, the results acquired from the small-scale tests were compared to the subsequent four tests with edge and central cracks in two common unstiffened and stiffened intermediate-scale structural configurations. The comparison showed that the crack pattern followed similar discrete steps proving that the scaling effect did not affect the crack pattern.

3 Numerical Simulations

The commercial code, ABAQUS/Explicit^{TM}, was used to carry out all the numerical simulations. This code offers an element library for a wide variety of geometric models. The preliminary analysis is conducted with a relatively coarse mesh.

However, increasing the number of elements also increases the required computational time to reach a solution and hence it increases cost. The meshing techniques implemented in this study require from 10 min to 20 h of computational time using powerful personal computers already installed at the industry. The numerical simulations of this research were performed on a HP ProLiant DL 145 G3 workstation (2-CPUs, 64-digit and 6 GB-memory) with Linux environment.

Prior to investigating the damage parameters that can be calculated from the calibration procedure, it is crucial to compare the deformed shape produced by the numerical simulations to the fracture mode of the experimental results.

This deformed shape is obtained from numerical simulations with the different types of meshes while acquiring the equivalent plastic strain to fracture from the corresponding displacement to fracture obtained from the experiment ($\delta_f = 4.385$ mm). The fine mesh captures very closely the failure mode from the test in both cases.

4 Analysis and Results

Marine vessels are always subjected to external or internal dynamic loading, operating in extreme environmental and functional conditions that can lead to the loss of the entire structure and human lives. Stiffened plates are the most commonly used built-up structural elements in marine vessels, which form the backbone of ships, appearing in decks, bottoms, bulkheads, and side shell. The damage of stiffened plates will result in the collapse of overall system structures. An increasingly popular approach to undertake ship hull ultimate strength analysis is to consider the failure of the individual stiffened plates and combine these to determine the failure load of the entire hull cross-section.

The results of the experiments on aluminum plates and panels reveal that the crack propagation and arrest phenomena highly depend on the presence of stiffeners, the type of stiffener and the stiffening configuration. Observations of the crack pattern in both experimental and numerical analyses show that the type of stiffener and its configuration can significantly affect the impact of fracture. A careful selection of stiffener type and its geometry can improve the global behavior of a marine structure based on the requirements set by the customer. Nevertheless, this selection should always take place in conjunction and with respect to the other types of loading that a vessel is designed to withstand.

5 Experimental Program

The present study investigates the characteristics of tensile strength of aluminum alloy base metal and fillet welded stiffeners for marine applications, concerning welded configurations. A series of tensile tests in a quasi-static condition are carried out on base metal and fillet welded stiffeners of aluminum alloys, varying the type of stiffener. The results and insights developed in this study will be useful for designing marine aluminum structures. These results are compared with T-type extruded aluminum stiffened panels to evaluate the effect of different manufacturing and design methods. As a last step, comparisons between stiffened and unstiffened structural configurations representing marine structures are performed. Additionally, the effect of stiffener dimensions on the crack propagation and arrest is examined. A series of experiments is performed, by changing the length and height of the web of a flatbar stiffener and then the classical T-type stiffener is examined.

5.1 Group I: Coupon Uniaxial Tensile Tests

The tests were conducted at room temperature and at atmospheric pressure on "dogbone" specimens, acquiring load vs. displacement data. These tests were conducted to verify the material properties and to examine the failure mode (slant fracture) of this marine aluminum alloy The results obtained from the coupon tensile testing on six specimens, Fig. 2, (two pairs taken from longitudinal and transverse directions with respect to the rolling of the plate) were almost identical, meaning that the required number of Group II tests (CT type of specimens) could be kept to a representative number of one test per different type of parameters examined, Figs. 3 and 4. In addition, the scope of the test was to get a macroscopic view of the crack pattern, rather than investigate microscopically the crack propagation phenomenon. These results were also used for calibrating the finite element model, as part of numerical simulations.

Fig. 2 Final stage (*left*) of a "dogbone" specimen (3AX), closer view (*center*) showing the mode of fracture (slant fracture) observed at the experimental tests, and schematic of directions that the specimens were cut with respect to the rolling direction (*right*)

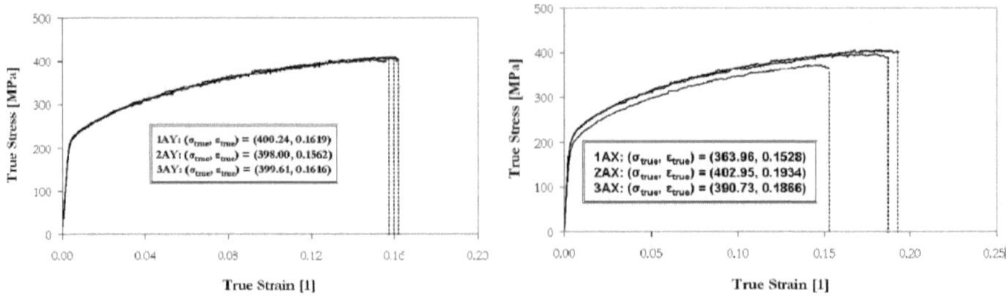

Fig. 3 True stress–strain curves of the "dogbone" specimens tested cut from two different material orientations (transverse [AY] and longitudinal [AX] direction, *left* and *right* respectively)

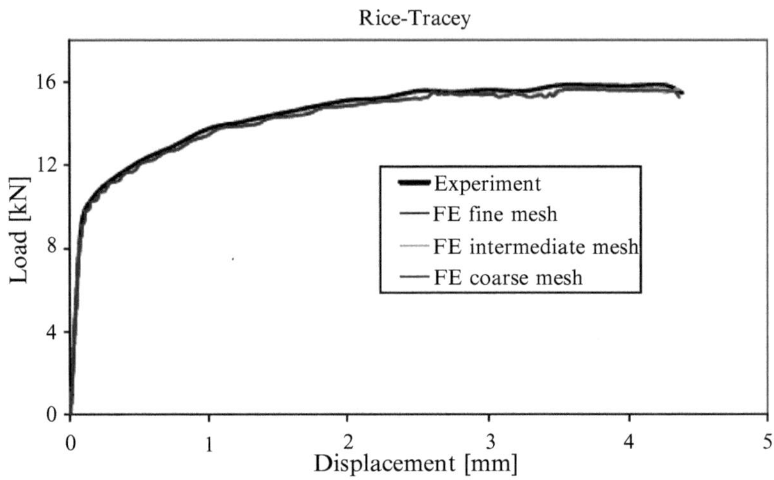

Fig. 4 Load vs. displacement graph obtained from experiment and numerical simulations using finite element code and the Rice and Tracey fracture criterion

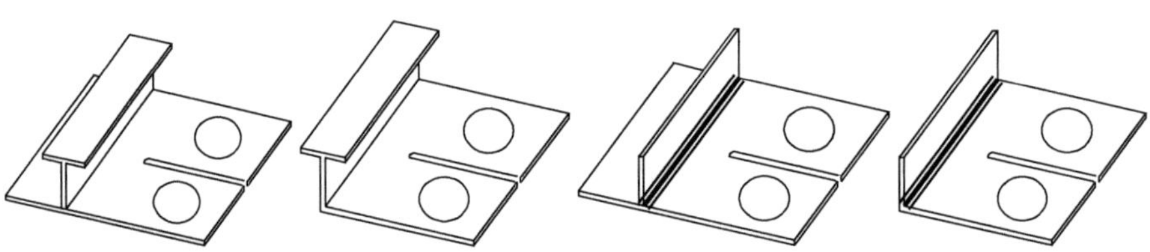

Fig. 5 Schematics of the stiffened CT specimens (CTE, CTEB, CTW and CTWB from *left* to *right*)

5.2 *Group II: Small-Scale Tests*

The Group II small-scale CT specimens, Fig. 5, were subsequently tested by varying several parameters, such as the structural configuration, the geometries of the specimen and crack tip, the displacement rate, the stiffener type and configuration (extruded and welded) and the crack length, as described below.

Fig. 6 Unstiffened CT specimens tested under varied displacement rate and crack length and tip geometry

5.2.1 Unstiffened Compact Tension Specimens

Introduction of a notch into the gage length of a metallic tensile test specimen increases the tensile strength above that measured on an identical metallic test specimen without a notch. The increased tensile strength of the notched specimen is caused by a subtle distortion of the applied uniaxial stress, resulting in a localized triaxial stress state at the root of the notch.

Quasi-static tests were conducted on notched CT specimens of 4 mm thickness (with no precracking) [13]. The tests were conducted at room temperature and at atmospheric pressure, using the previously mentioned MTS testing machine. During testing the load–displacement relationship was measured through a computerized data acquisition system, whereas at the same time the crack propagation pattern was traced through consecutive photographs taken at predetermined short intervals.

Thirteen unstiffened CT specimens were tested and in all cases the crack propagated perpendicular to the loading direction. It was observed that the crack initiated at a 45° angle at the notch and that the parameters examined (notch length and tip geometry, displacement rate) did not affect the crack direction, Fig. 6. Figure 7 present the good correlation between the experimental and numerical result with respect to the crack path.

5.2.2 Stiffened Compact Tension Specimens

Even though the main obstacle of the crack propagation in a steel plate is the welded stiffener, there has not been any significant work published on this area with the exception of that of Dexter and Pilarski [14] and Dexter et al. [15]. The presence of stiffeners on the crack propagation path can be a significant parameter for the crack arrest phenomenon, as initially observed by Rodd [9], adding value to the fracture toughness of the material, a fact which has been thoroughly investigated by Sielski [16] and Khan at al. [17]. Therefore, experimental and numerical work is

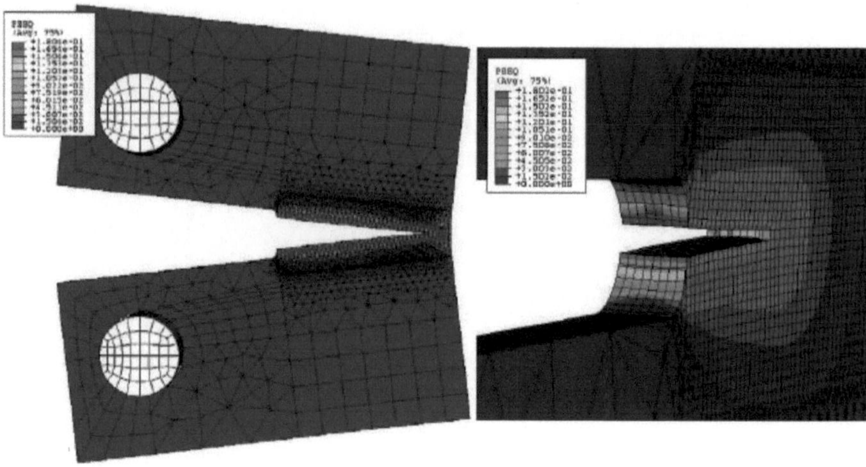

Fig. 7 CT specimen with initial notch length of 71 mm and blunt crack tip: crack path (*left*) and crack tip initiation obtained from the FE simulation (*right*)

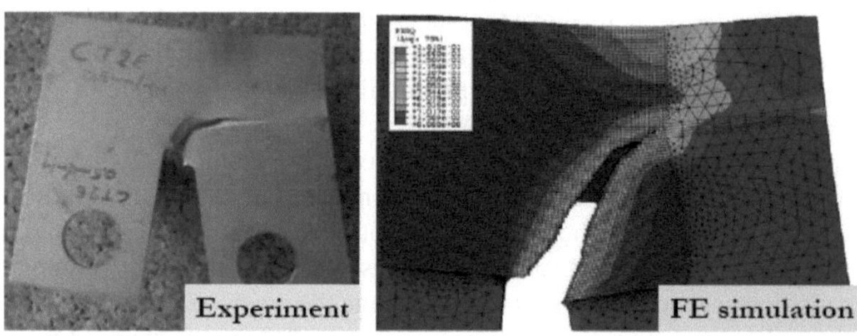

Fig. 8 Comparison between the crack pattern obtained from the experiment (*left*) and the numerical simulation (*right*) for the T-type extruded CT specimen

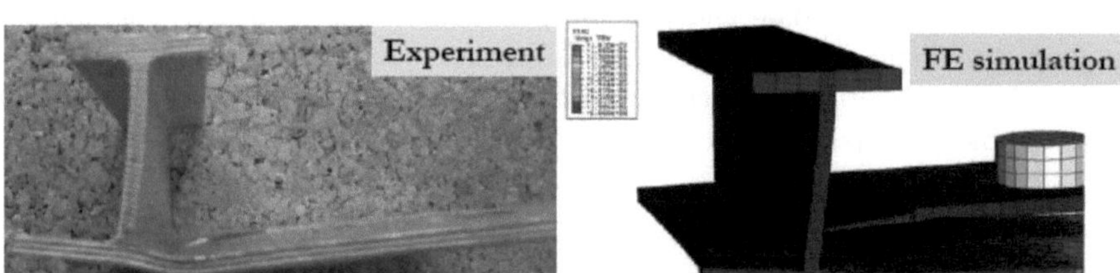

Fig. 9 Comparison between experimental result (*left*) and numerical simulation (*right*) with respect to the behavior of the T-type stiffener during the loading

performed for the evaluation and assessment of the stiffener dimensions and design, concerning the direction of the crack, Figs. 8 and 9. The comparison between theoretical predictions and experimental results will ultimately lead to the design of an alternative more efficient structure.

Ship structures are composed almost entirely of orthogonally (longitudinally and transversely) stiffened plating. Both the plating and stiffeners must be designed to sustain the working loads. Plate panels in the ship are welded around their periphery to stiffeners or adjacent panels. Fatigue tests are addressing the properties of local structural issues, such as capacity of plating between stiffeners, the proportions of the flange and web elements of stiffeners, and the fatigue strength of a welded detail. Local structural issues like these should, however, be understood before the larger, general structure can be designed. As the overall structure increases in size, the size, shape, and length of the members may be limited in order to control buckling of the main load carrying members. Therefore, grillage test structures, containing multiple longitudinal and transverse stiffeners, have to be evaluated for catastrophic modes of buckling failure and to define margins of safety for design against ultimate failure.

The most common techniques used in the aluminum shipbuilding industry for marine panel construction is either welded stiffeners or extruded ones. In the tests performed (again on CT specimens of the same dimensions as before, but now having a welded or extruded stiffener), it was observed that the crack did not propagate through the stiffener at the T-type extruded specimens, Fig. 10.

Additionally, and based on the base plate material behind the stiffener web, a "zig-zag" phenomenon was observed at around one third of the distance between the

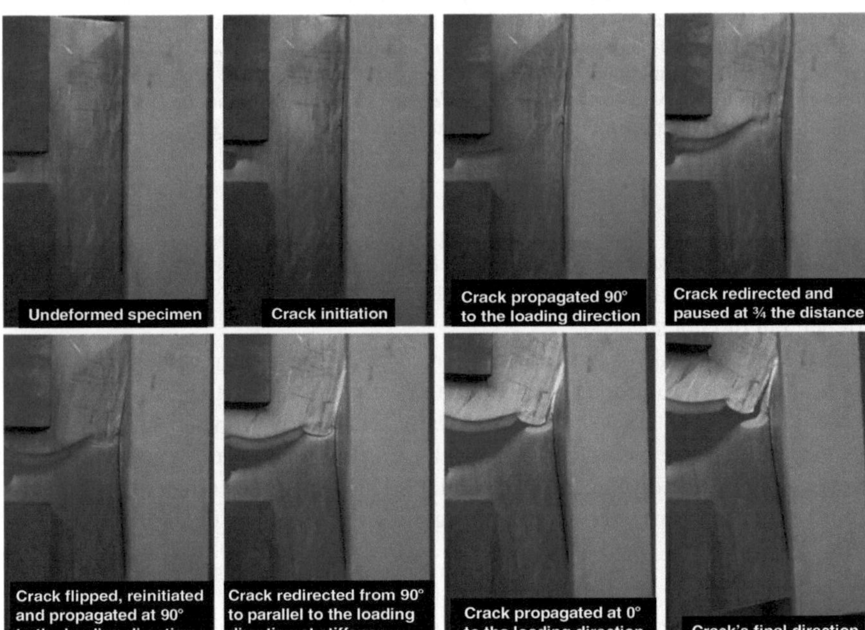

Fig. 10 Stepwise crack propagation in T-type extruded stiffened CTEB specimen (*top*-view, where the clevis are on the *left* and the flange on the *right* of each photograph) that has no material at the rear part of the stiffener showing the phenomena observed: crack propagation with 30° angle with respect to the loading direction, pause, flip, re-initiation and re-direction to 90° to the loading direction, and propagation parallel to the loading direction

Fig. 11 Stepwise crack propagation showing the phenomena observed at flatbar type welded stiffened CT2 W specimen tested: crack propagation with 30° angle with respect to the loading direction, crack kinking to a direction perpendicular to the loading, propagation through the weld and the stiffener, and, simultaneous propagation on the stiffener and the rear plate material with similar growth rate. *Top* view with clevis on the *left* and the flatbar stiffener on the *right* (first four photographs beginning upper *left* to the *right*) and *side* view (three photographs at the second row starting from the second on the *left*) of the stiffened CT specimen

initial notch and the stiffener web. While the crack reached the extruded stiffener's web, a delay was noticed, meaning that the crack "tried" to propagate through the stiffener, but did not succeed.

On the other hand, in two of the four cases of the flatbar welded specimens, the crack managed to propagate through the web and behind the stiffener, Fig. 11. The sequence of the crack propagation can be categorized in the following three steps: (a) through the weld, (b) on the web of the stiffener, and (c) at the rear plate material. Finally, the displacement rate and the initial notch geometry (crack length and different crack tip) did not seem to significantly affect the crack path.

5.3 Group III: Intermediate-Scale Tests

In contrast to fluid mechanics, the problems of scaling and size in solid mechanics have not come to the forefront of attention until late in the previous century. The classical view that any observed size effect was statistical was reversed during the

1980s. Ductile materials might exhibit size effects on the mean structural strength that are deterministic in nature, being caused by stress redistribution and energy release associated with stable propagation of large fractures or with formation of large zones of distributed cracking.

The intermediate-scale specimens were designed to meet the limitations of the experimental apparatus. The outcome of the calculations yielded specimens that could be called full-scale. It was decided instead to categorize them in the intermediate-scale, compared to the large size of marine structures, such as the size of a vessel. Due to the scarcity of material, we examined only the extruded panels. This group of experiments is divided in two sets. The first one is consisted of a pair of large "dogbone" specimens (600 mm length, 300 mm width and neck of 210 mm) with two edge cracks, one unstiffened, named DENT, and the other stiffened with an extruded stiffener in the center, named SSDENT. The second set is consisted of a pair of large "dogbone" specimens with similar dimensions to the previous set with one central crack, one unstiffened, named CCT, and the other stiffened with two extruded stiffeners equally spaced at each side of the crack, named TSCCT.

5.3.1 Double Edge Notched Tension (DENT) Specimen

The first type of specimen tested was a version of the "dogbone" specimen used in Group I of the experiments, but with increased dimensions and two artificial edge notches, each one having initial length of 22.5 mm, which were machined prior to testing. The specimen was mounted on the MTS device using a fixture designed specifically for the intermediate-scale specimens. Two cracks were initiated at both notches but it was observed that one of the cracks propagated at a greater distance compared to the opposite crack and they finally coalesced. The crack propagated with a very high growth rate. The path created can be simulated as a line perpendicular to the loading directions without any inclinations. The DENT specimen prior to failure and the results from the respective numerical simulations are shown in Fig. 12.

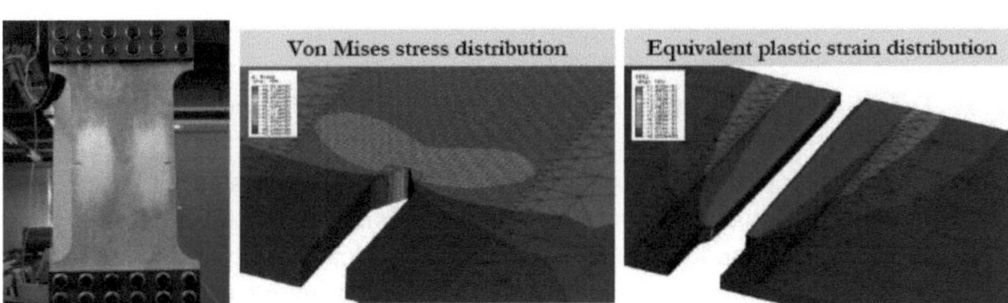

Fig. 12 DENT specimen mounted on the MTS device (*left*) and results from FE simulation showing the distributions of the von Mises stress at the crack initiation step (*center*) and the equivalent plastic strain (*right*) for the failed case

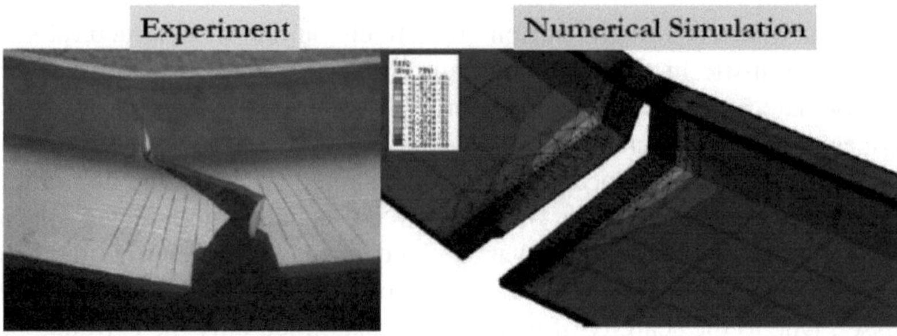

Fig. 13 Crack pattern observed in the experimental program (*left*) and the FE code (*right*). Note that minor deformations or crack kinking can alternate the crack path which is expected to be perpendicular to the loading direction at the "perfect" case

5.3.2 Single Stiffened Double Edge Notched Tension (SSDENT) Specimen

The phenomena observed during the test on the SSDENT specimen can be summarized at the following phases: onset of fracture at the artificial crack on the right after the completion of the out-of-plane deformation, crack on the right flipped and redirected itself, onset of fracture on the artificial crack on the left, flipping and redirection of the crack on the left, both cracks propagated in order to coalesce at the web of the stiffener, and, cracks coalescence, propagation on the web of the stiffener and onset of fracture at the flange of the stiffener. The correlation between the experiment and the numerical simulation was very good, as shown in Fig. 13.

5.3.3 Center Cracked Tension (CCT) Specimen

The next test examined the crack propagation of an unstiffened center cracked tension (CCT) specimen with similar dimensions to the DENT one, tested under displacement-control conditions and simulated numerically, Fig. 14. During this test the only phenomenon observed was that the crack initiated at both ends of the artificial crack of 44.5 mm initial length and propagated at both sides.

5.3.4 Twin Stiffened Center Cracked Tension (TSCCT) Specimen

In this test, a similar to the CCT specimen was used with the existence of two T-type extruded stiffeners located symmetrically at each side of the crack, tested under displacement-control conditions and simulated numerically, Fig. 15. Prior to the onset of fracture, the specimen presented out-of-plane deformation due to the existence of the stiffeners. As a second step, a dishing out-of-plane deformation was observed around the region of the artificial crack in the center of the specimen, with direction opposite to the one occurred in the whole specimen. Then the cracks propagated towards the web of the stiffeners and finally the flange of one of the stiffeners failed and the experiment was terminated.

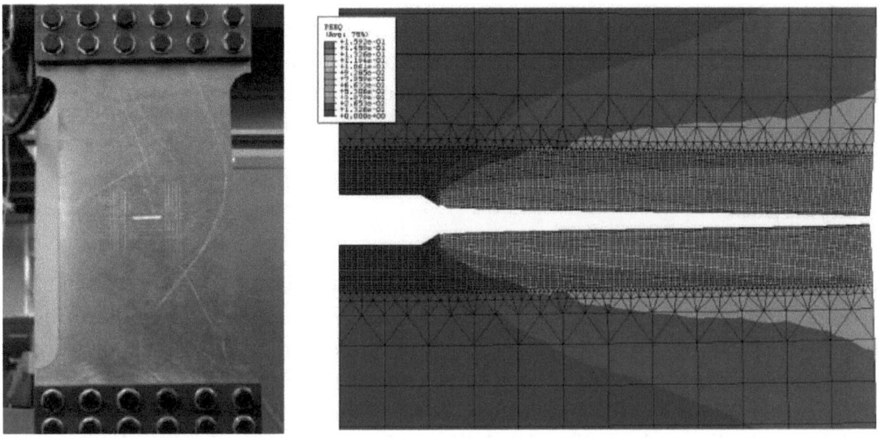

Fig. 14 Comparison between the results obtained at the experimental program (prior to failure) and the FE simulation (failure) for the CCT model

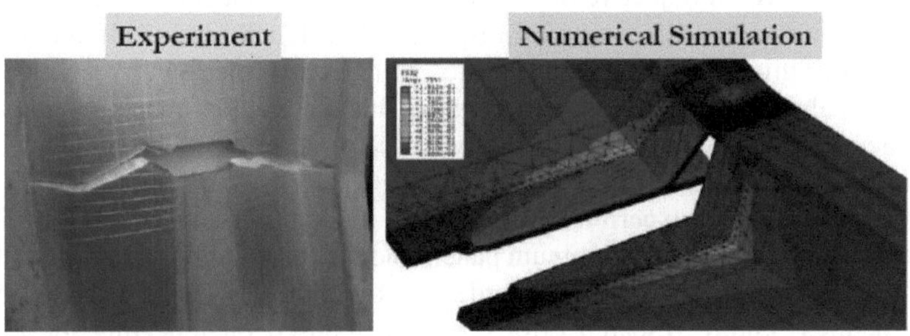

Fig. 15 Comparison between the results obtained at the experimental program and the FE simulation for the TSCCT model

The intermediate-scale experiments revealed a strong relationship between crack path and the presence of stiffeners. The thickness of the plates was 4.1 mm. For the extruded T-type stiffeners the web height was 4.1 mm, the web thickness was 3 mm, the flange length was 3 mm and the flange thickness was 4 mm. For the I-welded stiffeners the web height was 4.1 mm and the web thickness was 3 mm. It was observed that the major parameter affecting the crack pattern is the plate material at the rear part of the stiffener.

The crack managed to propagate through the web of the stiffener in two cases and failed to do so in all other tests including both the extruded and welded CT stiffened specimens. The main phenomena observed for the extruded CT specimens are the following: (a) the crack delayed its propagation the closer it was reaching the stiffener and accelerated while it turned parallel to the stiffener, (b) the crack followed the "zig-zag" pattern at half the distance between the crack initiation and the stiffener, and (c) the crack failed to propagate through the stiffener in all cases examined. The displacement rate did not seem to significantly affect the crack path in the cases examined. Finally, a visual examination of the intermediate-scale specimens indicated that the crack initiated and propagated from both directions independent of the crack location. Bending of the specimens preceded fracture for the stiffened ones.

6 Synopsis

Modern large-scale structures must be designed to survive exposure to extreme loading conditions from impact and explosions. Various structural failure modes contribute to the loss of integrity of structures subject to blast loading; these being dependent on material selection and structural configuration. Modeling the structure's response encompasses material constitutive equations, fracture and damage mechanics, nonlinear dynamics simulation codes and structural finite element analyses. The advancement and integration of these disciplines are necessary in order to provide a design framework for developing optimum structural configurations and materials. The overall objective of this study has been to understand the mechanics of fracture that govern the structural response of stiffened marine aluminum panels and apply this method to other type of structures. Experiments conducted on small-scale aluminum plates and panels reveal that the crack propagation and arrest phenomena are highly depended on the presence of stiffeners, the type of stiffener and the stiffening configuration.

The results of these experiments is expected to lead to the creation of a fracture test database on continuous aluminum panels and the formulation of appropriate criteria for cracked aluminum panels in order to extend the service of marine vehicles and structures. Future research efforts will be focused on repeating the experiments and on performing similar ones on intermediate size specimens to examine the scaling effect. Finally, calibrated numerical simulation provided comparable results emphasizing the effect of each parameter on the crack propagation and arrest phenomena.

The level of damage tolerance is empirically set and requires accurate prediction of crack growth under operational conditions. Accurate prediction of crack growth in marine aluminum structures plays a key role in the structural analysis of a vessel. In the advancing marine high speed market the evolution tends towards larger and faster vessels and this makes it difficult to base new designs on past experience. The result is an intensified need for rational procedures for design of both main structural parts and details in such high speed vessels.

This method should be applied to other metal structures, with emphasis on steel. Various structural failure modes contribute to the loss of integrity of marine vessels subject to blast loading; these being dependent on material selection and structural configuration. Modeling structural response encompasses material constitutive equations, fracture and damage mechanics, nonlinear dynamics simulation codes and structural finite element analyses. The advancement and integration of these disciplines are necessary in order to provide a design framework for developing optimum structural configurations and materials.

References

1. J.K. Paik, S. van der Veen, A. Duran, M. Collette, Considering aluminum welded panel structures for aerospace, marine and land-based applications: a comparison of ultimate compressive strength design methods, In: 9th Symposium on Practical Design of Ships and Other Floating Structures, Luebeck-Travemuende, Germany, (2004) 727–735.
2. J.K. Paik, S. van der Veen, A. Duran, M. Collette, Ultimate compressive strength design methods of aluminum welded stiffened panel structures for aerospace, marine and land-based applications: A benchmark study, Thin-Walled Structures 43 (10) (2005) 1550–1566.
3. T. Wierzbicki, P.F. Thomas, Closed-form solution for wedge cutting force through thin metal sheets, International Journal of Mechanical Sciences 35 (3–4) (1993) 209–229.
4. T. Wierzbicki, M. Yahia Yahiaoui, M. Sinmao, Research activities within the Joint MIT Industry Project on Tanker Safety, Joint MIT - Industry Program on Tanker Safety, Report No. 37 (1995).
5. L.B. Freund, V.C.F. Li, D.M. Parks, Analysis of a wire-wrapped mechanical crack arrester for pressurized pipelines, Journal of Pressure Vessel Technology-Transactions of the ASME 101 (1) (1979) 51–58.
6. Z. Zhuang, P.E. O'Donoghue, The recent development of analysis methodology for rapid crack propagation and arrest in gas pipelines, International Journal of Fracture 101 (3) (2000) 269–290.
7. J. Beach, Advanced Surface Ship Hull Technology – Cluster B, Naval Engineers Journal 103 (6) (1991) 27–37.
8. J. Sikora, J.M. Grassman, P. Sensharma, J. Watts, T.R. McNatt, Advanced double hull structural design technology, Naval Engineering Journal 109 (5) (1997) 117–128.
9. J.L. Rodd, M.P. Phillips, E.D. Anderson, Stranding Experiments on Double Hull Tanker Structures, In: Proceedings of the Advanced Double Hull (ADH) Technical Symposium Paper No. 2, Gaithersburg, MD (1994).
10. K.P. Galanis, V.J. Papazoglou, Effect of stiffening configurations on fracture of aluminium structures. "Early View" of Journal of Strain since May 2008.
11. K.P. Paik, A.K. Thayamballi, Ultimate Limit State Design of Steel-Plated Structures, Wiley, Chichester (2003).
12. International Standard ISO 12135: 2002(E) Metallic materials - unified method of test for the determination of quasistatic fracture toughness (2002).
13. K.P. Galanis, V.J. Papazoglou, T. Wierzbicki, Crack propagation in stiffened panels, In: International Conference in Collision and Grounding, Hamburg, Germany, September 9–12 (2007) 163–169.
14. R.J. Dexter, P.J. Pilarski, Crack propagation in welded stiffened panels, Journal of Constructional Steel Research 58 (5–8) (2002) 1081–1102.
15. R.J. Dexter, P.J. Pilarski, H.N. Mahmoud, Analysis of crack propagation in welded stiffened panels, International Journal of Fatigue 25 (9–11) (2003) 1169–1174.
16. R.A. Sielski, Fracture-mechanics of ship structures, Naval Engineers Journal 104 (3) (1992) 36–45.
17. M.Z.S. Khan, D.S. Saunders, I.A. Burch, A.P. Mouritz, Materials aspects of damage tolerance and reliability of ship structures and components, Naval Engineers Journal 106 (4) (1994) 192–207.

BEM Prediction of TBC Fracture Resistance

Loukas K. Keppas and Nick K. Anifantis

Abstract The present study investigates the transient behavior of interfacial cracks in thermal barrier coatings (TBC). It is assumed that these material systems withstand thermal shock in the presence of external operating pressure load acting on the coating surface. Partial of full crack closure takes place and the thermal contact resistance as well the friction between the crack faces is considered. The dependence of the thermal resistance on the contact pressure leads to a coupled the thermal–mechanical analysis. An appropriate boundary element formulation based on two-dimensional time-dependent thermoelasticity, implemented in a verified computer code, is utilized for the numerical solution. A series of parametric analyses examines the impact of coefficient of friction and the level of loading and thermal contact resistance on crack severity.

Keywords Boundary elements · Thermal barrier coatings · Interfacial crack · Thermal contact resistance

1 Introduction

The demand for endurance of metallic components that undergo extreme thermal conditions in applications such as combustion chambers of diesel engines and gas turbine blades has established the use of thermal barrier coatings (TBC) on the surface of such components. Operation under high thermal gradients and cycling loading can cause delaminations in these material systems. Several studies related to the thermal fracture of TBC have been conducted utilizing experimental techniques and finite elements [2, 6, 7]. The numerical prediction of the failure resistance of TBCs is very attractive for the design or the fracture assessment of TBCs. Therefore

L.K. Keppas and N.K. Anifantis (✉)
Machine Design Laboratory, Mechanical and Aeronautics Engineering Department, University of Patras, Rion 26500, Greece
e-mail: nanif@mech.upatras.gr

S. Pantelakis, C. Rodopoulos (eds.), *Engineering Against Fracture: Proceedings of the 1st Conference*,

the present study is focused on the application of a boundary element procedure, developed by the authors, for the solution of time-dependent coupled thermoelastic problems [4]. A simple case of a metal-to-ceramic bond with an existing interfacial crack is examined. It is assumed that coating surface is subjected to thermal shock combined with a uniform constant external pressure. The thermal contact resistance (TCR) between the crack faces depends on the normal contact traction. The assumption for the level of the (TCR) and the coefficient of friction has a significant impact on the resulting strain energy release rate (SERR). The parametric analyses provide valuable information for the failure assessment of such materials and at the same time reveal the robustness of the proposed boundary element methodology to treat this class of problems.

2 Boundary Element Analysis of Time Dependent Thermo-elasticity

It is assumed that the boundary Γ which surrounds a two dimensional body is discretized in a number of elements and after the application of 2-D boundary integral equations of time-dependent thermo-elasticity the matricial form the corresponding equations [1] is derived:

$$[Q]\{\theta^F\} = [\Theta^1]\{q^F\} + \{B^F\} \tag{1}$$

$$[T]\{u^F\} = [U]\{t^F\} + \{\overline{B}^F\} \tag{2}$$

where F is the current time point while θ^f, q^f are the nodal temperatures and heat fluxes at the time instant $f = 1, F$, respectively. The vectors $\{B^F\}$, $\{\overline{B}^F\}$ are defined as:

$$\{B^F\} = \sum_{f=1}^{F} -1\left([\Theta^{F+1-f}]\{q^f\} - [\Theta^{F+1-f}]\{\theta^f\}\right) \tag{3}$$

$$\{\overline{B}^F\} = \sum_{f=1}^{F} \left([\overline{Q}^{F+1-f}]\{\theta^f\} - [\overline{\Theta}^{f+1-f}]\{q^f\}\right) \tag{4}$$

The quantities $[Q]$, $[\Theta]$, $[T]$, $[U]$, $[\bar{Q}]$, $[\bar{\Theta}]$ are coefficient matrices expressing the contributions from Q_i, Θ_i, T_{ij}, U_{ij}, $\bar{Q}_i$, $\bar{\Theta}_i$, respectively. The vectors $\{\theta\}$, $\{q\}$, $\{u\}$, $\{t\}$ represent nodal boundary values of temperatures, heat fluxes, displacements and tractions, respectively.

The domain of the problem is divided into the sub-domains ${}^{I}\Omega$ and ${}^{II}\Omega$, such that $\Omega(\Gamma) =^{I} \Omega({}^{I}\Gamma) +^{II} \Omega({}^{II}\Gamma)$ (Fig. 1). A part of the boundary of the bodies ${}^{I}\Omega$ and ${}^{II}\Omega$, have the possibility to come into contact and the discretization in boundary elements creates a number of master-slave node pairs. Three possible contact states can be occur, namely open, adhesion and slip state described by the corresponding

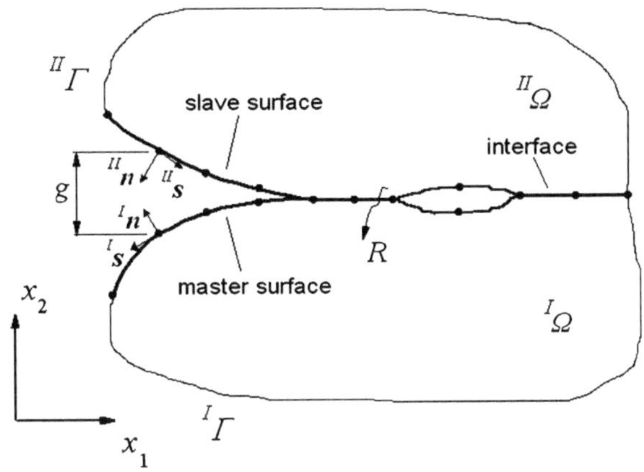

Fig. 1 General thermomechanical contact of two bodies

Table 1 Thermal contact conditions

Adhesion–slip	Open	Interface
$^{I}q^F = -^{II}q^F$	$^{I}q^F = -^{II}q^F$	$^{I}q^F = -^{II}q^F$
$^{I}\theta^F = ^{II}\theta^F - R(^{I}t_n^F)^{I}q^F$	$^{I}q^F = 0$	$^{I}\theta^F = ^{II}\theta^F$

Table 2 Mechanical contact conditions

Adhesion	Slip	Open	Interface
$^{I}t_t^F = -^{II}t_t^F$	$^{I}t_t^F = -^{II}t_t^F$	$^{I}t_t^F = -^{II}t_t^F$	$^{I}t_t^F = -^{II}t_t^F$
$^{I}t_n^F = -^{II}t_n^F$	$^{I}t_n^F = -^{II}t_n^F$	$^{I}t_n^F = -^{II}t_n^F$	$^{I}t_n^F = -^{II}t_n^F$
$^{I}u_t^F = ^{II}u_t^F$	$^{I}t_t^F = \pm\mu^{I}t_n^F$	$^{I}t_t^F = 0$	$^{I}u_t^F = ^{II}u_t^F$
$^{I}u_n^F = ^{II}u_n^F - g_n^F$	$^{I}u_n^F = ^{II}u_n^F - g_n^F$	$^{I}t_n^F = 0$	$^{I}u_n^F = ^{II}u_n^F$

boundary contact zones $\Gamma_o, \Gamma_\alpha, \Gamma_s$. The indices o, α, s represent the open, adhesion and slip state, respectively. The thermal and mechanical contact conditions at a time instant F for a node-pair in adhesion, slip, open state, as well as along the interface are provided in Tables 1 and 2, respectively. In these tables μ is the coefficient of friction while n,t are the local tangential and normal direction and g_n^0 expresses the possible initial normal gap between the nodes of the node-pair under consideration. The TCR R between the contacting faces depends on the contact pressure i.e. the normal traction. Finally, If Eqs. (1)–(2) are applied separately to both subregions $^{I}\Omega$ and $^{II}\Omega$ and the thermal and mechanical contact conditions of Tables 1 and 2 are substituted into Eqs. (1)–(2) the resulting expressions can be assembled to obtain the final equations for the thermal and elastic part of the problem [4]. The equation for the thermal part is:

$$\begin{bmatrix} {}^{I}Q & 0 & {}^{I}Q_i & {}^{I}Q_a & {}^{I}Q_s & {}^{I}Q_o & 0 \\ 0 & {}^{II}Q & {}^{II}Q_i & {}^{II}Q_a & {}^{II}Q_s & 0 & {}^{II}Q_o \end{bmatrix} \left\{ {}^{I,II}\theta^F \right\}^T$$
$$= \begin{bmatrix} {}^{I}\Theta & 0 & {}^{I}\Theta_i & {}^{I}\Theta_\alpha & {}^{I}\Theta_s & {}^{I}\Theta_o & 0 \\ 0 & {}^{II}\Theta & -{}^{II}\Theta_i & -{}^{II}\Theta_\alpha - R({}^{I}t^F_{n,\alpha}){}^{II}Q_\alpha & -{}^{II}\Theta_s - R({}^{I}t^F_{n,s}){}^{II}Q_s & 0 & {}^{II}\Theta_o \end{bmatrix} \left\{ {}^{I,II}q^F \right\}^T$$
$$+ \left\{ B^F \right\} \tag{5}$$

where:

$$\left\{ {}^{I,II}\theta^F \right\} = \left\{ {}^{I}\theta^F \; {}^{II}\theta^F \; {}^{I}\theta^F_i \; {}^{I}\theta^F_\alpha \; {}^{I}\theta^F_s \; {}^{I}\theta^F_o \; {}^{II}\theta^F_o \right\} \tag{6}$$
$$\left\{ {}^{I,II}q^F \right\} = \left\{ {}^{I}q^F \; {}^{II}q^F \; {}^{I}q^F_i \; {}^{I}q^F_\alpha \; {}^{I}q^F_s \; -{}^{I}q^F_o \; {}^{I}q^F_o \right\} \tag{7}$$

and for the elastic solution:

$$\begin{bmatrix} {}^{I}T & 0 & {}^{I}T_i & {}^{I}T_a & {}^{I}T_{t,s} & {}^{I}T_{n,s} & 0 & {}^{I}T_o & 0 \\ 0 & {}^{II}T & {}^{II}T_i & {}^{II}T_a & 0 & {}^{II}T_{n,s} & {}^{II}T_{t,s} & 0 & {}^{II}T_o \end{bmatrix} \left\{ {}^{I,II}u^F \right\}^T$$
$$= \begin{bmatrix} {}^{I}U & 0 & {}^{I}U_i & {}^{I}U_a & {}^{I}U_{n,s} \pm \mu {}^{I}U_{t,s} & {}^{I}U_o & 0 \\ 0 & {}^{II}U & -{}^{II}U_i & -{}^{I}U_a & -({}^{II}U_{n,s} \pm \mu {}^{II}U_{t,s}) & 0 & {}^{II}U_o \end{bmatrix} \left\{ {}^{I,II}t^F \right\}^T + \left\{ \overline{B}^F \right\} \tag{8}$$

where:

$$\left\{ {}^{I,II}u^F \right\} = \left\{ {}^{I}u^F \; {}^{II}u^F \; {}^{\mathrm{I}}u^F_i \; {}^{\mathrm{I}}u^F_a \; {}^{\mathrm{I}}u^F_{t,s} \; {}^{\mathrm{I}}u^F_{n,s} \; {}^{\mathrm{II}}u^F_{t,s} \; {}^{\mathrm{I}}u^F_o \; {}^{\mathrm{II}}u^F_o \right\} \tag{9}$$
$$\left\{ {}^{I,II}t^F \right\} = \left\{ {}^{I}t^F \; {}^{II}t^F \; {}^{\mathrm{I}}t^F_i \; {}^{I}t^F_a \; {}^{I}t^F_{n,s} \; -{}^{I}t^F_o \; {}^{II}t^F_o \right\} \tag{10}$$

The existence of the TCR R as function of normal traction t_n in Eq. (5) lead to a coupled problem demanding simultaneous solution of the Eqs. (5), (8). Every node-pair is examined according to Table 3 in order to check if any change in the contact status has occurred. The Eqs. (5), (8) are solved repeatedly during the solution procedure until the contact status does not change and the calculated TCR of all adhesion-slip node-pairs of the last iteration R_{last} is approximately equal to that of the previous iteration R_{prev}.

Table 3 Definition of contact status

Assumption	Decision	
	Open	Contact
Open	$\left\lvert {}^{II}u^F_n - {}^{I}u^F_n \right\rvert > g^{F-1}_n$	$\left\lvert {}^{II}u^F_n - {}^{I}u^F_n \right\rvert \leq g^{F-1}_n$
Contact	${}^{I}t^F_n \geq 0$	${}^{I}t^F_n < 0$
	Adhesion	Slip
Adhesion	$\left\lvert {}^{I}t^F_t \right\rvert < \left\lvert \mu({}^{I}t^F_n) \right\rvert$	$\left\lvert {}^{I}t^F_t \right\rvert \geq \left\lvert \mu({}^{I}t^F_n) \right\rvert$
Slip	${}^{I}t^F_t \cdot ({}^{I}u^F_t - {}^{II}u^F_t) > 0$	${}^{I}t^F_t \cdot ({}^{I}u^F_t - {}^{II}u^F_t) \leq 0$

3 Calculation of Fracture Parameters

The existence of an interfacial crack in a bimaterial solid justifies the choice of the strain energy release rate (SERR) as fracture assessment parameter. Singular quarter-point elements (QPEs) are used for the representation of the behavior of temperature and displacement field near the crack tip and traction singular quarter-point elements (TSQPEs) are utilized to account for the heat flux and traction field. The SERR G is related to the magnitude of complex stress intensity factor K_0 according to the equation:

$$G = \frac{(K_0)^2}{4}\left(\frac{1-\nu_C}{\mu_C} + \frac{1-\nu_S}{\mu_S}\right) \tag{11}$$

where ν_i, μ_i, represent the Poisson's ratio and shear modulus respectively while the index i denotes the coating c or the substrate s. The magnitude of K_0 is computed by the traction formula [3]:

$$K_0 = \frac{\sqrt{2\pi\ell}}{cosh(\pi\varepsilon)}[(t_s^A)^2 + (t_n^A)^2]^{1/2} \tag{12}$$

In the above equation t_s^A and t_s^A refer to the tangential and normal component of traction at the crack tip A and ℓ is the length of the crack tip element. The quantities ζ, ε are defined as $\zeta = (\mu_c + \kappa_c\mu_s)/(\mu_s + \kappa_s\mu_c)$ and $\varepsilon = 0.5\pi\ln\zeta$ with $\kappa_d = 3-4\nu_d$ for plane strain or $\kappa_d = (3-4\nu_d)/(1+4\nu_d)$ for plane stress and $d = c, s$.

4 Numerical Results and Discussion

The problem under consideration is illustrated in Fig. 2. There are two subdomains, a coating layer of thickness H_c and a substrate of thickness $H_s = 10H_c$ with an interfacial central crack of length $2a = 2(H_s/8)$ between the two layers. Nodular cast

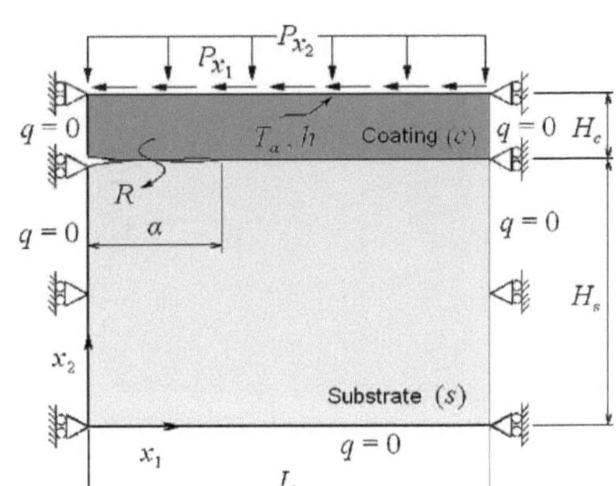

Fig. 2 Geometry and boundary conditions of the problem

iron is used as substrate and the coating material is Zirconia (ZrO_2) with properties provided in [6]. The free surface of the coating experiences a thermal shock by the ambient temperature $T_\alpha \gg T_0$, with T_0 being the initial uniform temperature, considering a coefficient of heat convection $h = 50\,\mathrm{W/m^2\,^\circ C}$. During the thermal shock a constant pressure Px_2 acts on the surface of the coating which may be combined with a constant shear stress Px_1. Plane strain conditions are assumed. Figure 2 shows the boundary conditions and geometrical aspects of the BEM model. Due to the lack of appropriate data from the literature for the relation of TCR with the contact pressure for the specific materials used in TBC systems Eqs. (13) for imperfect thermal contact [5] are considered.

$$R_1(t_2) = 0.0002 \times \exp(10^{-8} t_2),\ R_2(t_2) = 0.0002 \times \exp(10^{-8} t_2) \tag{13}$$

The function $R_1(t_2)$ gives very low resistance values and corresponds to perfect thermal contact [4]. The non-dimensional SERR and time are used for the presentation of results:

$$G^* = \frac{G}{E_C \alpha_C^2 (\Delta T)^2 H_S},\ \mathbf{t}^* = \frac{k_C \mathrm{t}}{\rho_C c_C H_S^2} \tag{14}$$

The evolution of non-dimensional SERR G^* is presented in Fig. 3 for perfect (R_1), imperfect (R_2) and adiabatic contact (insulation) conditions. The case considered here represents thermal shock without the presence of load on the coating surface and frictionless contact ($\mu = 0$). The results of the present analysis are compared to that of [6] which correspond to frictionless and insulated contact.

Apparently, the consideration of pressure dependent TCR has a noticeable impact on the predicted G^*. The extreme cases of thermal insulation and perfect thermal contact give a maximum difference in G^* about one order of magnitude. Low TCR values reduce the jump in temperature at the interface and the stress distribution

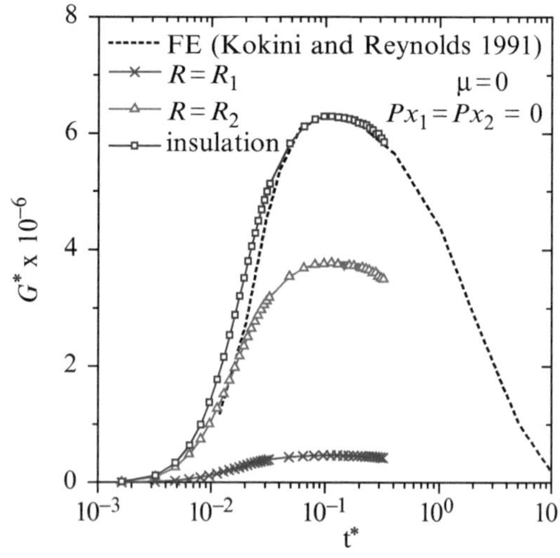

Fig. 3 G^* versus non-dimensional time for several levels of TCR

Fig. 4 Transient G^* for $R = R_1$

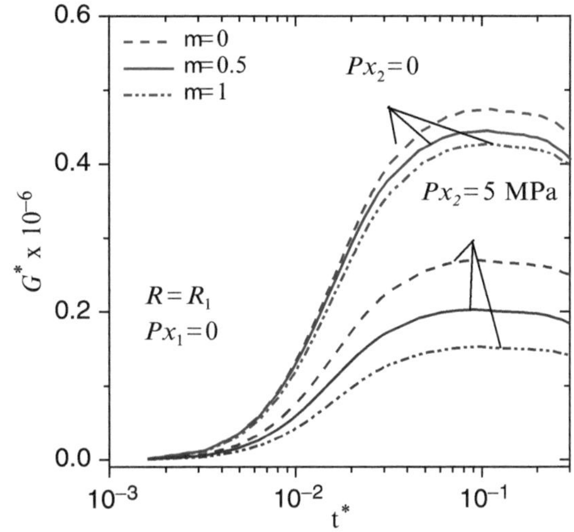

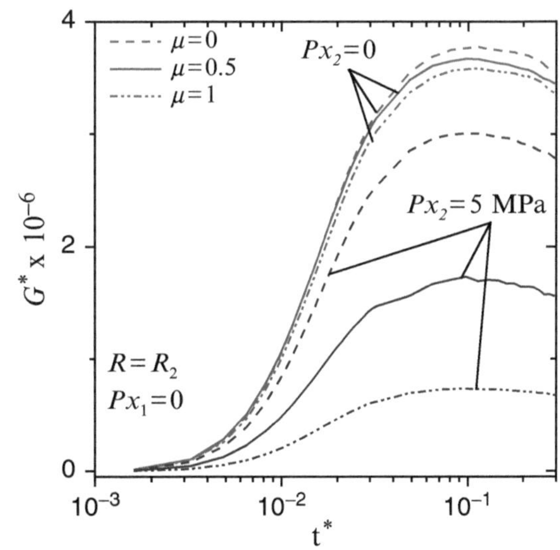

Fig. 5 Transient G^* for $R = R_2$

derives lower strain energy release rates. When thermal shock is combined with constant pressure Px_2 the crack closure is better pronounced. In this case the effect of coefficient of friction is not negligible according to Figs. 4 and 5 for $R = R_1$ and $R = R_2$, respectively. Both graphs indicate that assumption for frictionless contact can be justified for thermal loading only. Higher values of μ result to lower G^* when a pressure load is applied during a thermal shock. Additionally, it is obvious in Figs. 3–5 that the time points at which G^* attains its peak value is independent from the level of TCR, coefficient of friction and pressure.

Fig. 6 Effect of coefficient of friction and pressure Px_2 on G^*_{peak} for $R = R_1$

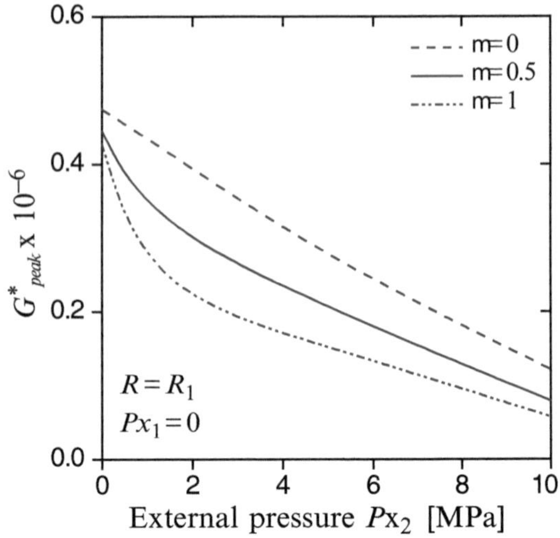

Fig. 7 Effect of coefficient of friction and pressure Px_2 on G^*_{peak} for $R = R_2$

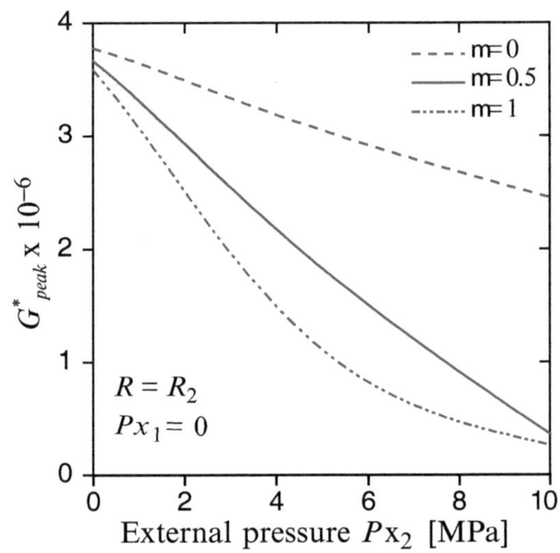

The fracture mode II is predominant for the cases under consideration where partial or approximately full closure of the crack occurs. Therefore, a reduction of the peak value of SERR G^*_{peak} is expected when the load Px_2 and the coefficient of friction are increased causing restriction of the sliding between crack faces. This is evident in Figs. 6 and 7 for $R = R_1$ and $R = R_2$, respectively. The relation between G^*_{peak} and Px_2 is linear for frictionless contact and exhibits non-linear behavior mostly when $\mu = 1$ for any level of TCR. In addition, the slope of $G^*_{peak} = G^*_{peak}(Px_2)$ for $\mu = 1$ is very intense initially and it is reduced at $Px_2 = 2\,\text{MPa}$ for $R = R_1$ and $Px_2 = 5\,\text{MPa}$ $R = R_2$.

Fig. 8 Effect of shear stress Px_1 on G^*_{peak} for $R = R_2$

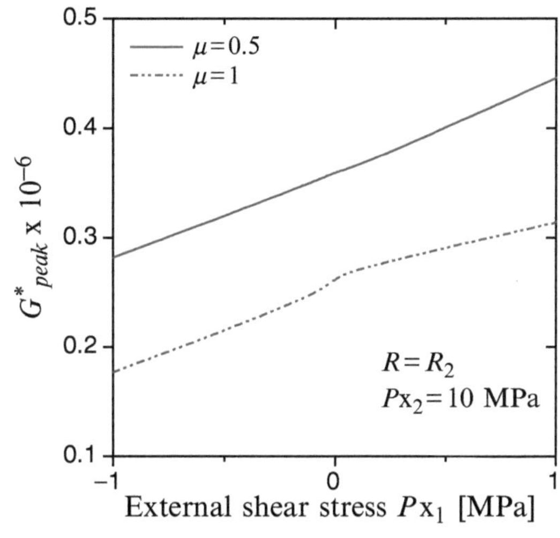

In some cases the components protected by TBCs may withstand a combined action of thermal shock, pressure and shear stress. This case was examined in the present analysis where frictional contact and $Px_2 = 10\,\text{MPa}$ was assumed and the magnitude of shear stress Px_1 takes values between $-1/10$ and of Px_2. The results are presented in Fig. 8. For both coefficients of friction G^*_{peak} exhibits a linear variation with the magnitude of Px_2 and it is reduced considerably when Px_2 directs towards the fractured edge of TBC (see Fig. 2).

5 Conclusions

The present study tried to emphasize the impact of thermal contact resistance on the fracture assessment of fractured thermal barrier coating systems under combined thermal and mechanical loading. The results show that disregarding the pressure-dependent contact resistance between the crack faces the prediction leads to conservative estimation of their failure resistance. The level also of the thermal resistance is important for the prediction. The higher the thermal contact resistance is, the higher the severity of the crack. It was found that the consideration of frictional contact and the level of coefficient of friction are critical for the computation of the strain energy release rate in the presence of pressure loads where intense crack closure occurs. Finally it can be concluded that numerical fracture analysis in cases of thermo-mechanical loading utilizing boundary element formulations is very efficient when transient coupled solutions are required.

References

1. Brebbia CA, Telles JC, Wrobel LC (1984) Boundary Element Techniques. Springer, Berlin
2. Evans AG, Hutchinson JW (2007) The mechanics of coating delamination in thermal gradients. Surface & Coatings Technology 201: 7905–7916
3. Gao YL, Tan CL (1992) Determination of characterizing parameters for bimaterial interface cracks using the boundary element method. Engineering Fracture Mechanics 41: 779–784
4. Keppas LK, Giannopoulos GI, Anifantis NK (2008) Transient coupled thermoelastic contact problems incorporating thermal resistance: A BEM approach. Computer Modelling in Engineering & Sciences (accepted)
5. Kishimoto K, Inoue H, Shibuya T (1995) A boundary element analysis of thermoelastic contact problems. Engineering Analysis with Boundary Elements 15: 329–337
6. Kokini K, Reynolds RR (1991) Transient heating vs cooling of interfacial cracks in ceramic-to-metal bonds. Engineering Fracture Mechanics 38: 371–383
7. Rangaraj S, Kokini K (2003) Interface thermal fracture in functionally graded zirconia–mullite–bond coat alloy thermal barrier coatings. Acta Materialia 51: 251–267

Session XII
Structural Analysis Composites

A Numerical Investigation of Fractured Sandwich Composites under Flexural Loading

Efstathios E. Theotokoglou and Ioannis I. Tourlomousis

Abstract Finite element analyses have been carried out for fractured sandwich composites loaded in flexure. Core skin debonds parallel to the beam axis are considered at different distances from the upper skin interface. Static non-linear elastic two-dimensional finite element analyses of cracked sandwich beams are accounted for the evaluation of stress intensity factors at the crack tips.

Keywords Sandwich composites · Shear stresses · Interfacial defects

1 Introduction

In sandwich structures, low density foam core are receiving increasing attention. Mechanical properties of cellular foams have been summarized by Gibson and Ashby [1]. Ashby et al. [2] investigated the fracture mechanisms of linear elastic foam. Zenkert and Backlund [3] studied the mode-II and mixed-mode crack propagation of PVC cellular foam. Harte and Fleck [4] investigated failure modes in aluminium foam core sandwich panels loaded in cyclic flexure fatigue. Burchardt [5] numerically and Burman and Zenkert [6] numerically and experimentally have investigated fatigue characteristics of foam. Noury et al. [7] studied fatigue crack growth in rigid PVC cellular foam under combined mode-I and mode-II loadings.

In sandwich structures the foam is typically the weakest part and is the first to fail under static or cyclic loading because it transfers the applied loads as shear stresses. Interface debond damages between the face and the core can originate either from the manufacturing process or during service. In sandwich structures interfacial defects have been studied by Triantafillou and Gibson [8], Carlsson et al. [9–11] and Zenkert et al. [12, 13]. Berggreen [14] and Berggreen et al. [15] have studied the ultimate failure of debond damaged sandwich panels loaded with lateral pressure.

E.E. Theotokoglou (✉) and I.I. Tourlomousis
Faculty of Applied Sciences, Department of Mechanics-Lab. of Strength Materials,
The National Technical University of Athens, Zographou Campus, Theocaris Bld., GR-0157 73, Athens, Greece
e-mail: stathis@central.ntua.gr

S. Pantelakis, C. Rodopoulos (eds.), *Engineering Against Fracture: Proceedings of the 1st Conference*,

In this paper a core-skin debond a parallel to the beam axis is considered (Fig. 1), in accordance with experimental data [16, 17]. When subjected to flexural loading, this debond will propagate slowly along the top interface and eventually kinked into the core as a shear crack. Stress intensity factors are calculated using the Finite Element Method and assuming linear fracture mechanics and plane strain, plane stress conditions for the cracked sandwich beam. Results from the finite element analysis show the influence of the upper face sheet to the value of the stress intensity factors when the distance of the crack from the upper skin interface diminishes.

2 Nature of the Problem

The sandwich considered is depicted in Fig. 1. It is composed from PVC-core, R-75 with properties given in Table 1. (Technical data from DIAB Inc.) and face sheets from 240F S2-glass fiber with epoxy resin. Face sheet was separated from the sandwich beam using cutting and polishing machine. The face sheet was loaded at different support spans and its tensile modulus was found 16,300 MPa with Poisson's ratio 0.3 in the longitudinal direction [17]. The dimensions of the test specimen were $L = 228.6$ mm (support span) and $b = 63.5$ mm (width). The core thickness was $t_2 = 12.7$ mm and the face sheet thickness was $t_1 = 2.28$ mm. The overall thickness was 17.26 mm.

In the experimental investigation [16, 17], the specimens were supported on two rollers with appropriate overhang. A crack initiated on the compression side just below the top face sheet/core interface. It was noticed that the crack always initiated at the sub-interface created by the resin soaked, and dry cells, below the

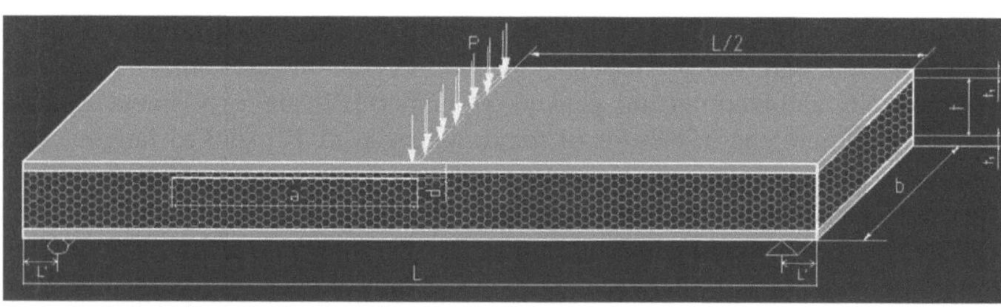

Fig. 1 A cracked sandwich beam under flexural loading

Table 1 Mechanical properties of R-75 [17]

Density (kg/m^3)	Compressive strength (MPa)	Compressive modulus (MPa)	Tensile strength (MPa)	Tensile modulus (MPa)	Shear strength (MPa)	Shear modulus (MPa)
75.2	1.1	38.0	2.0	62.0	0.9	29.0

actual top core-skin interface. This debonding crack was about 1–1.5 mm below the interface. The crack runs parallel to the beam axis from the point of initiation towards the end support (Fig. 1).

3 Numerical Investigation

The cracked sandwich beams are analyzed using the finite element method. Because of high stress gradients around the interface, different finite element meshes consisting of two dimensional plane strain and plane stress elements are used.

The finite element analysis is performed by the use of the general purpose finite element program ANSYS [18]. The six node two-dimensional plane strain triangular elements (PLANE 82) were used in order to model the beam. The frictionless contact area at the crack surfaces in the cracked sandwich beam, was modeled with two-node linear contact elements (CONTACT 178) [18], in order to prevent one surface from entering into the other in the load transfer region. The contact friction is assumed zero since the friction properties on the crack surfaces can not be reliably measured and since the effect of friction on the closed part of crack surfaces is small [6, 13, 15, 17]. Singular elements (mid-side nodes at 1/4) were used at the two crack tips in order to simulate the singular stress behaviour.

In order to analyze the event-1 crack propagation, a small crack is considered at different distances d (Fig. 1) below the interface immediately under the central load introduction and parallel to the beam axis. This small crack is considered propagating under the interface. For the different crack lengths and for the applied load:

$$P = rP_{ult} = 943.27N, \tag{1}$$

Where P_{ult} (= 1347.52N) the average static failure load [16, 17] of the sandwich beam and r (= 0.70) the lower stress level from the experimental investigation (the stress level r for a particular set of cycling loading defines the values of the maximum and minimum stresses).

The opening K_I and shear K_{II} mode stress intensity factors calculated from the nodal displacement on the crack lips, are [18]:

$$u = \frac{K_I}{4G}\sqrt{\frac{r}{2\pi}}\left((2\kappa-1)\cos\frac{\theta}{2}-\cos\frac{3\theta}{2}\right)-\frac{K_{II}}{4G}\sqrt{\frac{r}{2\pi}}\left((2\kappa+3)\sin\frac{\theta}{2}+\sin\frac{3\theta}{2}\right),$$

$$\upsilon = \frac{K_I}{4G}\sqrt{\frac{r}{2\pi}}\left((2\kappa-1)\sin\frac{\theta}{2}-\sin\frac{3\theta}{2}\right)-\frac{K_{II}}{4G}\sqrt{\frac{r}{2\pi}}\left((2\kappa+3)\cos\frac{\theta}{2}+\cos\frac{3\theta}{2}\right), \tag{2}$$

where u and υ the sliding and opening displacements respectively (r, θ) the polar-local coordinate system at the crack tip, G the shear modulus and $\kappa = 3-4\nu$ for plane strain, $\kappa = (3-\nu)(1+\nu)$ for generalized plane stress.

In case that there isn't any symmetry in the cracked model and $\theta = \pm 180^\circ$, we have:

$$K_I = \sqrt{2\pi}\frac{G}{(1+\kappa)}\frac{(\Delta\upsilon)}{\sqrt{r}}, \quad K_{II} = \sqrt{2\pi}\frac{G}{(1+\kappa)}\frac{(\Delta u)}{\sqrt{r}} \tag{3}$$

where Δu, $\Delta\upsilon$ the difference in displacements on the crack lips.

A very fine finite element mesh is considered in order to simulate the crack lengths. The finite element mesh consists of about 82,000 elements PLANE 82 in the case that $d = 1.5$ mm, 64,000 elements in the case that $d = 1.0$ mm and 42,000 elements in the case that $d = 0.5$ mm. We have used different finite element meshes in the core as the distance of the crack from the upper skin interface changes but without changing the number of elements in the upper and lower skin laminate (about 4,782 elements in the top laminate and 2,290 elements in the bottom laminate). In order to build the finite element meshes in the core we have considered different parameters concerning the crack length, the "fine" mesh around the crack tip and the distances of the crack from the upper skin interface. Twelve singular elements have been used around the crack tips for every crack length. The number of contact elements changes according to the crack length and in the finite element program ANSYS a code in APDL [18], is created in order to insert the contact elements for every crack length.

4 Numerical Results

After the formation of a small crack, the cracked sandwich beam is studied. The crack is considered at different distances $d\,(= 1.5, 1.0, 0.5$ mm) from the upper skin interface and just below the concentrated force. The position of the right crack tip is considered constant, immediately under the concentrated from whereas the left crack tip moves to the left, as it was observed from the experimental investigation [16, 17]. The results for the opening mode K_I and the shear mode K_{II} stress intensity factors based on relations (2) and (3) and for the plane strain and plane stress conditions, are given in Figs. 2–6 for the left and the right crack tips.

From the numerical investigation it is seen that as the crack propagates, K_I and K_{II} values increase at the left crack tip (Figs. 2, 3). K_{II} increases very faster when comparing with K_I. It is observed that the value K_I-stress intensity factors are very lower comparing with the K_{II}-stress intensity factors. It is also observed that as the distance between the crack and the upper skin interface decreases the value of the K_I and most apparently of the K_{II} also decrease but the percentage differences between the values of K_{II} for $d = 1.5$ mm and $d = 1.0$ mm and for $d = 1.0$ mm and $d = 0.5$ mm are rather small. This is due for K_I to the fact that the stress field in the core, where the shear stresses are dominated, between the crack and the interface is rather complex because of the small distance from the interface to the crack. On the other hand for K_{II} the shear stresses in the core become higher as we move from the upper skin interface (see also [14,15]). In case that plane stress is considered instead of plane strain for $d = 1.0$ mm, there are differences between the values of K_I and K_{II} (Figs. 4, 5). These differences increase as the crack length increases. At the right

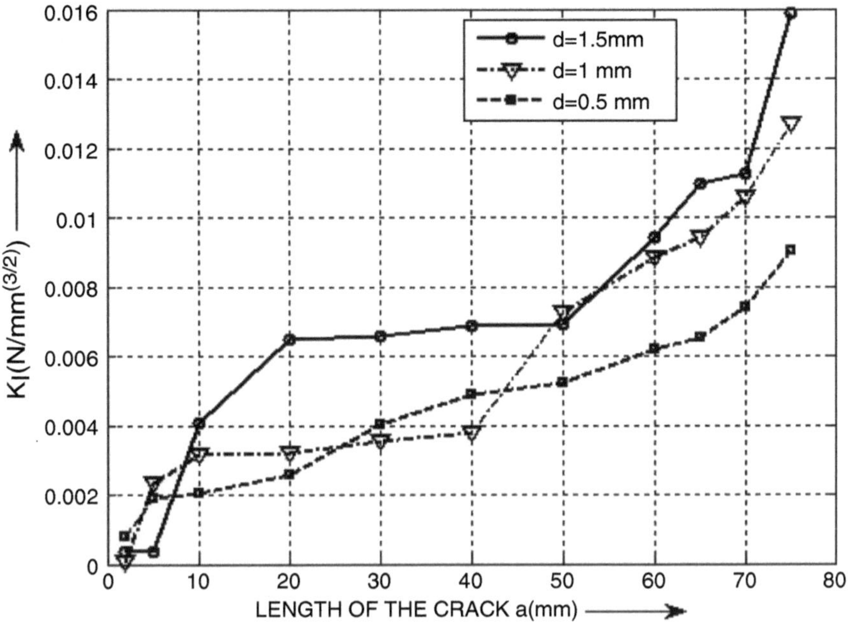

Fig. 2 Values of the stress intensity factors K_I for the left crack tip and for different distances d from the upper skin interface

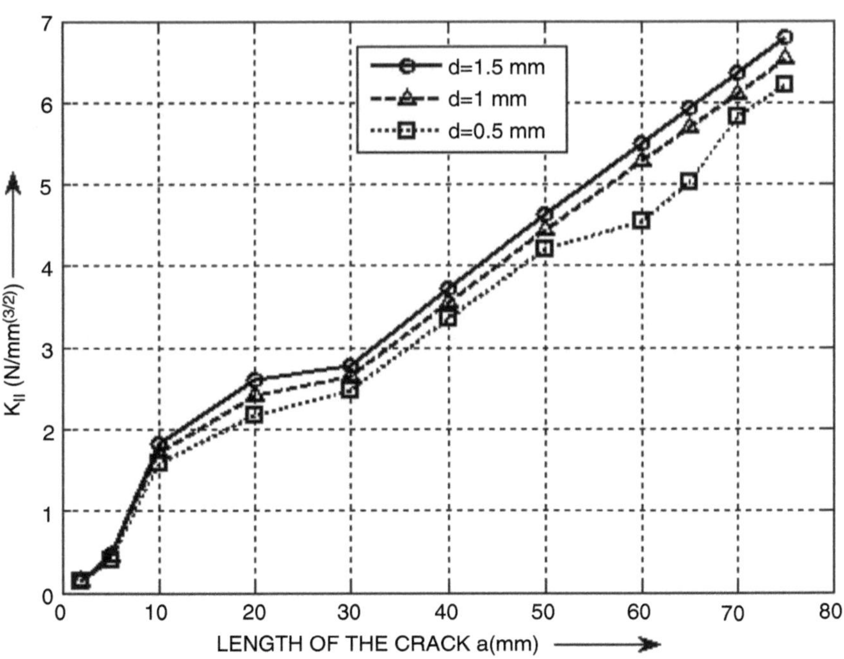

Fig. 3 Values of the stress intensity factors K_{II} for the left crack tip and for different distances d from the upper skin interface

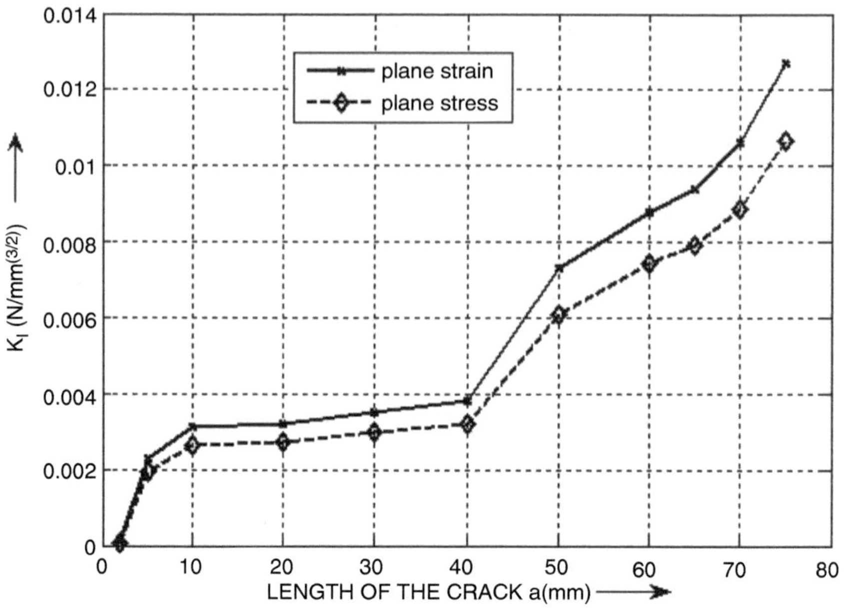

Fig. 4 Differences between the plane strain and plane stress cases for the values of the left crack tip stress intensity factors K_I and for $d = 1$ mm

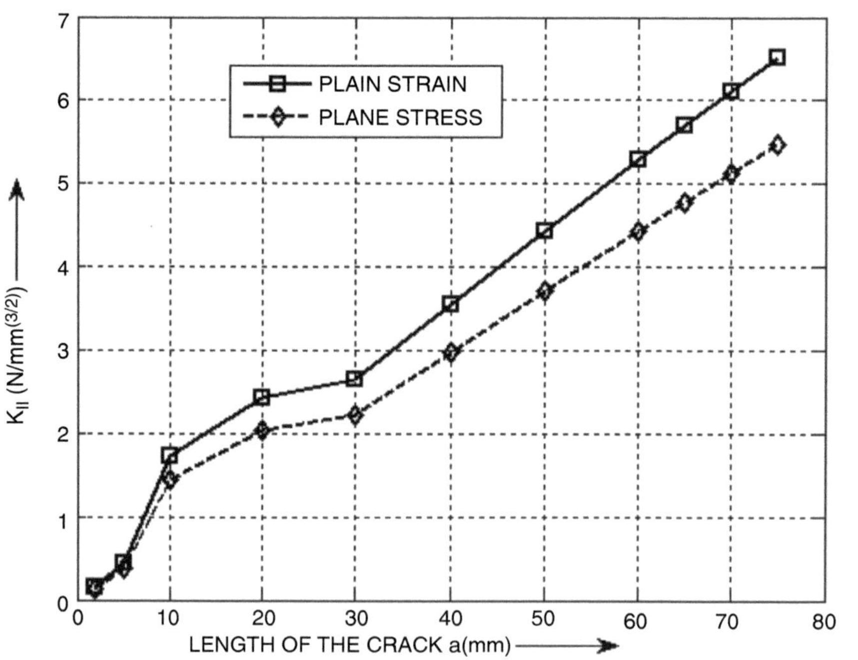

Fig. 5 Differences between the plane strain and the plane stress cases for the values of the left crack tip stress intensity factors K_{II} and for $d = 1$ mm

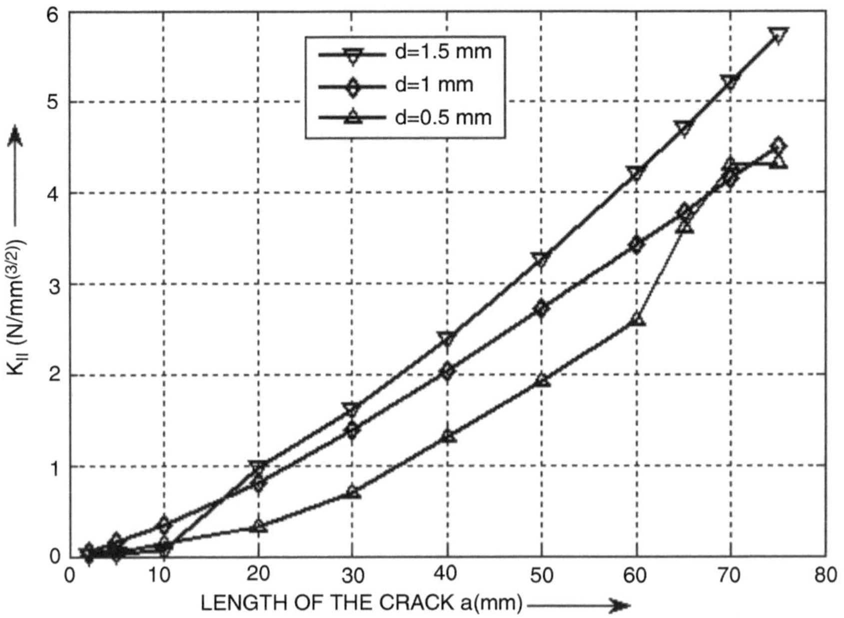

Fig. 6 Values of the stress intensity factors K_{II} for the right crack tip and for different distances d from the upper skin interface

crack tip the K_{II} values (Fig. 6) increase in a slow rate when comparing with the left crack tip. For the right crack tip it is observed that, there are percentage differences between the plane stress and the plane strain cases.

From the above analysis it results that a crack propagating into the core close to the upper skin interface, is subjected to mixed-mode loading conditions where Mode II is the dominating mode.

5 Summary

The behavior of fractured sandwich composites loaded, in three-point flexure was examined numerically. An "extra fine" finite element mesh was used in this study better than those used in [19–21].

The crack propagation process was simulated numerically via the finite element method using different finite element meshes for the different crack lengths, in the core of a sandwich beam very close to the upper core-skin interface and parallel to the beam axis. The finite element analyses were static and non-linear elastic. The non-linearity concerns contact elements used at the crack surfaces. The results for the stress intensity factors have big deviations from those derived in [19] using a "coarse" finite elements mesh. The results for the left and right crack tips are comparable with those derived in [21] for $d = 1.5\,\mathrm{mm}$.

The stress intensity factors at the vicinity of the crack tips were calculated for the different crack lengths using the linear fracture mechanics approach. The crack lengths considered in the core were in accordance with the crack initiation and propagation in the core of the sandwich beam just below the top face in the sub-interface, observed in the experiments [16,17]. From the numerical investigation it follows that the core is mainly subjected to shear. The crack propagation on the compression side is generally mode II dominated. The differences in the K_I values at the right crack tip and for the different distances d from the upper skin interface, were due to the complex stress situation immediately down the concentrated load [22–25].

The above numerical investigation gives a first indication of the crack growth behaviour in sandwich beams under flexural loading. Although the percentage differences between the plane strain and plain stress cases in the value of stress intensity factors were rather small, a three dimensional finite element analysis is further needed in order to verify our results.

References

1. L. H. Gibson, M. F. Ashby, "Cellular solid-structure and properties", 2nd ed. Cambridge: Cambridge University Press (1997).
2. M. F. Ashby, L. J. Gibson, S. K. Maiti, "Fracture toughness of brittle cellular solids", *Scripta Metallurgica*, 18, 213–217 (1984).
3. D. Zenkert, J. Backlund, "Poly(vinlchloride) sandwich core materials: fracture behaviour under mode ii loading and mixed-mode condition", *Materials Science and Engineering*, A108, 233–240 (1989).
4. A. M. Harte, N. A. Fleck, "The fatigue strength of sandwich beams with alloy foam core", *International Journal of Fatigue*, 23, 499–507 (2001).
5. C. Burchardt, "Bonded sandwich T-joint for maritime applications", Ph.D. dissertation, special Report No. 32, Institute of Mechanical Engineering, Aalborg University (1996).
6. M. Burman, "Fatigue crack initiation and propagation in sandwich structures", Doctoral thesis, Report 98–29, Department of Aeronautics, Royal Institute of Technology (1998).
7. P. M. C. Noury, R. A. Shenoi, I. Sinclair, "Fatigue crack growth in rigid PVC cellular foam under combined mode-I and mode-II loading", in Proceedings of Fourth International Conference on Sandwich Construction, Editor, Karl-Axel O., EMAS Publishing, Warrington, UK, 2, 491–502 (1998).
8. T. C. Triantafillou, L. J. Gibson, "Debonding in foam core sandwich panels", *Materials and Structures*, 22, 64–89 (1989).
9. L. A. Carlsson, L. S. Sendlein, S. L. Merry, "Characterization of face sheet/core shear fracture of composite sandwich beams", *Journal of Composite Materials*, 25, 101–116 (1989).
10. L. A. Carlsson, "On the Design of the Cracked Sandwich Beam (CSB) specimen", *Journal of Reinforced Plastics and Composites*, 10, 434–444 (1991).
11. L. A. Carlsson, S. Prasad, "Interfacial fracture of sandwich beams", *Engineering Fracture Mechanics*, 44, 581–590 (1993).
12. D. Zenkert, "Damage tolerance of foam core sandwich constructions", Ph.D. thesis, *Report 90-8*, Department of Aeronautics, Royal Institute of Technology (1990).
13. D. Zenkert, "Damage of sandwich beams with interfacial debondings", *Composite Structures*, **17**, 331–350 (1991).

14. C. Berggreen, "Damage tolerance of debonded sandwich structures", Ph.D thesis. Department of Mechanical Engineering, Technical University of Denmark (2004).
15. C. Berggreen, B. C. Simonsen, K. K. Borum, "Prediction of debond propagation in sandwich beams under FE-bared Fracture Mechanics and NDI Techniques", *Journal of Composite Materials*, 41, 493–520 (2007).
16. N. Kulkarni, "Fatigue response and life prediction of foam-core sandwich composites under flexural loading", Master thesis, Teskegee University (2002).
17. N. Kulkarni, H. Mahfuz, S. Jeelani, L. A. Carlsson, "Fatigue crack growth and life prediction of foam core sandwich composites under flexural loading", *Composite Structures*, 59, 499–50 (2003).
18. ANSYS Engineering Analysis System User's Manual, Swanson Analysis Systems, Inc. (1992).
19. E. E. Theotokoglou, L. A. Carlsson, H. Mahfuz, "Numerical study of fractured sandwich composites under flexural loading" in Proceedings of the 7th International Conference on Sandwich Structures, Editors, Thomsen, O.T., Bozhevolnaya, E., Lyckegaard, A., Springer, Amsterdam, Holland, 423–431 (2005).
20. P. P. L. Matos, R. M. McMeeking, P. G. Charalambides, M. D. Drory, "A method for calculating stress intensities in bimaterial fracture", *International Journal of Fracture*, 40, 235–254 (1989).
21. E. E. Theotokoglou, D. Hortis, L. A. Carlsson, H. Mahfuz, "Numerical study of fractured sandwich composites under flexural loading", *Journal of Sandwich Structures and Materials*, 10, 75–94 (2008).
22. Y. Frostig, "On stress concentration in the bending of sandwich beams with transversely flexible core", *Composite Structures*, 5, 405–414 (1996).
23. E. E. Theotokoglou, "Analytical determination of the ultimate strength of sandwich beams", *Applied Composite Materials*, 3, 345–353 (1996).
24. J. Kim, S. R. Swanson, "Design of sandwich structures for concentrated loading", *Composite Structures*, 52, 365–373 (2001).
25. Y. Frostig, T. O. Thomsen, "Localized effects in the non-linear behavior of sandwich panels with a transversely flexible core", *Journal of Sandwich Structures and Materials*, 7, 53–75 (2005).

Mechanical Properties and Failure Investigation of Metallic Open Lattice Cellular Structures

George N. Labeas, Milan M. Sunaric, and Vangelis P. Ptochos

Abstract A generic open lattice cellular structure under compressive loading is investigated by means of numerical analysis. The cellular structure is produced of stainless steel sheets using the punching and folding technique. The Finite Element method is used to derive the most important structural core properties, i.e. elasticity modulus, plateau stress and compaction strain, as well as to study the non-linear response and complex failure processes that occur as the core collapses in compression. Two concurrent Finite Element models are developed, a simple beam element model and a more complicated shell element model. The results obtained from simulations of the structure under compressive loading using the two versions of the FE modeling are compared to results of quasi – static compression experimental tests. The influence of geometrical parameters and strut cross section parameters on the core structure mechanical properties is parametrically studied.

Keywords Cellular structures · Mechanical properties · FE modelling · Non-linear response · Static compression

1 Introduction

Sandwich structures made of metallic or plastic core and composite skins have attracted the attention of the engineering industry since many years. The applications of cellular structures are widespread. Thermal insulation, packaging, structural, buoyancy and many other engineering sectors are the most common fields for the utilization of cellular structures. There is a wide variety of cores with various unit cell geometries, density and production technologies. The usual design

G.N. Labeas (✉), M.M. Sunaric, and V.P. Ptochos
Laboratory of Technology and Strength of Materials, University of Patras,
Panepistimioupolis, Rion 26500, Greece
e-mail: {labeas, msunaric, vptochos}@mech.upatras.gr

S. Pantelakis, C. Rodopoulos (eds.), *Engineering Against Fracture: Proceedings of the 1st Conference*,

parameters of sandwich constructions focuses on the determination of physical properties and linear elastic response (specific stiffness and strength) and more rarely include other behaviors such as fatigue behavior, crashworthiness and foreign object impact performance, etc. [1]. The mechanical behavior of sandwich structures is dependent on the mechanical properties of the core, as well as the metallic or composite skins. The core properties are dependent on the constitutive material and the cell geometry. Hence, for a given material, properties are adjusted by varying the parameters of the core geometry. The cellular cores may comprise closed cells, open cells or framework type geometries. A number of manufacturing processes are possible; they include investment casting, deformation forming, brazing etc. From the theoretical point of view, research is currently underway that addresses the analysis and optimization of lattice structures in the context of core materials in sandwich construction [2]. Already published results on truss-type cellular geometries by means of FE analysis for Kagome tetrahedral and Body Centered Cubic (BCC)-type geometries may be found in [3]. However, only few works compare the results of the FE analyses with the experimental data obtained on real structures [4].

1.1 Description of the Investigated Core Structure

The cellular core investigated in the present work is presented in Fig. 1a, as taken from [5]. The core is produced by ATECA from stainless steel sheets using the punching and folding technique, as described in [5]. The cellular structure has the dimensions $75 \times 75 \times 50\ \text{mm}^3$ (A, B and D dimensions in Fig. 1b respectively). The constitutive material for this core structure is stainless steel 304 L with elasticity modulus of 193 GPa and mass density of 8,000 kg/m^3. The structure comprises a regular generic unit-cell geometry, which is repeated periodically in all principal directions.

As it can be observed from Fig. 2a, the periodic unit-cell is of BCC (body centered cubic) type, i.e. its struts cross at the centroid of a cubic cell. The dimensions of the unit-cell in the investigated cores are $8.7 \times 10.7 \times 6.7\ \text{mm}^3$ (Fig. 2a), which

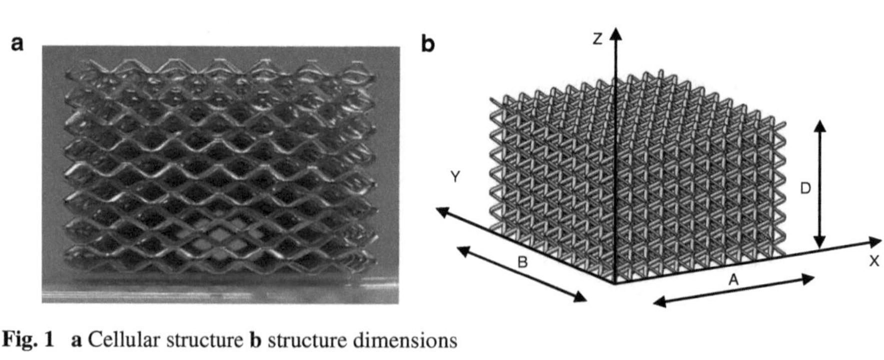

Fig. 1 **a** Cellular structure **b** structure dimensions

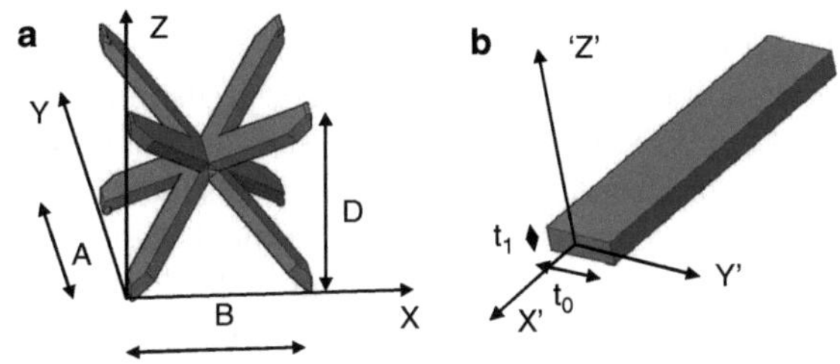

Fig. 2 **a** Unit-cell and **b** strut geometry

means that the core consists of $7 \times 9 \times 8$ cells. The struts comprising the unit-cell have rectangular cross-section of width $t_0 = 1.6$ mm and height $t_1 = 0.55$ mm (Fig. 2b). The structure is loaded in compression along the vertical z axis.

1.2 Finite Element Models

For simple two-dimensional or three-dimensional geometries, it is possible to calculate analytically the linear elastic response of cellular cores by using unit-cell approaches and mechanics of materials theory [6]. However, if the cell geometry becomes more complex, the analytical solutions may only roughly approximate the real unit-cell response; moreover, realistic loading or boundary conditions of cellular cores do not always lead to periodic boundary conditions at the unit-cell boundaries. The non-linear response and progressive failure analysis requires huge effort and still can yield only an estimation on collapse load, while different non-linear phenomena remain uncharacterized. In such cases, determination of the structural response requires extensive experimental testing. Hence, in the development and optimization phases of novel cellular structures, which require determination of the influence of cell-shape, size and constitutive material with respect to elasticity modulus, analytical and numerical prediction methodologies combined to some experimental testing result to a very efficient design procedure, as the rather expensive production and testing costs are minimized.

The present investigation is carried out by means of numerical Finite Element method. Two different FE modeling approaches are presented. The first approach uses the shell element formulation and models the entire structural geometry in detail, which results to a large number of elements thus high computational cost. The second approach uses a beam element formulation, which enables modeling of the entire core structure with less total number of nodes and elements.

For the development of the shell FE model, SHELL43 element of the ANSYS code is used. The SHELL43 element is well suited to model linear, warped, moderately-thick shell structures. The cellular core structure model built using this element type consists of 80,640 elements and 103,104 nodes. The beam element model consists of 24,192 BEAM188 elements and 45,576 nodes. The BEAM 188 element type is based on Timoshenko beam theory and is suitable for analyzing slender to moderately stubby/thick beam structures. It also includes stress stiffness terms, which enables analysis of lateral, flexural and torsional stability problems.

For both FE models the boundary conditions comprise loading on the top surface in the vertical direction and the constraining of vertical displacements on the bottom surface. A coupled displacement of all loaded nodes in the loading direction is enforced. There is no constraining for any lateral displacement of the structure. The comparison of the results obtained by the two approaches will result to a decision about which of the two approaches is better to be used in the analysis.

1.3 Prediction of Cellular Cores Linear Response

The linear response is mainly described by the elasticity modulus E_Z in the out-of plane direction, which is the most usual loading direction of cellular cores; E_Z value depends on the unit-cell geometry and size, strut aspect ratio and constitutive material properties. A linear static analysis is initially executed, for the prediction of the cellular core elasticity modulus. The elasticity modulus is estimated as the ratio of the computed global axial stress over the global axial strain. The global axial stress in this case is calculated as the overall axial load (in Z direction) over the loading surface (Fig. 3a). Also the global axial strain is calculated as the axial displacement of the loaded surface over the initial height of the structure (Fig. 3b).

The calculation of the elasticity modulus E_Z using the described methodology and the given geometrical parameters yields the value of 2,466 and 3,362 MPa, for the beam and the shell FE models, respectively. The experimental value of 2,641 MPa, which is measured in a test campaign performed by EADS IW [7], is closer to the calculated value of the beam model, although it is expected that the shell

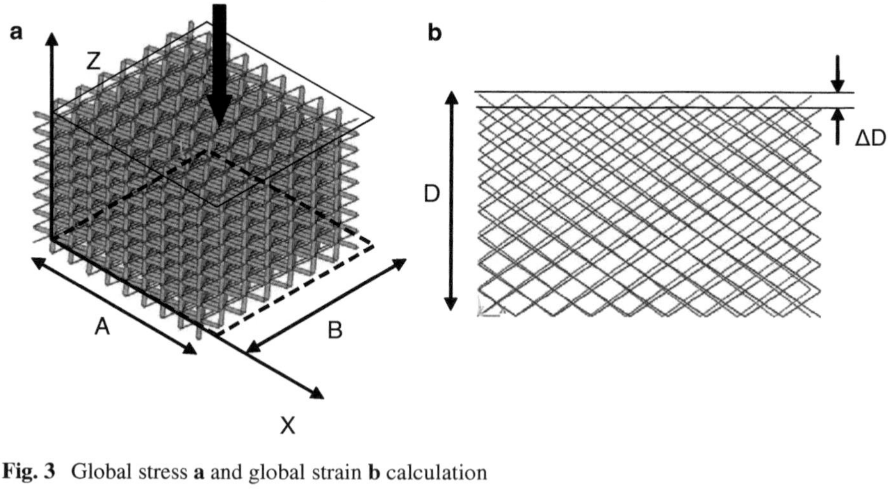

Fig. 3 Global stress **a** and global strain **b** calculation

model would yield better prediction due to a detailed modeling of the unit cell. This may be explained by existence of imperfections and irregularities in the core structure, which results from the process of joining the layers for the fabrication of the core structure. The possible imperfections in this type of cellular structure include truss layers misalignment, as well as layer debonding. The current FE modeling approach does not include any kind of imperfection. Moreover, the structure modeled with shell elements is shown to behave relatively stiffer compared to the beam element model, due to the fact that shell element succeed to model more accurately the junction of the struts which in reality is a complex surface (Fig. 4b); on the other hand this strut junction area which is of very high local stiffness is approximated by only a single node in the beam element model (Fig. 4a), the stiffness of which is lower compared to the reality.

A parametric study on the influence of the strut cross section dimensions on the elastic modulus E_Z is performed based on both beam and shell FE models. The influence of the strut ratio t_1/L (where t_1 is the cross-section height and L is the strut length, as illustrated in Fig. 2b) on the computed core elasticity modulus are presented in Fig. 5a. Also the influence of the strut ratio t_0/L (where t_0 is the cross-section width and L is the strut length as illustrated in Fig. 2b) on the computed

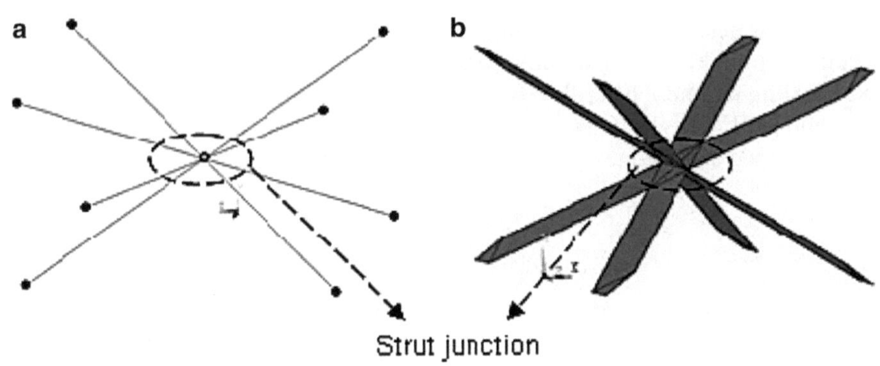

Fig. 4 Beam **a** and shell **b** unit-cell models

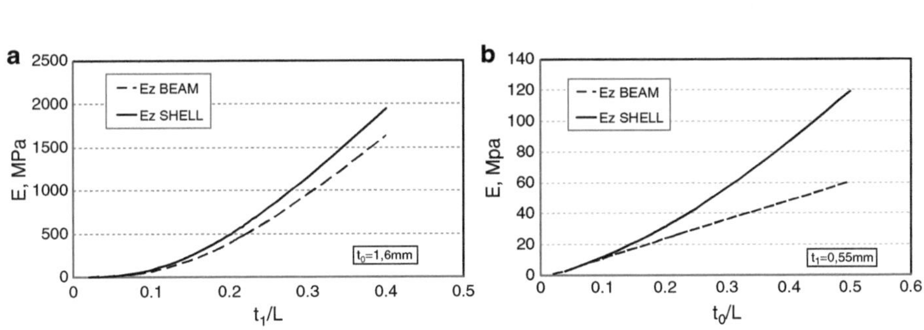

Fig. 5 Influence of strut cross-section height-over-length **a** and width-over-length **b** ratios on elastic modulus

core elasticity modulus are presented in Fig. 5b. It can be observed that elasticity modulus values increase when the value of either strut ratios increase, which can be explained by the increase of the struts second moment of inertia.

1.4 Prediction of Cellular Cores Non-linear Response

The non-linear response is driven by the failure mechanisms taking place during the cellular core compression. Typically the compression response can be described by the peak failure stress and the plateau stress. The peak failure stress corresponds to the initiation of core failure, which triggers the core progressive compaction. In general during the compaction phase the compressive stress does not grow significantly until the core stacking is finished.

For the prediction of the non-linear response, it is necessary to introduce failure mechanisms into the analysis model. The failure mechanisms that may possibly occur during core compaction are: (a) material plasticity, which mainly refers to the development of plastic hinges in the strut junction points, and (b) buckling of struts, which introduces geometrical non-linearity in the core behavior.

In the present analysis the critical elasto-plastic buckling load is analytically calculated, as described hereafter, in order to avoid the handling of geometric non-linearities in the FE model which could lead to numerical non-convergence in the post-buckling regime. The critical buckling load is calculated by analytical methods provided in [8]. The elastic buckling is based on the Euler formula, which for long beams clamped at the edges gives the critical elastic buckling load as:

$$P_{cr} = \frac{4\pi^2 E_0 I}{L^2} \tag{1}$$

In Eq. (1), L is the strut length, E_0 is the constitutive material elasticity modulus and I is the minimum principal second moment of inertia for the given strut cross-section. For inelastic buckling, the material modulus E_0 of Eq. (1) should be properly modified in order to take into account material plasticity. According to the 'tangential modulus' method, the modulus E_0 is replaced by the tangent modulus E_t, which is calculated from the slope of the stress–strain curve as:

$$E_t = \frac{d\sigma}{d\varepsilon} \tag{2}$$

The 'reduced modulus' method suggests replacing the modulus E_0 by the reduced modulus E_R, which is given as:

$$E_R = E_0 \frac{4(E_t/E_0)}{(1+\sqrt{E_t/E_0})^2} \tag{3}$$

The critical strut elasto-plastic buckling load is then calculated according to:

$$P_{cr} = \frac{4\pi^2 E^* I}{L^2} \tag{4}$$

In the case of 'tangential modulus' method, the modified modulus E^* equals to E_t, while in case of 'reduced modulus' method the modified modulus E^* equals to E_R.

The used critical buckling load takes a value between the theoretically calculated values of the 'tangential modulus' method and the 'reduced modulus method'. Comparing the results of both methods for the elasto-plastic buckling failure, it is estimated that the buckling in the core structure will occur at global compression stress of 2.57 MPa. From the experiments on core structures, it has been observed that the failure occurs at plateau stress of maximum 0.6 MPa. Therefore, the buckling failure is not the primary failure mode for this type of cellular cores.

The struts of the cellular core are mainly subjected to bending loads, which at critical level result in generation of plastic hinges in struts, in the regions near the strut junction points. In order to predict this type of failure, material plasticity needs to be introduced. Hence, a bi-linear isotropic elasto-plastic stress–strain material behaviour is introduced in the numerical model, with flow stress of 240 MPa, ultimate stress of 580 MPa and deformation at fracture of 50%.

The bending of the diagonal struts results in core compaction, which is finalized when physical contact occurs between the struts. The structure is then compacted and the deformation at which the compaction occurs is defined as compaction strain. From the physics of the compaction phenomenon, it is obvious that it can be simulated numerically only by introducing contact definitions in the numerical model. However, in order to avoid the increase of computational complexity, which is high even for the case of $75 \times 75 \times 50\ mm^3$ size core presently analyzed, a simple engineering approach is used for estimation of compaction strain. By this approach the final height of the structure at compaction is approximated by the overall thickness of the compacted struts, which for eight levels of unit cells at height gives the $H_f = 8.8$ mm. Using the formula

$$\varepsilon_{compact} = (H - H_f)/H \tag{5}$$

the compaction strain of this structure is estimated to be 82%.

1.5 *Cellular Core Plateau Stress and Compaction Strain Results*

Using the developed analysis methodology, numerical simulations are performed for the core structure. Predicted deformed shapes for a part of the core comprising beam elements and shell elements are presented in Fig. 6. It can be noticed that the dominant failure mode of this structure is plastic bending of struts and formation of plastic hinges around the strut junctions.

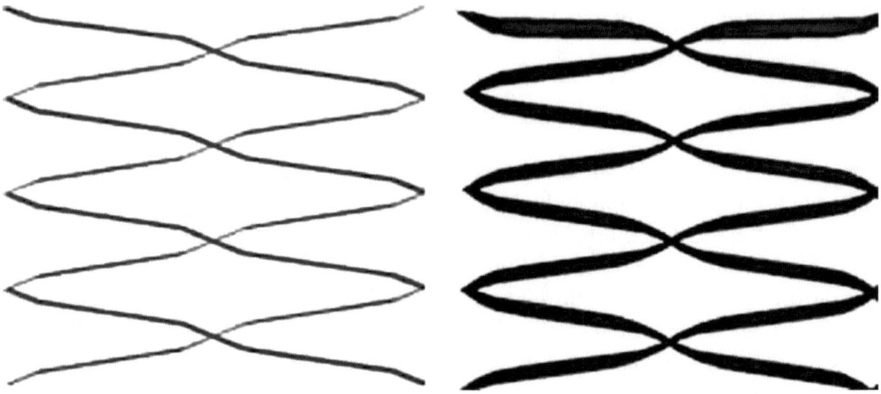

Fig. 6 The beam and shell element model deformed unit-cell

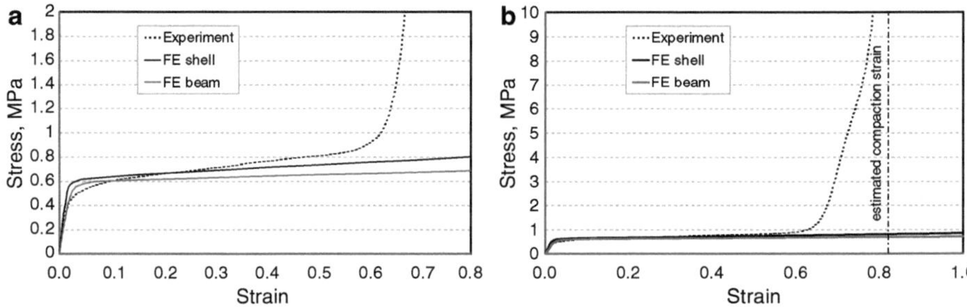

Fig. 7 **a** Comparison of measured and predicted non-linear response; **b** comparison of measured and estimated compaction strain

The calculated non-linear stress–strain response of the core is compared to the experimental curve in Fig. 7a. The numerical models approximate well the experimentally obtained response from [7]. There is a small difference in predicted plateau stress between the beam and the shell element model. The prediction of the compaction strain approximates well the values that are obtained in the compression experiments on the core structures, as can be observed in Fig. 7b.

2 Conclusions

An analysis methodology for the prediction of structural response of compressively loaded cellular cores, capable to calculate their elasticity modulus, plateau stress and compaction strain, is proposed. The cellular core elasticity modulus predicted from the linear static analysis correlates well with the experimentally obtained values. Numerical results have shown that the linear response is strongly influenced by the geometric parameters of the strut. The deviation of the numerically predicted and experimentally obtained structural responses may be explained by different types of imperfections that this type of core structure naturally includes. Thus, for the improvement of the numerical models, the most common types of imperfections should be identified and included in the modeling methodology.

For the non-linear response prediction, the failure mechanism of material plasticity is identified as the principal failure mechanism and is included in the model. Both the shell and the beam numerical models show good approximation of the core structure non-linear response, with the shell structure being somehow stiffer than the beam structure. The maximum compaction strain is successfully approximated analytically, in order to avoid the introduction of contact definitions in the numerical simulation. In general the performed analysis results approximate well the values obtained from the experimental compression tests.

Acknowledgements The results performed in this paper are partially obtained in the frame of the EU funded research project "CELPACT". The authors wish to acknowledge EU for financial support of this project, as well as Mrs. Yannick Girard EADS-IW-France and Mrs. Valia Fascio – ATECA-France for providing the experimental results and production data of the investigated core structures.

References

1. McKown S, Mines RAW (2006) Impact behavior of metal foam cored sandwich beams, European Conference on Fracture 16, Alexandroupolis, Greece, Paper No. 254.
2. Luxner MH, Stampfl J, Pettermann HE (2007) Numerical simulations of 3D open cell structures – influence of structural irregularities on elasto-plasticity and deformation localization, International Journal of Solids and Structures 44, pp. 2990–3003.
3. Deshpande VS, Fleck NA, Ashby MF (2001) Effective properties of the octet-truss lattice material, Journal of the Mechanics and Physics of Solids 49, pp. 1747–1769.
4. Wadley HNG, Fleck NA, Evans AG (2003) Fabrication and structural performance of periodic cellular metal sandwich structures, Composites Science and Technology 63(16), pp. 2331–2343.
5. Cellular Structures for Impact Performance – CELPACT Project, CELPACT 18M Interim Activity Report, 2008.
6. Gibson LJ, Ashby MF (1999) Cellular Solids, Cambridge Solid State Science Series, Cambridge University Press, Cambridge, United Kingdom.
7. Cellular Structures for Impact Performance – CELPACT Project, Experimental Results of Open Lattice Core, Internal Communication Report, 2008.
8. Curtis HD (1997) Fundamentals of Aircraft Structural Analysis, WCB/McGraw-Hill, Boston, MA, USA.

MIX
Papier aus verantwortungsvollen Quellen
Paper from responsible sources
FSC® C105338

If you have any concerns about our products,
you can contact us on
ProductSafety@springernature.com

In case Publisher is established outside the EU,
the EU authorized representative is:
Springer Nature Customer Service Center GmbH
Europaplatz 3, 69115 Heidelberg, Germany

Printed by Libri Plureos GmbH
in Hamburg, Germany